Apollo

The Definitive Sourcebook

Richard W. Orloff and David M. Harland

Apollo

The Definitive Sourcebook

Richard W. Orloff
Oakhurst
New Jersey
USA

David M. Harland
Space Historian
Kelvinbridge
Glasgow
UK

SPRINGER–PRAXIS BOOKS IN SPACE EXPLORATION
SUBJECT *ADVISORY EDITOR*: John Mason B.Sc., M.Sc., Ph.D.

ISBN 10: 0-387-30043-0 Springer Berlin Heidelberg New York
ISBN 13: 978-0-387-30043-6

Springer is a part of Springer Science + Business Media (*springeronline.com*)

Library of Congress Control Number: 2005936334

Cover design: Jim Wilkie
Copy editing: Alex Whyte
Typesetting: BookEns Ltd, Royston, Herts., UK

Printed in Germany on acid-free paper

"That the conquest of space is possible
must now be regarded as a matter beyond serious doubt"
Arthur C. Clarke, 1951

"... utter bilge"
Professor Richard van der Riet Woolley, Astronomer Royal,
commenting in 1956 on the prospects of space travel

"Beep, beep, beep ..."
Sputnik, 4 October 1957

"I believe that this nation should commit itself to achieving the goal, before this decade is out,
of landing a man on the Moon"
John F. Kennedy, 25 May 1961

"I think the most significant event that took place in the '50s was the launching of Sputnik"
John F. Kennedy, one week before his assassination in November 1963

"one giant leap for Mankind"
Neil Armstrong, as he stepped onto the lunar surface on 21 July 1969

"... America's challenge of tomorrow"
Gene Cernan as he departed the lunar surface on 14 December 1972

When Sir Arthur C. Clarke was asked in 2002 which event in the
twentieth century he would never have predicted, he replied,
"That we would have gone to the Moon and then stopped"
And, on reflection, he added,
"The space age hasn't begun yet"

This book is dedicated to
the 400,000 workers across America
who made Apollo a reality

A field geologist a long way from home.

Table of contents

List of illustrations

List of tables

List of appendices

Authors' preface

The word 'epic' is woefully inadequate to describe the Apollo program of the 1960s and early 1970s, during which time the United States of America landed a total of 12 men on the Moon.* Although politically initiated as a 'space race' in the context of the Cold War with the Soviet Union, the program became a great feat of exploration. Innumerable texts have been produced, describing the program from many points of view, but few provide comparative statistical data.

The aim of this book is to provide a comprehensive reference for *facts* about Apollo. As numerous NASA centers and contractors created post-mission reports, there are often differences in certain reported measurements. Additionally, as time passed, typographical errors crept into some Apollo-related publications. To resolve such conflicts, we turned to original documents, some of which were previously unavailable to the public, but have now been made available through the US Freedom of Information Act. In addition to an introduction to the program and individual mission narratives that include comprehensive timelines from just prior to liftoff to just after the recovery of the crew and spacecraft, we have included a series of statistical tables in which much of this data is listed in a form designed to enable missions to be compared. Prelaunch and countdown events are given in local US Eastern time, in which missions were launched, but mission events are in either Greenwich Mean Time (GMT) or Ground Elapsed Time (GET), the latter measured from 'Range Zero', which is defined as the last integral second before liftoff. The 'T–' and 'T+' style is used for events during countdown and ascent to orbit, and the time is in seconds. In general, however, our notation uses colon separators, and in order to avoid confusion with 24-hour GMT the format for GET is hhh:mm:ss.sss (i.e. with three digits in the hour, with preceding zeroes early on). To remain within our assigned number of pages, we have focused on the origins and operational aspects of the program, at the expense of its vehicles and facilities. Having documented Apollo, we look forward to a renewal of human lunar exploration in a decade or so.

Richard W. Orloff and David M. Harland
December 2005

* Throughout the Apollo era all US astronauts were male, and the terminology (which we unashamedly retain) reflected this.

Acknowledgments

The information contained in our mission logs was derived primarily from uncopyrighted NASA and contractor reports (see the Bibliography) and, in some cases, is used verbatim without attribution. In addition, we must acknowledge the following copyrighted works:

- The source for some of the astronaut biographical data is the excellent *Who's Who In Space: The International Space Year Edition*, by Michael Cassutt, although most information was derived from NASA biographies.
- The primary source for descriptions of the mission insignias is the official NASA text that accompanied each emblem, but additional information was obtained from *Space Patches from Mercury to the Space Shuttle*, by Judith Kaplan and Robert Muniz, 1986, and *"All we did was fly to the Moon": Astronaut Insignias and Callsigns* by Richard L. Lattimer, 1985.
- The COSPAR designations for spacecraft and launch vehicle stages assigned once in space are from *R.A.E. Table of Earth Satellites 1957–1986*, a compilation of installments originally issued by the Royal Aircraft Establishment in England.
- Artwork of the planned lunar traverses is courtesy NASA.
- The Apollo lunar surface traverse maps are courtesy the US Geological Survey.
- The cross-sectional geological interpretations of the Apollo landing sites are from the *Lunar Sourcebook: A User's Guide to the Moon*, edited by Grant H. Heiken, David T. Vaniman and Bevan M. French, and published by Cambridge University Press.

Over the years, many people helped us to locate original NASA documents, images and other information, and to check the manuscript for errors: Becky Fryday, formerly Media Services, Lyndon B. Johnson Space Center; Bunda L. Dean, formerly Lyndon B. Johnson Space Center; Dale Johnson, George C. Marshall Space Flight Center; Daryl L. Bahls, The Boeing Company; David H. Ransom Jr, Sedona, AZ; J.L. Pickering, Normal, IL; Ricky Lanclos, Nederland, TX; Dr Eric M. Jones, editor of the Apollo Lunar Surface Journal on the Internet; Dr John B. Charles, Lyndon B. Johnson Space Center; Florastela Luna, Lyndon B. Johnson Space Center; Gary Evans, TRW; Gordon Davie, Edinburgh, Scotland; W. David Woods, Glasgow, Scotland, editor of the Apollo Flight Journal on the Internet; Ken MacTaggart, Glasgow, Scotland; Dwayne A. Day, Space Policy Institute, Washington, DC; Robert Andrepont, Opelousas, LA; Ed Hengeveld, The Netherlands; Brian Lawrence, Wantage, England; Janet Kovacevich, formerly Lyndon B. Johnson Space Center; Joan Ferry and Lois Morris, Woodsen Research Center, Rice University;

Joey Pellarin Kuhlman, formerly Lyndon B. Johnson Space Center; Kenneth Nail, formerly John F. Kennedy Space Center; Kipp Teague, Lynchburg, VA; Donnis Willis, Satsuma, AL; Lee Saegesser, formerly NASA Headquarters; Lisa Vazquez, formerly Lyndon B. Johnson Space Center; Mike Gentry, Lyndon B. Johnson Space Center; Margaret Persinger, Kennedy Space Center; Oma Lou White, formerly George C. Marshall Space Flight Center; Paulo D'Angelo, Rome, Italy; Philip N. French and Jonathan Grant, NASA Center for Aerospace Information; Robert Sutton, Chantilly, VA; Robert W. Fricke Jr, Lockheed Martin; Ruud Kuik, Amsterdam, The Netherlands; Dr David R. Williams, National Space Data Center, Robert H. Goddard Space Flight Center; Hayes M. Harper, Downers Grove, IL; Stephen Tellier, Lunar and Planetary Institute, Houston, TX; Lieutenant Colonel George H. Orloff US Army (Retired), Oakhurst, NJ; Harald Kucharek, Karlsruhe, Germany; Kay Grinter, Kennedy Space Center; Lindsay Reed, Beachwood, NJ; Sabrina Echeverria, Bayville, NJ; and Louise Alstork, Stanley Artis, Steve Garber, Hope Kang, Roger Launius, Warren Owens, and Michael Walker, NASA Headquarters, Washington, DC.

Introduction

THE NATIONAL AERONAUTICS AND SPACE ADMINISTRATION

On 4 October 1957, to mark the International Geophysical Year, which ran from mid-1957 to the end of 1958, the Soviet Union launched the world's first artificial satellite, naming it 'Sputnik' for 'fellow traveler'. The United States, which had expected this technological milestone to be achieved by its own Vanguard project, was shocked. Worse was to come. The attempt to launch Vanguard on 6 December from Cape Canaveral on Florida's Atlantic coast was a disaster. The vehicle rose several feet, lost thrust, fell back and exploded. In an effort to recover from this humiliation, Wernher von Braun of the Army Ballistic Missile Agency (ABMA) at the Redstone Arsenal in Huntsville, Alabama, drew from storage a Redstone missile fitted with 14 solid rockets – 11 in the outer ring serving as the second stage and three in the center serving as the third stage. This variant, known as the Jupiter-C, had been developed to accelerate a payload into the atmosphere from the top of a ballistic arc in order to test a thermal shield for the warhead of the Jupiter missile. With a fourth stage formed from yet another solid rocket, the Jupiter-C was capable of putting a payload into orbit. This was done on 31 January 1958. The satellite, named Explorer, was supplied by the Jet Propulsion Laboratory (JPL) of the California Institute of Technology in Pasadena, which had developed the Sergeant rockets that formed the upper stages.[1] The payload, supplied by James van Allen of the University of Iowa, was integrated into the final stage and the whole thing went into orbit. In detecting the presence of energetic charged particles trapped in Earth's magnetic field, this instrument made the first significant discovery of the Space Age.

[1] In 1938 Caltech's Daniel Guggenheim Aeronautical Laboratory was contracted by the Army Air Corps to develop 'jet-assisted take-off' rockets for heavy bombers. In 1942 laboratory leader Theodore von Kármán set up the Aerojet Engineering Corporation to manufacture such 'JATO' rockets. By 1943, when the program was expanded to include the development of ballistic missiles, the name 'Jet Propulsion Laboratory' had been adopted even though, strictly speaking, it was not involved in research into jet engines. In quick succession, it created a series of solid-fuel missiles named Private, Corporal and Sergeant.

Although national honor had been restored, the Department of Defense was deeply concerned that the Soviet rocket was so much more powerful than its own. On 7 February President Dwight David 'Ike' Eisenhower created the Advanced Research Projects Agency (ARPA) to develop national goals and coordinate (although not conduct) the necessary research. On 1 April Roy W. Johnson was appointed director, reporting directly to the Secretary of Defense. On the recommendation of James Rhyne Killian, Eisenhower's special assistant for science and technology, an executive order on 2 April established the National Aeronautics and Space Administration (NASA) to manage a national civilian space program. Eisenhower also set up the National Aeronautics and Space Board to advise on policy. In June the National Academy of Sciences – which had been established in 1863 as a private institution to promote the advancement of science and, on being requested, advise the government on scientific matters – created a Space Science Board to advise on the scientific exploration of space. The National Aeronautics and Space Act was passed by Congress on 16 July and was signed into law on 29 July. Simultaneously, the National Advisory Committee for Aeronautics (NACA), created in 1915 to undertake basic research in aerodynamics, was wound up and its facilities transferred over to its successor: the Langley Aeronautical Laboratory at Langley Field, which was established in Hampton, Virginia, in 1917, together with its Pilotless Aircraft Research Station at Wallops Island; the Ames Aeronautical Laboratory at Moffett Field, which was established in 1939 in Mountain View, California; the Lewis Flight Propulsion Laboratory, which was established in 1941 in Cleveland, Ohio; and the High-Speed Flight Station, which was established in 1949 at Muroc Field in California, which was renamed Edwards Air Force Base in 1950.[2] Langley was pursuing aerodynamic studies related to manned spaceflight, Lewis was developing a rocket engine that burned hydrogen as its fuel, and Edwards was developing the X-15 rocket-powered aircraft capable of flying on a ballistic arc up to the 'fringe' of space. Although NASA's remit was significantly broader than that of its predecessor, it did *not* immediately gain control of either the rocketry expertise at ABMA or of JPL.[3] On 8 August Thomas Keith Glennan, president of the Case Institute of Technology in Cleveland, Ohio, was nominated as administrator, and Hugh Latimer Dryden, who for the past decade had been NACA director, was retained as Glennan's deputy, and their appointments were confirmed within days.

A MANNED SATELLITE

Six weeks after Explorer was launched, Maxime A. Faget at Langley started to write the aerodynamic specifications for a recoverable satellite. Contrary to the prevailing belief that a spacecraft would be a derivative of the ever-higher ever-faster trend in aircraft, he

[2] On their transfer to NASA, Langley became the Langley Research Center and Ames became the Ames Research Center.

[3] On 3 December 1958 Eisenhower ordered that JPL be transferred to NASA, and in September 1959 the Department of Defense voluntarily yielded ABMA.

decided to design a 'capsule' that would be launched vertically on a rocket. In space, its orientation (its 'attitude') would be controlled by thrusters firing cold gas. On leaving orbit, it would make a ballistic re-entry and land by parachute. The main prerequisites were a rocket with sufficient thrust to achieve orbital speed, and a thermal shield to protect the capsule during re-entry. Although powerful rockets were already under development, the thermal problem posed a technological challenge. Faget decided on a conical capsule that would enter the atmosphere 'blunt end forward'. The blunt shape would compress the air and generate a shock wave that would prevent most of the heat from impinging on the capsule, and the base would be protected by an ablative shield that would flake off into the slipstream to prevent heat from penetrating the capsule. As the capsule would create no aerodynamic lift, it would follow a ballistic trajectory. This outline was presented to an aerodynamics conference at Ames on 18 March in a paper entitled *Preliminary studies of manned satellites – wingless configuration, non-lifting*.

In September 1958 Glennan and Johnson (of ARPA) agreed to cooperate in the development of a "manned satellite". On 7 October Robert Rowe Gilruth formed an informal group at Langley to explore how to achieve "at the earliest practicable date, orbital flight and successful recovery, of a manned satellite". Left unsaid but understood, was that this must be achieved ahead of any Soviet effort to send a man into space. On 5 November this working party was formalized as the Space Task Group. NASA promptly invited concept proposals from industry for a spacecraft, and set out to recruit 'astronauts' to fly it. On 17 December Glennan announced that the project had been named 'Mercury'. Eleven companies submitted bids, and on 12 January 1959 the contract was awarded to the McDonnell Aircraft Company in St Louis, Missouri.

On 9 April 1959 Glennan announced to a packed press conference the names of the seven military test pilots who had been selected to be astronauts. They were Lieutenant Commander Alan Bartlett Shepard Jr, Lieutenant Malcolm Scott Carpenter and Lieutenant Commander Walter Marty Schirra Jr from the Navy; Lieutenant Colonel John Herschel Glenn Jr from the Marines; and Captain Virgil Ivan 'Gus' Grissom, Captain Donald Kent 'Deke' Slayton and Captain Leroy Gordon Cooper Jr from the Air Force.

THE SATURN LAUNCH VEHICLE

Following World War II, North American Aviation re-engineered the engine of the German V-2 missile, and ABMA incorporated it into the Redstone medium-range missile. When the company developed a more powerful engine, ABMA used it to substantially upgrade the Redstone to create the Jupiter intermediate-range missile. In 1955 North American Aviation created its Rocketdyne division in the West San Fernando Valley in California to design and manufacture engines. For the Atlas intercontinental-range missile for the Air Force, it devised a power plant with three engines that delivered a total of between 360,000 and 380,000 pounds of thrust. In parallel, the company assessed the feasibility of developing a single engine that would produce 1 million pounds of thrust. In April 1957 ABMA initiated a design study for an advanced rocket (dubbed the Super Jupiter) that would be powered by a cluster of E-1 engines, then being considered by

Rocketdyne, each of which could match the thrust of the Atlas trio. On finishing the study in December, the Army submitted to the Department of Defense *A National Integrated Missile and Space Vehicle Development Program*, in which it pointed out that clustering four such engines would yield a booster with 1.5 million pounds of thrust several years before a single engine of comparable power could be developed. On 23 June 1958 the Air Force contracted Rocketdyne to undertake the preliminary design work for a single-chamber engine capable of delivering between 1 and 1.5 million pounds of thrust, which Rocketdyne designated the F-1.

In July 1958 ARPA expressed interest in a booster with 1.5 million pounds of thrust using the clustering technique, but employing proven engines instead of the 'paper' E-1 in order to reduce the development timescale. In response, von Braun suggested clustering eight of the engines used on the Jupiter missile, and dubbed the result the Juno V.[4] In view of the state of the art in rocketry, skeptics doubted that simultaneously igniting eight primary engines would be feasible, and even if it was, combustion instabilities would cause engine shutdowns. Nevertheless, on 15 August ARPA issued Order 14, calling for ABMA to demonstrate the cluster on a static rig by the end of 1959. On 11 September 1958 Rocketdyne was contracted to upgrade the Jupiter engine to a configuration that would be suitable for clustering. Rocketdyne designated this engine the H-1. To save time and money in preparing the test rig, ABMA decided *not* to fabricate a single structure of 20-foot diameter in which large kerosene and liquid oxygen tanks were installed one above the other, but instead to stack eight Redstone tanks around a somewhat wider Jupiter tank. On 23 September ARPA expanded Order 14 to include four test *flights*, the first of which was to be in the fall of 1960, to provide "a reliable high-performance booster to serve as the first stage of a multistage carrier capable of performing advanced missions". ABMA decided to retain the *ad hoc* arrangement of tanks for the flight vehicle.

On 27 January 1959 NASA sent the White House *A National Space Vehicle Program*. Although it bore the name of Abraham Hyatt, chief of launch vehicles, the report had been written primarily by Milton W. Rosen, an engineer who would later take over Hyatt's position. It began by observing that the Vanguard, Juno I, Juno II and Thor–Able were all impromptu developments that could not serve as the basis for a national space program, and pointed out that since launch vehicles would take a long time to develop, it would be necessary to initiate development long before their missions and payloads were defined. Three developments were proposed: one using the Atlas missile, which could be utilized with or without an upper stage; another using the Juno V with a variety of upper stages; and, later, an enormous booster that clustered the F-1 engine. On 2 February the Juno V was renamed 'Saturn' – a name that had been introduced informally by ABMA some months earlier – and two days later ARPA called for the upper stages to be defined as a matter of urgency. Over the ensuing months, and in line with the off-the-shelf strategy pursued by ABMA, the options for adapting existing missiles were considered and

[4] For launching scientific satellites, the Jupiter-C was renamed Juno I. To launch heavier payloads, the upper stages were mated with a stretched Jupiter missile as Juno II. Follow-on concept studies for multistage launchers were labeled Juno III and Juno IV, so this new proposal became Juno V.

classified by an alphabetic nomenclature utilizing numerical subdivisions. The Saturn A-1 followed up a suggestion by Rosen to mount a two-stage Titan I missile on the booster. The Saturn A-2 was an ABMA proposal to create a second stage by clustering Jupiter missiles. The Saturn B-1 was a longer-term option that would involve the development of a completely new second stage. All of these proposals used kerosene-burning engines. Prior to being assigned in 1958 as director of the office of space flight development at NASA headquarters to manage a coordinated effort to devise boosters suitable for a variety of missions, Abraham Silverstein had been associate director at Lewis where, in the early 1950s, he had initiated the design of a hydrogen-burning engine, which was now in development to power the Centaur upper stage for the Atlas and Titan. In March ARPA proposed that when this stage became available it should be used with the Saturn. But by July ABMA had concluded that off-the-shelf upper stages would be too limiting: in particular, the narrowness of the missiles compared to the 20-foot-diameter Saturn first stage would severely restrict the payload. Meanwhile, the Air Force had concluded that it could achieve the goals of the Department of Defense using missile conversions, and therefore discussions were initiated to transfer von Braun's team and the Saturn launch vehicle to NASA, which was delighted to get this expertise 'in house', because this would enable it to configure the vehicle to its own requirements. The formal transfer was signed by Glennan on 21 October, but the transition of facilities and staff took six months.

On 27 November, while this handover was underway, Silverstein created the Saturn vehicle evaluation committee, which promptly reaffirmed that *new* stages would be required. The main issue was whether these should burn hydrogen rather than kerosene. For a specific propellant mass, hydrogen promised 40% greater lifting power than kerosene. If the upper stages were to burn kerosene, they would require more engines, more propellant and more tankage to make up this thrust – all of which was additional mass – and despite its unprecedented thrust the vehicle would soon reach its limit. ABMA would have preferred to use kerosene-burning engines, but acquiesced because the Centaur was scheduled to be flight tested long before the Saturn would be ready to accept upper stages, by which time it was fair to assume the engine would be reliable. By the end of 1959, therefore, the 'A' series using off-the-shelf conversions had been discarded as too limiting; the 'B', with its large new kerosene-burning second stage, had been ruled out by the decision to use hydrogen-burning engines; and a 'building block' approach had been devised for the development of the 'C' series, which was initially envisaged as having three variants. The largest variant, the C-3, would use the S-I, S-II, S-III, S-IV and S-V stages; the C-2 would be the C-3 without the S-III stage; and the C-1 would be the C-2 without the S-II stage. Thus, despite this eminently logical nomenclature, the first vehicle to be developed would comprise the S-I, S-IV and S-V stages! While the 20,000-pound-thrust hydrogen-burning engine then in development would be able to power the S-IV and S-V stages, the S-II and S-III would require an engine with 200,000 pounds of thrust (with four engines in the S-II and two in the S-III). Silverstein recommended an immediate start on the development of this new engine (which, on being contracted on 31 May 1960, was designated the J-2 by Rocketdyne). The Saturn C-1, C-2 and C-3 were expected to be capable of launching the components for a space station and the manned missions that would assemble it in space and operate it, and even be capable of launching a circumlunar mission. However, a lunar landing would require either several C-3 launches, each carrying a part of the spacecraft, or a booster powered by F-1 engines capable of dispatching the mission with a single vehicle. On 7 December ARPA accepted Silverstein's call and requested an engineering and cost study of (1) the S-IV stage with a cluster of four 20,000-pound-thrust hydrogen-burning engines, and (2) an S-V stage based on the version of the Centaur using two such engines. This study was submitted by NASA/ABMA on 28 December. A few days later, in line with the Saturn vehicle evaluation committee, NASA approved a research and development plan for 10 Saturn C-1 launchers. On 14 January 1960 Eisenhower urged NASA to "accelerate" work on "the super booster", and on 18 January gave the Saturn the government's highest national priority rating – known as 'DX' – for procurement of materials and payment of overtime rates. On 1 July 1960 Eisenhower attended a ceremony to rename the ABMA facility in Huntsville the George C. Marshall Space Flight Center, and NASA appointed von Braun as its director.

NASA'S LONG-TERM PLAN

At the start of 1959 the House select committee on astronautics and space exploration issued *The Next Ten Years in Space*, which was a staff study of a poll of the aerospace community. Prominent among the projected post-Mercury manned space programs was a circumlunar flight that would undertake a lunar reconnaissance as a precursor to sending a mission to land on and explore the surface. On 15 December 1958 von Braun had briefed Glennan on how a lunar landing might be mounted. The simplest strategy, known as 'direct ascent', had been depicted in the well-received 1950 movie *Destination Moon*, but von Braun calculated that it would require a rocket with a cluster of at least 10 of the most powerful engines then under consideration. A more sensible solution, he insisted, would be to assemble the spacecraft in Earth orbit from specialized components launched on a succession of smaller rockets. He outlined four options in which as many as 15 smaller rockets would launch the components of a 200-ton spacecraft into low orbit, where they would be linked up prior to departing for the Moon. In fact, he observed, it would be best to create a space station, at which missions to the Moon could be prepared. In view of this, on 7 April 1959 Glennan requested funding in Fiscal Year 1960 to conduct a feasibility study of rendezvous in space: specifically to identify referencing methods by which the relative positions of vehicles in space could be accurately measured; to develop apparatus to enable one vehicle to locate and track another; and to develop the guidance and control systems required to enable a vehicle to follow a precisely determined path.

The next day, 8 April, NASA established the research steering committee for manned space flight to study options for post-Mercury activities. It was chaired by Harry J. Goett, who was at that time at Ames but in September became director of the Robert H. Goddard Space Flight Center, which was opened by NASA on 1 May 1959 at Greenbelt, Maryland. Goett invited the various field centers to nominate representatives to the committee. It met on 25–26 May, and again on 25–26 June, and drew up its conclusions on 9 December. It studied a scenario outlined by von Braun, which envisaged placing nine Centaur stages into low orbit to serve as (and refuel) the 'escape' propulsion system, three more to serve as (and refuel) the stage that would perform the lunar orbit insertion maneuver and the descent to the surface, one to break out of lunar orbit, and one to insert the manned spacecraft into Earth orbit to initiate the mission – a total of 14 launches, all of which would have to be accomplished within the period of several weeks owing to the difficulty of storing cryogenic propellants in space. Although the more powerful Saturn variants then under consideration would ease the task, the committee concluded that a direct ascent would be simpler. However, as it would take many years to develop the requisite launch vehicle, the committee called for a strategy by which the Saturns would facilitate the assembly of a space station in the mid-1960s, circumlunar flights by the end of the decade, and a lunar landing in the 1970s. Homer J. Stewart, head of the office of program planning and evaluation, drew on this report in preparing the *Ten Year Plan* to be presented to Congress. On 29 December Silverstein established the Space Exploration Program Council under the chairmanship of associate administrator Richard E. Horner to oversee the development and implementation of programs. On 28 January 1960 Horner presented the *Ten Year Plan* to the House committee on science and astronautics, and on 28 March Glennan presented it to the Senate committee on aeronautical and space sciences.

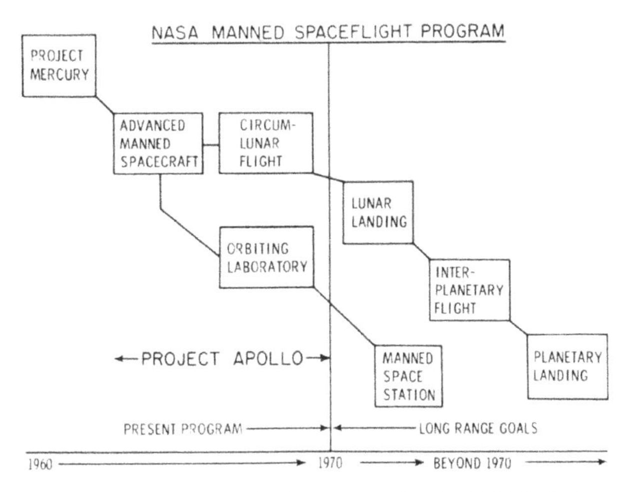

When NASA drew up its long-term plan in 1959, it did not envisage making a manned lunar landing until after 1970.

CONCEIVING THE APOLLO SPACECRAFT

Meanwhile, on 27 May 1959 Gilruth had directed the Space Task Group to study the possibility of "an advanced manned spacecraft". On 12 August the new-projects panel chaired by H. Kurt Strass was established to consider an advanced program involving automated satellites, probes and landing craft which would lead up to a manned lunar landing. On 18 August the panel recommended that the first major project to be investigated should be a "second generation" spacecraft with a three-man capacity, in-space and atmospheric maneuverability, advanced abort devices, potential for a near-lunar-return velocity of 36,000 feet per second, and advanced recovery techniques. In January 1960 Silverstein suggested to Gilruth that this new project should be named 'Apollo'. On 1 February von Braun gave the Saturn vehicle evaluation committee *A Lunar Exploration Program Based Upon Saturn-Boosted Systems*, refining how a lunar landing might be undertaken. On 5 March the Space Task Group was instructed to draw up a draft specification for the advanced manned spacecraft. The preliminary guidelines for its development were formulated on 8 March, refined in April and May during consultations with the other field centers, and published in June. It was to be able to support a three-man crew in a shirt-sleeved environment for up to 14 days; be compatible with both the Saturn C-1 and C-2 variants; be capable of operating with "a space laboratory" in orbit around the Earth and of undertaking a "lunar reconnaissance"; and hopefully be ready for manned qualification flights in 1966. On 16–17 May Langley held a NASA-wide meeting to review the issues relating to space rendezvous. The consensus was that funding should be sought to conduct experiments to investigate the practicalities of rendezvous. On 25 May the Space Task Group established the advanced vehicle team under the chairmanship of Robert O. Piland to undertake a preliminary design study of the three-manned spacecraft. On 5 July the House committee on science and astronautics declared that the *Ten Year Plan* was a good program as far as it went, but it did not go far enough; the committee recommended that a "high priority program" should be undertaken "to place a manned expedition on the Moon in this decade", and urged NASA to draw up a plan to achieve this. On 25 July Glennan approved the name Apollo, and on 28 July Dryden made the announcement at the start of a two-day NASA–industry program plans conference hosted by headquarters to appraise industry of the agency's *plans* for launch vehicle development and its *aspirations* for post-Mercury manned missions (aspirations, because as yet there was no funding commitment). On 30 August another session was held to outline the requirements for the spacecraft, as defined by the Space Task Group, emphasizing that it must not only operate in conjunction with a space station in Earth orbit but also make a circumlunar flight and be capable of directly supporting a lunar landing mission. On 1 September 1960 the Space Task Group appointed Faget chief of its flight systems division, and established within it the Apollo project office headed by Piland.[5]

On 12–13 September the Space Task Group briefed potential industrial bidders for the contract to develop the Apollo spacecraft, and released the formal request for proposals. By the closing date of 9 October, 14 bids had been received. On 25 October Convair/Astronautics of San Diego, General Electric of Philadelphia, and the Martin Company of Baltimore were each awarded six-month $250,000 study contracts to propose designs. By 21 October, the Space Task Group had decided that the spacecraft must have a modular design in which a 'service module' of subsystems could be jettisoned prior to re-entry, and that the 'command module' in which the crew would return to Earth should have a lift-to-drag ratio of 0.35 with sufficient roll capability to enable it to control its hypersonic trajectory in order to aim for a specific recovery point, incorporate an overall heat shield, and be rated for a nominal maximum load of 8 *g*. All this information was forwarded to the industrial studies as part of the documentation for the initiation of their studies in November. Meanwhile, on 30 September the Space Exploration Program Council decided that because the spacecraft was likely to be too heavy to be launched by the Saturn C-1 it would require the Saturn C-2, the development of which should proceed apace. However, on 5 October the Marshall Space Flight Center agreed to a Space Task Group request to launch "boilerplate" spacecraft on

[5] The Apollo project office in the flight systems division was raised to report directly to Gilruth in December 1961, and then superseded by the Apollo spacecraft project office in January 1962.

Saturn C-1's to test the jettisoning of the launch escape system, to supplement 'live' tests using smaller Little Joe II boosters.

After serving on the team that laid the groundwork for the creation of NASA in 1958, George Michael Low, then chief of the special projects branch at the Lewis Propulsion Laboratory, was reassigned as program chief for manned space flight in the office of space flight development at headquarters. In a reorganization on 8 December 1959, this became the office of space flight programs. On 17 October 1960 Low told Silverstein of his intention to form a committee to investigate the circumlunar objective in greater detail, in order to ensure that the Apollo spacecraft would be capable of supporting a landing mission.

THE WAY TO THE MOON – THE MISSION MODE

Robert Channing Seamans Jr, formerly a senior manager at the Radio Corporation of America, superseded Richard Horner as associate administrator on 1 September 1960. He saw his role as assisting Glennan and Hugh Dryden by serving as the agency's 'general manager'. In late 1959 Langley had established a lunar mission steering group to investigate issues related to landing on the Moon. One paper by William H. Michael Jr outlined how a specialized craft could be separated from its mother ship in lunar orbit, descend to the surface, and then rendezvous after lifting off. Michael noted that since less propellant would be required to land and return a small lander than the entire craft, a two-part spacecraft would significantly reduce the mass of the launch vehicle. The complication was the requirement to undertake a rendezvous in lunar orbit. On his first visit to Langley, Seamans was briefed on this by John Cornelius Houbolt. As a result of having been involved in a project which the Air Force hoped would enable a craft to intercept Soviet satellites, Seamans was familiar with the concept of rendezvous in space, and he asked the group to make a presentation at headquarters in December. On 10 December the group briefed the Space Task Group on the benefits of lunar orbit rendezvous, then on 14 December briefed managers in Washington on the assembly of a spacecraft in orbit, the savings in mass to be gained from lunar orbit rendezvous as compared to direct ascent, and the practicalities of rendezvous – which Houbolt said should be straightforward.

John C. Houbolt outlines the concept of lunar orbit rendezvous.

Soon after his 'get to know you' tour of the field centers, Seamans decided on some restructuring. He made the Space Task Group a separate 'field element' that reported to Silverstein, rather than through Langley director Floyd L. Thompson to the office of research and advanced technology. This move, which came into effect on 3 January 1961, reflected the role of the Space Task Group in managing project Mercury, which was at that time gearing up to its first manned launch, and defining the Apollo spacecraft – two functions that would be better managed by simplifying the link to Silverstein, who was in overall charge of manned space. Seamans also made Low director of manned space craft and flight missions, which put him directly under Silverstein, as opposed to being several levels down as a mere program chief.

On 5–6 January 1961, after receiving presentations on Earth orbit rendezvous by von Braun, lunar orbit rendezvous by Houbolt, and direct ascent by Melvyn Savage of headquarters, the Space Exploration Program Council formalized the *ad hoc* committee that Low had recently established, renamed it the Manned Lunar Landing Task Group, expanded its remit to investigate the options for the 'mission mode' for a lunar landing, and required it to supply cost and schedule estimates for inclusion in the Fiscal Year 1962 budget proposal. Concluding that rendezvous was a generic skill that the agency would have to perfect, Low recommended that funding be sought to make a start on developing this capability. Meanwhile, on 10 January the lunar mission steering group at Langley and the Space Task Group refined the savings in mass likely to be derived from using lunar orbit rendezvous for a manned lunar landing.

CHANGE OF ADMINISTRATION

Although Eisenhower had ordered the development of the Saturn booster and then accelerated the program, 1960 was an election year with John Fitzgerald Kennedy running against Richard Milhous Nixon, and NASA knew that there would be no commitment to develop the Apollo spacecraft until after the election in November. On 11 January 1961, shortly before his inauguration, Kennedy released a report by an *ad hoc* committee on space which had been chaired by Jerome B. Wiesner of the Massachusetts Institute of Technology, whom he intended to make his science adviser, recommending actions that his administration ought to take. During the campaign Kennedy had made an issue of Eisenhower's space policy, although his criticism had focused on what he referred to as the "missile gap" rather than space exploration as such. Wiesner's report had been prepared during the transition and, its title notwithstanding, it addressed both the space and missile programs. In fact, in urging more effort on large launch vehicles and downplaying the significance of the manned space program, the report was consistent with Eisenhower's view that large boosters were a national priority and that project Mercury represented a sufficient response to the perceived insult of Sputnik. On 20 January Kennedy was sworn in as the 35th President of the United States of America. In view of the change of administration, Glennan submitted his resignation. On the advice of Vice President Lyndon Baines Johnson, on 30 January Kennedy nominated James Edwin Webb to supersede him. Dryden agreed to remain in place as deputy. On 9 February the Senate approved Webb's appointment, and on 15 February he was sworn in.

SEEKING FUNDING FOR APOLLO

On 7 February 1961 George Low's Manned Lunar Landing Task Group sent *A Plan for a Manned Lunar Landing* to Seamans. Apollo would serve multiple roles, including servicing a space station in low Earth orbit. Although the Saturn C-2 would be able to launch the spacecraft on a lunar reconnaissance, a larger launch vehicle would be required for a direct ascent landing mission. The Saturn C-2 was expected to be capable of injecting 8 tons onto a translunar trajectory, whereas a landing mission might require in excess of 36 tons, and this would involve either multiple Saturn C-2 launches and assembly of the spacecraft in Earth orbit prior to setting off (this being von Braun's preferred mode) or the development of a single launcher (which von Braun warned might pose too great a leap beyond the state of the art). The committee noted that regardless of which mode was chosen, 'orbital operations' (which was von Braun's term for space rendezvous) would be required for the assembly of the space station and, given the existence of the space station, this might well host the assembly of a lunar spacecraft. Wiesner, who had already criticized NASA's interest in expanding the manned program, favoured the use of satellites for scientific research. Although new to the agency, Webb was firmly of the opinion that getting men into space would be the primary means by which the nation would advance its capacity to operate in space in whatever manner it might later choose, and therefore the agency's objective should be to provide this capability.

On 20 March the Marshall Space Flight Center recommended revisions to the configurations of the Saturn variants. Firstly, it recommended eliminating the S-V stage from the Saturn C-1 in order to reduce the development time. The two-stage version would still be able to insert 10 tons into low orbit, which represented a significant capability for launching satellites. It also recommended upgrading the S-IV stage by switching from four LR-119 engines with a total of 70,000 pounds of thrust to six LR-115 engines with 90,000 pounds of thrust. On 31 March NASA approved the development of the Saturn C-2, which would require specifying the new second stage and appointing the contractor to supply it.

In late March, in order to make a start on developing the Apollo spacecraft in line with the *Ten Year Plan*, Webb met Kennedy to discuss whether the White House would support a NASA request for the money Eisenhower had denied. The next day Kennedy accepted Wiesner's advice that boosters were the 'pacing item', and on 28 March sent Congress an amended budget request that was $125.67 million more than Eisenhower had sought, but he opted to postpone making a decision on Apollo until NASA's long-term plans had been evaluated.

KENNEDY'S CHALLENGE

On 12 April Yuri Alexeyevich Gagarin flew the first manned orbital mission for the Soviet Union. At his regular press conference later that day, Kennedy said, "no one is more tired than I am" of seeing the United States second in space. The Soviets had "secured large boosters which have led to their being first in Sputnik, and led to their first putting their man in space. We are, I hope, going to be able to carry out our efforts, with due regard to the problem of the life of the man involved, this year." He expected that "the news will be worse before it is better, and it will be some time before we catch up". That same day, Seamans established the Saturn program requirements committee chaired by William A. Fleming, an acting assistant administrator, to review on an ongoing basis the plans for missions that were to use the Saturn, and correlate them with the development and

A Redstone missile lifts off with Al Shepard's Mercury spacecraft.

procurement of launch vehicles. On the evening of 13 April, Kennedy asked Webb and Dryden to the White House to discuss how NASA might catch up with the Soviet Union in space. On 10 April the dormant National Aeronautics and Space Council had been revived, and Lyndon Johnson was appointed its chairman. On 20 April Kennedy asked the Space Council to canvass opinion and suggest a long-term strategy. "Do we have a chance of beating the Soviets by putting a laboratory in space, or by a trip around the Moon, or by a rocket to land on the Moon, or by a rocket to go around the Moon and back with a man? Is there any other space program which promises dramatic results in which we could win?" He demanded a response "at the earliest possible moment". At a press conference on 21 April Kennedy pointed out that his administration was considering the options and cost of space. "If we can get to the Moon before the Russians, then we should." At a meeting of the Space Council on 24 April, Lyndon Johnson, as Wiesner later related, "went around the room saying, 'We've got a terribly important decision to make. Shall we put a man on the Moon?' And everybody said 'yes'. And he said 'thank you' and reported to the president that the panel said we should put a man on the Moon." On 28 April Johnson submitted the Space Council's recommendation: "Largely due to their concerted efforts and their earlier emphasis upon the development of large rocket engines, the Soviets are ahead of the United States in world prestige attained through impressive technological accomplishments in space. The US has greater resources than the USSR, etc. The country should be realistic and recognize that other nations, regardless of their appreciation of our idealistic values, will tend to align themselves with a country which they believe will be the world leader. The US can, if it will firm up its objectives and employ its resources, have a reasonable chance of attaining world leadership in space. If we don't make a strong effort now, the time will soon be reached when the margin of control over space and other men's minds through space accomplishment will have swung so far on the Russian side that we will not be able to catch up. Even in those areas in which the Soviets already have the capability to be first and are likely to improve upon such capability, the United States should make aggressive efforts, as the technological gains as well as the international rewards are essential steps in gaining leadership. Manned exploration of the Moon, for example, is not only an achievement with great national propaganda value, but is essential as an objective, whether or not we are first in its accomplishment..." Kennedy was receptive, but postponed making a formal decision until after the first manned Mercury mission, which was scheduled for early May.

On 25 May 1961 John F. Kennedy challenged his nation to land a man on the Moon before the decade was out.

On 2 May, Seamans, alerted to Kennedy's thinking, asked Fleming to chair an *ad hoc* task group for a manned lunar landing study and "detail a feasible and complete approach" for an "early" landing. The word "early" was in contrast to the previous expectation that a landing would not occur until some time in the 1970s. In fact, the date in the guidelines for this study was 1967 because it was thought that the Soviets might make a landing that year in order to mark the 50th anniversary of the Bolshevik Revolution. Fleming drew his committee primarily from headquarters and the Space Task Group. After assessing a scenario that Houbolt had recently outlined involving two Saturn C-2 launches, the committee dismissed rendezvous as an unjustifiable risk, which left only direct ascent. Consequently, in its report on 16 June, the committee recommended the development of a massive booster.

Meanwhile, on 5 May Alan Bartlett Shepard flew the first manned Mercury mission. Riding a Redstone missile, his 15-minute ballistic arc down the Atlantic Missile Range was less impressive than a full orbit, but NASA was finally in the manned spaceflight business. At a press conference later that day, Kennedy said that he was minded to undertake "a substantially larger effort in space". And on cue, that same day, the Space Task Group completed *Project Apollo, Phase A: General Requirements for a Proposal for a Manned Space Vehicle and System – Statement of Work*, this being the draft specification for the Apollo spacecraft. The basis for this planning was a circumlunar mission. On 8 May Webb and Robert S. McNamara, the Secretary of Defense, sent Lyndon Johnson their joint recommendation on how America should challenge the Soviet Union in the space arena. Several days later, the White House let it be known that Kennedy would soon address a joint session of Congress. On reviewing a draft speech calling for a manned lunar landing in 1967, Webb urged that the phrase "by the end of the decade" be substituted as a more realistic date.

On 25 May 1961 Kennedy gave a speech to Congress on the theme of *Urgent National Needs* that was billed as his second State of the Union message. In view of recent space achievements by the Soviets, he proclaimed, "Now it is time to take longer strides, time for a great new American enterprise, time for this nation to take a clearly leading role in space achievement, which in many ways may hold the key to our future on Earth." Having outlined the political background, he laid down the gauntlet. "I believe that this nation should commit itself to achieving the goal, before this decade is out, of landing a man on the Moon, and returning him, safely, to the Earth." He had opted for a lunar landing precisely because it posed a great technical challenge. By literally 'shooting for the Moon', he was betting that America would not only be able to catch up with the Soviets in space, but forge ahead. Having concluded that space was the arena of superpower politics, he was challenging his rival, Nikita Khrushchev, for world leadership. He had imposed the deadline to ensure that reaching the Moon was perceived as a 'race'. To reinforce the magnitude of the task, he noted, "No single space project in this period will be more impressive to mankind, or more important for the long-range exploration of space; and none will be so difficult or expensive to accomplish." Sending a man to the Moon was to be the modern form of the ancient practice of 'single combat', by which opposing armies lined up and each dispatched a single warrior to decide the issue. Kennedy made sure everyone understood this analogy, "In a very real sense, it will not be one man going to the Moon, if we make this judgment affirmatively it will be an entire nation, for all of us must work to put him there." Finally, to emphasize what was at stake, he warned, "If we are to go only halfway, or reduce our sights in the face of difficulty, in my judgment it would be better not to go at all."

THE MODE DEBATE CONTINUES

NASA thus had its post-Mercury manned program, which was neither to build a space station, nor to fly a lunar reconnaissance, but to make a landing – and do it within the decade! On the day of Kennedy's speech, 25 May 1961, Seamans asked Don Ostrander, director of launch vehicle programs, to set up a committee "to assess a wider variety" of options. Ostrander appointed Bruce T. Lundin of Lewis as chair. This *Broad Study of Feasible Ways for Accomplishing a Manned Lunar Landing Mission* was to focus on the launch vehicle issues. In contrast to Fleming, Lundin drew his membership primarily from the field centers. One new factor was that the Marshall Space Flight Center had just concluded that a circumlunar mission would require a more powerful vehicle than the Saturn C-2. On 1 June, following up on the recommendations of March, the S-V stage was deleted from the Saturn C-1, but a mock up was to be retained for the Block I test flights of this vehicle, the first of which was imminent.

On 10 June the Lundin committee reported with *A Survey of Various Vehicle Systems for the Manned Lunar Landing Mission*. As direct ascent was deemed to be impractical in the 1967–1970 timescale, the committee investigated rendezvous around the Earth and rendezvous around the Moon, together with the possibility of using both, and recommended using two or three Saturn C-3 launches to assemble the spacecraft in Earth orbit prior to setting off for the Moon, with the spacecraft entering lunar orbit as a preliminary to landing. To opt for lunar orbit rendezvous would require the development of a larger launcher that might serve no other role, whereas the Saturn C-3 was better able to undertake a variety of types of mission. On 16 June the Fleming committee submitted *A Feasible Approach for an Early Manned Lunar Landing*, and recommended direct ascent. Noting that the 'pacing item' in an attempt to achieve a lunar landing within the decade was likely to be the development of the launch vehicle and its static test and launch facilities, the report recommended that the construction of a new NASA launch site be given a high national priority. On 18 June Seamans decided that it would be necessary to even out the level of detail in the mode studies before a decision could be made. While

Ostrander worked with von Braun to develop an end-to-end mission plan, Ostrander's deputy, Donald H. Heaton, was asked on 20 June to establish an *ad hoc* task group to explore the program requirements for a range of rendezvous-based modes, in particular the configurations of the upper stages of the 'advanced Saturns', and report in August. In order to develop a NASA-wide viewpoint, this committee drew its membership both from headquarters and the field centers. In the meantime, on 22 June Webb chaired a meeting at headquarters that decided to terminate work on the Saturn C-2, to authorize the preliminary design of the C-3 and to undertake studies of more powerful variants in order to provide necessary input to the decision on the mission mode. By this time von Braun was considering a C-3 with a pair of F-1 engines in its first stage delivering a total of 3 million pounds of thrust; a C-4 with four F-1 engines; a C-5 with five F-1 engines on the first stage, four J-2 engines on the second stage and a single J-2 on the third stage; and a C-8 with eight F-1 engines in the first stage, eight J-2 engines in the second stage and a single J-2 on the third stage in a configuration that some referred to as the 'Nova' booster. On 23 June Seamans followed up on Fleming's warning about facilities by asking Kurt Heinrich Debus, the director of NASA's launch operations directorate, and Major General Leighton Ira Davis, the commander of the Air Force Missile Test Center at Cape Canaveral,[6] to report on the launch requirements, methods and procedures for a manned lunar landing by various mission modes, specifying respective range responsibilities, authority and management. On 30 June Seamans clarified that they were to focus on possible launch sites, issues of land acquisition, facilities for spacecraft and launch vehicle preparation, and the launch facilities, but *not* to consider tracking and command stations. They reported on 31 July that they had assessed eight possible sites: Cape Canaveral (on-shore); Cape Canaveral (off-shore); Mayaguana Island (downrange on the Atlantic Missile Range); Cumberland Island, Georgia; Brownsville, Texas; White Sands Missile Range, New Mexico; Christmas Island in the Pacific Ocean; and South Point on Hawaii. On the basis of minimizing costs and allowing for the tight timescale, they had shortlisted Cape Canaveral (on-shore) and White Sands, both of which had established tracking ranges, but the need to preclude spent first stages falling on populated areas imposed severe constraints on the range of launch azimuths at White Sands. On 24 August, therefore, NASA announced that it would build the Apollo launch facilities on Merritt Island, several miles up the coast from Cape Canaveral.

On 20 July Seamans asked Nicholas Erasmus Golovin, one of his assistants, to chair a large launch vehicle joint planning committee, which was to recommend launch vehicle configurations required not only for Apollo but also the activities of the Department of Defense. This group rapidly decided that, in the case of Apollo, the requirements of the launch vehicle were "inextricably linked" to the mission mode, and set out to compare direct ascent with methods of rendezvous. The office of launch vehicle programs described direct ascent. The Marshall Space Flight Center described Earth orbit rendezvous. At the request of the chairman, a study of 'lunar surface rendezvous' was

[6] On 15 May 1964 the Air Force Missile Test Center was redesignated the Air Force Eastern Test Range.

presented by JPL. Houbolt described lunar orbit rendezvous on 29 August, and on 31 October sent a two-volume *Manned Lunar-Landing Through the Use of Lunar Orbit Rendezvous*, which detailed contributions by the entire Langley team. At this time, the Space Task Group was wary of the unproven task of rendezvous, and therefore argued for direct ascent. After three months of debate, the committee found itself unable to recommend a mode, but it did conclude that the F-1 engine *would* be required, which at this time meant either the Saturn C-4 or C-5. Meanwhile, in August the Heaton committee had submitted its report, *Earth Orbital Rendezvous for an Early Manned Lunar Landing*, stating that "rendezvous offered the earliest possibility for a successful manned landing", and envisaging a dual launch in which the spacecraft would rendezvous and dock with a propulsion module prior to heading for the Moon. It also opined that while the Saturn C-3 with the eight-engine H-1 cluster *might* be adequate, a wiser choice would be a variant utilizing the more powerful F-1 engine. Faced with committees reaching contradictory conclusions, Seamans realized that the matter of the mode rested on two imponderables: (1) the viability of orbital operations, and (2) the engineering challenge of developing a "superbooster". Put simply, if rendezvous proved impracticable there would be no option but to develop the superbooster, but if the challenges of rendezvous had been overstated then the development of such an enormous launcher would be unnecessary.

SATURN TESTING STARTS

Rocketdyne successfully test fired the H-1 prototype at full power on 17 December 1958. The first production engine was delivered to Huntsville on 3 May 1959, where it was installed in a static rig and test-ignited on 21 May, before being fired for 80 seconds on 29 May. On 3 June construction of Launch Complex 34 began at Cape Canaveral to facilitate flight testing. On 28 March 1960 two of the engines of the S-I stage test article were simultaneously fired for 8 seconds; on 6 April four were fired; and on 29 April all eight were fired. In a public demonstration on 26 May, the stage fired for 35 seconds and achieved 1.3 million pounds of thrust. That same day, assembly of the first flight article started. On 15 June the test stage was fired for the full mission duration of 122 seconds. On 7 March 1961 the first flight article was installed for tests. On 5 June Pad 34 was declared operational. At 2,800 tons, the 310-foot-tall gantry was the largest movable land structure in North America. The first test flight occurred on 27 October 1961, with the S-IV and S-V stages 'boilerplate' mockups ballasted with water. The perfect flight confounded the skeptics who had doubted that the engines would all ignite, or, if they did, that premature shutdowns would terminate the flight at an early stage.

The eight-engine first stage of the Saturn C-1 launch vehicle.

MERCURY MARK II?

On 20 January 1961, with Apollo not yet funded, Robert Gilruth convened a panel of the Space Task Group to discuss immediate post-Mercury options, focusing primarily on the hardware requirements for the spacecraft and the launch vehicle. Max Faget outlined two broad classes of mission, one to demonstrate rendezvous in space and the other to pursue extended mission duration. The publicly stated goal of Mercury was a three-orbit flight, but it was evident that with supplemental consumables the spacecraft would be able to operate for a day or so. However, the addition of either a rocket engine to conduct orbital maneuvering or the stores to sustain a week-long mission would push the spacecraft over the mass that the Atlas could place in orbit, in which case it would be necessary either to add an upper stage or to switch to a more powerful launcher. Faget was also concerned that the task of rendezvousing in space might prove to be "too hazardous for a one-man operation", and so he proposed making it into a two-man spacecraft, in which case a more powerful vehicle would certainly be required. And if the spacecraft was to be adapted for more sophisticated missions, he had a list of "refinements" intended to make it easier to assemble, check out and service. On 1 February Gilruth told James Arthur Chamberlin, the project's engineering director, to work with the McDonnell Aircraft Company to draw up the specifications for a 'Mercury Mark II' spacecraft. Although Chamberlin set out simply "to make a better mechanical design" without changing the overall form or the function of the spacecraft, he soon concluded that it would require to be completely redesigned, and reported this to Gilruth on 17 March. Prior to joining NASA in 1959, Chamberlin had designed fire-control systems for fighter aircraft, and these systems were built to facilitate frequent maintenance. The flaw with Mercury, he said, was that its electrical systems were stacked inside the capsule in the manner of a multi-tiered cake, whereas they should be directly accessible from outside, as on an aircraft. He proposed stripping the spacecraft's systems down and repackaging them so that the spacecraft would not only be simpler to assemble but also easier to service on the launch pad. Like Faget, he wanted to transform an engineering test vehicle into an operational spacecraft. On 12 April McDonnell submitted a proposal to provide a detailed engineering study of such a redesign, and this was approved two days later. Furthermore, the company was authorized to purchase long-lead-time items, so that, if NASA decided to pursue the idea, half a dozen of the 'new' spacecraft could be built without inordinate delay.

In St Louis, Chamberlin methodically worked out how "to increase component and system accessibility" in the Mercury spacecraft, and when he reported to the Capsule Review Board on 9 June, Gilruth, Low and Faget were astonished by the scope of the changes. In "dissociating" what he saw as overly integrated systems Chamberlin wanted to repackage – or modularize – almost every system and place it outside rather than inside the pressure shell. Furthermore, in a reversal of the original design requirement that the spacecraft be capable of flying automatically, Chamberlin wanted to delete the automatic sequencer that had proven so difficult to certify – "the root of all evil" he called it – and rely instead on the astronaut to control the vehicle. Only the external shell of the spacecraft had been sacrosanct, because he was not permitted to change the aerodynamics. Chamberlin had taken a particular dislike to the launch escape system. The Atlas burned kerosene with oxygen, but this was such a volatile mixture that a tower with an escape rocket had been placed on the nose the Mercury spacecraft to pull it clear of the fireball of an exploding booster. Quite apart from the fact that this represented dead weight, the sequencer was a nightmare to check out. In Chamberlin's view, if the spacecraft's functionality was to be upgraded, then items such as the automatic sequencer would have to be deleted to accommodate the limitations of the Atlas. The initial decision to use the Atlas for Mercury had been straightforward, because it had been the most powerful missile in the inventory at that time. But on 8 May, Albert C. Hall, the general manager of the Martin Company's Baltimore division, had briefed Silverstein on its new Titan II intercontinental missile, which was a two-stage vehicle with much greater capacity than the Atlas. Silverstein suggested that Gilruth should consider 'man rating' it, and Chamberlin seized on this as a means of escaping from the Mercury spacecraft's mass limit. It was not only the raw performance that was attractive: the Titan II had a rugged design that would make it straightforward to erect and check out. It also had one major advantage: although the hypergolic propellants burned on coming into contact, eliminating the need for an ignition system, they were less explosive, which meant that in a catastrophic malfunction there would be time for an ejection seat to carry the occupant clear. The heavy launch escape system with its infernal sequencer could now be dispensed with, and as the ejection seat would require the installation of a larger hatch, this could be designed to hinge open instead of being bolted on, which would not only simplify the preparations for launch but would also offer the prospect of the astronaut opening the hatch in space to venture outside. When the Capsule Review Board met again on 12 June it was concluded that if the spacecraft was to be made *operational* in the long-term, the investment in such an extensive redesign would pay off. However, lacking the necessary funding, the Board restricted Chamberlin to developing the "minimum modifications" to facilitate a 24-hour flight, which was something that seemed likely to be required once the first orbital mission had been achieved.

Nevertheless, the Space Task Group set out to work through the implications of Chamberlin's full proposal. On 7 July Walter Burke of McDonnell gave a briefing in which he outlined three options. The first (which is what the company had been authorized to do) was the "minimum modifications" for a 24-hour flight. The second was to "reconfigure" the spacecraft as Chamberlin had outlined. As a third enhancement, Burke proposed adding another seat. In fact, Faget had raised this possibility at the Capsule Review Board on 9 June, arguing that if they were going to overhaul the spacecraft to the degree proposed by Chamberlin, they might as well make it a two-man vehicle. He had already suggested that if rendezvous proved to be as tricky as some feared, it would be prudent to include a second pilot to share the load. And if the hinged hatch was to facilitate external activity, then the second pilot would be able to look after the spacecraft and render assistance as necessary. There was a good case to be made for modularizing the spacecraft and upgrading its capabilities, but even after Kennedy directed NASA to shoot for the Moon there was no *requirement* to do so. However, when Gilruth and Silverstein went to St Louis on 27 July to review progress on the "minimum modifications", they were shown a wooden mockup of the two-man spacecraft, which, on a panel between the seats, had an integrated hand-controller for the orbital maneuvering system. Although not much larger than Mercury, it had a spectacularly different 'look and feel'. Wally Schirra, one of the astronauts, was impressed. The next day, Silverstein told the company to concentrate exclusively on the two-man variant, as it was simply too good a spacecraft to ignore.

ISSUING CONTRACTS

While the mission mode was debated, NASA, with its audacious mandate in place, moved rapidly to place contracts. On 22 May 1961 the Space Task Group finished the second-draft *Statement of Work* for the specifications of the spacecraft, and in June factored into this the conclusions drawn from the three six-month conceptual studies submitted earlier in the month. On 18–20 July a NASA–industry Apollo technical conference held at headquarters briefed potential bidders. On 28 July Seamans approved procurement, appointed the source evaluation board, and gave it authority to form the requisite assessment teams.[7] The request for proposals was sent to 12 companies. A three-phase program was specified: Phase A called for the Saturn C-1 to be used for a variety of unmanned tests in Earth orbit, leading to a 14-day manned mission; Phase B was to use the Saturn C-3 (or, if necessary, a more powerful variant) to conduct a circumlunar flight; and Phase C was to make the lunar landing. It was stipulated that the spacecraft must be capable of supporting the landing mission, but what this might actually involve could not be determined until the mode issue had been resolved. On 9 October five companies submitted bids.

On 7 September NASA announced its decision to fabricate and assemble the first stages of the Saturns at the government-owned Michoud Ordnance Plant near New Orleans, Louisiana. This sprawling facility had been commissioned early in World War II as a yard for making Liberty ships. It had been converted in 1943 to make cargo aircraft, was decommissioned at a later date, then converted to make tanks for the Army during the Korean War. Although it was declared as surplus, it ideally suited NASA's needs because the intracoastal waterways of the Tennessee Valley Authority provided access to Huntsville by barge, and its Gulf coast location enabled onward shipment of finished stages to the Cape. On 17 September NASA invited 36 companies to bid, and a NASA–industry conference was held on 26 September, with the closing date for bids being 16 October. On 17 November NASA announced that the Chrysler Corporation had been chosen to manufacture the first stage of the Saturn C-1, with an order for 20 units, the first of which was to be delivered in early 1964. In the meantime, on 11 September, North American Aviation had been selected to supply the S-II stage, although its development had to remain 'on hold' while the mission mode was debated.[8] On 3 October the space vehicle board was established to coordinate joint activities by the Space Task Group and

[7] The Apollo spacecraft source evaluation board was initially chaired by Walter C. Williams, the associate director of the Space Task Group, but when Williams was reassigned on 1 November 1961 Faget took over.

[8] On 6 November 1961 the Marshall Space Flight Center decided to add a fifth J-2 engine to the S-II for a total of 1 million pounds of thrust. North American Aviation did not receive the go-ahead to start work until 30 October 1962.

the Marshall Space Flight Center. One of its first acts was to initiate a study (to be completed by March 1962) of a proposal to replace the cluster of six LR-115 engines on the S-IV stage by a single J-2 engine with 200,000 pounds of thrust, in order to create the S-IVB as the third stage of the 'advanced Saturn'. NASA announced on 25 October that it was to buy a 13,500-acre tract on the Pearl River, some 35 miles northeast of Michoud, where it would build the test facilities for the large stages of the new launch vehicle. This became the Mississippi Test Facility.

In its report submitted on 24 November the source evaluation board rated the Martin Company the highest on its 'scorecard' scheme, and recommended it as "the outstanding source for the Apollo prime contractorship". After reviewing the Board's reasoning, Webb, Dryden and Seamans consulted Gilruth, who observed that although Martin was experienced in making launch vehicles, North American Aviation, which had scored a close second, had considerably greater experience in designing "flying machines" and had excelled on experimental aircraft such as the X-15 rocket plane. Although it already had won the contract for the second stage of the 'advanced Saturn', Webb decided that North American Aviation was capable of managing both projects, and the decision was announced on 28 November. By this time, the *Statement of Work* of 28 July had been refined and expanded. One item required the spacecraft to have a single-engine service propulsion system that used storable hypergolic propellants and, in addition to performing all the necessary maneuvers in space, would also provide an abort contingency for the remainder of the ascent to orbit after the launch escape system had been jettisoned. If the mission mode were to require the entire spacecraft to land on the Moon, a lunar landing

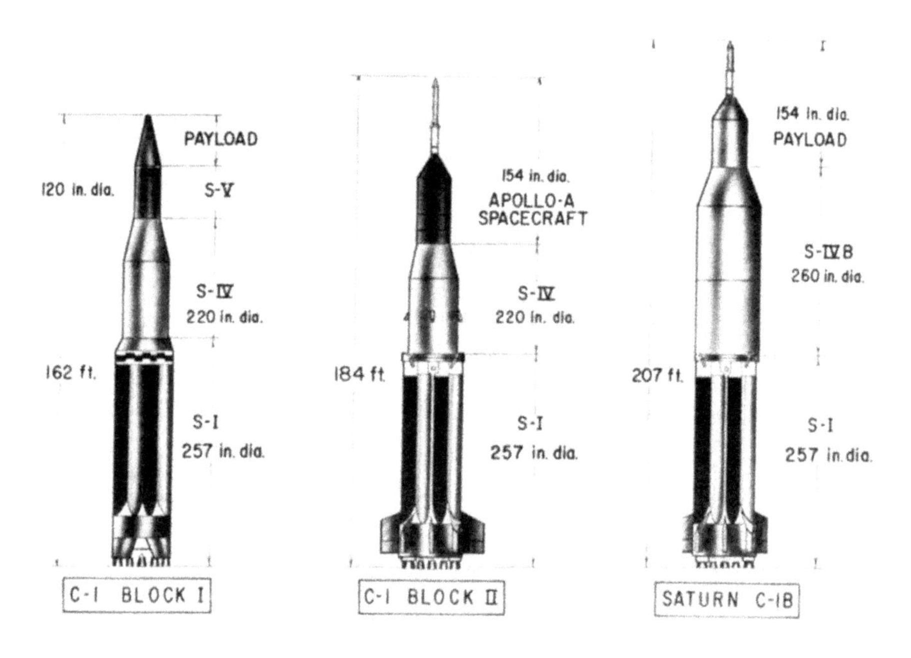

By the end of 1961, three versions of the Saturn C-1 launch vehicle were envisaged.

propulsion module would perform the descent, and the service propulsion system would execute the liftoff. This contract was for the command and service modules (CSMs). In addition to 'real' spacecraft, two series of 'boilerplates' were ordered, one of a simple cold-rolled steel construction to be used in drop/impact tests, and others of higher fidelity for a variety of abort tests. Lewis already had an outline design for a lunar landing propulsion module appropriate for either direct ascent or Earth orbit rendezvous. If a lunar lander was required, that contract would be offered separately.

REORGANIZATION

On 24 September 1961 NASA announced a major restructuring of its headquarters management by creating four program offices: advanced research and technology; space science; applications; and manned space flight. The effect was to place the administration, if not the funding, of all its activities (some of which were aeronautical) on a par with the office of manned spaceflight, since fully three-quarters of the agency's budget went to that office. In effect, Webb had drawn the power of decision-making to headquarters, since the directors of the 'offices' and 'centers' reported to Seamans, who had budgetary control. The obvious candidates for the directorship of the office of manned space flight were Abraham Silverstein and Wernher von Braun, but because their relationship was stormy Webb decided to seek an outsider. On 21 September Dyer Brainerd Holmes, an executive of the Radio Corporation of America, who had managed the development of the Ballistic Missile Early Warning System, had been hired. His position abolished, Silverstein returned to Lewis as its director. In May 1961 the Space Task Group had proposed the creation of a 'Manned Spacecraft Development Center', but Webb had decided to rename the Space Task Group the Manned Spacecraft Center and charge it with not only the design, development, evaluation and testing of manned spacecraft, but also the training of astronauts and the management of mission operations. On 14 September it was decided that this new facility should be built on 1,600 acres of pastureland some 22 miles southeast of Houston, Texas, which had been donated for the purpose by Rice University.

DECIDING THE LAUNCH VEHICLE

On 6 November 1961 Holmes asked Milton Rosen, who was now the director of launch vehicles and propulsion in the office of manned space flight, to establish a working group to further investigate the issues raised by the committees that had addressed the mission mode, and to recommend a launch vehicle program that would satisfy not only manned spaceflight but also the agency's other potential future programs. Although the committee recognized the advantages to be derived from the use of a small lander and lunar orbital rendezvous, it rejected this since a failed rendezvous in lunar orbit would be irrecoverable. Reporting on 20 November, the committee recommended making an urgent effort to demonstrate rendezvous in Earth orbit. As regards launch vehicles, a lunar landing involving a rendezvous mode would require a Saturn C-5 that had five F-1 engines in its first stage, five J-2 engines in its second stage, and a single J-2 engine in its third stage, whereas direct ascent would require a superbooster with eight F-1 engines in its first stage, four 'M-1' engines (still to be specified) in its second stage, and a single J-2 engine in its third stage. The committee recommended that since the S-IVB was common to both, its development should be authorized with a view to flight trials as the second stage of the Saturn C-1 some time in 1964. In view of the fact that a rendezvous rehearsal, and the proof of its viability, would not be feasible until at least 1964, the *primary* effort should be directed toward developing the superbooster required for the direct ascent mode. On 15 December Boeing was hired to manufacture the first stage of the 'advanced Saturn' at Michoud, and on 20 December the Douglas Aircraft Corporation was authorized to start planning the refit of the S-IV stage to use a single J-2 engine. On 21 December Holmes created the Manned Space Flight Management Council. Drawing on senior managers at headquarters and the field centers, this would set policy for manned space planning. At its first meeting the Council decided in favor of a launch vehicle with five F-1 engines in the first stage, five J-2 engines in the second stage and a single J-2 in the third stage. The Saturn C-5 offered flexibility *vis-à-vis* the mission mode, and would be able to support a variety of other programs with heavy payloads. With a capacity to insert 113 tons into low orbit or send 41 tons toward the Moon, it was certainly capable of dispatching a circumlunar mission, indicating that a landing mission by Earth orbit rendezvous could be done with just two launches. The savings in mass resulting from the use of a small lunar lander and lunar orbit rendezvous *might* enable such a landing mission to be dispatched by a single launcher, but if necessary this two-part spacecraft could be assembled in space prior to setting off to the Moon. If rendezvous was shown to be infeasible, the launcher would be scaled up to the Saturn C-8 configuration. When on 25 January 1962 NASA formally approved the development of the Saturn C-5, the development of the Saturn C-1 was sustained to facilitate early testing of the Apollo spacecraft but work on all other variants was terminated. On 18 March the Marshall Space Flight Center issued a schedule for Saturn C-5 development which envisaged the first test launch in late 1965, attaining 'man-rated' status by the eighth flight in mid 1967, and the first manned flight later that year leading to the lunar landing within the deadline set by Kennedy.

Meanwhile, on 15 January 1962 the Apollo project office was subsumed into the Apollo Spacecraft Project Office (ASPO), which was charged with providing technical direction to North American Aviation and its subcontractors. Charles W. Frick, formerly of Convair's General Dynamics, was appointed manager, reporting to Gilruth, and Robert Piland, who had headed the Apollo project office, was retained as Frick's deputy. On 9 February Gilruth established an *ad hoc* lunar landing module working group which, under the direction of ASPO, was to draw up the specifications for the lunar landing propulsion module that Lewis was designing. As North American Aviation commenced work on the main Apollo spacecraft, it subcontracted the Lockheed Propulsion Company on 13 February to develop the solid rocket for the launch escape system; the Marquardt Corporation on 2 March to develop the attitude control engines of the service module; the Aerojet–General Corporation on 3 March to develop the service propulsion system; the Pratt & Whitney division of the United Aircraft Corporation on 8 March to develop the fuel cell power system; the Avco Corporation on 23 March for the ablative heat shield of the command module; and the Thiokol Chemical Corporation on 6 April to build the solid rocket that would jettison the launch escape system, either at the required point during a normal launch or after an abort.

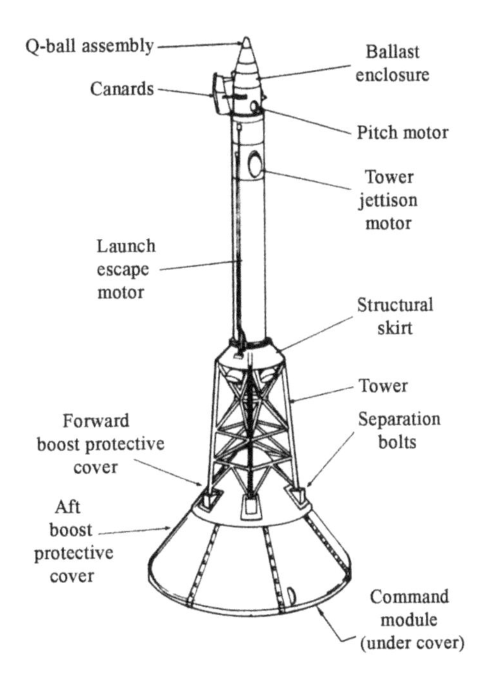

The launch escape system was designed to draw the Apollo command module away from a malfunctioning booster.

Meanwhile, on 12 March the advance elements of ASPO relocated to the Manned Spacecraft Center facility in Houston. The next day Webb requested that Apollo be given the government's highest national priority rating ('DX') for procurement of materials and payment of overtime rates, and after an endorsement of this request by the Space Council, Kennedy agreed on 11 April.

THE DECISION FOR GEMINI

After presenting the case for lunar orbit rendezvous to the committees chaired by Lundin, Heaton and Golovin, to no avail, on 15 November 1961 Houbolt sent a letter to Seamans, who had requested to be kept informed. Explaining that he was writing "somewhat as a voice in the wilderness", Houbolt presented a critique of Apollo planning. A superbooster would take so long to develop that it offered little prospect of achieving the lunar landing within Kennedy's deadline. He argued for using the Space Task Group's proposed 'Mercury Mark II' spacecraft to prove the feasibility of rendezvous in Earth orbit, and then using the Saturn C-3 to assemble a spacecraft that would enter lunar orbit, where the lander would detach, land, lift off and rendezvous. When Seamans circulated this letter around headquarters, Low was supportive but Rosen was dismissive. On 4 December Seamans told Holmes he was "essentially in accord" with the recommendations of the Rosen committee. Two days later, Holmes told Seamans that since rendezvous offered the prospect of achieving the lunar landing sooner than by waiting for the superbooster, it was "essential" to develop a rendezvous capability. He recommended that this be the primary objective for 'Mercury Mark II', with a key secondary objective being to demonstrate a man could survive in space for at least as long as required on a lunar landing mission. On 7 December Seamans approved this recommendation, and later that day Gilruth announced that the two-man spacecraft would be built by McDonnell – since it was regarded as an upgrade of the Mercury spacecraft, the contract was signed on 22 December without competition. The spacecraft would be launched by a 'man-rated' Titan II. That same day, Seamans recommended that the Atlas be used to place an Agena stage into orbit to serve as the rendezvous and docking target. At the suggestion of Alex Nagy at headquarters, on 3 January 1962 this interim program was named 'Gemini'. At the beginning of 1961 NASA had been concerned that it might not secure funding for *any* manned program to follow

An Atlas missile lifts off on 20 February 1962 with John Glenn's Mercury spacecraft.

Mercury, but by the end of the year it had acquired *two* new programs and a mandate of staggering audacity. Nevertheless, it was not until 20 February 1962 that the agency was able to launch John Herschel Glenn on an Atlas missile for its first orbital mission.

DECIDING THE MISSION MODE

On 30 December 1961, in an effort to resolve the mode question, Holmes appointed Joseph F. Shea to conduct a review of the alternatives and make a recommendation. Shea was to undertake this review in such a way that all the field centers could sign up to the chosen mode, to ensure that the agency proceeded with Apollo in a unified manner. Shea was an excellent choice for this appointment because, having recently been recruited from the Space Technology Laboratories in Los Angeles to be Holmes's deputy, he was sufficiently 'neutral' to overcome field center dogma. On visiting the field centers in mid-January 1962, Shea realized that the Marshall Space Flight Center and Manned Spacecraft Center were each viewing the issue from its own perspective without really appreciating the other's point of view, and each was convinced that it was 'right'. Von Braun was in favor of Earth orbit rendezvous because he had faith that rendezvous was viable and because he considered the superbooster with a first stage of eight F-1 engines, providing a total of 12 million pounds of thrust, to be too great a technological leap. Gilruth had initially been in favor of direct ascent because it did not require rendezvous. However, as the Langley studies had demonstrated, resolving the mission mode was not simply an issue of launch vehicles; the act of landing on the Moon had also to be considered, and indeed was likely to be the deciding factor. Direct ascent and Earth orbit rendezvous would both involve landing the entire spacecraft on the Moon, which was not only inefficient in terms of propellant but also technically difficult. By late 1961, however, as the Manned Spacecraft Center started to plan 'Mercury Mark II' with the primary objective of demonstrating the feasibility of rendezvous, Gilruth was won over by the mass-savings to be gained by using a separate lander and gave his support to lunar orbit rendezvous. On 6 February 1962 Houbolt and Charles W. Mathews of the Manned Spacecraft Center updated the Manned Space Flight Management Council on the relative merits of rendezvous around the Earth and around the Moon. Then Shea – in an effort to encourage the field centers to see the issue from their opponents' viewpoint, not just their own – instituted a series of workshops at headquarters.

The first workshop on 13–15 February was devoted to Earth orbit rendezvous, and reaffirmed that Gemini must provide "substantial experience with rendezvous techniques pertinent to Apollo". On 2–3 April the second workshop explored lunar orbit rendezvous in detail. It was assumed that the mission would require a single Saturn C-5, which would be able to insert the S-IVB directly into low orbit and thereby obviate a restart capability. In this scenario, after the S-IVB had made the translunar injection, the Apollo mother ship would retrieve the 'Lunar Excursion Vehicle' from the adapter on top of the S-IVB, but how this was to be done remained to be specified. The early proposal by the Langley team for an open-deck lander was rejected in favor of a pressurized cabin. Lunar orbit insertion was to be by the service propulsion system. Two of the three astronauts would transfer to the lander by a means that had to be specified, but might involve an axial airlock in the command module. The descent propulsion system would require to be throttleable in order to enable the lander to hover prior to touchdown. It remained to be decided whether the same engine would be used for liftoff. After rendezvous, docking, and crew transfer, the lander would be discarded, the service propulsion system would put the spacecraft into a return trajectory to Earth, and the service module would be jettisoned shortly before the command module entered the atmosphere. Shea's strategy was succeeding, because in working out the detail of lunar orbit rendezvous, von Braun began to accept its merits. As indeed did Rosen, who had dismissed Houbolt's missive when shown it by Seamans. This was just as well, since Walter C. Williams, associate director of the Manned Spacecraft Center, warned the Manned Space Flight Management Council on 24 April that the protracted debate was delaying key decisions for the development of the Apollo spacecraft. On 3 May ASPO gave a presentation to Holmes on lunar orbit rendezvous. Two days later, the preliminary *Statement of Work* for the 'Lunar Excursion Module' (LEM) was completed, in case lunar orbit rendezvous was chosen, and on 29 May, with a consensus developing in favor of this mode, the Council approved the schedule for letting the development contract.

By this point, Shea had concluded that both Earth orbit rendezvous and lunar orbit rendezvous were *feasible*, but the *better* mode was still at issue in view of the constraints. On 7 June von Braun told Shea that he accepted Gilruth's rationale for a specialized lander, which was the factor that *required* lunar orbit rendezvous. On 22 June Shea formally recommended lunar orbit rendezvous to the Manned Space Flight Management Council, which concurred, reaffirmed its decision to utilize the Saturn C-5, and endorsed Rosen's proposal to use the first stage of the Saturn C-1 to test the S-IVB and then use this combination to facilitate early testing of the two spacecraft. On 28 June Seamans, Holmes and Shea briefed Webb and Dryden, who accepted the recommendation. The formal decision was announced on 11 July 1962.

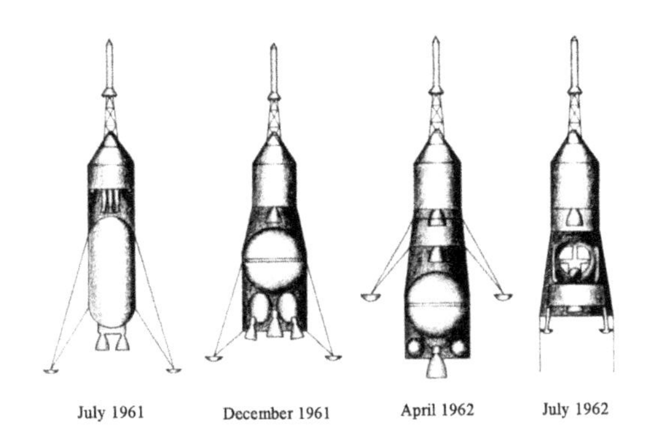

As the Apollo 'mission mode' was debated, a number of spacecraft configurations were proposed.

On 25 July the Manned Spacecraft Center invited 11 companies to bid for the contract to develop the LEM. Nine proposals were submitted on 4 September, and after the presentations on 13–14 September the evaluation teams set to work. On 8 August NASA gave the Douglas Aircraft Company permission to proceed with the S-IVB stage. On 14 August the *Statement of Work* to North American Aviation was updated to include the LEM requirements, and that company promptly issued the structural requirements for the adapter that would mate the 154-inch-diameter service module of the spacecraft to the 260-inch-diameter S-IVB, within which the LEM would be carried during launch. On 5 September ASPO asked the company to investigate docking systems and crew transfer methods. The terms of reference stipulated that the retrieval of the LEM from the S-IVB must take place as soon as possible after translunar injection. The options for doing so included the spacecraft making a free fly-around, a tethered fly-around, and being repositioned utilizing a mechanical means. On completing its evaluation in November, the company recommended a free fly-around involving the following stages: (1) as soon as possible after the S-IVB made translunar injection, the spacecraft would separate from the Spacecraft/LEM Adapter (SLA); (2) the SLA would split into four sections and splay out at an angle to ensure that the engine bell of the service propulsion system made a clean exit; (3) the spacecraft would fire its thrusters to move clear, rotate through 180 degrees, return, and dock with the roof of the LEM; (4) the LEM would then be released from the spent stage.

Although NASA considered the mode question to be closed, the choice was attracting criticism from knowledgeable outsiders. While von Braun was leading Kennedy, Johnson and McNamara around a stored Saturn C-1 at Huntsville on 11 September, he began to describe lunar orbit rendezvous. Kennedy interjected: “I understand Dr Wiesner doesn’t agree with this?” In offering his criticisms Wiesner started a heated debate that Webb pre-empted by pointing out that the request for proposals to build the LEM had already been invited from industry. Kennedy, having listening intently, nodded his agreement, but this did not end the matter. When Wiesner continued to criticize the mode decision, Webb told him on 24 October that the contract for the LEM would be awarded unless the president ordered the mode to be reviewed. Wiesner backed down, and on 7 November it was announced that Grumman Aircraft Engineering Corporation of Bethpage, New York, had won the contract with a design for a two-stage vehicle.

Meanwhile, on 17 September 1962 NASA announced the selection of nine additional astronauts: Lieutenant Charles ‘Pete’ Conrad Jr, Lieutenant Commander James Arthur Lovell Jr, and Lieutenant Commander John Watts Young from the Navy; Major Frank Frederick Borman II, Captain James Alton McDivitt, Captain Thomas Patten Stafford, and Captain Edward Higgins White II from the Air Force; and Neil Alden Armstrong (a former naval aviator, now working as a civilian test pilot for NASA) and Elliot McKay See Jr (a civilian test pilot for the General Electric Company). At the same time, Slayton was appointed as coordinator of astronaut activities, reporting to Gilruth.[9] In addition to

[9] When the Manned Spacecraft Center was reorganized in 1963 this post was renamed assistant director for flight crew operations.

the administrative tasks associated with the astronaut office, which he managed in the manner of a military unit, Slayton was responsible for flight crew assignments. On 30 October Webb once again adjusted the top management. In addition to being director of the office of manned space flight, Holmes was made a deputy associate administrator, in which capacity he took direct responsibility for the field centers primarily engaged in manned space projects (i.e. the Marshall Space Flight Center, Manned Spacecraft Center and Launch Operations Center) that had previously reported to Seamans.

MISSION PLANNING ISSUES

By the end of 1962, NASA had taken all the crucial decisions that defined how it would address Kennedy's challenge. On 16 January 1963 the mission analysis branch of the flight operations division at the Manned Spacecraft Center set out to study operational procedures for lunar orbit prior to the LEM initiating the powered descent. It was presumed that the vehicle would be released in a circular orbit at an altitude of 80 nautical miles above the mean lunar surface. One option was for the vehicle to transfer into an elliptical descent orbit of the same period but with a 50,000-foot perilune just short of the landing site, make a low pass for a final reconnaissance, and initiate the powered descent on the next revolution. The possibility of having the mother ship use its service propulsion system to enter the descent orbit prior to releasing the LEM was ruled out by the amount of propellant it would consume. The most severe constraint was that the LEM be able to initiate an abort late in the descent, and in particular during the terminal hover phase. The recommendation was that the main spacecraft remain in a circular orbit at 80 nautical miles. Meanwhile, the Manned Spacecraft Center set up the Apollo mission planning panel on 27 February to ensure that the spacecraft would be able to achieve the program's objectives, to design the lunar landing mission, and to develop contingency plans for all manned missions. On 27 March the panel stipulated two requirements for the landing mission: (1) both astronauts must exit the LEM and work on the surface together; (2) duration limits must be established. Based on the 48-hour operational requirement of the LEM's systems, these limits varied from 24 hours in flight and 24 hours on the surface at one extreme, to 3 hours in flight and 45 hours on the surface at the other. Grumman was instructed to design the vehicle to support the entire range of options. As the 'reference mission' was refined in ever greater detail, the time that the S-IVB would spend in the Earth 'parking orbit' was discussed by the Apollo spacecraft mission trajectory sub-panel on 30 April. Since the 4.5-hour operating limit of the S-IVB stage's sequencer dictated the maximum number of revolutions, the translunar injection was scheduled for the second revolution. In June the mission analysis branch investigated a 'skipping' profile for atmospheric entry following a high-speed lunar return in order to minimize to the heating rate. After the entry interface at 400,000 feet the command module was to penetrate the atmosphere to a height of 300,000 feet (the figure chosen for the study) and then skip back out to pursue a ballistic arc that would lead to second contact and landing. Because there would be no active control of the trajectory during this part of the flight, the length of the skip would be determined by the angle and speed at the onset of the ballistic arc, which would, in turn, determine the landing site. The accuracy of the landing would therefore be

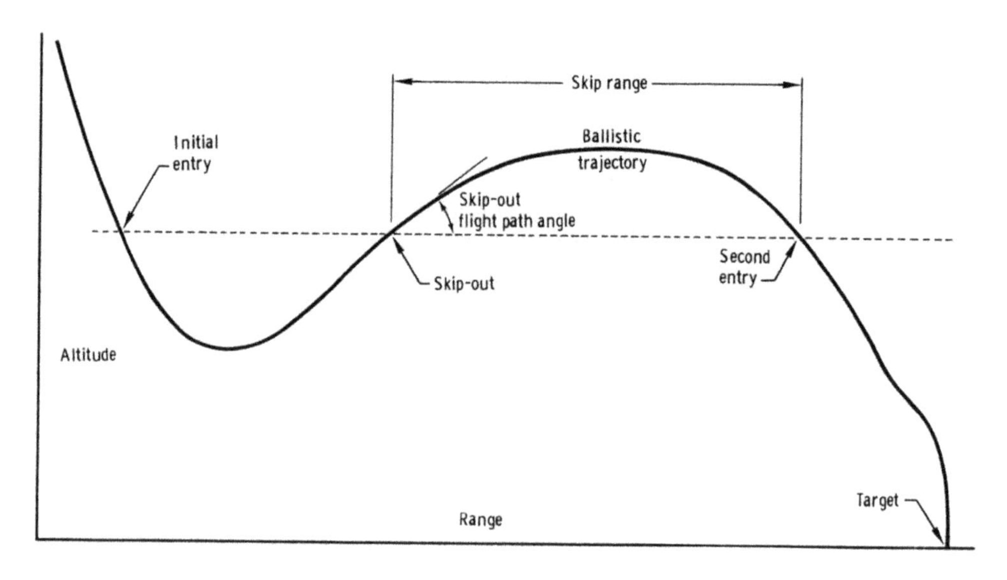

On concluding a high-speed return from the Moon, an Apollo command module was to fly a 'skipping' atmospheric re-entry, with the first encounter reducing its energy sufficiently to make the second comparable to that from a low Earth orbit.

directly related to 'dispersions' in both speed and angle at this crucial moment. When calculations showed that the profile was more sensitive for the shallower trajectories, it was decided to perform the skip at the steepest possible angle consistent with the greatest permissible *g*-forces. On 1 August NASA, North American Aviation and Grumman reviewed the relative merits of having the LEM enter an equi-period orbit with its apolune above the 80-mile orbit and its perilune at 50,000 feet, versus a Hohmann transfer in which the apolune would remain at the altitude of the parking orbit. Grumman argued for the Hohmann transfer as this was the minimum-energy option, and would maximize the propellant available for the powered descent, but because there was no consensus NASA called for further investigation of the reliability, propellant consumption, and operational flexibility of each option.

On 21 August 1963 the mission planning branch reported an analysis of how mission rules affected the degree of flexibility in launch operations. These rules specified a free-return trajectory, a predetermined lunar landing site, and requisite illumination at various points of the mission. As John P. Bryant, the author of the report, expressed it, the illumination constraints were "by far the most restrictive". These included a minimum of three successive daily launch windows, a minimum daily window of 3 hours, a daylight launch, and the retrieval of the LEM from the S-IVB stage in sunlight. Bryant pointed out that the inevitable result "of imposing an ever-increasing number of flight restrictions [was] the eventual loss of almost all operational flexibility" and recommended that each restriction be evaluated with a view to eliminating "as many as possible". On 16–23 October, in investigating ways of reducing the weight of the two Apollo vehicles, the spacecraft technology division of the Manned Spacecraft Center

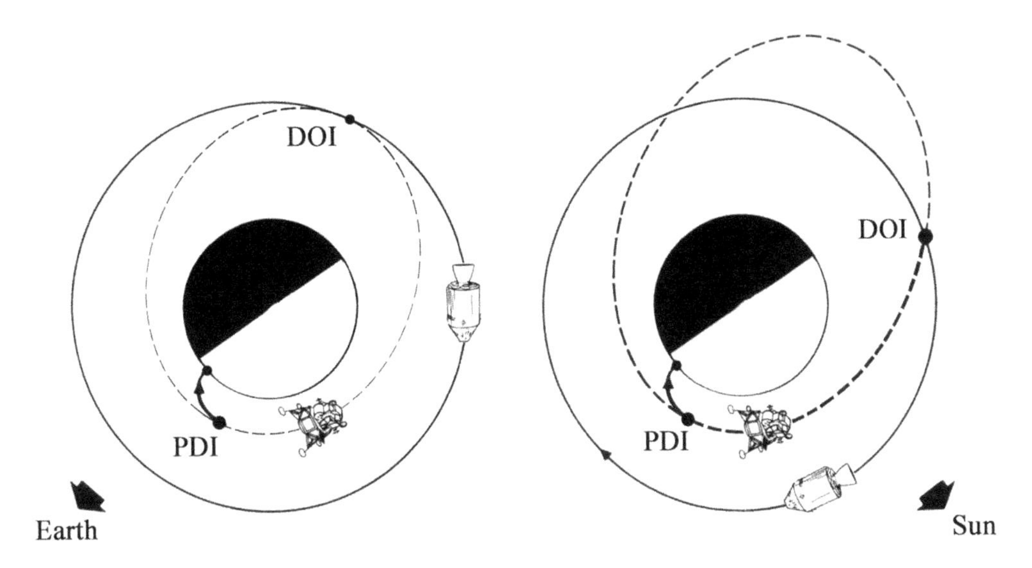

If, on separating from the CSM, the LEM adopted a descent orbit with the same orbital period and a perilune at an altitude of 50,000 ft, at which point it would initiate the powered descent, the CSM would be able to maintain line-of-sight contact throughout the descent (right). The maneuvers for a 'Hohmann transfer' would be more propellant-efficient, but the mother ship, remaining in the higher and hence slower orbit, would not be able to maintain line-of-sight contact (left). Note that the Sun has just risen over the target. [Not to scale. DOI: descent orbit insertion; PDI: powered descent initiation.]

considered: (1) eliminating the free-return trajectory; (2) having the LEM utilize a Hohmann transfer orbit in order to enable it to carry less propellant; (3) reducing the time available to rendezvous; and (4) reducing the operating life of the LEM. On 27 November Shea directed that even if it was decided to use a Hohmann transfer, the tanks of the LEM must be designed to accommodate the propellant that would be required for the equi-period mode, in order to preserve that option. (This would also provide a margin of propellant to facilitate the more demanding missions that would follow.) Another possible compromise was related to the fact that the Moon would rotate beneath the orbital plane of the mother ship while the LEM was on the surface. By deleting the 'reconnaissance' pass over the landing site, restricting the site to within 5 degrees of the lunar equator, and limiting the duration of independent activity to less than 48 hours, the plane change that the ascent stage would require to make during the rendezvous would be able to be reduced to 0.5 degree, which would further reduce the amount of propellant it was required to carry.

LEM DESIGN ISSUES

One early decision for Grumman was the number of legs on the descent stage of the LEM. As tests showed three to be inadequate and five to be unstable, it was decided to use four. In the early months, the weight and height of the vehicle increased, which raised the center of mass and required the span of the legs to be widened, and since the SLA would not be able to accommodate this span the legs had to be made deployable. On 13 February 1963 Grumman hired Rocketdyne to develop the descent propulsion system (DPS) and the Bell Aerospace Company to develop the ascent propulsion system (APS), and on 11 March the Marquardt Corporation was hired to supply the reaction control system. Rocketdyne proposed to achieve the desired throttleability across a 10:1 range of thrust by the controlled injection of gaseous helium into the hypergolic propellants. On realizing that throttleability constituted what Frank Canning, Grumman's assistant project manager, described as the "biggest development problem", it was decided to order a mechanical throttle as a contingency. On 14 March Grumman hosted a conference for potential bidders, and early in May chose the Space Technology Laboratories. On 3 July Bell was authorized to start work on the APS, the design of which was complicated by the fact that its nozzle would be embedded in the descent stage and would be required to ignite as pyrotechnics severed the connections between the two stages in a mode referred to as 'fire in the hole'. With the lander rapidly putting on weight, the LEM flight technology systems panel was established on 10 May by the Manned Spacecraft Center to coordinate all issues pertaining to weight control, engineering simulation and environment. The initial configuration of the LEM had the crew in seats and four curved windows, two of which were low to give a view of the ground during the final phase of the landing. On 16 June Grumman was instructed to consider having the crew fly standing up in order to be able to eliminate not only the weight of the seats but also most of the glass. On 16 July the company recommended two small flat triangular windows set at an angle. This revision greatly improved crew mobility, visibility, control accessibility and egress–ingress. After reviewing a mockup in which the flight controls were modified for standing crewmen, the Manned Spacecraft Center accepted these changes on 18 September.

ONGOING CHANGES

Having reduced to three the number of variants of the Saturn launch vehicle, on 7 February 1963 the Saturn C-1 with the S-IV became the Saturn I, the Saturn C-1 with the S-IVB became the Saturn IB, and the Saturn C-5 became the Saturn V. On 13 February the Manned Spacecraft Center rearranged the Apollo spacecraft project office by appointing two deputy managers: James L. Decker for the CSM and Robert O. Piland for the LEM. On 20 February Holmes, in Washington, appointed Joseph F. Shea deputy director for systems and George M. Low deputy director for programs, to give them increased authority over the directorates of the office of manned space flight. On 10 May Gilruth revised his management by splitting development from operations: as deputy director for development and programs James C. Elms was to manage manned space flight projects and plan, organize and direct administrative and technical support; and as deputy director for mission requirements and flight operations Walter C. Williams was to manage the drawing up of mission plans and rules, the training of crews, and the provision of ground support and mission control facilities.

In mid-1963 NASA hoped to start manned Gemini missions in 1964 and finish the series in 1965. After the final Mercury mission on 15 May 1963, the astronauts argued to continue flying until Gemini was ready, but Webb refused. The Apollo schedule called for four manned Block I CSMs using the Saturn I, starting in 1965; up to four manned flights using the Saturn IB, starting in 1966; eight test flights to 'man rate' the Saturn V, starting in late 1965; at least six missions in Earth and lunar orbit using the Saturn V, starting in 1967; and a landing in 1968 or 1969. On 12 June, Holmes, who judged Apollo to be falling behind schedule and had lost a dispute with Webb over budgetary priorities, submitted his resignation. On 23 July NASA announced that George E. Mueller, vice president of the Space Technology Laboratories, would take over as director of the office of manned space flight on 1 September.

When Charles W. Frick resigned as ASPO manager on 3 April 1963, Piland served as acting manager until Shea took over on 22 October. Low took on Shea's duties as deputy director of systems in addition to his own post of deputy director of programs at the office of manned space flight. On 27 August the Manned Space Flight Management Council decided to create a deputy associate administrator for manned space flight operations to enable Mueller to divest himself of this subsidiary role, which he had inherited from Holmes, and on 22 October Mueller drew Williams from the Manned Spacecraft Center for the job.[10] Webb announced a reorganization of headquarters on 9 October in which he created three associate administrators under Seamans. Thus, in addition to being the director of the office of manned space flight, Mueller became the associate administrator for manned space flight with responsibility for the three field centers mainly involved in manned programs. The Goddard Space Flight Center, JPL and related facilities went to the office of space science and applications (which merged two offices) under associate administrator Homer E. Newell, and the former NACA facilities went to the office of advanced research and technology under associate administrator Raymond L. Bisplinghoff.

MUELLER TAKES CHARGE

In view of the reason for his predecessor's resignation, Mueller instigated a review, and this confirmed Apollo to be slipping. On 29 October 1963 he announced to the Manned Space Flight Management Council that the only way to recover time would be to reduce the number of development flights. The plan drawn up by the Marshall Space Flight Center in March 1962 envisaged a series of launches of the Saturn V in which the stages were tested in sequence, with the first test, on which only the first stage would be 'live', in late 1965, with a view to 'man rating' the launch vehicle by mid 1967. Mueller said he proposed to reduce this eight-launch research and development phase by adopting 'all up' testing with all 'live' stages, modules and systems, including both of the spacecraft. Gilruth

[10] On 24 April 1964 Williams, frustrated at headquarters, resigned. The recently created post of deputy associate administrator for manned space flight operations was initially left vacant, and then abolished in December 1964.

On 7 November 1963 the launch escape system lifts Apollo spacecraft BP-6 clear in its first 'pad abort' test (PA-1).

and von Braun objected, but Mueller insisted, and Webb concurred. In addition, a recent study by Bellcomm Inc. had recommended reassigning the early tests of the Apollo Block I spacecraft from the Saturn I to the Saturn IB.[11] On 30 October Mueller cancelled the four manned test flights with the Saturn I that had been scheduled for 1965. The development of the Saturn IB for manned missions would be accelerated and the 'all up' testing strategy employed in this case too. After coming to terms with this, Gilruth asked von Braun whether the Saturn IB could lift both the CSM and LEM, and was told that it would be feasible only if their weights were controlled. At the White Sands Missile Range in New Mexico on 7 November, the Apollo launch escape system executed its first 'pad abort' test. On 18 November Mueller directed that if the LEM was not ready, the early Saturn IB flights would fly without it, but it must be phased into the test program as soon as possible. Furthermore, he directed that two successful development flights of the Saturn IB and Saturn V would be sufficient to 'man rate' them. The schedule that he issued on 31 December 1963 envisaged the tenth and final Saturn I mission in June 1965, the first Saturn IB test in the first quarter of 1966 and the first manned mission later in the year. The first Saturn V test flight was to be in the first quarter of 1967, with the first manned mission (hopefully on the third flight) scheduled for later that year. The same day (31 December) Mueller established the Apollo program office, appointing himself as director and recruiting Brigadier General Samuel C. Phillips, who had managed the development of the Minuteman missile, as his deputy.

Meanwhile, on 18 October 1963 NASA announced the recruitment of a further 14 astronauts: Major Edwin Eugene 'Buzz' Aldrin Jr, Captain William Alison Anders, Captain Charles Arthur Bassett II, Captain Michael Collins, Captain Donn Fulton Eisele, Captain Theodore Cordy Freeman, and Captain David Randolph Scott from the Air Force; Lieutenant Alan LaVern Bean, Lieutenant Eugene Andrew Cernan, Lieutenant Roger Bruce Chaffee, and Lieutenant Commander Richard Francis Gordon Jr from the Navy; Captain Clifton Curtis Williams from the Marines; and Ronnie Walter Cunningham (a research scientists at the Rand Corporation) and Russell Louis 'Rusty' Schweickart (a research scientist at the Massachusetts Institute of Technology). On 5 November Gilruth placed assistant directors under Elms, by now deputy director of the Manned Spacecraft Center, to strengthen the local management of Gemini and Apollo:

[11] Bellcomm Inc. was a systems-engineering group created in 1962 by the American Telephone & Telegraph Company at NASA's request to conduct independent analyses of aspects of the Apollo program.

The astronauts recruited by NASA to fly the Mercury, Gemini and early Apollo missions.

lead flight director Christopher Columbus Kraft Jr was made assistant director for flight operations; Faget assistant director for engineering and development; and Slayton, assistant director for flight crew operations. In addition, G. Merritt Preston was appointed to manage Manned Spacecraft Center operations in Florida. On 17 January 1964 Elms resigned, and two days later Low was assigned in his place.

KENNEDY'S ASSASSINATION

On 16 November 1963 Kennedy flew to Cape Canaveral to inspect the 'moonport' that NASA was in the early stage of building on nearby Merritt Island. Von Braun gave him a guided tour of Pad 37B, which was being prepared for the fifth Saturn I, which was to be the first to fly a 'live' S-IV stage. On display were scale models of various launch vehicles to depict the great size of the Saturn V. On 22 November Kennedy was assassinated while on a visit to Dallas, Texas, and later that day Lyndon Baines Johnson was sworn in as his successor. On 28 November Johnson directed during an address on television that Cape Canaveral be renamed Cape Kennedy. The next day he signed an executive order renaming NASA's Launch Operations Center the John F. Kennedy Space Center.[12] However, the City of Cape Canaveral, incorporated in 1962 and sandwiched between Port Canaveral to the north and Cocoa Beach to the south, decided by city council vote *not* to change its name.

FIRMING UP THE SCHEDULE

On 15 January 1964 the Manned Spacecraft Center proposed to Joe Shea of ASPO that two of the Saturn IB 'all up' development flights be used to test the heatshield of the Apollo command module, to enable the early 'all up' tests of the Saturn V to be described as 'demonstration' rather than 'development' for the spacecraft.[13] On 7 February ASPO directed Grumman to provide two LEM test articles (LTA) and 11 mission-capable LEMs, the first three of which were to be capable of either manned or unmanned operation. On 23 March Mueller directed that if the first two unmanned CSM test flights were successful, the next mission would be a long-duration manned flight, after which there would be two tests of the LEM, the first one unmanned and the second in conjunction with a manned CSM – providing that it proved possible for the Saturn IB to lift both vehicles together.

[12] This name change officially took effect on 20 December 1963.

[13] On 26 October 1962 a nomenclature was introduced by which the pad abort tests were to run in sequence from PA-1; the Little Joe II flights were to start at A-001; missions using the Saturn I were to start at A-101; missions using the Saturn IB were to start at A-201; and missions using the Saturn V were to start at A-501, with the 'A' standing for 'Apollo'. As a refinement, the 'SA' prefix was used by the Marshall Space Flight Center (giving precedence to its launch vehicle) and the 'AS' prefix was used by the Manned Spacecraft Center (giving precedence to its spacecraft). Also the term 'space vehicle' was introduced to describe the integrated 'launch vehicle' and 'spacecraft', with the 'spacecraft' being the CSM, the LM (if present) and the SLA.

On 16 November 1963 NASA administrator James E. Webb greets John F. Kennedy on his arrival at Cape Canaveral (top). In the blockhouse of Pad 37 (left) George E. Mueller briefs (seated left to right) George M. Low, Kurt H. Debus, Robert C. Seamans, Webb, Kennedy, Hugh L. Dryden, Wernher von Braun, Major General Leighton I. Davis and Senator George Smathers of Florida. Having inspected a model of SA-5, the first Block II Saturn I, Kennedy and von Braun look up (right) at the structure on which it would be erected for launch on 29 January 1964.

REFINING THE DOCKING SYSTEM

Since the specifications for the Apollo spacecraft had been drawn up prior to the decision to use lunar orbit rendezvous, the Block I was not equipped to operate in conjunction with the LEM. It lacked facilities to enable it to rendezvous and dock – not even with a space station – restricting it to flying solo missions. However, its development went ahead while the specifications for the Block II were drawn up to enable it to test the systems that were to carry over to its successor.

When North American Aviation initially drew up the requirements for how the Apollo spacecraft would dock with another vehicle in May 1962, the company had envisaged a spacesuited crewman entering the airlock that was to be on the axis of the command module and, once the two vehicles were several feet apart with zero relative motion, opening the outer hatch to grasp the other vehicle and "maneuver it into the sealing faces for final clamp". But in July the company started a study of mechanical docking systems. In August NASA began to discuss with the company a number of possible configurations for the command module airlock as a means of transferring crewmembers to the LEM after that vehicle had been retrieved from the S-IVB. One suggestion was to design the docking system to become a pressurized tunnel. This was accepted, and in September the airlock was deleted. On 16 July 1963 the company was instructed to halt work on an 'impact' type of docking system, and instead to concentrate on the 'extendable boom' concept. On 19–20 November Grumman and North American Aviation agreed to employ a 'probe and drogue' docking system. A probe on the nose of the command module would be extended early in the approach, and once a set of 'capture latches' on the tip of the probe had engaged with the conical drogue on LEM ('soft docking'), the probe would be retracted to draw the vehicles together to enable a dozen latches in their docking collars to establish a rigid connection ('hard docking'). In accepting this recommendation in December, ASPO directed that either vehicle must be capable of playing the active role during the rendezvous. However, with the docking system on the 'roof' of the LEM to enable the CSM to retrieve that vehicle from the S-IVB, the requirement that the LEM be able to play the active role in a docking posed a problem, as the crew, facing 'forward', would not be able see the target. Grumman therefore designed the hatch on the front, by which the crew would access the lunar surface, to serve as the docking system for the

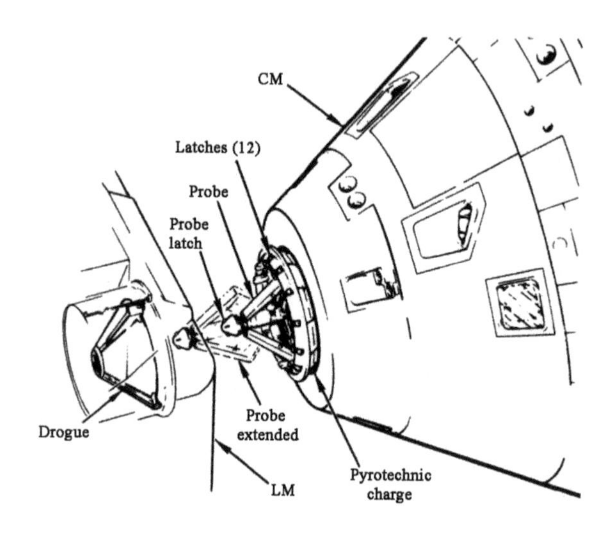

For Apollo a 'probe and drogue' docking assembly was selected.

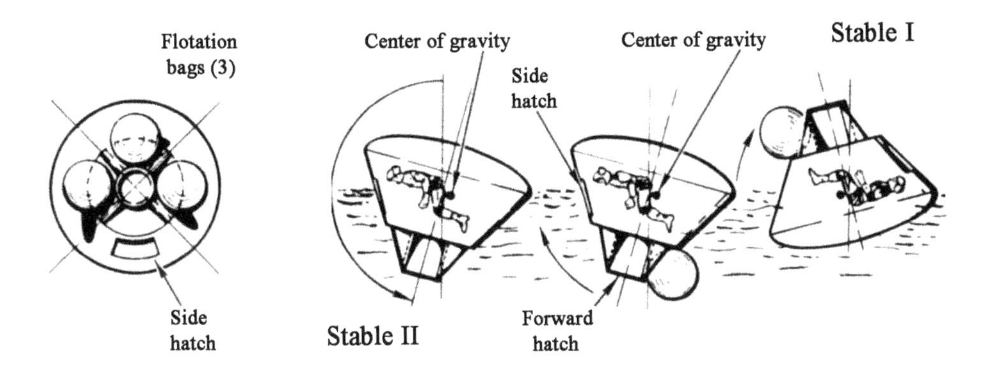

Inflatable bags were to ensure that the Apollo command module would be self-righting in water.

rendezvous. That is, if the CSM was the active vehicle then it would dock at the roof hatch, and if the LEM was active it would use its forward hatch. On 26 February 1964 ASPO directed that in the event of the tunnel failing to pressurize, the returning astronauts would spacewalk and enter the command module using its side hatch. The hatch on the Block I CSM was not designed for spacewalking, but this rule required the hatch on the Block II to be capable of being swung open in space. On 15 March North American Aviation completed a crew transfer study using a mockup of the tunnel in which the hatches and docking mechanisms were set on counterbalances. Engineers in unpressurized suits suspended from rigs to simulate weightlessness showed that it would be possible to remove and reinstall the hatches and docking system components in space. Another test on 24 March confirmed that a crewman wearing a pressurized suit and a life-support backpack could enter the command module through its side hatch. In late June these results were confirmed by zero-*g* trials on board a KC-135 training aircraft.

On 15 April 1964 ASPO told North American Aviation to design the command module with an 'upright' stable flotation attitude. On 15 September the company recommended the installation of airbags in the apex compartment which would be inflated by an electric pump as seawater was permitted to flood the aft compartment of the module. Tests established that two 2-cubic-foot airbags were sufficient for the purpose, but a third was added for redundancy.[14] ASPO ordered this system to be installed on both Block I and Block II variants.

REFINING THE LEM

At a meeting on 29 November 1962, after Grumman had been chosen to build the LEM but before the *Statement of Work* was refined NASA had directed that all manned Saturn

[14] The apex-up flotation attitude was defined as 'Stable I' and the undesirable apex-down attitude was 'Stable II'.

V missions must carry a LEM and that this would provide backup propulsion in the event of a service propulsion system failure. In particular, during the translunar coast the LEM was to be able to make such mid-course corrections as required to return to Earth. On 12 March 1964 ASPO directed North American Aviation to verify that the docking system would be sufficiently rigid to withstand the stresses imparted during such maneuvers.

Meanwhile, on 11 February 1964 Grumman was told to add an abort guidance system (AGS) using a simple inertial reference system affixed to the structure of the vehicle to facilitate the rendezvous in the event of the failure of the primary guidance, navigation and control system (PGNCS). This would eliminate the need to build redundancy into the primary, and would also save weight. On 24–26 March ASPO formally reviewed the test mockup at Grumman, a full-scale replica of the LEM that included representations of the hatches, windows, cabin equipment arrangement, illumination, crew stations, display panels and instrument layout, and crew support and restraint systems. It gave an early assessment of crew mobility, egress and ingress. By now the weight-saving effort had prompted the deletion of the second drogue, and if the LEM was to be active in the docking procedure the cabin would require to be depressurized to transfer the drogue from its original position in the upper hatch to the forward hatch. As a follow-up to the crew mobility trials, on 24 April ASPO initiated a study to recommend modifications to enable an active LEM to dock using its upper hatch. It was determined that this would require the installation of a small window set above the commander's head. A Grumman study identified the best location, size and reticulation of this window, and on 22 May ASPO directed the company to install the window and delete the forward docking system and tunnel. Now that the forward hatch was no longer required to house the drogue, it could be enlarged to improve passage by a crewmember wearing a bulky life-support backpack, while on the lunar surface or during an emergency transfer in space. On 5 June Grumman was instructed to add a narrow platform in front of the egress hatch, promptly dubbed the 'front porch', to improve access to the ladder. On 5–8 October the mockup was inspected to assess the recent alterations and recommend further modifications. Then in January 1965 Grumman rejected the circular forward hatch for a hinged trapezoidal door. After a review on 1–8 April, ASPO deemed this hatch to be "inherently highly reliable", with the only "remotely possible" problem being the failure of the release latch, in which case the surface excursion would have to be cancelled.

Meanwhile, in September 1963 Grumman had instructed the Bell Aerosystems Company to develop an ablative nozzle extension for the APS engine, and had initiated an internal study of a radiatively cooled design as a weight-saving alternative. But on 11 May 1964 ASPO told Grumman to end work on the radiatively cooled nozzle as this was drawing people away and hindering development of the primary option. The design of the throttle for the DPS was resolved on 28 January 1965, when it was decided to adopt the mechanical throttle of Space Technology Laboratories; the contract with Rocketdyne for a throttle involving the injection of helium into the propellant was cancelled. On 15 April Grumman began firing trials of the APS engine at White Sands.

On 12–19 November 1964, at a workshop to further reduce the weight of the LEM, ASPO asked Grumman to consider replacing the fuel cells with chemical storage batteries. On 10 February 1965 the company submitted its proposal for a battery powered descent stage. When it was decided on 2 March to use batteries in both stages, Grumman cancelled

After the design with which Grumman won the LEM contract was refined, a full-sized mockup was built.

the contract with Pratt & Whitney for fuel cells. This greatly simplified the pyrotechnics for staging by eliminating the requirement to sever cryogenic umbilicals. On 15 April Grumman's design for two batteries in the ascent stage and four in the descent stage was approved.

During joint meetings on 19–26 November 1964, North American Aviation and Grumman agreed the alignment of the vehicles during docking – the overhead window of the LEM would face the right-hand docking window of the command module. On 7 December ASPO told Grumman that in order to conserve the CSM's propellant, it had been decided that the LEM would nominally play the active role in the rendezvous. Then on 8 February 1965 ASPO deleted the radar from the Block II spacecraft, and on 17 February North American Aviation was directed to install a transponder for the LEM's radar. At the same time, Grumman was told to modify the VHF system to enable the CSM to infer range and range-rate, as a contingency against the failure of the radar, and to fit a strobe light to enable the CSM to track the LEM optically using its telescope/sextant. By 25 June simulations had identified the best position for the transponder. In mid-March, Cline W. Frasier of the guidance and control division at the Manned Spacecraft Center had raised the issue of deleting the radar from the LEM and instead having it conduct the rendezvous using a star tracker, a xenon strobe light on the CSM, and a hand-held sextant for the pilot of the LEM. In mid-April Grumman was told to design the LEM in such a manner as to be able to use either system. In October Slayton argued for the radar, as this would more readily provide range and range-rate than would the optical tracker. At the end of 1965 Mueller and Shea hatched the 'rendezvous sensor olympics', which was to be undertaken in the spring of 1966 to provide the basis for making this decision. William A. Lee of ASPO objected to pausing RCA's development of the radar pending the decision, because if the issue was resolved in favor of the radar then development time would have been lost and the delay would most likely preclude installing the radar on LEM-1 and LEM-2, in turn undermining Mueller's call for 'all up' testing. On 1 February 1966 Slayton wrote to the chief of the guidance and control division, saying, "The question is not which system can be manufactured, packaged, and qualified as flight hardware at the earliest date; it is which design is most operationally suited to accomplishing the lunar mission." On 25 May the Manned Spacecraft Center recommended that the primary effort be devoted to the radar because the 'olympics' had shown it to offer sufficiently greater flexibility to justify its weight, and that Hughes continue with the development of the optical tracker at a reduced level of funding as a contingency. Phillips and Mueller concurred.

SUIT ISSUES

On 21 April 1964 ASPO directed that all three Apollo crewman must wear their suits from launch through translunar injection, and during lunar operations and re-entry, but providing that at least one man was suited at all times the others could take off their suits while coasting to and from the Moon. When a review of the mockup of the Block I CSM at North American Aviation on 28–30 April revealed that three suited crewmen in couches had their mobility restricted as a result of overlap at the shoulders and elbows, it was decided to adapt the less-bulky Gemini suit in order not to delay the Block I missions while the Apollo suit was modified for use in the Block II. On 30 June the company was ordered to implement such changes as necessary to make the Block I command module compatible with the Gemini suit. On 16 July the David Clark Company of Worcester, Massachusetts, which was making the Gemini suit, was told to modify the design for use on Apollo.

GEARING UP

In November 1964 Shea, Mueller and Phillips drew up an outline schedule to test the Apollo hardware in advance of the introduction of the Saturn V, but it was still unclear whether the weights of the two spacecraft were sufficiently constrained to enable them to be lifted together by a Saturn IB for the joint mission. On 16 December Shea directed that the Block I manned missions should use low orbits from which, in the event of the failure of the service propulsion system, the spacecraft should be able to de-orbit itself using its reaction control system thrusters, and in the event of these too failing, the orbit must decay naturally and result in re-entry within an acceptable duration.[15]

Outline schedule for Apollo drawn up in November 1964

Mission	Payload	Launch date
AS-201	CSM-009 (unmanned)	1965
AS-202	CSM-011 (unmanned)	1966
AS-203	No spacecraft (S-IVB development flight)	Jul 1966
AS-204	CSM-012 (manned)	Oct 1966
AS-205	CSM-014 (manned)	Jan 1967
AS-206	LEM-1 (unmanned)	Apr 1967
AS-207	CSM-101 (manned) and LEM-2	Jul 1967

On 31 August 1964, lead flight director Christopher Columbus Kraft Jr had appointed John D. Hodge, Eugene F. Kranz and Glynn S. Lunney to share round-the-clock flight operations once the Gemini missions began. On 24 December Everett E. Christensen was made mission operations director, a position that effectively superseded deputy associate administrator for manned space flight operations, which had been vacant since Williams' departure in April. At the same time, two posts of mission director were also created, with the intention that the appointees would serve on alternate missions. In addition, activities at the Cape were consolidated, with Kurt Debus being made director of launch operations and Merritt Preston, who had been managing the Manned Spacecraft Center's activities at the Cape, becoming his deputy.

[15] This precaution was reputedly a headquarters response to the situation depicted by Martin Caidin in his recent novel *Marooned.*

Shea, Kraft and Slayton were briefed on 18 January 1965 by the mission planning and analysis division of the Manned Spacecraft Center on the Saturn IB and early Saturn V flights. On 21 January, in response to a query by Phillips, Shea pointed out that the current estimate was that the Saturn IB would be able to insert 35,500 pounds into a circular orbit at 105 nautical miles. This, however, was less than the combined 'control weights' of the CSM and LEM by 870 pounds, and the two vehicles were currently above their control weights. Shea said that in view of the difficulty in constraining the weights, the best solution would be to find a way of increasing the launcher's capacity by 1,000 pounds – and while this increase in performance was being investigated, it would be possible to delete from the service module one set of propellant and helium pressurant tanks without seriously affecting the plans for the Block I missions. In fact, the launcher had a 'control payload', which was the specified minimum mass that it was to be capable of placing into the reference orbit, and a 'design goal', which exceeded this. On 23 February Phillips told Shea that the Marshall Space Flight Center would endeavor to increase the payload to 36,500 pounds. The development version of the eight-engine cluster had generated 1.3 million pounds of thrust, but the fifth flight had introduced an upgraded cluster that finally achieved its specification of 1.5 million pounds of thrust. In August 1963 Rocketdyne had proposed a further upgrade that would increase the power to 1.6 million pounds of thrust, and on 8 November NASA had ordered this be done. By 23 April 1965 the uprated engine had completed its qualification testing at Huntsville, and was declared fit for use with the S-IB stage of the Saturn IB. On 12 May Huntsville reported that it would be possible to uprate the engine by a further 5,000 pounds of thrust, to increase the S-IB to 1.64 million pounds of thrust. The rocketeers were nevertheless fighting a losing battle since, by then, both spacecraft had put on even more weight.

Shea had established the configuration control board, with himself in the chair, on 13 January 1965 in order to rule on all proposals for engineering changes to the spacecraft. On 10 May he faced a dilemma: the 'all up' testing regime required that all spacecraft incorporate a full set of subsystems, but it had been suggested that the landing radars be omitted from LEM-1 and LEM-2 on the basis that a radar would serve no function on an Earth orbital mission. Omitting the radar on these early test flights would save money and assist in the effort to trim the weight of the vehicle at this critical juncture, but to do so would set the precedent for a series of one-of-a-kind spacecraft, each of which was tailored to specific development objectives, and none of which demonstrated all the systems in conjunction. Furthermore, by relieving the pressure on the call to cut weight in the short term, such compromises might undermine that effort in the long term. On 27 May the Manned Spacecraft Center reaffirmed that LEM-1 must test the radar. However, on 25 June ASPO assistant manager Harry L. Reynolds warned Owen E. Maynard, the chief of the systems engineering division at the Manned Spacecraft Center, that it was "becoming increasingly clear that we are going to have a difficult job keeping the LEM weight below the control weight". On 6 July Grumman requested that it be allowed to deliver the early LEMs without some subsystems, but Shea insisted that they must all leave the factory in a fully functional condition. At that time, LEM-1 was to be delivered to the Kennedy Space Center in November 1966, with the next five vehicles following in 1967, but it was becoming increasingly evident that this schedule would be difficult to meet. On 13 September 1965 Shea established the weight control board to enable subsystem managers to meet on a weekly basis and report progress in controlling the weights of the spacecraft, and when appropriate to create *ad hoc* task forces to chase up specific issues and report back.

Meanwhile, on 17 February 1965, Shea clarified for North American Aviation the Block I schedule. CSM-009 and CSM-011 were to fly on AS-201 and AS-202 missions, and be configured for unmanned use. CSM-012 and CSM-014 were to be delivered for manned missions, but be capable of being adapted at the Cape for unmanned flight. The decision for CSM-012 would be made six months before the scheduled launch date for AS-204, and if it were to be flown unmanned this would be either to gain additional data on the spacecraft's characteristics or to provide additional time for the Marshall Space Flight Center to obtain orbital data on the S-IVB stage. CSM-017 and CSM-020, which were assigned to the early tests of the Saturn V launch vehicle, need not be capable of manned use. The first manned Block II was to be CSM-101, in conjunction with LEM-2. On 22 March Glynn S. Lunney, chief of the flight dynamics branch of the flight control division at the Manned Spacecraft Center, was appointed the assistant flight director for AS-201 and AS-202. Once the spacecraft had been released on a high ballistic arc it was to fire its service propulsion system to accelerate to 29,000 feet per second, significantly faster than a normal orbital entry but not as fast as a lunar return. CSM-009 was to test the performance of the ablator, and CSM-011 was to produce a high total heat load and assess the ablator's interaction with its support structure. On 25 June Carroll H. Bolender was appointed as deputy director of mission operations at the office of manned space flight, his first task being to plan these two missions. On 10 August ASPO issued the flight assignments: LEM-1 (AS-206), LEM-2 (AS-207), LEM-3 (AS-503), LEM-4 (AS-504), LEM-5 (AS-505) and LEM-6 (AS-506). Six test articles had been ordered for ground tests: LTA-1 was retained at Bethpage to resolve problems encountered during the initial fabrication, assembly and checkout procedures; LTA-2 went to the Marshall Space Flight Center for launch vibration tests; LTA-3 and LTA-5 were to assess the structural effects of firing the engines; LTA-8 went to the Manned Spacecraft Center for thermal–vacuum environmental tests; and LTA-10 went to the North American Aviation factory in Tulsa, Oklahoma, where the SLA was being manufactured, for fit checks. In July 1965, in an effort to cut costs, Grumman had been told to delete LTA-4 (meant for vibration tests), the ascent stage of LTA-5 and the two flight test articles, and instead to refurbish two of the test articles for flight once their ground testing was over. The company said it would refurbish LTA-10 and LTA-2 for the first two Saturn V flights. The first three LEMs were to incorporate development flight instrumentation to record conditions during launch: the first two on the Saturn IB and the last on a Saturn V. A requirement was to minimize the difference between LEM-3 and LEM-4, and to make all subsequent production vehicles identical.

The emblem designed by NASA for the Apollo program.

On 30 August 1965 Mueller announced the choice of the emblem to represent the Apollo Program – it had a large letter 'A' superimposed on a black sky which depicted the constellation of Orion, an Earth and a Moon, and trajectory arcs. The astronauts were represented by the three stars of Orion's belt, which lay along the 'bar' of the 'A'.

On 21 October Phillips postponed AS-201 to January 1966 and AS-202 to June 1966 in order to reflect the delivery dates for CSM-009 and CSM-011, but the dates of the other early flights remained unchanged, with AS-203 in July 1966, AS-204 in October 1966 and AS-205 in January 1967.

DEVELOPING THE SATURN V

On 22 January 1965 Phillips forecast a busy year for Apollo, with a great deal of ground testing, the development testing of the CSM for the first manned mission, flight qualification of the Saturn IB and initial testing of the Saturn V. Three days later Seamans ventured to a congressional budget committee that a lunar landing "in early 1970" was "conceivable". And indeed, following years of development, the strands of the program were coming together. Rocketdyne had started work on the 200,000-pound-thrust J-2 engine in May 1960 to power the hydrogen-burning upper stages of the 'advanced Saturn', whose configuration was at that time open to debate. On 27 November 1963 it made its first full-duration (500-second) test. On 4 December 1964 the Douglas Aircraft Corporation fired an S-IVB on a static rig at its Sacramento test site, running it at full power for 10 seconds. This was a 'battleship' version (i.e. the equivalent of a 'boilerplate') using thick stainless steel propellant tanks rather than the lightweight aluminum of the operational version. On 7 December 1964 the first S-IVB mockup – which was accurate in terms of mass, center of gravity and structural stiffness but had models of the engine and other systems – arrived in Huntsville to undergo stress testing. Meanwhile, on 19 January 1959 NASA had taken over from the Air Force the contract with Rocketdyne for the F-1 engine. On 10 February 1961 the prototype was test fired for the first time, and sustained 1.55 million pounds of thrust for several seconds, thereby setting a record for a single-chamber engine by a considerable margin. On 9 April 1961 it was announced that the engine had achieved 1.64 million pounds of thrust. On 26 May 1962 the engine was successfully fired at full power for its intended operating duration of 2.5 minutes. On 16 April 1965 the first S-IC stage of the Saturn V was test ignited for a few seconds at the Mississippi Test Facility. On 24 April the S-II was first tested at Rocketdyne's facility at Santa Susana in California, delivering the requisite 1 million pounds of thrust. On 21 June Rocketdyne made the one-thousandth test firing of the F-1 engine. At Huntsville on 5 August the S-IC made a "perfect" full-duration test, during which it responded to steering commands provided by the blockhouse. On 9 August the S-II made its first full-duration firing. That same day,

Rocketdyne's mighty F-1 engine.

the first production version of the S-IVB was tested, and on 20 August it was fired for 3 minutes, shut down for half an hour, then reignited for almost 6 minutes in a simulation of its role on a lunar flight. Meanwhile, on 27 August North American Aviation reported that it had finished ground testing the service propulsion system of the spacecraft. Unfortunately, by early 1966 the development of the S-II was slipping behind schedule. In an effort to recover, North American Aviation recruited a new manager, Robert E. Greer, a retired Air Force general, who took the company team to the Mississippi Test Facility. On 23 April 1966 the S-II was successfully fired for 15 seconds, but faulty instrumentation caused firings on 10, 11 and 16 May to cut off early. In tests on 17 and 20 May, the stage fired for 150 and 350 seconds, but fires broke out in two places on the vehicle during a test on 25 May. As the stage was being removed from the stand on 28 May its hydrogen tank exploded, damaging the facility and injuring five people. Mueller began to send weekly progress reports on the S-II to company president John Leland Atwood, and warned that the S-II stood an excellent chance of replacing the LEM as the 'pacing item' in the program; not that this was news to Atwood. North American Aviation was also experiencing difficulties with the subcontractors for the spacecraft, both for the Block I and Block II.

PREPARING APOLLO

On 2 December 1965 Hugh Dryden died of cancer, and a week later Webb announced that the president had agreed to promote Seamans to deputy administrator.[16] Although Seamans was sworn in on 21 December, he retained the responsibilities of associate administrator until Newell took over that post in August 1967, whereupon John E. Naugle took over as associate administrator for space science and applications.

A significant operational step toward lunar orbit rendezvous was accomplished on 15 December 1965, when Gemini 6 rendezvoused with Gemini 7. The straightforward manner in which this was done raised the prospect of undertaking the manned test of the LEM without the requirement to reduce the weights of the CSM and LEM to enable the Saturn IB to carry them together. On 28 January 1966 Phillips asked ASPO to determine the impact, including the effects on ground support equipment and mission control, of a *dual* AS-207/208 mission as early as the scheduled date for AS-207, the Saturn IB that was to have orbited them together. The idea was for the near-simultaneous launches of AS-207 with CSM-101 and AS-208 with LEM-2 to lead to a rendezvous and docking, at which point the mission would unfold as planned. On 2 February John P. Mayer, chief of the mission planning and analysis division, told Kraft that the principal scheduling constraint would be to prepare the Real-Time Computer Complex in Houston to plan and support such a mission, in which case the decision on whether to attempt it would have to be taken very soon. Mayer also urged that if the IBM personnel who had worked

[16] The High-Speed Flight Station at Edwards Air Force Base was renamed the H.L. Dryden Flight Research Facility.

on the Gemini 6/7 rendezvous could be spared, they should be reassigned to help to plan the new dual mission. On 4 February John D. Hodge, chief of the flight control division, pointed out that some of the operational issues associated with near-simultaneous launches would be obviated if there was a delay, the duration of which would require to be determined. On 24 February Howard W. 'Bill' Tindall Jr, assistant chief of the mission planning and analysis division, recommended launching the CSM first into a circular orbit at 260 nautical miles and, 24 hours later, launching the LEM into a circular orbit at 110 nautical miles. There would be two follow-up launch windows of about 3 minutes each day that would yield ideal in-plane and phasing conditions for the CSM to make the rendezvous at the end of the third orbit. On 1 March Shea urged that this mission should be attempted. On 8 March Phillips told the Manned Spacecraft Center, Marshall Space Flight Center and Kennedy Space Center to initiate planning, with a view to launching one month later than had been set for AS-207 on the previous schedule.

Meanwhile, AS-201 lifted off from Pad 34 on 26 February 1966 and released CSM-009 on a ballistic arc to test its heat shield. On 16 March Gemini 8 achieved the first docking with an Agena target vehicle, thereby boosting confidence in the decision to attempt the AS-207/208 dual mission. On 21 March NASA announced that Gus Grissom was to be 'command pilot' for the first Apollo mission, with Ed White as 'senior pilot' and Roger Chaffee as 'pilot'.[17] They were to be backed up respectively by James McDivitt, David Scott and Rusty Schweickart. In each case, there was a Gemini commander, a Gemini pilot, and a rookie. Slayton nominated Grissom to this role immediately following his Gemini 3 test flight in March 1965. After commanding Gemini 4 in June 1965, McDivitt had been reassigned to back up Grissom. Gemini 4 pilot White first backed up Gemini 7 and was then assigned to Grissom's crew. Although Slayton had introduced a 'rotation' in which a pilot progressed to command a Gemini mission, following Gemini 8 Scott was immediately added to McDivitt's crew to give them early experience of Apollo training prior to taking on the AS-207/208 dual mission. The performance of the service propulsion system of CSM-009 had raised some issues, but as long as CSM-011 succeeded, AS-204 would launch the manned mission with CSM-012 in the fourth quarter of 1966, otherwise that pair would be used for a third unmanned test. In contrast to the conservative three-orbit Gemini 'shakedown' flight, the first Apollo mission was to be an 'open-ended' flight of up to 14 days "to demonstrate spacecraft and crew operations and evaluate spacecraft hardware performance in Earth orbit".

On 4 April 1966 the Manned Spacecraft Center revised its senior management job titles, with 'assistant director for' becoming 'director of' in order to make more explicit the fact that the post had *primary* rather than subordinate responsibility for that activity: as a result, for example, Kraft ceased to be the assistant director for flight operations and became the director of flight operations. On 12 May NASA deleted 'Excursion'

[17] The 1961 crew terminology had envisaged a commander/pilot, a navigator/copilot, and a flight engineer/scientist.

from 'LEM', renaming the lander the Lunar Module 'LM'. In June the 'Apollo 1' crew patch was approved. On 13 July Slayton and Kraft jointly wrote to Shea: "A comprehensive examination of the Apollo missions leading to the lunar landing indicates there is a considerable discontinuity between the missions AS-205 and AS-207/208. Both missions AS-204 and AS-205 are essentially long-duration system validation flights. AS-207/208 is the first of a series of very complicated missions. A valid operational requirement exists to include an optical equal-period rendezvous on AS-205." A rendezvous with the spent S-IVB stage on one of the Block I missions would provide an opportunity to examine the control dynamics, visibility, and piloting techniques that would be required to perform the basic AS-207/208 mission. In July every spacecraft on Grumman's production line through LM-4 was late. The focus, of course, was on LM-1, but late shipments by subcontractors had created a bottleneck in its assembly. However, the 'rate of slippage' was slowing, and on 6 October Shea reported his belief that Grumman would be able to deliver LM-1 early in 1967. By the end of 1966 LM-1 and LM-2 were in test stands, and LM-3 through LM-7 were in various stages of production, but by the end of January 1967 it was evident that LM-1 would not be able to be shipped in February after all.

AS-202 was launched from Pad 34 on 25 August 1966, with CSM-011 which, in successfully accomplishing its mission, concluded the unmanned testing of the Block I CSM. Because 80 percent of the objectives for CSM-002, CSM-009 and CSM-011 had (between them) been met, AS-204 was now released for the manned Apollo 1.[18] On 26 August the command module of CSM-012 was delivered to the Kennedy Space Center in a conical container prominently labeled 'Apollo One'. North American Aviation was to have sent it several weeks earlier, but the failure of the glycol pump in the environmental control unit led to the exchange of this unit with its CSM-014 counterpart. Although the customer acceptance review identified other "eleventh-hour problems" associated with the environmental control system, NASA authorized its shipment. On 7 October the office of manned space flight held the AS-204 design certification review, and declared that the launch vehicle and spacecraft conformed to the design requirements and would be flightworthy as soon as a number of deficiencies had been rectified. Phillips sent a list of deficiencies to Lee B. James at the Marshall Space Flight Center, Shea at the Manned Spacecraft Center, and John G. Shinkle, the Apollo program manager at the Kennedy Space Center, ordering speedy compliance. On 11 October Phillips was informed by Bolender of a report which he had received the previous day from Shinkle, detailing mounting delays in the preparation of CSM-012. When the spacecraft was delivered, 164 'engineering orders' had been identified as 'open work', even though the data package had identified only 126 such items. By 24 September the list had grown to 377, and Shinkle deemed that some 150 of the 213 additional orders should have been identifiable by North American Aviation prior to the customer acceptance review. The main issues included the

[18] CSM-002 was launched on 20 January 1966 by a Little Joe II for the A-004 high-altitude abort test at the White Sands Missile Range.

malfunctioning environmental control unit, which had failed again, problems with the reaction control system, a leak in the service propulsion system, and even design deficiencies with the couches that had required engineers to be dispatched to the Cape. On 12 October Phillips wrote to Mark E. Bradley, vice president of the Garrett Group – whose AiResearch division had supplied the environmental control unit under subcontract to North American Aviation – saying that its reliability was threatening a "major delay" to the AS-204 mission. To Phillips, the problems seemed "to lie in two categories: those arising from inadequate development testing, and those related to poor workmanship". On 25 October the propellant tanks of the service module for CSM-017 (assigned to AS-501) failed catastrophically during a test at North American Aviation. The normal pressure was 175 pounds per square inch, but it failed after 100 minutes at the maximum specified operating pressure of 240 pounds per square inch. In fact, the test of SM-017 had been ordered after the discovery of cracks in the tanks of SM-101 (assigned to AS-207). The failure was especially mystifying because, in 'proof testing', the tanks of SM-017 had been subjected to 320 pounds per square inch for several minutes. ASPO established an investigation, which was to report by 4 November. Because SM-012 had been through the same test regime, Shea grounded it pending the investigation. The problem was determined to be stress corrosion (i.e. cracking) in the titanium resulting from the use of methyl alcohol as a test liquid; the purpose of the test was to verify the integrity of the tanks, and the hydrazine and nitrogen tetroxide propellants were toxic. Using a liquid that was compatible with titanium would eliminate the problem, and it was decided to use freon and isopropyl alcohol for the oxidizer and fuel tanks respectively, with the further proviso that the systems had not been previously exposed to propellant and would be purged by gaseous nitrogen after the test. With the issue understood, the tanks of SM-012 were removed for inspection, and verified free of cracks. A replacement environmental control unit was delivered on 2 November and testing resumed once this had been installed, but the unit had to be returned to AiResearch and was not reinstalled in the command module until mid-December.

On 29 September NASA named the crew for the CSM-014 mission as Wally Schirra (command pilot), Donn Eisele (senior pilot) and Walt Cunningham (pilot), who would be backed up by Frank Borman, Tom Stafford and Michael Collins respectively. Schirra was the only member of the prime crew to have flown in space, but all members of the backup crew had flown previously. In fact, Slayton had given Schirra and Borman these assignments in March, upon their return from the global 'goodwill tour' following Gemini 6/7. Stafford and Collins had been assigned following Gemini 9 in June and Gemini 10 in July, respectively. Slayton had earmarked the rookies Eisele and Chaffee to Grissom's crew, but in late 1965 Eisele had injured his shoulder in weightlessness training on a KC-135 aircraft and had to withdraw from training for several months, prompting Slayton to move him to Schirra's crew instead of White, whom Slayton had planned to assign to Schirra after backing up Gemini 7. The Apollo 2 mission was to be a straightforward rerun of Apollo 1, to further evaluate the spacecraft's basic systems prior to advancing to the Block II. In early December, accepting that it would be impossible to launch Apollo 1 before the end of the year, Mueller rescheduled it for February 1967 and deleted the Block I reflight in order to preclude the delay in flying CSM-012 from impacting the missions already scheduled for that year. Much to Schirra's dismay, his crew was reassigned to back up Apollo 1.[19] As soon as possible after AS-204, AS-206 would launch LM-1 for an unmanned test, and if this proved satisfactory McDivitt's crew would fly the dual mission (which was now AS-205/208 because the deletion of CSM-014 had freed AS-205) as the new Apollo 2 in August. On 22 December NASA announced this change, together with the fact that McDivitt's crew would be backed up by Thomas Stafford, John Young and Eugene Cernan. Furthermore, if two unmanned tests proved sufficient to 'man rate' the Saturn V, it might be possible to launch AS-503 before the end of 1967 with a CSM and LM on a high-apogee mission. The crew was named as Frank Borman (CDR), Michael Collins (CMP) and William Anders (LMP), backed up by Pete Conrad, Dick Gordon and Clifton Williams.[20] These assignments were made after Stafford flew Gemini 9 in June, Young and Collins flew Gemini 10 in July, Conrad and Gordon flew Gemini 11 in September, and Cernan backed up Gemini 12 in November. The schedule in force in December 1966 therefore envisaged AS-503 in October 1967, AS-504 in December 1967, and the lunar landing in late 1968 or early 1969.

The Gemini missions had demonstrated that if an astronaut on a spacewalk is to be able to work effectively, he must be provided with a variety of mobility and stability aids. On 6 December 1966 Slayton warned Shea that without handholds and tethering points, an emergency transfer from the forward hatch of the LM to the command module would not be feasible. On 26 December he recommended that a spacewalk should be scheduled 100 hours into the AS-503 mission, following the two DPS firings but before the descent stage (with its porch) was jettisoned. One of the two astronauts would egress and ingress through the forward hatch to assess the environmental control system in the LM during depressurization, the operation of the forward hatch, the Apollo suit with its life-support backpack, and the egress procedure for the emergency external transfer. While outside, the spacewalker was to comprehensively photograph the exterior of the LM in order to verify that it had not been damaged during its retrieval from the S-IVB/SLA.

SETBACK AND RECOVERY

With the launch of AS-204 set for 21 February 1967, a 'flash fire' broke out in CM-012 while it was pressurized with pure oxygen for a 'plugs out' countdown test on 27 January, killing Grissom, White and Chaffee, and threatening to halt the program in its tracks. Nevertheless, on 31 January the Manned Spacecraft Center, Marshall Space Flight Center and Kennedy Space Center were directed to proceed as planned with preparations for AS-

[19] Having lobbied for the cancellation of the Block I reflight, Schirra had hoped to put himself first in line for the dual mission, but Slayton required the command module pilot of a Block II to have prior experience of rendezvous, which Eisele lacked.

[20] On 29 November 1966 Slayton revised the crew nomenclature to commander, command module pilot and lunar module pilot. On 14 June 1967 the acronyms CDR, CMP and LMP were announced by Low, who was by then ASPO manager. Because Slayton required the Block II CMP to have had experience of rendezvous, the only way he could introduce a rookie was as the LMP; this constraint would be relaxed later.

501 with CSM-017 and LTA-10R, except that during tests the command module was not to be pressurized with pure oxygen without specific authorization. On 2 February CSM-014 was delivered to the Kennedy Space Center, shipped by North American Aviation to assist in the training of the technicians who were to disassemble CM-012 during the investigation of the fire. On 3 February Mueller announced that although manned flights were grounded indefinitely, the unmanned AS-206, AS-501 and AS-502 missions were to proceed as soon as delivery of the hardware allowed. During the investigation of the fire, Mueller suggested that when the Block II spacecraft became available the CSM-only mission should be deleted and the effort switched to combined testing with the LM, but Gilruth said that it would not be wise to test *two* new vehicles at once. In March it was decided to fly an 11-day CSM-only mission, in effect to perform Grissom's mission with the upgraded model, and Slayton tipped off Schirra that his crew would fly it, backed up by Stafford's crew. On 21 February, the day that Apollo 1 had been scheduled to launch, Floyd Thompson, leading the investigation, gave a preliminary briefing to Mueller on the findings, and on 25 February Seamans sent a memo to Webb listing Thompson's early recommendations.[21] On 15 March Slayton requested that one of the primary objectives of Schirra's flight – which had become the third mission to be named Apollo 2 – be a rendezvous with the spent S-IVB stage, and proposed that this occur "after the third period of orbital darkness". Because the flight was to focus on evaluating the spacecraft's systems, Kraft pointed out on 18 April that if a problem were to arise that would require the cancellation of the rendezvous, then any preliminary maneuvers would complicate the preplanned contingency deorbit procedures. The rendezvous should not be initiated until "after a minimum of one day of orbital flight", and it should be "limited to a simple equi-period exercise with a target carried into orbit by the spacecraft". On 5 April Phillips told the Manned Spacecraft Center, Marshall Space Flight Center and Kennedy Space Center that the mission profile for the first manned flight would be based on that developed for Grissom's flight, dated November 1966. As the complexity of the mission was to be limited to that previously planned, and as no rendezvous had been planned, the rendezvous exercise would be assessed in terms of how it would complicate the mission. The double-hatch of the Block I had been replaced on the Block II by a single 'unified hatch' on a hinge that swung outward. It had a manual release for either internal or external operation, and could be opened in 60 seconds irrespective of whether the cabin exceeded ambient pressure. It had a mechanism and seal similar to that of the Gemini spacecraft to protect against being accidentally opened in the vacuum of space, but was capable of being opened for a spacewalk. Nevertheless, Phillips directed that there "be no additions that require major new commitments such as opening the CM hatch in space or exercising the docking subsystem". On 2 June Phillips agreed with Low that there should be a rendezvous using a small transponder 'pod' (as developed for an early Gemini rendezvous test) but Phillips insisted that this should not be listed as a primary objective.

NASA announced on 20 March 1967 that the unmanned LM-1 flight would be switched from AS-206 to AS-204, which had become available. The rationale for the AS-

[21] The Apollo 204 Review Board submitted its final report on 5 April 1967.

205/208 dual mission with CSM-101 and LM-2 was to ensure that testing the LM would not be held up by slippage in the development of the Saturn V, but by this time it was evident that a Saturn V launch would soon be feasible. The first two development flights were to carry refurbished LM test articles, but unless the pace of LM development dramatically picked up, the heavy launch vehicle would become available before the LM, and hence eliminate the need for the *ad hoc* dual mission. It was decided that if the LM-1 flight proved unsatifactory, LM-2 would be launched unmanned by AS-206 to address the remaining test objectives. On 25 March Mueller directed that missions be numbered in the sequence of their being flown, regardless of whether they employed the Saturn IB or Saturn V and were manned or unmanned. In the old scheme only manned missions had been counted: CSM-012 had been assigned to Apollo 1 and, as the plan was repeatedly revised, Apollo 2 was assigned in turn to CSM-014 (Schirra), CSM-101 (McDivitt) and CSM-101 (Schirra). Mueller named AS-501, the first scheduled mission, Apollo 4 without explaining his accounting. On 30 March Low suggested that the AS-201 and AS-202 CSM test flights be retrospectively assigned the designations Apollo 2 and Apollo 3 in order to lead into the forthcoming Apollo 4 mission, but this was rejected by Mueller on 24 April.[22] On 7 April Joe Shea became deputy associate administrator for manned space flight with responsibility for technical aspects of the program, and George Low succeeded him as ASPO manager at the Manned Spacecraft Center. A few days later, Everett Christensen resigned as mission director in Washington. On 17 April a joint meeting by the mission operations division and flight operations directorate at the Manned Spacecraft Center announced that: (1) successful firings by the descent and ascent stages of an unmanned LM, including the 'fire in the hole' separation, should be regarded as prerequisites to a manned LM attempting these functions; (2) a demonstration of EVA transfer should *not* be a prerequisite to manned independent flight of the LM; (3) the Saturn V should be 'man rated' as rapidly as possible; (4) three manned Earth orbit flights involving the CSM in conjunction with the LM should be considered the *minimum* number of missions prior to a lunar landing; and (5) although a lunar orbit mission should *not* be a formal step in the program, the necessary planning should be undertaken as a contingency in the event of the CSM achieving lunar-mission capability ahead of the LM. On 27 April ASPO sent the Block II Redefinition Task Team, headed by Frank Borman, to North American Aviation. This 'Tiger Team' had the authority to make on-the-spot decisions that previously would have required judgment by the configuration control board. It was to oversee the 'redefinition' of the Block II spacecraft, to provide a rapid response to questions regarding detail design, quality and reliability, test and checkout, baseline specifications, configuration control, and schedules. Meanwhile, North American Aviation had hired William Bergman from the Martin Company to supersede Harrison Storms as project manager. On 8 May Low confirmed that CSM-101 would be launched by AS-205 on an open-ended mission of up to 11 days in order to evaluate its systems, and the following day Webb informed the Senate space committee that the mission would be flown by Schirra, Eisele and Cunningham. Webb had canvassed suggestions for how best

[22] AS-203 was not included because it did not carry a spacecraft.

to impress upon Congress that the program was recovering from the setback of the capsule fire, and Mueller had urged flying the Saturn V as soon as possible. On 9 May, therefore, it was announced that AS-501 would be scheduled for early in the autumn.

PICKING UP THE PACE

In May 1967 the Manned Spacecraft Center had proposed three manned Saturn V missions *prior to* attempting the lunar landing. When Mueller advocated a landing *on* the third mission, Kraft warned Low that a landing should not be tried "on the first flight which leaves the Earth's gravitational field" because flying to the Moon was such a great step forward in terms of operational capability that this should be demonstrated separately, in order to enable the landing crew to focus on activities associated with landing. Having accepted Kraft's argument that the lunar landing crew should not be the first to venture out to the Moon, on 20 September Low took a team to the office of manned space flight. Owen E. Maynard, chief of the systems engineering division in Houston outlined the proposed sequence: (A) Saturn V and unmanned CSM development; (B) Saturn IB and unmanned LM development; (C) Saturn IB and manned CSM evaluation; (D) Saturn V and manned CSM/LM joint development; (E) CSM/LM trials in an Earth orbit possessing a 'high' apogee; (F) CSM/LM trials in lunar orbit; (G) the first lunar landing; (H) further 'minimalist' landings; (I) lunar orbital surveys; and (J) 'enhanced capability' landings.[23] This alphabetically labeled series was not a list of *flights*, since several flights might be required to achieve one *mission*. Two Saturn V development flights were already scheduled as Apollo 4 and Apollo 6, and the LM-1 test as Apollo 5. Phillips asked whether a second Saturn V test was really necessary, and von Braun and Low said it was, as the second would confirm the data from the first. If the development of the Saturn V were to prove to be protracted then the 'D' mission would be addressed by reverting to the Saturn IB plan in which the CSM and LM would be launched separately and rendezvous in orbit. Most of the discussion focused on the proposal to fly a lunar orbital flight "to evaluate the deep space environment and to develop procedures for the entire lunar landing mission short of LM descent, ascent and surface operations". When Mueller argued that "Apollo should not go to the Moon to develop procedures", Low replied that developing crew operations would not be the main reason for the mission; there was still a lot to be learned about navigation, thermal control and communications in deep space. Although the meeting left this matter undecided, the alphabetic designators soon became common shorthand.

On 2 October 1967 Phillips confirmed that LM-2 should be configured for an unmanned test flight, and directed that LM-3 be paired with CSM-103 for the first manned mission of the complete Apollo spacecraft. Grumman's latest schedule called for LM-2 to be shipped in

[23] The last two categories represented the lunar phase of the so-called Apollo applications program, and when this fell by the wayside the main program was expanded to include 'enhanced capability' landings.

February 1968, LM-3 in April, and LM-4 in June. On 4 November Mueller announced the sequence of flights for 1968: AS-204 with LM-1, then AS-502 (the second unmanned test), AS-503 (the third unmanned test, if required), AS-206 (LM-2, if required), AS-205 (CSM-101, manned) and AS-504 (CSM-103 and LM-3, manned). On 15 November Low said that in the event of AS-503 being unmanned, the payload should be BP-30 and LTA-B.

THE FIRST SATURN V FLIGHT

On 25 May 1966, five years to the day after Kennedy's speech calling for a lunar landing, the 2,700-ton diesel-powered crawler drove a full-scale engineering model of the Apollo–Saturn V from the Vehicle Assembly Building to Launch Complex 39 in order to verify the ground facilities and assist in the development of training procedures. It was a striking demonstration of the 'mobile launcher' concept. The schedule called for the launch of AS-501 early in 1967, but few believed that this would be feasible because the S-II stage had become the 'pacing item'. In fact, the delivery of the first 'live' S-II to the Kennedy Space Center had already slipped from July to October 1966, and when it arrived at the Mississippi Test Facility on 13 August the inspectors found a number of cracks, which delayed the start of its acceptance firings. In November Phillips revised the schedule to call for the S-II to be delivered to the Kennedy Space Center on 9 January 1967 for launch in April. Meanwhile, the S-IC stage had been erected on the mobile launch platform in the VAB on 27 October. In order not to delay the checkout, it was decided to stack a bobbin-shaped 'spacer' instead of the S-II to support the S-IVB, and on 12 January 1967 the spacecraft was added to begin its own checkout. On the S-II's delivery on 21 January, a number of faults were found. By now the launch had slipped to May. On 14 February the spacecraft was retrieved and transferred to the Operations and Checkout Building for examination as part of the investigation of the Apollo 1 fire, and so many wiring discrepancies were identified that repairs were not completed until June. Meanwhile, the S-IVB was de-mated, and on 23 February the spacer was replaced by the S-II. However, on 24 May the S-II was de-mated for inspection after factory inspectors discovered cracks in another S-II which was being readied for shipment, and was not restacked until mid-June. Once the S-IVB had been added, the revised spacecraft was installed on 20 June. The vehicle was finally rolled out to Pad A on 26 August. It was announced that the six-day countdown demonstration test would start on 20 September, but this slipped to 27 September, and a number of problems delayed its completion to 13 October; but, to be fair, this *was* the first Saturn V and the launch operations team was learning about the vehicle. The countdown began on 6 November, proceeded smoothly, and the first Saturn V lifted off on time on 9 November for its 'all up' demonstration flight. Suddenly, there was a real prospect of achieving Kennedy's challenge.

TESTING THE LM

On 20 November NASA announced that in view of the Apollo 4 success, McDivitt, Scott and Schweickart, who had been expecting to fly the Saturn IB version of the 'D' mission, would be launched on the *first* manned Saturn V, and be backed up by Conrad, Gordon and Al Bean (who had been assigned in place of Williams upon the latter's death in an aircraft crash on 5 October 1967). Borman, Collins and Anders would still fly the high-apogee 'E' mission, but be launched on the *second* manned Saturn V, and be backed up by Armstrong, Lovell and Aldrin. If the lunar orbit 'F' mission were to be deleted, then the *next* crew would be the one that attempted the first landing. Slayton is reputed to have planned to rotate Grissom from Apollo 1 to command the first landing. Schirra's position in the sequence mirrored that of Grissom, but while Schirra's spacecraft was a Block II, the fact that the mission did not have a LM meant that the CMP was not required to have prior experience of rendezvous. The big question was whether Slayton would rotate a backup crew for the 'F' mission or establish a 'new' crew with Schirra in command. By backing up Schirra, Stafford was first in line for rotation and it was noted that his crew was by far the most experienced yet formed.

In mid-1966 Phillips had hoped that AS-206 would be able to launch LM-1 in April 1967, and Debus, allowing six months to check out the spacecraft, had requested Grumman to ship it to the Kennedy Space Center in September 1966. However, manufacturing issues and instabilities in the ascent engine had intervened. When AS-206 was erected on Pad 37B in January 1967 the delivery date of its payload was still undefined. The AS-204 launch vehicle had been delivered to the Cape in August 1966 and erected on Pad 34 for Apollo 1. As it was undamaged by the fire that destroyed the spacecraft on 27 January, it had been reassigned to LM-1. It was de-erected and inspected in March, and on 12 April was re-erected on Pad 37B as 'AS-204R'. In the absence of the spacecraft, Grumman built a plywood mockup to verify the pad facilities. On 12 May Low told headquarters that while Grumman had promised to deliver LM-1 in June, he was skeptical. John J. Williams headed a 400-man spacecraft operations team at the Cape. When LM-1 arrived on 27 June, the preliminary examination identified a significant number of departures from its specifications. The previous year, NASA had decided to install a refurbished LM test article as a mass-model on the first Saturn V test. If LM-1 had been ready for AS-501 it *might* have been installed to further the 'all up' testing strategy, but the spacecraft's protracted development had precluded this. On 26 July 1967 C.H. Bolender was reassigned as ASPO program manager for the LM at the Manned Spacecraft Center. On 19 November LM-1 was mated with its launch vehicle and, because no CSM was present, the SLA was topped by a nose cone. The mission was launched as Apollo 5 on 22 January 1968, and four days later the LM-2 flight requirements meeting determined that: (1) apart from minor anomalies, LM-1 had achieved all its flight objectives; (2) it should be possible to achieve the objectives for LM-2 either by additional ground testing or on a manned mission; and (3) it was not necessary to undertake additional unmanned flights to 'man rate' the LM. Grumman's long-established view had been that there should be a second test, but the company relented after the review of the LM-1 data by the Manned Space Flight Management Council on 6 February. On 6 March NASA cancelled the shipment of LM-2 to the Cape. If AS-502 repeated the success of Apollo 4, then AS-503 would indeed be manned, hopefully before the end of the year.

Robert Seamans had resigned on 2 October 1967, giving three months' notice. On 5 February 1968 the Senate had accepted the nomination of Thomas O. Paine of General Electric, and on 25 March he was sworn in as deputy administrator.

OVERCOMING PROBLEMS WITH DARING PLANS

AS-502 was launched with CSM-020 and LTA-2R on 4 April 1968 as Apollo 6. In contrast to the flawless inaugural flight, this time the S-IC suffered a longitudinal 'pogo' oscillation that would have been intolerable to a crew, two of the engines of the S-II cut off early and the S-IVB failed to restart. Nevertheless, the spacecraft was inserted into an acceptable parking orbit and functioned well, with its service propulsion system making a rather longer burn than planned in order to recover the high apogee required to subject the heat shield to re-entry at near-lunar speed. On 23 April Mueller urged working to man AS-503. Although Phillips directed the next day that this vehicle be prepared with CSM-103 and LM-3, he also demanded contingency planning to reconfigure it with BP-30 and LTA-B if it was decided to make a third unmanned test. The Kennedy Space Center pointed out that so long as it was given sufficient notice of the configuration, the boilerplates would be able to be launched in mid-October but the manned mission would not be ready until late November at the earliest. On 26 April Webb approved this planning for a manned mission in the fourth quarter of the year, subject to a resolution of the anomalies suffered by AS-502. In seeking to overcome the pogo, the Marshall Space Flight Center enquired whether the emergency detection system could be configured to trigger an automatic abort. When Slayton argued against doing this, Low ordered the development of a 'pogo abort sensor' with a display in the command module to enable the crew to judge whether to initiate an abort. On 17 August, by which time it was clear that the pogo would be able to be eliminated, Low recommended that work on this sensor be terminated, and one week later Phillips concurred.

Meanwhile, when combustion instability caused by the fuel injector of the LM's ascent propulsion system continued into the summer of 1967, NASA got Rocketdyne to develop an alternative injector as a contingency measure. In April 1968 Grumman was instructed to coordinate the testing of Rocketdyne's injector in Bell's engine. In May, Low decided to utilize this hybrid, but told Rocketdyne to undertake the integration work. By mid-August it was evident from qualification testing that the modified engine was free of instabilities, and that there was no requirement to mount another mission specifically to demonstrate it in space. On 13 May Low met Kraft, Slayton and Faget to discuss whether the 'fire in the hole' staging test of the ascent propulsion system should be demonstrated on the 'D' or 'E' missions. One factor was that LM-3 would be the last to have the development flight instrumentation. Faget argued that while such data was desirable, it was not essential; more important, he said, would be photographs of the base of the ascent engine after the burn. In view of this line of argument, and the fact that the test would significantly increase the complexity of the 'D' mission, Low postponed the issue of whether to make the 'fire in the hole' test on the 'E' mission until the performance of the engine on LM-1 had been thoroughly analyzed. On 17 May Kraft told Low that the 'E' mission was already a complex affair, and that as additional objectives were assigned the probability of accomplishing them diminished. In particular, Kraft saw little need for a 'fire in the hole' staging demonstration. He understood the engineer's desire to test all the systems in space, in both normal and backup modes, but the first 'fire in the hole' test at the White Sands Test Facility on 22 December 1967 had achieved all its objectives and further ground testing would provide the data to enable the pressure and temperature transients of lunar lift off to be calculated.[24] Also in the spring of 1968 the Manned Spacecraft Center began a study of extending the apogee of the 'E' mission out to almost lunar distance to investigate navigation, communications and thermal control in the event of the lunar orbital 'F' mission being deleted; this alternative mission was labeled 'E-prime'.

On 7 May 1968 CSM-101 passed its final customer acceptance review, and at the end of the month was delivered to the Kennedy Space Center. The inspectors were delighted to find fewer discrepancies than on any previous spacecraft. Schirra, however, did not accept it as being flightworthy until the completion of the altitude-chamber tests in June; but when these had been done he was upbeat: "We're on a high-speed track, and this train is moving out." In contrast, when LM-3 was delivered on 14 June, the inspectors found over 100 deficiencies, many of which were classified as major. One month later George C. White, Mueller's chief of reliability and quality assurance, briefed him on the issues that the certification review board would need to address. Charles W. Mathews, now working for the office of manned space flight, was sent to assess the situation, and reported that the spacecraft was unlikely to be ready for AS-503.

On 7 August Low told the Manned Space Flight Management Council that LM-3 was unlikely to be ready for launch until February 1969. The following day, Low flew to the Cape with C.H. Bolender (his LM manager), Scott H. Simpkinson (ASPO's test division chief) and Owen G. Morris (of the LM project engineering division in Houston) to discuss the status of AS-503 with Phillips, Debus (director at the Kennedy Space Center), Roderick O. Middleton (Apollo program manager at the Cape) and Rocco A. Petrone (director of launch operations). On 9 August, in a memo entitled *Special Notes*, Low said that if Apollo was to achieve a lunar landing in 1969, it was essential to fly the first manned Saturn V mission before the end of 1968. By this point, the pogo problem was heading toward resolution and the other issues that marred AS-502 had been overcome. The delivery of CSM-103 was imminent, but LM-3 was seriously late. In April 1967, the Manned Spacecraft Center had outlined a lunar-orbit contingency mission, and on 7 August Low had asked Kraft to look into the possibility of sending CSM-103 to the Moon without a LM. At 08.45 in Houston on 9 August, Low met Gilruth and Kraft to discuss this option. Gilruth was enthusiastic; and Kraft said it would be technically feasible, largely because several months earlier it had been decided to put the 'Colossus' program in the spacecraft computer for CM-103. At 09:30 Slayton was called in and gave his support. Low then telephoned Phillips, who was still in Florida, and a meeting was scheduled in Huntsville at 14:30. In addition to Gilruth, Low, Kraft, Slayton and Phillips, also in attendance were Debus, Petrone, George H. Hage (Phillips's deputy), Wernher von Braun, Eberhard F.M. Rees (von Braun's deputy), Lee B. James (Saturn V program manager at the Marshall Space Flight Center) and Ludie G. Richard (of the Marshall Space Flight Center) – making 12 in all. Low said that if the Apollo 7 evaluation of CSM-101 went well, it would be technically possible to send CSM-103 out to the Moon in December, but if CSM-101 was flawed then CSM-103 would require to be confined to Earth orbit to continue the evaluation of the spacecraft's systems. Kraft pointed out that a 'loop' around

[24] In the event, neither Apollo 9 nor Apollo 10 made a 'fire in the hole' separation in space.

the back of the Moon and straight back to Earth would be insufficient; the spacecraft would require to enter lunar orbit to contribute significant insight to the lunar landing mission (which may well be the next mission to venture to the Moon, depending on whether the 'F' mission was undertaken). There was general agreement to start planning for this contingency, and that it be kept secret, pending the final decision and public announcement. The meeting broke up at 17:00. On returning to Houston, at 20:30 Low briefed George W.S. Abbey (his technical assistant), Kenneth S. Kleinknecht (his CSM manager), C.H. Bolender (his LM manager) and Dale D. Myers (North American Aviation's Apollo program director). On 10 August Slayton offered this 'new' mission to McDivitt, who opted to wait for LM-3 in order to retain the 'D' mission for which his crew had been training for so long. Borman, who had followed the discussions to extend the apogee of his 'E' mission out to lunar distance, readily accepted. On 12 August Kraft told Low that if they required a daylight launch for an Atlantic recovery after an abort, the lunar launch window would open on 20 December. As the substitute for LM-3, Low selected LTA-B as this had been assigned to accompany BP-30 on the unmanned mission and was already in preparation.

On 14 August the original 12 conferees, minus Rees, were joined by William C. Schneider, Julian B. Bowman (both of the office of manned space flight) and Thomas O. Paine (deputy administrator) for a meeting at headquarters to draw up a recommendation. During the meeting, Mueller called from Vienna in Austria where he and Webb were at a United Nations Conference on the Exploration and Peaceful Uses of Outer Space. Mueller was skeptical, and said he would not be able to discuss it until 22 August. Paine, playing the devil's advocate, pointed out to the conferees that until recently there had been doubts about whether the Saturn V was safe for manned flight, yet now they were considering having it send a spacecraft on an impromptu mission to orbit the Moon. Paine asked for comments. Von Braun pointed out that once it was decided to man AS-503, it did not matter how far the spacecraft went. Hage noted that there were a number of points in the mission where go/no-go decisions could be made, minimizing the risk. Slayton opined that not to pursue this option would significantly diminish the chance of achieving Kennedy's deadline. Debus had no technical reservations. Nor did Petrone. Bowman argued that it would be a 'shot in the arm' for the program. James said doing it would enhance the safety of later flights. Richard said it would enhance lunar capability. Schneider endorsed the proposal wholeheartedly. Gilruth pointed out that although it was an impromptu mission, it would enhance the chance of being able to achieve the overall goal of the program. Kraft reiterated that it should be a lunar orbital rather than a circumlunar mission. Low said that if Apollo 7 was a success, then they could either fly this impromptu mission or await LM-3 and launch in February/March – and given the deadline for the lunar landing the decision was obvious. Paine concurred. Phillips accepted the recommendation and ordered planning to continue. Phillips and Paine discussed it with Mueller and Webb on the telephone the next day. Mueller had warmed to the idea overnight. Webb was "fairly negative" (as Phillips later put it) but asked for information to be sent by telegram. On 16 August Webb called Paine and agreed to mission planning, with the proviso that there was no public announcement. On 17 August Phillips told Low that although Webb had authorized preparations for a launch in December, there must be no 'leak' that the spacecraft might go to the Moon; the decision was contingent on the outcome of Apollo 7.

Meanwhile, CSM-103 had arrived at the Kennedy Space Center, and work had started on the modifications needed to send it to the Moon. On 19 August Phillips directed that, irrespective of whether CSM-103 was confined to low Earth orbit or went to the Moon, if AS-503 was manned the mission was to be designated as 'C-prime'. If so, then McDivitt's crew would fly the 'D' mission riding AS-504 with CSM-104 and LM-3. The 'E' mission had been deleted. That same day, Phillips told the press that if the flight of CSM-101 was successful, AS-503 would be manned; but since the LM would not be ready this would be a CSM-only mission. By not mentioning the possibility of leaving Earth orbit, he readily conveyed the impression that this would be an Earth orbit mission. On 3 September Low directed that in the event of the 'C-prime' mission being confined to Earth orbit, it would undertake the parking orbit preparations for the translunar injection and then fly one of a number of alternative missions. In one, the spacecraft would separate without the S-IVB having restarted, and then fire its service propulsion system to set up a 6,440-km apogee. Alternatively, this apogee could be achieved by the spacecraft remaining with the S-IVB, which would make a brief burn. In the lunar option, the S-IVB would make the full burn. No matter which mission was executed, once the spacecraft was free it would simulate the transposition, docking and extraction of the LM. If it went to the Moon, it would circle the Moon 10 times over a period of 20 hours before heading home. On 9 September Borman's crew began training in the simulator at the Kennedy Space Center for the lunar mission, and on 19 September Mueller declared the Saturn V fit for manned flight. CSM-103 was mated with the launch vehicle on 7 October, the launch escape system was added on 8 October, and the next day the space vehicle was driven out to Pad 39A.

On 31 March 1968 Lyndon Johnson had announced that he would not stand for re-election. After informing Johnson on 16 September, James Webb announced his resignation, effective 6 October, his 62nd birthday, and Paine was promoted to acting administrator. Meanwhile, on 20 September CSM-101 passed its flight readiness review, and later in the day Schirra announced that Apollo 7 would be his final mission as he intended to retire from NASA, thus ending the public speculation about whether Schirra would command the lunar landing mission.

On 11 October AS-205 lifted off from Pad 34 with CSM-101 for Apollo 7 with the primary objectives of demonstrating CSM/crew performance; demonstrating crew/space vehicle/mission support facilities performance during a manned CSM mission; and demonstrating the CSM rendezvous capability using the S-IVB stage as the target. These objectives were to be addressed during in the first three days, but the flight was to be open-ended in order to assess the sustained performance of the spacecraft's systems, with a maximum duration of 11 days. Although Schirra was disinclined to load up the flight plan with secondary tasks which, if they proved problematical, might create a sense of 'failure' despite the fact that the primary objectives were achieved, the extended mission provided ample opportunity to conduct 'experiments' for the science community. Several hours into the mission, the spacecraft separated from the S-IVB, moved clear, turned and moved back in as if to retrieve the LM (which was absent). When Cunningham reported that the panels of the SLA had not fully deployed, it was decided that on future missions the panels would be jettisoned rather than hinged open. The rendezvous rehearsal was accomplished the following day. At Schirra's insistence, one man was awake at all times to monitor the systems. All three men developed head colds early in the flight, making them grumpy, and the provision of in-flight TV, which was one of the secondary objectives, provided a focus for their frustration, but when this was finally supplied the quality of the monochrome camera was excellent. The service propulsion system fired perfectly eight times. Schirra later described the flight as a "101% success".

On 7 November Mueller declared that AS-503 was fit for a mission to the Moon. On 11 November Phillips recommended to the Manned Space Flight Management Council that Apollo 8 enter lunar orbit. Later that day, Mueller told Paine that he had discussed the mission with the science and technology advisory committee and with the president's science advisory committee, both of which had endorsed the proposal, and he recommended that it should be attempted. After speaking to Borman by telephone, who confirmed his willingness to fly the mission, Paine gave the formal go ahead and told Phillips to make the necessary preparations. The next day NASA announced that Apollo 8 would be launched on 21 December and would be a lunar orbital mission. On 13 November it was announced that Stafford, Young and Cernan, backed up by Cooper, Eisele and Mitchell, had been assigned to Apollo 10. This established the precedent for a crew backing up one mission, skipping two, then 'rotating' to be the prime crew of the next mission. It had yet to be decided, however, whether Apollo 10 would fly the 'F' or 'G' missions.[25]

On 21 December AS-503 launched Apollo 8, which pursued the full mission. An important assignment while in lunar orbit was to take overlapping vertical and oblique pictures to determine the locations and elevations of landmarks on the far side of the Moon near the eastern limb, to enable a lander to perform navigational checks shortly prior to initiating its powered descent. With this audacious mission, NASA took a great step towards achieving its objective of a lunar landing before the decade was out.

The crew named in 1966 for the high-apogee 'E' mission was Borman, Collins and Anders, backed up by Conrad, Gordon and Williams. The flight assignments were revised following the Apollo 1 fire, and although Borman retained his crew and mission, he lost the first manned Saturn V and his backup crew became Armstrong, Lovell and Aldrin. In the summer of 1968, however, Collins had to stand down to have a bone spur removed from his spine, and Lovell was moved up to replace him. In appointing Fred Haise, a LM-specialist, to the backup crew, Slayton had to switch Aldrin from the LMP to CMP assignment because Haise was a rookie and the rule required the CMP to have rendezvous experience. In the event, by accepting the prospect of a lunar orbital mission Borman regained the first manned Saturn V. As a side effect of this change in the launch order, Slayton's crew rotation scheme put Armstrong instead of Conrad in line to command Apollo 11, which would attempt the first lunar landing if Apollo 10 flew the 'F' mission.

On 10 January 1969 John D. Stevenson, the director of mission operations in Washington, circulated a revised form of the tentative schedule for 1969 that had been

[25] CSM-101 had flown on Apollo 7, CSM-102 had been retained by North American Aviation for ground testing, CSM-103 had been assigned to the Apollo 8 'C-prime' mission, CSM-104 was to fly the Apollo 9 'D' mission, CSM-105 was for ground testing, and CSM-106, which was delivered to the Kennedy Space Center on 25 November 1968, was assigned to Apollo 10.

issued a year earlier. This called for launching the delayed 'D' mission on 28 February. As the 'E' mission had been rendered irrelevant by Apollo 8, this meant that if the 'F' mission lifted off on 17 May the launch window with the requisite illumination of the primary landing site would open on 15 July. The rationale for the 'F' mission had been to gain experience of conditions in deep space, but as this had been provided by Apollo 8, was another reconnaissance needed? The decision on the 'F' mission was postponed until LM-3 had been put through its paces. The launch of Apollo 9 was slipped to 3 March to enable the crew to recover from a mild respiratory infection. The fact that Schweickart suffered space sickness at an early stage resulted in his external transfer being cancelled, but he was able to perform an excursion by restricting himself to the LM's porch and so make the only test of the life-support backpack planned prior to the lunar landing. When the LM undocked to undertake its independent flight, this marked the first time that a crew flew in space in a vehicle that was incapable of entering the atmosphere; if it got into trouble, the only hope of rescue would be for the CSM to rendezvous and dock. On 5 March, while the mission was underway, president Richard Milhaus Nixon nominated Paine to be NASA administrator.

DRESS REHEARSAL

On 24 March 1969 NASA reported that Apollo 10 would fly the 'F' mission. The original concept had called for the LM merely to undock, enter a slightly different orbit, and return and redock, but in December 1968 the mission planning and analysis division at the Manned Spacecraft Center had urged putting the descent propulsion system through a realistic rehearsal in which the LM would lower its perilune sufficiently to test the ability of the landing radar to detect and lock onto the surface, with the illumination for the low passes exactly as it would be on the landing mission, to enable the primary site to be documented and to identify landmarks on the approach route. Howard Tindall also suggested that the LM should initiate the powered descent and then execute an early abort by 'fire-in-the-hole' staging, but his colleagues convinced him that this would be overly adventurous. One aspect of the decision to proceed with the 'F' mission was to assess the operation, tracking and communications of two vehicles in lunar orbit. As Apollo 8 had confirmed that 'mascons' (discovered by tracking Lunar Orbiter 4) made the lunar gravitational field uneven, it was necessary to evaluate how the guidance and navigation system of the LM coped while making the low passes of the descent orbit. In other words, it had been decided to exploit the fortuitous relaxation in schedule pressure, take the next step beyond what had been achieved by Apollo 8, and perform a full-dress rehearsal to the point of the LM initiating the powered descent. The LM was to separate from the CSM in the lunar parking orbit, enter the descent orbit with a 50,000-foot perilune, make two low passes over candidate landing sites, then simulate an abort and execute the rendezvous. In fact, it would have been impracticable for LM-4 to attempt the 'G' mission because it was incapable of landing on the Moon as it did not have the requisite 'software' for the powered descent – at that time, computer memory was a physically wired technology. Furthermore, owing to propellant limitations in the ascent stage on this somewhat overweight LM, if it had landed it would have been unable to lift off and rendezvous. If

Apollo 10 had been assigned the first landing, it would necessarily have had to wait for LM-5, but to do this would have wasted the opportunity for a rehearsal, and it was deemed better to seek more information to ease the way for its successor. Apollo 10 lifted off on 18 May, and on the fifth day, with the LM in the descent orbit, the ability of the landing radar to lock onto the surface was established,. The prime candidate landing site (initially designated II-P-6 and later ALS-2), which was located just north of the lunar equator in the southwestern part of Mare Tranquilitatis, was photographed at optimum illumination and at much higher resolution than had been achieved by the Lunar Orbiters, with the advantage that the film was able to be returned to Earth for processing rather than being scanned and transmitted. A visual inspection indicated that the eastern end of the target ellipse was smooth and level, but a landing in the west might require some maneuvering to select an acceptable place among small craters and boulders. Passes were also made over ALS-1 in eastern Tranquilitatis at a higher illumination than desired for a landing, and over ALS-3 in Sinus Medii at a lower illumination.

THE LANDING CREW

On 16 July 1969 Apollo 11 lifted off to attempt to land on the Moon. By late 1968 Collins had recovered from his spinal surgery, so when Slayton rotated Armstrong from backing up Apollo 8 to commanding Apollo 11 he reinstated Collins as CMP and put Aldrin back in the LMP slot. Lovell and Anders were rotated into backup, with Lovell commanding, Anders as CMP and Haise as LMP. Consequently, as had been the case for Apollo 10, all members of the prime crew for this mission had previous flight experience.

Saturn I

Development program

1961–1965

BACKGROUND

The first stage of the Saturn I launch vehicle was powered by a cluster of eight H-1 engines with an inner quartet that were fixed and four peripheral engines that were mounted on gimbals to provide control of roll, pitch and yaw. In order to reduce the tooling costs, the vehicle was made by stacking tanks from the Redstone and Jupiter missiles in parallel, with a 267-cm-diameter Jupiter tank being surrounded by eight 178-cm-diameter Redstone tanks. The central tank held oxidizer, and those around it alternated fuel and oxidizer. As the cryogenic liquid oxygen would cause the oxidizer tanks to shrink when loaded, these were the main structural elements, and the kerosene tanks had slip joints at the top. To prevent the vehicle becoming unbalanced in flight, there were interconnecting pipes to maintain an even level in each set of tanks. Consequently, in addition to each of the four fuel tanks feeding a pair of engines (one inner and one outer), the fact that it was interconnected meant that if one of the engines shut down prematurely the fuel in the tank above it would automatically be distributed to the remaining engines. In addition to each of the peripheral oxidizer tanks feeding a pair of engines and being interconnected for purposes of level balancing, it was also linked to the central oxidizer tank. At the top of the stage were 48 spheres containing nitrogen to pressurize the fuel tanks. The oxidizer was pressurized by a heat-exchanger that volatized the liquid oxygen to the gaseous phase. The redundant design incorporated 16 pumps. The resulting pipework made the S-I a plumber's nightmare, but the H-1 was a reliable engine. The tankage was held together at the base by the 'tail section', and at the top by a 'spider beam'. The tail section contained the structural element against which the engines thrusted, and the hold down points. The spider consisted of an axial hub with eight radial I-beams and eight outer sections connected in the manner of the chords of a circle. It was mated with the outboard oxidizer tanks, and supported the upper stages. A major aspect of the design of the engine cluster was the thermal shielding to protect the engine components from a phenomenon known as 'base heating', caused when the exhaust interacted with the shock waves that trailed the vehicle and created pockets of dead air and zones of turbulent mixing that heated the base of the vehicle; skeptics doubted that this would be able to be overcome, but static testing enabled an iterative approach to succeed.

SA-1: FIRST FLIGHT TEST

SA-1 lifted off from Launch Complex 34 at Cape Canaveral, Florida, at a Range Zero time of 15:06:04 GMT (10:06:04 EST) on Friday, 27 October 1961, and flew a ballistic arc over the Atlantic Ocean to verify the propulsion system, control system, and structural integrity of the 1.3×10^6-lb-thrust S-I stage.

A dummy third stage (S-V) arrived at Cape Canaveral by barge on 1 May 1961. The first stage (S-I) arrived on 15 August, together with the dummy second stage (S-IV). The S-IV held a tank with 90,000 lb of water ballast in order to simulate the mass and aerodynamic characteristics of the actual stage. The S-V was just a tank of 102,000 lb of water mated using a short conical adapter skirt and topped with a nose cone of a Jupiter missile. The dummy units were taken to Hangar D to be inspected. The S-I was transferred to Launch Complex 34, and erected on the pedestal on 20 August. The upper stages had been erected by 25 August. The 10-hr countdown was started at 23:00 EST on 26 October. Although no technical difficulties imposed delays, two holds were called to await more favorable cloud conditions for photographic coverage of the ascent; the first hold at T–120 min (07:00 EST) for 34 min, and the second at T–20 (09:14) for 32 min. The automated sequencing proceeded satisfactorily, and the compatibility of the ground support equipment with the vehicle was confirmed. Because this was the first attempt to launch this complex vehicle, the engineers estimated that it had only a 30% chance of completing a nominal flight. The condition of the ground support equipment after the launch was better than feared.

SA-1 preparation event	Date
Dummy S-V stage delivered by barge to Cape Canaveral.	01 May 1961
Operational S-I stage and dummy S-IV stage and payload delivered by barge.	15 Aug 1961
S-I stage erected on pedestal at Launch Complex 34.	20 Aug 1961
Dummy stages and payload erected on booster.	23 Aug 1961
Service structure removed for radio-frequency tests.	15 Sep 1961
Fuel test completed.	17 Sep 1961
Liquid oxygen loading test completed.	04 Oct 1961
Final overall test completed.	10 Oct 1961
Engine gimbal checks completed.	13 Oct 1961
Simulated flight test performed.	16 Oct 1961
Simulated flight test repeated.	23 Oct 1961
RP-1 fuel loaded.	26 Oct 1961
Countdown started at T–10 hours.	26 Oct 1961
Launch.	27 Oct 1961

The ignition sequence lit the H-1 engines in matching pairs at 100-millisec intervals, and once they were all confirmed to be running smoothly the hold-down arms were released. After 17.89 sec of vertical rise, the vehicle started to pitch over in order to fly out over the Atlantic Ocean. It attained Mach 1 at T+49 at an altitude of 3.6 nmi, and the maximum dynamic pressure of 5.28 psi at T + 61 occurred at an altitude of 5.9 nmi. The pitching ended at T + 100.49. The center engines were shut down at T + 110.1, the greatest acceleration of 137 ft/sec^2 was at T + 110.25, the outer engines were cut off by a fuel depletion sensor at T + 116.08, which was some 1.6 sec earlier than predicted: for this test, only 83% of the capacity had been used and 880 lb less fuel had been loaded than first believed. The maximum Earth-fixed velocity of 5,297 ft/sec occurred at T + 116.3. No staging operation was attempted: the entire vehicle followed an arc that peaked at an altitude of 74 nmi, 90 nmi downrange, at T + 250. Telemetry ceased at T + 409, probably when aerodynamic stresses caused structural failure of the adapter between the S-I and the S-IV stages, where the telemetry system was located. At that time, the vehicle was some 180 nmi downrange and descending through an altitude of 10.6 nmi.

The S-I was more heavily instrumented than earlier rockets in order to produce a detailed record of its performance so that, in the event of its loss, the cause of the failure could be confidently diagnosed. Although this vehicle flew without active guidance, it incorporated hardware that was to be made operational on later flights.

No major malfunctions or deviations that could be considered a serious system failure or design deficiency occurred, and all of the test objectives were achieved. The engines proved to be slightly more powerful than predicted, and the trajectory was higher than intended due to the greater acceleration. Sloshing of propellants induced roll oscillations late in the boost phase, but the control system was able to overcome this by gimbaling the outer engines. The objectives of the mission were:

1. Flight test the cluster of eight 165,000-lb-thrust H-1 kerosene and liquid oxygen engines. *Achieved.*
2. Flight test the S-I stage clustered propellant tanks structure. *Achieved.*
3. Flight test the S-I stage control system. *Achieved.*
4. Flight test the four support arms and four holddown arms on the launch pad. *Achieved.*
5. System flight test:
   (a) Bending and flutter. *Achieved.*
   (b) Sloshing. *Achieved.*
   (c) Base heating. *Achieved.*
   (d) Aerodynamic engine torques. *Achieved.*
   (e) Airframe aerodynamic heating. *Achieved.*

COMPLETING PHASE I

SA-2 on 25 April 1962 and SA-3 on 16 November 1962 also flew ballistic arcs without active guidance, but after the S-I had shut down explosive charges were detonated to simulate the catastrophic failure of the second stage, and the manner in which the water ballast dispersed in the upper atmosphere was studied for Project Highwater. For SA-3, four solid fuel retrorockets were mounted in the adapter between the S-I and S-IV stages and fired 12 sec after S-I shutdown to qualify them for future use. In the case of SA-4 on 28 March 1963, one inboard engine (#5) was cut off by a timer at T + 100.62 sec in order to

SA-1 mission event	GET (h:m:s)	Date (GMT)	Time (h:m:s)
Launch vehicle transfers to internal power.	–000:00:35	27 Oct 1961	15:05:29
Ignition command.	–000:00:03.03	27 Oct 1961	15:06:00
Range zero.	000:00:00.00	27 Oct 1961	15:06:04
Thrust commit.	000:00:00.23	27 Oct 1961	15:06:04
Launch commit.	000:00:00.59	27 Oct 1961	15:06:04
First motion.	000:00:00.75	27 Oct 1961	15:06:04
Liftoff signal (start program device).	000:00:00.89	27 Oct 1961	15:06:04
Initiate (pitch) tilt.	000:00:17.89	27 Oct 1961	15:06:21
Mach 1.	000:00:49.00	27 Oct 1961	15:06:53
Maximum dynamic pressure.	000:01:01.00	27 Oct 1961	15:07:05
Terminate tilt.	000:01:40.19	27 Oct 1961	15:07:44
S-I center engine cutoff.	000:01:50.10	27 Oct 1961	15:07:54
Maximum acceleration.	000:01:50.25	27 Oct 1961	15:07:54
S-I outer engine cutoff.	000:01:56.08	27 Oct 1961	15:08:00
Maximum Earth-fixed velocity.	000:01:56.3	27 Oct 1961	15:08:00
End of second thrust decay.	000:01:59.00	27 Oct 1961	15:08:03
Apex.	000:04:09.24	27 Oct 1961	15:10:13
Loss of telemetry.	000:06:49.35	27 Oct 1961	15:12:53

demonstrate that the loss of one engine would not cause instabilities that would disrupt the others. By now, the stabilized platform for the active guidance system had been proved capable of establishing a space-fixed coordinate reference against which to accurately determine the attitude and trajectory of the vehicle.

PHASE II TESTING

On 29 January 1964, SA-5 introduced the Block II S-I stage with uprated engines delivering 1.5×10^6 lb of thrust, stretched tankage for the extra propellant required to satisfy their thirst, and large tail fins for greater stability; the S-IV stage powered by a cluster of six RL-10 engines that burned hydrogen in oxygen; active guidance; and a camera that was to make a movie of the staging sequence and then be ejected for recovery from the sea. In this case the vehicle was capped by a nose cone. SA-5 also marked the transfer of Saturn I operations to Pad B of Launch Complex 37.* In contrast to the previous test flights, after rising vertically for 8 sec the S-I made a 3.5-deg/sec roll in order to align the guidance system with the flight azimuth of 105°E of N prior to starting the pitch maneuver. The staging sequence was executed without incident. Hydrogen had already been vented through the RL-10 engines in order to "condition" them. At S-IB shutdown, the S-IV was released and, as the spent stage was retarded by its retrorockets,

* Pad A at Launch Complex 37 was not built.

small rockets on the second stage fired to provide ullage to settle its propellants before ignition. With oxidizer depletion triggering S-IV shutdown, the total Earth-fixed velocity was 88.6 ft/sec greater than nominal, the altitude was 3.3 nmi higher, and the surface range was 10 nmi longer. The anomalous cross-range deviation of 11.5 nmi was attributed to thrust vector misalignments. Nevertheless, the orbit ranging between a perigee of 141.6 nmi and an apogee of 424.3 nmi was acceptable. The objectives of the mission were:

1. Launch vehicle propulsion, structural and control flight test. *Achieved.*
2. First live test of S-IV stage. *Achieved.*
3. First flight test of instrument unit. *Achieved.*
4. S-I/S-IV 'single plane' separation. *Achieved.*
5. Recovery of movie camera. *Achieved.*
6. Flight control utilization of S-I stage tail fins. *Achieved.*
7. The venting of hydrogen to 'chilldown' the RL-10 engines prior to S-IV ignition. *Achieved.*
8. Separation initiated by timer. *Achieved.*
9. Place 38,000 lb into orbit. *Achieved.*

SA-5 mission event	GET (h:m:s)	Date (GMT)	Time (h:m:s)
Ignition command.	–000:00:02.99	29 Jan 1964	16:24:58
Range zero.	000:00:00.00	29 Jan 1964	16:25:01
First motion.	000:00.00.31	29 Jan 1964	16:25:01
Liftoff signal (start program device).	000:00:00.41	29 Jan 1964	16:25:01
Initiate roll maneuver.	000:00:08.78	29 Jan 1964	16:25:09
Terminate roll maneuver.	000:00:13.05	29 Jan 1964	16:25:14
Initiate (pitch) tilt.	000:00:15.71	29 Jan 1964	16:25:16
S-IV liquid hydrogen prestart (venting through the RL-10 engines) and start movie cameras.	000:01:47.66	29 Jan 1964	16:26:48
Terminate tilt.	000:02:13.31	29 Jan 1964	16:27:14
S-IV liquid oxygen pre-start (ports blown open).	000:02:19.75	29 Jan 1964	16:27:20
S-I center engine cutoff.	000:02:20.75	29 Jan 1964	16:27:21
S-I outer engine cutoff.	000:02:26.73	29 Jan 1964	16:27:27
S-IV ullage rocket ignition.	000:02:27.02	29 Jan 1964	16:27:28
S-I/S-IV 'single plane' separation.	000:02:27.14	29 Jan 1964	16:27:28
S-I retrorocket ignition.	000:02:27.16	29 Jan 1964	16:27:28
S-IV ignition.	000:02:28.84	29 Jan 1964	16:27:29
Jettison ullage rockets.	000:02:47.14	29 Jan 1964	16:27:48
S-IV cutoff.	000:10:30.97	29 Jan 1964	16:35:31
Orbital insertion.	000:10:39.97	29 Jan 1964	16:35:40

SA-6 on 28 May 1964 was the first Saturn I to carry a prototype ("boilerplate") of the Apollo command/service modules to evaluate the structural compatibility of the launch vehicle and spacecraft. The command module of CSM-013 (BP-13) had a crew compartment, side hatch, forward access way, forward compartment cover, aft heat shield, communications and instrumentation, and an environmental control system to maintain an internal temperature suitable for the electronics. The service module had dummy reaction control system quadrant mounts, and was mated by a cylindrical adapter to the Instrument Unit of the S-IV stage. As the first test flight to verify the compatibility of the spacecraft with this launch vehicle, the mission was designated A-101.

Although one of the inboard engines cut off prematurely at T + 116.8 sec because the teeth were stripped from one of the gears in a turbopump, the control system compensated by extending the S-I boost phase by 2.7 sec. Ten seconds after staging, the launch escape system was jettisoned from the spacecraft. On this occasion, the guidance system was permitted to control the shutdown of the S-IV. The total velocity at engine #8 cutoff was 36.5 ft/sec slower than nominal and 328 ft/sec slower at shutdown, but the S-IV was able to recover, and when it was shut down the velocity was 10.2 ft/sec greater than nominal, the altitude was 1.3 nmi lower and the range was 22.5 nmi short. The insertion orbit ranging between 98.4 and 123 nmi was retarded by air drag and the stage and its payload re-entered over the western Pacific on 2 June.

SA-6 mission event	GET (h:m:s)	Date (GMT)	Time (h:m:s)
Ignition command.	–000:00:02.95	28 May 1964	17:00:57
Range zero.	000:00:00.00	28 May 1964	17:01:00
First motion.	000:00:00.17	28 May 1964	17:01:00
Liftoff signal (start program device).	000:00:00.40	28 May 1964	17:01:00
Guidance computer release.	000:00:00.43	28 May 1964	17:01:00
Initiate roll maneuver.	000:00:08.40	28 May 1964	17:01:08
Terminate roll maneuver.	000:00:12.70	28 May 1964	17:01:12
Initiate (pitch) tilt.	000:00:15.55	28 May 1964	17:01:15
S-IV liquid hydrogen pre-start (venting through the RL-10 engines) and start movie cameras.	000:01:47.01	28 May 1964	17:02:47
Engine #8 premature shutdown.	000:01:57.28	28 May 1964	17:02:57
Terminate tilt.	000:02:14.55	28 May 1964	17:03:14
S-IV liquid oxygen pre-start (ports blown open).	000:02:21.21	28 May 1964	17:03:21
S-I center engine cutoff.	000:02:23.23	28 May 1964	17:03:23
S-I outer engine cutoff.	000:02:29.23	28 May 1964	17:03:29
S-IV ullage rocket ignition.	000:02:29.50	28 May 1964	17:03:29
S-I/S-IV 'single plane' separation.	000:02:29.62	28 May 1964	17:03:29
S-I retrorocket ignition.	000:02:29.68	28 May 1964	17:03:29
S-IV ignition.	000:02:31.31	28 May 1964	17:03:31
Jettison ullage rockets.	000:02:41.62	28 May 1964	17:03:41
Jettison launch escape tower.	000:02:41.62	28 May 1964	17:03:41
Initiate S-IV guidance commands.	000:02:48.23	28 May 1964	17:03:48
Terminate S-IV guidance corrections.	000:10:22.13	28 May 1964	17:11:22
S-IV cutoff on guidance command.	000:10:24.86	28 May 1964	17:11:24
Orbital insertion.	000:10:34.86	28 May 1964	17:11:34
Orbital decay re-entry.	–	02 Jun 1964	–

SA-7 on 18 September 1964 repeated the Apollo boilerplate mission as A-102, except that in this case the launch escape system was discarded by firing its escape and pitch motors in order to demonstrate this alternative method. On this occasion, the camera pods ejected by the first stage fell further downrange than expected, and an approaching hurricane prevented their recovery. When the pods washed ashore seven weeks later in outwardly poor condition, the film was found to be undamaged.

Ten launch vehicles had been built in the expectation of failures, but all of the objectives had been achieved in seven flights, and so NASA decided to build real payloads into the final three Apollo boilerplates. In orbit, the shell of the CSM was jettisoned to expose a structure that unfolded a pair of panels in concertina-fashion to a total span of 98 ft. These panels contained detectors to register the impact of micrometeoroids, to provide data to determine whether the shielding of the Apollo spacecraft would be adequate. The wing-like configuration of the resulting vehicle prompted the mission name of 'Pegasus'.

Saturn I flight	Launch (GMT)	Time (h:m:s)	Remarks
SA-1	27 Oct 1961	15:06:04	S-I stage flight on ballistic arc.
SA-2	25 Apr 1962	14:00:34	S-I stage flight on ballistic arc.
SA-3	16 Nov 1962	17:45:02	S-I stage flight on ballistic arc.
SA-4	28 Mar 1963	20:11:55	S-I stage flight on ballistic arc.
SA-5	29 Jan 1964	16:25:01	S-I and S-IV stages on orbital trajectory.
SA-6	28 May 1964	17:07:00	S-I, S-IV and boilerplate Apollo (BP-13).
SA-7	18 Sep 1964	16:22:43	S-I, S-IV and boilerplate Apollo (BP-15).
SA-9*	16 Feb 1965	14:37	S-I, S-IV and boilerplate Apollo (BP-16) as Pegasus 1.
SA-8*	25 May 1965	07:35:01	S-I, S-IV and boilerplate Apollo (BP-26) as Pegasus 2.
SA-10	30 Jul 1965	13:00:00	S-I, S-IV and boilerplate Apollo (BP-9) as Pegasus 3.

* Note that SA-8 and SA-9 were flown out of sequence.

CONCLUSIONS

Contrary to early expectations, the Saturn I had an unprecedented 100% record of success. Only one of the 80 first-stage engines used on the 10 flights had required to be shut down, and in that case the vehicle had compensated for its loss. But despite its rapid development and flight record the Saturn I had been overtaken by events and was declared obsolete.

THE SATURN IB

By further uprating the H-1 engine cluster to deliver 1.6×10^6 lb thrust, the S-I became the S-IB. In addition, the spider beam was enlarged and reinforced to accommodate the S-

IVB second stage, which was wider and heavier than the S-IV, the tail fins were redesigned to reflect the fact that the vehicle would be longer and heavier, and the performance was further raised by eliminating or redesigning components to a less conservative margin – as the experience with the S-I had shown to be feasible – in order to shed 9 tons.

Testing the first stage for SA-1 at Huntsville.

With its dummy upper stages, SA-1 lifts off from Pad 34 on 27 October 1961.

The final Block I version of the Saturn I was SA-4.

The first Block II Saturn I to carry a "boilerplate" Apollo spacecraft and launch escape system was SA-6.

A Pegasus micrometeoroid payload is stowed within a "boilerplate" spacecraft.

AS-201

The first flight of the Saturn IB: a CSM on a ballistic arc

26 February 1966

BACKGROUND

AS-201 was the first in a series of test flights to 'man rate' the Saturn IB launch vehicle and Apollo spacecraft. The Saturn IB used an upgraded version of the first stage of the Saturn I, of which 10 had successfully launched between 27 October 1961 and 30 July 1965, and the new S-IVB upper stage. Prior to this, 10 boilerplate Apollo spacecraft and one production-line model (CSM-002) had been launched for a variety of purposes. The payload was a Block I spacecraft (CSM-009) that had neither a guidance and navigation system nor an S-band transmission system, was powered by batteries instead of fuel cells, and had an *ad hoc* flight-control sequencer. The plan was to release the spacecraft on a high ballistic arc, and then have it fire its service propulsion system to dive back into the atmosphere in order to test the heat shield.

AS-201 was launched from Launch Complex 34 at Cape Kennedy, Florida, at a Range Zero time of 16:12:01 GMT (11:12:01 EST) on Saturday, 26 February 1966, and 37 min later the command module splashed into the South Atlantic.

LAUNCH PREPARATIONS

The countdown started at 00:00 EST on Sunday, 20 February 1966, with the clock at T–52 hr 30 min. Weather prompted recycling to T–13 hr at 18:30 on 22 February. On 23 February, the weather and a break in a submarine cable to one of the downrange stations caused a 24-hr delay, and high winds, heavy overcast, low ceiling, intermittent rain and high seas in the recovery area prompted a further 24-hr slippage on 24 February. The countdown resumed at 17:15 on 25 February, with the clock at T–780 min. The 60 min available for holds that would not impact the launch time were consumed in two 30-min segments. The first, at T–266 min, was to catch up with liquid oxygen tanking. The other, at T–90 min, was to complete the liquid hydrogen tanking, which had been delayed by work on a helium regulator in the ground support equipment. Three unplanned holds and recycles delayed the launch by a total of 207 min. A hold called at T–30 min to complete the closeout of the spacecraft lasted 78 min. An attempt to top up a gaseous nitrogen tank in the S-IB stage at T–35 sec was foiled by a balky valve, and the clock had to be held at T–4 sec. The clock was recycled to T–15 min while this issue was investigated, and was then resumed. But at T–5 min 34 sec it had to be held again because the nitrogen problem persisted. After 31 min in hold, the problem was finally resolved and the clock was again recycled to T–15 min. At 11:12:01 EST the vehicle was finally able to lift off into a bright clear sky with a visibility of 8.65 nmi. This was the first time that Launch Complex 34 had been used since a Saturn I launch in 1963. At T+3 sec the vibration shook free a high-voltage fuse in the electrical substation and disabled the water-quenching system, with the result that the flame and undamped vibration caused substantial damage to the pad and its associated systems.

ASCENT PHASE

The ignition sequence started the eight H-1 engines of the S-IB stage in matched pairs at 100-millisec intervals. The vehicle was launched on a pad azimuth of 100°E of N, but at T+11.2 sec it began to pitch over and initiated a 9.35-sec roll to align its inertial guidance system with the flight azimuth of 105°E of N. It achieved Mach 1 at T+64.5, and passed through maximum dynamic pressure at T+76.0. The pitch maneuver was terminated at T+134.39 in order to stabilize the vehicle in readiness for staging. At T+137.0 the level sensors in the propellant tanks were enabled. The shutdown of the four center engines at T+141.46 was 0.89 sec later than planned. The outer engines were cut off by a fuel depletion sensor at T+146.94, which was 0.37 sec later than nominal. The space-fixed velocity was 26.4 ft/sec faster than nominal.

Next came the task of separating the S-IVB, which was on its maiden flight. In rapid succession over an interval of 5 sec three solid rocket motors around the rear rim of the S-IVB that were canted out at an angle of 9.5 deg fired to provide ullage to settle the liquid propellants in the stage's tanks; pyrotechnics detonated to separate the stages; four solid rockets fired to retard the S-IB; within 0.97 sec of 'first motion' the J-2 engine cleared the cylindrical interstage, and the J-2 was commanded to ignite. As a result of the slow opening of the main oxidizer valve, the buildup to 90% thrust took longer than expected: 3.6 sec after the engine start command, rather than 2.9 sec. Their job done, the ullage motors were jettisoned. At T+172.62 a small solid rocket drew the tower of the launch escape system from the nose of the spacecraft, and several seconds later the two cameras in the interstage that had filmed the separation of the S-IVB were ejected. Although the drogue parachutes failed to deploy properly, one camera was able to be recovered from the Atlantic to provide visual proof of the clean separation. Some 90.68 sec into the S-IVB burn, and 14.22 sec ahead of schedule, the propellant utilization system activated a valve to adjust the initial 5.50:1 mixture ratio. However, instead of the planned 5.21:1 this gave 5.1:1. This reduced the thrust and obliged the stage to extend its burn by 9.76 sec in order to attain the requisite velocity and altitude. The guidance system cut off the J-2 at T+602.86, at which time the space-fixed velocity was 1.65 ft/sec below nominal, the altitude was 0.40 nmi higher than nominal, and the surface range was 16.76 nmi longer, but the range excess was explained by the prolonged burn. In this case, the trajectory was a high suborbital arc.

AS-201 preparation event	Date
North American Aviation begins assembly of CSM-009.	Oct 1963
S-IB stage arrived at KSC by barge and was offloaded and moved into hanger AF.	14 Aug 1965
S-IB stage transported to LC 34 and erected on the pad. It was used as a spacer for the ground test S-IVB-F to verify the cryogenic propellant loading systems.	18 Aug 1965
The forward bulkhead on fuel tank F1 of the S-IB was inadvertently reversed by overpressurization during instrument compartment leak tests.	10 Sep 1965
S-IVB stage arrived at KSC by barge and was taken to hangar AF for inspection and modification.	19 Sep 1965
S-IVB-F stage de-erected.	28 Sep 1965
Fuel tank F1 of the S-IB stage was exchanged.	29 Sep 1965
S-IVB stage erected.	01 Oct 1965
IU arrived at KSC by barge and was offloaded and transported to hanger AF.	20 Oct 1965
NASA approves shipment of CSM-009 to KSC.	20 Oct 1965
Initial power applied to the S-IB stage, with interim operating system programs for the RCA-110A computer.	21 Oct 1965
IU erected.	25 Oct 1965
CM-009 arrived at KSC.	25 Oct 1965
Initial power applied to the S-IVB stage.	26 Oct 1965
SM arrived at KSC.	27 Oct 1965
Launch vehicle electrical mate accomplished.	10 Nov 1965
Primary RCA-110A ground computer operating tape arrived.	12 Nov 1965
Command guidance and control system checkout commenced.	29 Nov 1965
Spacecraft erected.	26 Dec 1965
CSM/SLA mating completed.	27 Dec 1965
Exploding bridge wire test completed.	31 Dec 1965
Holddown arm qualification test completed.	03 Jan 1966
Launch vehicle/spacecraft electrical mate completed.	18 Jan 1966
Flight readiness review completed.	21 Jan 1966
Plugs-in overall test completed.	24 Jan 1966
Launch escape system mate and thrust vector alignment completed.	24 Jan 1966
Space vehicle flight electrical mate completed.	28 Jan 1966
Plugs-out overall test completed.	02 Feb 1966
Dry countdown demonstration test (CDDT) completed.	08 Feb 1966
Wet CDDT completed.	09 Feb 1966
Flight readiness test completed.	12 Feb 1966
Commenced RP-1 tanking.	19 Feb 1966
T–52 hr 30 min launch countdown commenced.	20 Feb 1966
Terminal countdown commenced.	25 Feb 1966
Liftoff.	26 Feb 1966

Meanwhile, at T + 533 sec the S-IB had impacted 251 nmi downrange, in the Atlantic at 27.30°N, 76.04°W.

AS-201 ascent phase event	GET (h:m:s)	Altitude (nmi)	Range (nmi)	Earth-fixed velocity (ft/sec)	Space-fixed velocity (ft/sec)	Event duration (sec)	Geocentric latitude (°N)	Longitude (°E)	Space-fixed flight path angle (deg)	Space-fixed heading angle (°E of N)
Liftoff	000:00:00.37	0.0	0.0	–	–	–	–	–	–	–
Mach 1 achieved	000:01:04.5	4.24	–	–	–	–	–	–	–	–
Maximum dynamic pressure	000:01:16.0	6.08	–	–	–	–	–	–	–	–
S-IB center engine cutoff	000:02:21.46	28.54	29.14	6,026.05	7,187.75	–	–	–	–	–
S-IB outboard engine cutoff	000:02:26.94	31.34	33.92	6,324.19	7,498.13	–	–	–	–	–
S-IB/S-IVB separation	000:02:27.76	–	–	–	–	–	–	–	–	–
S-IVB cutoff	000:10:02.86	141.08	875.96	21,439.86	22,768.63	–	–	–	–	–

POST-BOOST PHASE

The spacecraft control programmer was activated at T + 663 sec. At T + 845 the S-IVB released the CSM and splayed the four panels of the SLA to an angle of 45 deg to ensure a clean extraction of the SPS engine bell. A moment later, the spacecraft fired the thrusters in the 'quads' of its SM-RCS for 18 sec to move clear. At T + 1,084 the vehicle reached the peak of its ballistic arc, 2,415.2 nmi downrange at an altitude of 226.1 nmi. At T + 1,181 the thrusters were fired again, this time to provide ullage to settle the propellants in their tanks, and 30 sec later the SPS was ignited for a 184-sec burn. The SM tanks could accommodate 18 tons of propellants, but in this case had been loaded with just 3.35 tons. After 80 sec the thrust chamber pressure began to decline as a result of the oxidizer tank having ingested helium during a pre-flight test, and by cutoff the pressure had decreased by 30%. Once the SPS had shut down, the thrusters were fired again to reinstate ullage and the main engine was reignited for a 10-sec burn, during which the chamber pressure oscillated from 70% down to 12%. Several seconds later, the thrusters began a pitch maneuver at a rate of 5 deg/sec for a duration of 18 sec in order to produce a 90 deg change in attitude. After separating, the command module used its own thrusters to continue this pitch rotation for another 82.5 deg, then rolled 180 deg in order to orient its heat shield for atmospheric entry. However, a fault in the electrical power system caused a loss of steering control that resulted in a rolling entry. The discarded service module burned up and was destroyed. The S-IVB impacted at T + 1,916 sec, 4,677 nmi downrange in the South Atlantic at 9.66°S, 10.08°W.

RECOVERY

The suborbital arc with a steep entry was to subject the heat shield to a high heating rate – that is, a high temperature for a comparatively short period. However, since the velocity at entry was 782 ft/sec slower than planned and the flight path was 0.44 degree shallower, the heating rate and *g*-load were less than planned. The onset of the radio blackout was at T + 1,580 sec. The maximum heating rate occurred at T + 1,631. The temperature at the base of the heat shield peaked at about 2,000°C. At T + 1,640 the load peaked at 14.3 *g* instead of the planned 16.0 *g*, but this was still much greater than would occur on a manned mission. At T + 1,695 the blackout ended. The three drogue parachutes were deployed at T + 1,840, and the main parachutes followed at T + 1,906. At T + 2,240 the CM splashed into the South Atlantic at 8.18°S, 11.15°W, 40 nmi uprange of the target. It adopted the apex-up flotation attitude. The recovery aids deployed, including the high-frequency antennas, but no transmissions were received. As the automated disconnect failed, the parachutes had to be cut free by swimmers. The CM was recovered by the USS *Boxer* 2.5 hours after splashdown, and later returned to North American Aviation in Downey, California, for post-flight testing.

CONCLUSIONS

The launch vehicle performed satisfactorily. Although all spacecraft test objectives were achieved, the entry heating rate was less than planned and some of the flight measurements were corrupted by an electrical short circuit. In order to provide the time to diagnose and rectify the fault in the SPS engine, AS-202 was rescheduled after AS-203, which was to be an S-IVB development flight that would not carry a spacecraft.

MISSION OBJECTIVES

Launch Vehicle Primary Objectives

1. Demonstrate structural integrity and compatibility of the launch vehicle and spacecraft and determine launch loads. Specifically:
   (a) Demonstrate structural integrity and compatibility of the space vehicle during S-IB powered flight and confirm structural loads and dynamic characteristics. *Achieved.*
   (b) Demonstrate structural integrity and compatibility of the S-IVB space vehicle during powered flight and coast. *Achieved.*
2. Demonstrate separation of:
   (a) S-IVB from S-IB. *Achieved.*
   (b) CSM from S-IVB/IU/SLA. *Achieved.*
3. Verify operation of propulsion, guidance, and control systems. Specifically:
   (a) Demonstrate S-IVB propulsion system including program mixture ratio shift and determine system performance parameters. *Achieved.*
   (b) Demonstrate S-IB propulsion system and evaluate subsystem performance parameters. *Achieved.*
   (c) Demonstrate launch vehicle guidance system, achieve guidance cutoff, and evaluate system accuracy. *Achieved.*
   (d) Demonstrate launch vehicle control system during S-IB powered flight, S-IVB powered flight, S-IVB coasting, and evaluate performance characteristics. *Achieved.*
   (e) Demonstrate launch vehicle sequencing system. *Achieved.*
4. Evaluate performance of the space vehicle emergency detection system in an open-loop configuration. *Achieved.*
5. Demonstrate the mission support facilities required for launch, mission operations, and recovery. *Achieved.*

Launch Vehicle Secondary Objectives

1. Confirm launch vehicle powered flight external environment. *Achieved.*
2. Evaluate launch vehicle internal environment. *Achieved.*
3. Evaluate S-IVB/IU in-flight thermal conditioning system. *Achieved.*
4. Demonstrate adequacy of S-IVB residual propellant venting system. *Achieved.*

AS-201, the first Saturn IB, lifts off with the Block I spacecraft CSM-009.

Spacecraft Primary Objectives

1. Demonstrate separation of:
   (a) Launch escape system and boost protective cover from the spacecraft during powered flight. *Achieved.*
   (b) CSM from S-IVB/IU/SLA. *Achieved.*
   (c) CM from SM. *Achieved.*
2. Verify the adequacy of the CM heat shield for entry from low orbit at an entry speed of 28,000 ft/sec. *Achieved.*
3. Verify operation of the SPS engine, including restart. *Achieved.*
4. Verify pressure and temperature control by the environmental control system. *Achieved.*
5. Verify communications. *Partially achieved.*
6. Verify the CM-RCS. *Achieved.*
7. Verify the SM-RCS. *Achieved.*
8. Verify the spacecraft control system. *Achieved.*
9. Verify the Earth landing system. *Achieved.*
10. Verify the electrical power system. *Partially achieved.*

MISSION TIMELINE

AS-201 mission event	GET (h:m:s)	Date (GMT)	Time (h:m:s)
Guidance reference release.	–000:00:05.038	26 Feb 1966	16:11:55
Range zero.	000:00:00:00	26 Feb 1966	16:12:01
First motion.	000:00:00.11	26 Feb 1966	16:12:01
Liftoff.	000:00:00.37	26 Feb 1966	16:12:01
Start tilt (pitch) maneuver.	000:00:11.20	26 Feb 1966	16:12:12
Start roll maneuver.	000:00:11.20	26 Feb 1966	16:12:12
Complete roll maneuver.	000:00:20.55	26 Feb 1966	16:12:21
Mach 1.	000:01:04.5	26 Feb 1966	16:13:05
Maximum dynamic pressure.	000:01:16.0	26 Feb 1966	16:13:17
Tilt arrest.	000:02:14.39	26 Feb 1966	16:14:15
Enable S-IB propellant level sensors (to generate the cutoff signal).	000:02:17.00	26 Feb 1966	16:14:18
S-IB center engine cutoff.	000:02:21.46	26 Feb 1966	16:14:22
S-IB outer engine cutoff.	000:02:26.94	26 Feb 1966	16:14:27
S-IVB ullage ignition.	000:02:27.52	26 Feb 1966	16:14:28
S-IB/S-IVB separation.	000:02:27.76	26 Feb 1966	16:14:28
S-IB retro ignition.	000:02:27.81	26 Feb 1966	16:14:28
First motion of S-IVB stage.	000:02:27.88	26 Feb 1966	16:14:28
J-2 clears interstage.	000:02:28.89	26 Feb 1966	16:14:29
J-2 ignition command.	000:02:29.35	26 Feb 1966	16:14:30
S-IB retro at 90% thrust.	000:02:29.40	26 Feb 1966	16:14:30
Ullage off.	000:02:31.50	26 Feb 1966	16:14:32

AS-201 mission event - *continued*	GET (h:m:s)	Date (GMT)	Time (h:m:s)
J-2 at 90% thrust.	000:02:32.90	26 Feb 1966	16:14:33
Jettison ullage motor casings.	000:02:39.77	26 Feb 1966	16:14:40
Jettison the launch escape tower.	000:02:52.64	26 Feb 1966	16:14:53
Eject S-IB film camera pods.	000:02:54.3	26 Feb 1966	16:14:55
Start iterative guidance mode.	000:02:56.11	26 Feb 1966	16:14:57
Initiate mixture ratio shift.	000:04:00.00	26 Feb 1966	16:16:01
S-IB impact.	000:09:13	26 Feb 1966	16:21:14
S-IVB cutoff.	000:10:02.86	26 Feb 1966	16:22:03
Start spacecraft control programmer.	000:11:03.1	26 Feb 1966	16:23:04
S-IVB/CSM separation command.	000:14:03.2	26 Feb 1966	16:26:04
S-IVB/CSM separation complete.	000:14:04.95	26 Feb 1966	16:26:05
Initiate 1st SM-RCS +X translation to move clear.	000:14:06.7	26 Feb 1966	16:26:07
Finish 1st SM-RCS +X translation.	000:14:24.7	26 Feb 1966	16:26:25
Peak altitude.	000:18:04.4	26 Feb 1966	16:30:05
Initiate 2nd SM-RCS +X translation for ullage.	000:19:41.2	26 Feb 1966	16:31:42
SPS 1st ignition.	000:20:11.2	26 Feb 1966	16:32:12
Finish 2nd SM-RCS +X translation.	000:20:12.2	26 Feb 1966	16:32:13
SPS 1st cutoff.	000:23:15.2	26 Feb 1966	16:35:16
Initiate 3rd SM-RCS +X translation.	000:23:15.7	26 Feb 1966	16:35:16
SPS 2nd ignition.	000:23:30.7	26 Feb 1966	16:35:31
Finish 3rd SM-RCS +X translation.	000:23:40.7	26 Feb 1966	16:35:41
SPS 2nd cutoff.	000:23:40.7	26 Feb 1966	16:35:41
SM-RCS initiates 5-deg/sec pitch rate.	000:23:44.1	26 Feb 1966	16:35:45
Terminate pitch (after 90-deg rotation).	000:24:02.0	26 Feb 1966	16:36:03
SM/CM separation.	000:24:15.0	26 Feb 1966	16:36:16
CM initiates 5-deg/sec pitch rate.	000:24:22.6	26 Feb 1966	16:36:23
Terminate pitch (after 82.5-deg rotation).	000:24:39.1	26 Feb 1966	16:36:40
CM initiates 5-deg/sec roll rate.	000:24:39.1	26 Feb 1966	16:36:40
Terminate roll (after 180-deg rotation).	000:25:15.1	26 Feb 1966	16:37:16
Enter radio blackout.	000:26:20	26 Feb 1966	16:38:21
Maximum entry heating rate.	000:27:11	26 Feb 1966	16:39:12
Maximum *g*-load.	000:27:20	26 Feb 1966	16:39:21
End radio blackout.	000:28:15	26 Feb 1966	16:40:16
Drogue parachute deployment.	000:30:53.9	26 Feb 1966	16:42:54
Main parachute deployment.	000:31:45.7	26 Feb 1966	16:43:46
S-IVB impact.	000:31:56	26 Feb 1966	16:43:57
Splashdown.	000:37:19.7	26 Feb 1966	16:49:20

AS-203

The second flight of the Saturn IB: evaluating the S-IVB

5 July 1966

BACKGROUND

Although planned as the third of a series of tests of the Saturn IB launch vehicle, when the second mission was delayed NASA opted to proceed with AS-203 out of sequence. The objectives were to evaluate the performance of the Instrument Unit in orbital flight, evaluate the attitude control and continuous venting systems of the S-IVB, demonstrate the chilldown systems in a simulated restart of the S-IVB, and determine the heat transfer characteristics of the propellant tanks. Because this mission did not involve an Apollo spacecraft, the IU was capped by an aerodynamic nose cone, inside of which was located the subcritical cryogenic nitrogen experiment. In orbit, television cameras were to observe the behavior of the liquid hydrogen fuel in conditions approximating those of a Saturn V flight, for which the S-IVB was to be the third stage.

AS-203 was launched from Pad B of Launch Complex 37 at Cape Kennedy, Florida, at a Range Zero time of 14:53:17 GMT (10:53:17 EDT) on Tuesday, 5 July 1966.

LAUNCH PREPARATIONS

As the preliminary preparations for launch had been accomplished by the Countdown Demonstration Test completed early on 4 July, it was decided to pick up the count at T–11 hr 30 min that evening at 21:30 EDT. It ran smoothly until 08:45 on 5 July, at T–15 min, when a hold was called to catch up with tasks that had to be achieved by that time, and to investigate the failure to transmit of TV camera #2 for the liquid hydrogen experiment. After a delay of 12 min 40 sec the count was resumed, and continued to T–5 min, when it was held for further investigation of the camera issue. On deciding to proceed with only camera #1 available for the experiment, having delayed 88 min 37 sec, the count was recycled to T–15 min. After a 2-min hold at T–3 min to verify the status of the radar at the Bermuda tracking station, the vehicle was launched at 10:53:17, some 113 min 17 sec later than planned. The visibility was 8.6 nmi, and the winds were light from the west-northwest, with a few scattered clouds.

AS-203 preparation event	Date
S-IB stage arrived at KSC by barge and was offloaded and moved into hanger AF.	12 Apr 1966
S-IB stage erected on LC 37B.	18 Apr 1966
S-IB fin installation completed.	19 Apr 1966
S-IVB stage erected.	21 Apr 1966
IU erected.	21 Apr 1966
Nose cone erected.	21 Apr 1966
Vehicle mechanical and propulsion checkout started.	22 Apr 1966
Launch vehicle electrical mate accomplished.	23 Apr 1966
Initial power applied to S-IB stage.	25 Apr 1966
Initial power applied to IU.	28 Apr 1966
Initial power applied to S-IVB stage.	28 Apr 1966
Launch vehicle switch selector functional.	02 May 1966
Propellant dispersion test accomplished.	25 May 1966
Launch vehicle completed full pressure test.	26 May 1966
Launch vehicle liquid oxygen simulate and malfunction test accomplished.	03 Jun 1966
RP-1 simulate and malfunction test accomplished.	06 Jun 1966
Liquid oxygen and liquid hydrogen loading test accomplished.	07 Jun 1966
Launch vehicle flight sequencer exploding bridge wire test completed.	09 Jun 1966
Launch vehicle sequencer malfunction test completed.	09 Jun 1966
Launch vehicle plugs-in overall test #1 accomplished.	11 Jun 1966
Launch vehicle plugs-in overall test #2 accomplished.	13 Jun 1966
Launch vehicle plugs-out overall test completed.	20 Jun 1966
Space vehicle flight sequence test completed.	22 Jun 1966
Flight readiness test completed.	27 Jun 1966
Commenced RP-1 tanking operations.	28 Jun 1966
S-IVB stage auxiliary propulsion system load and fire accomplished.	29 Jun 1966
Countdown demonstration test completed.	01 Jul 1966
Launch countdown started at T–11 hr 30 min.	04 Jul 1966

ASCENT PHASE

The vehicle was launched on a pad azimuth of 90°E of N, but at T + 12.2 sec it began to pitch over, and initiated an 18-sec roll to align its inertial guidance system with the flight azimuth of 72°E of N. It achieved Mach 1 at T + 51.64 and passed through maximum dynamic pressure at T + 70.0. The pitch maneuver was terminated at T + 133.9 to stabilize the vehicle in readiness for staging. At T + 136.3 the level sensors in the fuel tanks were enabled. The center engine cutoff occurred 1.20 sec earlier than predicted, and 3.44 sec later a fuel depletion sensor cut off the outer engines, which was 0.44 sec later than predicted. The altitude was 1.20 nmi higher than nominal and the space-fixed velocity 76 ft/sec greater. Both of the cameras that filmed the separation of the S-IVB were ejected, but only one could be recovered. The propellant utilization system of the S-IVB operated in open-loop mode with an average mixture ratio of 4.95:1 maintained throughout the burn. On determining that it had achieved the desired velocity and altitude, the guidance

AS-203 ascent phase event	GET (h:m:s)	Altitude (nmi)	Range (nmi)	Earth-fixed velocity (ft/sec)	Space-fixed velocity (ft/sec)	Event duration (sec)	Geocentric latitude (°N)	Longitude (°E)	Space-fixed flight path angle (deg)	Space-fixed heading angle (°E of N)
Liftoff	000:00:00.86	0.0	0.0	–	–	–	–	–	–	–
Mach 1 achieved	000:00:51.64	3.60	–	–	–	–	–	–	–	–
Maximum dynamic pressure	000:01:10.00	7.12	–	–	–	–	–	–	–	–
S-IB center engine cutoff	000:02:19.24	33.66	46.18	8,595.9	9,797.7	–	–	–	–	–
S-IB outboard engine cutoff	000:02:22.68	35.64	50.69	8,901.9	10,109.9	–	–	–	–	–
S-IB/S-IVB separation	000:02:23.44	36.08	51.70	10,122.7	–	–	–	–	20.18	–
S-IVB cutoff	000:07:13.35	103.14	721.85	24,202.1	25,539.7	–	–	–	–	–
Earth orbit insertion	000:07:23.348	–	–	–	25,569.9	–	–	–	–	–

system issued the cut off signal at T + 433.35 sec, some 2.90 sec earlier than predicted. The space-fixed velocity was 2.0 ft/sec less than nominal, the altitude was 0.065 nmi higher and the surface range was 3.75 nmi shorter, but considering that this was the first time an S-IVB had put itself into orbit it was extremely satisfactory. Its international designation upon achieving orbit was 1966-059A.[1]

After attaining its peak altitude of 72.4 nmi at T + 271.5 sec, the S-IB fell into the Atlantic at T + 584, some 437 nmi downrange at 30.46°N, 72.52°W.

EARTH ORBIT PHASE

At orbital insertion, the liquid hydrogen was controlled by a combination of tank baffles and deflectors and by ullage created by venting liquid oxygen. Successfully maintaining the fluid in an essentially settled condition at this critical time enabled continuous venting of liquid hydrogen to maintain it in this condition during the coasting phase. A simulated J-2 restart verified the fuel recirculation chilldown, the fuel antivortex screen, the liquid oxygen recirculation chilldown, and the storage bottles. The fact that the rise in the liquid hydrogen pressure while in orbit was greater than predicted gave data on the heat transfer characteristic of the S-IVB tank that would assist in planning Saturn V missions. Radar tracking by ground stations around the world monitored how the parameters of the orbit were changed by the thrusting effect of continuous venting.

AS-203 S-IVB orbit evolution	Space-fixed velocity (ft/sec)	Apogee (nmi)	Perigee (nmi)	Period (min)	Inclination (deg)
Start of revolution 1	25,570	102.2	100.1	88.21	31.9824
Start of revolution 2	25,564	108.2	104.2	88.40	31.9827
Start of revolution 3	25,557	112.8	106.2	88.56	31.9761
Start of revolution 4	25,550	117.1	109.3	88.70	31.9863

The subcritical cryogenic nitrogen experiment demonstrated the feasibility of using such a system to store and deliver cryogenics in a weightless environment. It was able to maintain pressure control throughout, and a progressive decrease in the fluid quantity indicated that the vapor was uniformly delivered from a two-phase mixture. The TV study

[1] *RAE Table of Earth Satellites 1957–1986*, pages vii and viii. The international Committee on Space Research (COSPAR) assigns orbital objects a designation based on the year of launch (first four digits) and the number of successful launches so far in that year (next three digits). Objects on suborbital paths are not included. In this nomenclature, the letter 'A' usually refers to the instrumented spacecraft, 'B' to the rocket, and 'C', 'D', 'E', etc., to fragments. In this case, the S-IVB became item 'A' because it was the principal object.

of the liquid hydrogen tank was undertaken during the first revolution, and then the engine restart was simulated. After 6.3 hours, early in the fifth revolution, the vehicle re-entered the atmosphere over the Caribbean.

CONCLUSIONS

Despite the pre-launch failure of one of its two TV cameras, the observations of the liquid hydrogen experiment verified that continuous venting during coasting flight provided sufficient ullage to maintain the hydrogen settled in its tank, as would be required on a Saturn V dispatching a lunar mission. The simulated restart verified the chilldown and recirculation systems of the J-2 engine. AS-203 was therefore a successful test of the S-IVB stage.

MISSION OBJECTIVES

Launch Vehicle Primary Objectives

1. Evaluate the S-IVB liquid hydrogen continuous venting system in coasting flight. *Achieved.*
2. Evaluate the J-2 engine liquid hydrogen chilldown and recirculation system for simulated engine restart. *Achieved.*
3. Determine cryogenic liquid/vapor interface and fluid dynamics of hydrogen propellant in a near-weightless environment. *Achieved.*
4. Determine the heat transfer into liquid hydrogen through the wall of the tank and obtain data required to develop a thermodynamic model of the propellant. *Achieved.*
5. Evaluate the S-IVB and IU checkout in orbital flight. *Achieved.*
6. Demonstrate the S-IVB auxiliary propulsion system operation and evaluate performance parameters. *Achieved.*
7. Demonstrate the adequacy of the S-IVB/IU thermal control system. *Achieved.*
8. Demonstrate the launch vehicle guidance system operation, and its ability to achieve an accurate engine cutoff to insert a payload into orbit. *Achieved.*
9. Demonstrate the structural integrity and dynamic characteristics of the launch vehicle. *Achieved.*
10. Demonstrate the mission support facilities and operations required for launch and mission conduct. *Achieved.*

Launch Vehicle Secondary Objectives

1. Evaluate the launch vehicle powered flight external environment. *Achieved.*
2. Verify the launch vehicle sequencing system operation. *Achieved.*
3. Evaluate the performance of the emergency detection system of the launch vehicle in an open-loop configuration. *Achieved.*
4. Evaluate the separation of the S-IVB/IU/nose cone from the S-IB. *Achieved.*
5. Verify the operation of the launch vehicle propulsion system and evaluate system performance parameters. *Achieved.*

Lifting off without a spacecraft, AS-203 was to test the S-IVB stage in orbit.

6. Evaluate the subcritical cryogenic nitrogen experiment for the Manned Spacecraft Center. *Achieved.*

MISSION TIMELINE

AS-203 mission event	GET (h:m:s)	Date (GMT)	Time (h:m:s)
Guidance reference release.	–000:00:04.4	5 Jul 1966	14:53:12
Initiate ignition sequence.	–000:00:02.487	5 Jul 1966	14:53:14
Range zero.	000:00:00.00	5 Jul 1966	14:53:17
First motion.	000:00:00.63	5 Jul 1966	14:53:17
Liftoff.	000:00:00.86	5 Jul 1966	14:53:17
Start tilt (pitch) maneuver.	000:00:12.2	5 Jul 1966	14:53:29
Start roll maneuver.	000:00:12.2	5 Jul 1966	14:53:29
Complete roll maneuver.	000:00:30.1	5 Jul 1966	14:53:47
Mach 1.	000:00:51.64	5 Jul 1966	14:54:08
Maximum dynamic pressure.	000:01:10.00	5 Jul 1966	14:54:27
Tilt arrest.	000:02:13.9	5 Jul 1966	14:55:30
Enable propellant level sensors.	000:02:16.27	5 Jul 1966	14:55:33
Activate interstage film cameras.	000:02:18.63	5 Jul 1966	14:55:35
S-IB center engine cutoff.	000:02:19.24	5 Jul 1966	14:55:36
S-IB liquid oxygen depletion cutoff signal.	000:02:20.74	5 Jul 1966	14:55:37
S-IB fuel depletion cutoff signal.	000:02:21.73	5 Jul 1966	14:55:38
S-IB outer engine cutoff.	000:02:22.68	5 Jul 1966	14:55:39
S-IVB ullage ignition.	000:02:23.23	5 Jul 1966	14:55:40
S-IB/S-IVB separation.	000:02:23.44	5 Jul 1966	14:55:40
S-IB retro ignition.	000:02:23.53	5 Jul 1966	14:55:40
First motion.	000:02:23.55	5 Jul 1966	14:55:40
S-IB retro at 90% thrust.	000:02:23.59	5 Jul 1966	14:55:40
J-2 clears interstage.	000:02:24.48	5 Jul 1966	14:55:41
J-2 ignition command.	000:02:24.89	5 Jul 1966	14:55:41
Ullage off.	000:02:27	5 Jul 1966	14:55:44
J-2 at 90% thrust.	000:02:28.21	5 Jul 1966	14:55:45
Jettison ullage motor casings.	000:02:25.43	5 Jul 1966	14:55:42
Start iterative guidance mode.	000:02:38.49	5 Jul 1966	14:55:55
Eject S-IB film camera pods.	000:02:48	5 Jul 1966	14:56:05
Peak altitude of S-IB.	000:04:31.5	5 Jul 1966	14:57:48
S-IVB cutoff.	000:07:13.35	5 Jul 1966	15:00:30
Initiate S-IVB coast.	000:07:13.43	5 Jul 1966	15:00:30
Orbital insertion.	000:07:23.348	5 Jul 1966	15:00:40
Low-*g* experiment activated.	000:08:34.71	5 Jul 1966	15:01:51
S-IB impact.	000:09:44.0	5 Jul 1966	15:03:01
S-IVB re-entry.	006:21	5 Jul 1966	21:14

AS-202

The third flight of the Saturn IB: testing the CM heat shield

25 August 1966

BACKGROUND

AS-202 was the third in a series of test flights to 'man rate' the Saturn IB launch vehicle and Apollo spacecraft. The payload (CSM-011) was a fully functional Block I spacecraft as intended for the first manned mission but with the couches, crew equipment, and cabin post-landing ventilation omitted, and a flight-control sequencer (different from that used on AS-201), three auxiliary batteries, four cameras, and a variety of flight qualification instrumentation added. The service module was loaded with 10.56 tons of propellant to boost the spacecraft on a high ballistic arc that would lead to splashdown in the Pacific. The primary objectives were to verify the emergency detection system in closed-loop configuration, and to demonstrate the spacecraft subsystems and test the heat shield in a re-entry with a high heat load.

As its designation suggests, AS-202 was intended to be the second test flight, but it slipped behind AS-203 as a result of delays involving the spacecraft. It lifted off from Launch Complex 34 at Cape Kennedy, Florida, at a Range Zero time of 17:15:32 GMT (13:15:32 EDT) on Thursday, 25 August 1966.

LAUNCH PREPARATIONS

At 23:30 EDT on 24 August the countdown was picked up at T–12 hr, with two built-in 30-min holds at T–4 hr and T–30 min aiming for a planned launch time of 12:30 the next afternoon. However, the emergency detection system test that had been scheduled for T–11 hr 30 min was postponed for 1 hr, then had to be conducted without the normal data interface due to a problematic link between the RCA-110A ground computer and the spacecraft. At 02:25 on 25 August, a 1-hr hold was called at T–9 hr 5 min, when this computer encountered difficulty communicating with the launch vehicle digital computer in the IU. On determining this to be a procedural issue involving the RCA-110A, the count was resumed at 03:25, but when the problem recurred the count was halted again for 40 min at T–8 hr 24.5 min. As a result of the prospect of the station on the island of Antigua having to shut down due to the approach of Hurricane Faith, at 06:25 the clock was advanced to T–6 hr 5 min in order to re-establish the originally planned 12:30 liftoff time. This step was accomplished by eliminating the final emergency detection system test at T–3 hr and by doing spacecraft closeout in parallel with the launch vehicle pre-tanking operations. At 12:10, with the clock at T–20 min, a hold was called to overcome a computer update problem on the communications ship *Rose Knot Victor* stationed in the Atlantic. The count was resumed at 12:51, and ran down to T–3 min, at which time a hold was called after the auxiliary propulsion system on the S-IVB indicated loss of power, but this was resolved 4 min later and the count was resumed at 13:12, with liftoff at 13:15:32. In summary: although there had been a total of 145 min of unplanned holds, the fact that 60 min had been built into the countdown and 40 min had been regained by advancing the clock, the actual delay beyond the planned liftoff time was only 45 min. The temperature and humidity were moderate, the surface winds were light from the southwest, and although visibility was about 8.6 nmi with high broken cloud coverage there were cumulus clouds and rain showers in sight. Hurricane Faith obliged the Antigua site to shut down 45 min after launch.

AS-202 preparation event	Date
S-IB stage arrived at KSC by barge and was offloaded and moved into hanger AF.	07 Feb 1966
S-IB stage erected on LC 34.	04 Mar 1966
Propulsion check started.	07 Mar 1966
S-IVB stage erected and mated.	10 Mar 1966
IU erected and mated.	11 Mar 1966
Launch vehicle power application test.	17 Mar 1966
Launch vehicle electrical mate test.	21 Mar 1966
Launch vehicle selector functional test.	24 Mar 1966
Launch vehicle power transfer test.	25 Mar 1966
Launch vehicle propellant dispersion test.	18 Apr 1966
Propulsion check completed.	19 Apr 1966
Launch vehicle flight sequence and exploding bridge wire test.	23 Apr 1966
Launch vehicle sequence malfunction test.	27 Apr 1966
Launch vehicle plugs-in overall test #1.	29 Apr 1966
S-IB stage 22% liquid oxygen load test.	30 Apr 1966
CSM-011 erected and mated.	02 Jul 1966
Launch vehicle emergency detection system test.	13 Jul 1966
Launch vehicle sequence malfunction test.	14 Jul 1966
Launch vehicle plugs-in overall test #2.	15 Jul 1966
Space vehicle electrical mate test.	21 Jul 1966
Space vehicle emergency detection system test.	21 Jul 1966
Space vehicle plugs-in overall test.	23 Jul 1966
Space vehicle plugs-out overall test.	28 Jul 1966
Launch vehicle countdown demonstration test completed.	08 Aug 1966
Space vehicle systems flight readiness test.	16 Aug 1966
S-IB stage RP-1 loaded.	18 Aug 1966
S-IVB stage auxiliary propulsion system propellants loaded.	21 Aug 1966
Launch countdown started at T–12 hr.	24 Aug 1966
Liftoff.	25 Aug 1966

ASCENT PHASE

The vehicle was launched on a pad azimuth of 100°E of N, but at T + 10.3 sec it initiated a 6-sec roll to align its inertial guidance system on the flight azimuth of 105°E of N, and at T + 10.7 started the pitch maneuver. The damage to the pad facilities was no worse than expected. At T + 63.7 it achieved Mach 1, and at T + 79.5 passed through maximum dynamic pressure. At T + 136.52 the level sensors in the fuel tanks were enabled. The pitch maneuver was terminated at T + 138.0 to stabilize the vehicle in readiness for staging. The center engine cutoff at T + 143.47 was 2.03 sec earlier than predicted. A fuel sensor cut off the outer engines 3.90 sec later, which was 1.27 sec earlier than predicted. The space-fixed velocity was 68.2 ft/sec greater than nominal. Staging was nominal. The launch escape tower was jettisoned together with the boost protective cover of the command module at T + 168.93. Operating in closed-loop, the S-IVB propellant utilization system provided an average mixture ratio of 5.5:1 during the high thrust period, and adjusted this to 4.7:1 at T + 475.72, which was 24.12 sec later than predicted. The cutoff at T + 588.47 was 13.73 sec earlier than predicted, probably due to the late mixture ratio shift, but the iterative guidance had yielded a space-fixed velocity that was only 2 ft/sec greater than nominal and an altitude only 0.05 nmi lower. The 21-nmi deficit in surface range was due to the shorter-than-predicted burn.

Meanwhile, at T + 253 the S-IB trajectory peaked at an altitude of 60.4 nmi at a range of 121.4 nmi, and at about T + 537 the stage impacted 243.2 nmi downrange at 27.39°N, 76.12°W. Although one of the cameras that filmed the staging was lost, the recovery aircraft detected the radio beacon of the other at T + 13 min and at T + 26 min observed the green dye marker at the predicted point, 2 nmi from the site of the S-IB impact.

POST-BOOST PHASE

At cutoff, the S-IVB was at an altitude of 120 nmi and climbing on a ballistic arc. The spacecraft separated at T + 598.7, and 11 sec later fired its SPS engine to place itself on a higher trajectory that would result in entry over the Pacific. This sequence was observed in real-time by a TV camera mounted in the SLA structure. When the efflux from the SPS engine impinged on the splayed SLA panels it torqued the S-IVB, but this was soon restabilized by its auxiliary propulsion system. The SPS burn was terminated after 216.7 sec, once it had achieved the velocity increment. As a thermal test, the CSM then turned to face the apex of the command module Earthward, which attitude was maintained as the spacecraft flew through the peak altitude of 618 nmi above Africa. Descending over the Indian Ocean, the spacecraft realigned itself with the velocity vector, the SPS was fired for 89.2 sec to accelerate for atmospheric entry, and then fired briefly twice more in rapid succession as a demonstration.

Meanwhile the S-IVB had performed the common bulkhead reversal test. In an effort to save weight, the propellant tanks shared a bulkhead. It was a sophisticated structure as it had to cope with the difference in pressure between the tanks, and insulate the liquid oxygen at –172°C from the liquid hydrogen at –253°C, lest the oxygen froze solid. The test provided data on the strength of the bulkhead by stressing it to destruction. At T + 800 the

AS-202 ascent phase event	GET (h:m:s)	Altitude (nmi)	Range (nmi)	Earth-fixed velocity (ft/sec)	Space-fixed velocity (ft/sec)	Event duration (sec)	Geocentric latitude (°N)	Longitude (°E)	Space-fixed flight path angle (deg)	Space-fixed heading angle (°E of N)
Liftoff	–	–	–	–	–	–	–	–	–	–
Mach 1 achieved	000:01:03.69	4.2	–	–	–	–	–	–	–	–
Maximum dynamic pressure	000:01:19.5	7.3	–	–	–	–	–	–	–	–
S-IB center engine cutoff	000:02:19.57	29.4	27.1	5,952.9	7,096.6	–	–	–	–	–
S-IB outboard engine cutoff	000:02:23.47	31.4	30.4	6,162.5	7,317.6	–	28.40	–80.00	26.43	101.83
S-IB/S-IVB separation	000:02:24.43	31.9	31.0	–	7,321.6	–	–	–	26.32	–
S-IVB cutoff	000:09:48.47	117.3	840.7	20,979.7	22,311.5	–	23.76	–65.88	3.99	111.84

AS-202 post-boost phase event	GET (h:m:s)	Altitude (nmi)	Space-fixed velocity (ft/sec)	Event duration (sec)	Geocentric latitude (deg N)	Longitude (deg E)	Space-fixed flight path angle (deg)	Space-fixed heading angle (°E of N)
CSM separation	000:09:58.70	120.0	–	–	–	–	3.82	–
Start 1st SPS burn	000:10:09.7	122.5	22,297	–	23.28	295.26	3.58	112.35
End 1st SPS burn	000:13:46.4	182.7	25,501	216.7	17.73	307.49	5.71	116.42
Peak altitude	000:41.14	617.1	22,664	–	–26.58	26.48	0.00	107.55
Start 2nd SPS burn	001:05:56.1	247.0	25,071	–	–18.62	112.20	–5.84	64.11
End 2nd SPS burn	001:07:25.3	201.8	27,443	89.2	–15.92	117.48	–7.35	62.62
Start 3rd SPS burn	001:07:34.5	197.2	27,477	–	–15.66	117.95	–7.27	62.48
End 3rd SPS burn	001:07:38.3	194.9	27,576	3.8	–15.51	118.18	–7.29	62.41
Start 4th SPS burn	001:07:47.5	189.1	27,624	–	–15.18	118.80	–7.18	62.24
End 4th SPS burn	001:07:51.3	186.8	27,719	3.8	–15.08	119.05	–7.19	62.17
CM/SM separation	001:11:04.0	94.1	28,315	–	–8.15	130.81	–4.71	59.64
Entry interface	001:12:28	65.8	28,512	–	–4.86	136.00	–3.53	59.01

stage reached the peak of its trajectory, at an altitude of 145 nmi, 1,543 nmi downrange, and at T + 941 lost structural integrity and ceased to transmit telemetry.

RECOVERY

At T + 4,188 sec the spacecraft began to reorient itself to jettison the service module, which was completed at T + 4,264, then the command module oriented itself for re-entry. In contrast to the 'rolling' entry on AS-201, this time the CM controlled its attitude to fly a trajectory that 'skipped' off the atmosphere and pursued an arc that led to a second contact and final entry. This profile produced a double peak in the heating rate, and was designed to expose the heat shield to low heat rates with high heat loads – that is, lower temperatures but over considerably longer periods than in a direct entry. A similar profile was to be employed after a lunar return. Although the maximum temperature of the surface of the capsule was calculated to have reached 2,700°F, the interior did not exceed 70°F.

At 18:48 GMT, after its 93-min flight, and 7.5 sec later than predicted, the CM splashed down in the Pacific Ocean at 16.11°N, 168.90°E, 435 nmi southwest of Wake Island. It adopted the apex-up flotation attitude, the parachutes were released and the recovery aids deployed. Because the –3.53-deg flight path angle at entry was steeper than the predicted value of –3.48 deg and the 0.28 (±0.02) lift-to-drag ratio was less than the predicted 0.33 (±0.04), the site was 205 nmi short of the aim point. Once a recovery aircraft had spotted the CM, swimmers were delivered to attach the flotation collar. The capsule was recovery by USS *Hornet* some 10 hours after the mission began.

CONCLUSIONS

In its closed-loop configuration, the emergency detection system monitored for either multiple H-1 engine failures or excessive angular rates on the vehicle. The engine failure mode was enabled at T + 10.89 sec, shortly after the vehicle began to pitch over and while it was rolling onto the flight azimuth. The angular rates mode was inhibited at T + 134.68 and the engine failure mode at T + 135.08, shortly before the level sensors were activated. If the system had detected a cause for abort during this interval it would have activated the launch escape system. The ability of the solid rocket in the tower to pull a command module clear in a variety of conditions had been demonstrated by a series of tests using Little Joe II rockets. The Q-ball (the sensor on top of the tower which measured angular rates) was powered off at T + 139.77, as the S-IB was shutting down, and the tower was jettisoned soon after the S-IVB ignited. The emergency detection system monitored the performance of the stage for the CSM which, if an abort was required, would separate and fire its SPS engine to draw clear, preparatory to a 'turnabout' maneuver to release the command module to make an emergency descent.

Other than the outlet temperature of the glycol evaporator in the environmental control system briefly exceeding its specified limit, and the exhaust temperature of the condenser of the fuel cells in the electrical power system rising toward its upper limit, the spacecraft systems operated satisfactorily in flight. This mission qualified the heat shield for orbital missions, but it would require additional testing for a lunar return. Planning for future missions would take into account the lower than expected lift-to-drag ratio of the command module. Both the Saturn IB launch vehicle and the Block I spacecraft were declared ready for the first manned mission.

MISSION OBJECTIVES

Launch Vehicle Primary Objectives

1. Demonstrate structural integrity and compatibility of the space vehicle during S-IB stage powered flight and confirm structural loads and dynamic characteristics.
2. Demonstrate structural integrity and compatibility of the space vehicle during S-IVB stage powered flight and orbital coast.
3. Demonstrate separation of the S-IVB/IU/spacecraft from the S-IB.
4. Demonstrate separation of the CSM from the S-IVB/IU/SLA.
5. Demonstrate S-IB propulsion system operation and evaluate system performance and parameters.
6. Demonstrate S-IVB propulsion system operation, including propellant mixture ratio shift, and evaluate system performance and parameters.
7. Demonstrate launch vehicle guidance system operation, achieve guidance cutoff, and evaluate system accuracy.
8. Demonstrate launch vehicle control system operation during S-IB powered flight, S-IVB powered flight and orbital coast, and evaluate performance characteristics.
9. Demonstrate launch vehicle sequencing system operations.

10. Demonstrate the in-flight performance of the S-IB and S-IVB secure range command systems.
11. Evaluate performance of the space vehicle emergency detection system in a closed-loop configuration.
12. Demonstrate the mission support facilities and operations required for launch, mission operations, and CM recovery.

Launch Vehicle Secondary Objectives

1. Confirm launch vehicle powered flight external environment.
2. Evaluate S-IVB/IU in-flight thermal conditioning system.
3. Verify adequacy of the S-IVB residual propellant venting.
4. Evaluate the S-IVB common bulkhead reversal test.

Spacecraft Primary Test Objectives

1. Demonstrate structural integrity and compatibility of the launch vehicle and spacecraft, and confirm launch loads including:
   (a) Demonstrate capability and structural integrity of the Saturn IB/CSM. *Achieved.*
   (b) Determine structural loading of the SLA when subjected to the Saturn IB launch environment. *Achieved.*
2. Demonstrate separation of the launch escape system and boost protective cover from the spacecraft, the separation of the spacecraft from the S-IVB/IU/SLA, and the separation of the command module from the service module. *Achieved.*
3. Verify the standpipe fix (requiring a minimum burn time of 198 sec) and the ability of the SPS engine to make multiple restarts (at least three burns of at least 3-sec duration at 10-sec intervals). *Achieved.*
4. Verify the guidance and navigation subsystem. *Achieved.*
5. Verify the ability of the environmental subsystem to control cabin pressure and temperature. *The glycol evaporator ceased to function between T+840 and T+4,080 sec, allowing the outlet temperature to exceed 75°F.*
6. Verify the unified S-band communication subsystem between the spacecraft and the Manned Space Flight Network in downlink mode for telemetry and simulated voice and in turnaround ranging mode. *Because the signal levels were weaker than expected, telemetry was not continuous.*
7. Verify the SM-RCS. *Achieved.*
8. Verify the CM-RCS. *Achieved.*
9. Verify the stabilization and control subsystem. *Achieved.*
10. Verify the electrical power subsystem. *The condenser exhaust temperature of the fuel cells approached its maximum limit.*
11. Demonstrate the Earth landing system, involving the parachute subsystem and recovery aids. *Achieved.*
12. Evaluate the space vehicle emergency detection subsystem in closed-loop configuration. *Achieved.*
13. Evaluate the heat shield at high heat load during entry at approximately 28,000 ft/sec. *Achieved.*

The Saturn IB is prepared for launch with the Block I spacecraft CSM-011.

CM-011 awaits recovery.

When the Little Joe II booster broke up during abort test A-003 on 19 May 1965, the launch escape system self-triggered to save the "boilerplate" spacecraft BP-22.

14. Demonstrate the mission support facilities required for launch, mission operations, and command module recovery. *Achieved.*

Spacecraft Secondary Test Objective

1. Demonstrate subsystem performance other than the minimum required for manned orbital capability, including:
   (a) Long-duration (approximately 200 sec) SPS performance, including shutdown characteristics. *Achieved.*
   (b) Obtain data on SPS burn stability. *Achieved.*

MISSION TIMELINE

AS-202 mission event	GET (h:m:s)	Date (GMT)	Time (h:m:s)
Guidance reference release.	–000:00:04.468	25 Aug 1966	17:15:27
Ignition sequence command.	–000:00:02.478	25 Aug 1966	17:15:29
Range zero.	000:00:00.00	25 Aug 1966	17:15:32
First motion.	000:00:00.73	25 Aug 1966	17:15:32
Liftoff.	000:00:00.93	25 Aug 1966	17:15:32
Start roll maneuver.	000:00:10.30	25 Aug 1966	17:15:42
Start tilt (pitch) maneuver.	000:00:10.70	25 Aug 1966	17:15:42
Enable multiple-engine cutoff monitor.	000:00:10.89	25 Aug 1966	17:15:42
Complete roll maneuver.	000:00:16.30	25 Aug 1966	17:15:48
Mach 1.	000:01:03.69	25 Aug 1966	17:16:35
Maximum dynamic pressure.	000:01:19.5	25 Aug 1966	17:16:51
Enable S-IB propellant level sensors (to generate the cutoff signal).	000:02:16.52	25 Aug 1966	17:17:48
Tilt arrest.	000:02:18.00	25 Aug 1966	17:17:50
Start film cameras in interstage.	000:02:19.28	25 Aug 1966	17:17:51
S-IB center engine cutoff.	000:02:19.57	25 Aug 1966	17:17:51
Q-ball power off.	000:02:19.77	25 Aug 1966	17:17:51
S-IB outer engine cutoff.	000:02:23.47	25 Aug 1966	17:17:55
S-IVB ullage ignition.	000:02:24.03	25 Aug 1966	17:17:56
S-IB/S-IVB separation.	000:02:24.23	25 Aug 1966	17:17:56
First motion.	000:02:24.33	25 Aug 1966	17:17:56
J-2 clears interstage.	000:02:25.30	25 Aug 1966	17:17:57
J-2 ignition command.	000:02:25.62	25 Aug 1966	17:17:57
Jettison ullage motor casings.	000:02:36.22	25 Aug 1966	17:18:08
Jettison the launch escape tower.	000:02:48.93	25 Aug 1966	17:18:20
Start iterative guidance mode.	000:02:52.40	25 Aug 1966	17:18:24
Peak of S-IB trajectory.	000:04:13.0	25 Aug 1966	17:19:45
Initiate mixture ratio shift.	000:08:18.15	25 Aug 1966	17:23:50
S-IB impact.	000:08:56:8	25 Aug 1966	17:23:34
S-IVB cutoff.	000:09:48.47	25 Aug 1966	17:25:20
S-IVB/CSM separation.	000:09:58.70	25 Aug 1966	17:25:30

AS-202 mission event – *continued*	GET (h:m:s)	Date (GMT)	Time (h:m:s)
SM-RCS +X translation.	000:10.00.0	25 Aug 1966	17:25:32
SPS 1st ignition.	000:10:09.7	25 Aug 1966	17:25:41
Peak of S-IVB trajectory.	000:13:21	25 Aug 1966	17:28:53
SPS 1st cutoff.	000:13:46.4	25 Aug 1966	17:29:18
Peak of CSM trajectory.	000:41:14	25 Aug 1966	17:56:46
Bulkhead test induces S-IVB structural failure.	000:15:41.2	25 Aug 1966	17:31:13
SPS 2nd ignition.	001:05:56.1	25 Aug 1966	18:21:28
SPS 2nd cutoff.	001:07:25.3	25 Aug 1966	18:22:57
SPS 3rd ignition.	001:07:34.5	25 Aug 1966	18:23:06
SPS 3rd cutoff.	001:07:38.3	25 Aug 1966	18:23:10
SPS 4th ignition.	001:07:47.5	25 Aug 1966	18:23:19
SPS 4th cutoff.	001:07:51.3	25 Aug 1966	18:23:23
Start maneuver to separation attitude.	001:09:48.2	25 Aug 1966	18:25:20
SM/CM separation.	001:11:04.0	25 Aug 1966	18:26:36
CM start maneuver to entry attitude.	001:11:12.2	25 Aug 1966	18:26:44
Entry interface (400,000 ft).	001:12:28.0	25 Aug 1966	18:28:00
Enter radio blackout.	001:13:36.0	25 Aug 1966	18:29:08
End radio blackout.	001:23:28.0	25 Aug 1966	18:39:00
Jettison CM apex.	001:26:58.3	25 Aug 1966	18:42:30
Drogue parachutes deployment.	001:26:59.9	25 Aug 1966	18:42:31
Main parachutes deployment.	001:27:48.2	25 Aug 1966	18:43:20
Splashdown.	001:33:02.2	25 Aug 1966	18:48:34

Apollo 1

Cabin fire

27 January 1967

BACKGROUND

As the first manned mission of the Apollo Program, Apollo 1 was scheduled for launch on 21 February 1967 from Launch Complex 34 at Cape Kennedy, Florida, by the fourth Saturn IB (i.e. AS-204). However, the prime crew were killed in a command module fire during a routine countdown rehearsal on 27 January.

The crew consisted of Lieutenant Colonel Virgil Ivan 'Gus' Grissom (USAF), command pilot; Lieutenant Colonel Edward Higgins White II (USAF), senior pilot; and Lieutenant Commander Roger Bruce Chaffee (USN), pilot. Selected in the astronaut group of 1959, Grissom had been pilot of MR-4, America's second and last suborbital flight, and command pilot of the first two-man flight, Gemini 3. Born on 3 April 1926 in Mitchell, Indiana, Grissom was 40 years old on the day of the Apollo 1 fire. Grissom received a BS in mechanical engineering from Purdue University in 1950. His backup for the mission was Captain Walter Marty 'Wally' Schirra [shi-RAH] (USN). White had been pilot for the Gemini 4 mission, during which he became the first American to walk in space. He was born on 14 November 1930 in San Antonio, Texas, and was 36 years old on the day of the Apollo 1 fire. He received a BS from the US Military Academy at West Point in 1952, an MS in aeronautical engineering from the University of Michigan in 1959, and was selected as an astronaut in 1962. His backup was Major Donn Fulton Eisele [EYES-lee] (USAF). Chaffee was training for his first spaceflight. He was born on 15 February 1935 in Grand Rapids, Michigan, and was 31 years old on the day of the Apollo 1 fire. He received a BS in aeronautical engineering from Purdue University in 1957, and was selected as an astronaut in 1963. His backup was Ronnie Walter 'Walt' Cunningham.

VEHICLE HISTORY

Apollo 1 preparation event	Date
Fabrication of CM-012 at North American Aviation, Downey, CA.	Aug 1964
Basic structure completed.	Sep 1965
Installation and final assembly of subsystems completed. Critical design reviews completed. Checkout of all subsystems initiated, followed by integrated testing of all spacecraft subsystems.	Mar 1966
Customer acceptance readiness review completed. NASA issued certificate of flightworthiness and authorized spacecraft to be shipped to KSC.	Aug 1966
Command module received at KSC.	26 Aug 1966
CM-012 mated with SM-012 in altitude chamber. Alignment, subsystems and system certification tests and functional checks performed.	Sep 1966
First combined systems tests completed.	01 Oct 1966
Design certification document issued which certified design as flight worthy, pending satisfactory resolution of open items.	07 Oct 1966
First manned test at sea level pressure to verify total spacecraft system operation completed.	13 Oct 1966
Unmanned test at altitude pressures using oxygen to verify spacecraft system operation.	15 Oct 1966
Manned test with flight crew completed.	19 Oct 1966
Second manned altitude test with backup crew initiated, but discontinued when failure occurred in oxygen system regulator in spacecraft environmental control system. Regulator removed and found to have design deficiency.	21 Oct 1966
Apollo program director conducted recertification review that closed out majority of open items remaining from previous reviews.	21 Dec 1966
Sea level and unmanned altitude tests completed.	28 Dec 1966
Manned altitude test with backup flight crew completed.	30 Dec 1966
Command module removed from altitude chamber.	03 Jan 1967
Spacecraft mated to launch vehicle at Launch Complex 34. Various tests and equipment installations and replacements performed.	06 Jan 1967

ACCIDENT

The accident occurred during the Plugs-Out Integrated Test. The purpose of this test was to demonstrate all space vehicle systems and operational procedures in as near a flight configuration as practical and to verify systems capability in a simulated launch. The test began at 12:55 GMT on 27 January 1967. After initial system tests were completed, the flight crew entered the command module at 18:00. The command pilot noted an odor in the CM-012 environmental control system suit oxygen loop and the count was held at 18:20 while a sample of the oxygen in this system was taken. The count was resumed at 19:42 with hatch installation and subsequent cabin purge with oxygen beginning at 19:45. The odor was later determined to be unrelated to the fire. Communication difficulties were encountered and the count was held at approximately 22:40 to troubleshoot the problem. The problem consisted of a continuously live microphone that could not be turned off by the crew. Various final countdown functions were still performed during the hold as communications permitted. By 23:20, all final countdown functions up to the transfer to simulated fuel cell power were completed and the count was held at T–10 min pending resolution of the communications problems. Over the next 10 min there were no events that appeared to be related to the fire. The major activity during this period was routine troubleshooting of the communications problem – all of the other systems were operating normally. There were no voice transmissions from the spacecraft from 23:30:14 until the transmission reporting the fire, which began at 23:31:04.7. During the period beginning about 30 sec prior to the report, there were indications of crew movement. These were provided by the data from the biomedical sensors, the command pilot's live microphone, the guidance and navigation system, and the environmental control system. There was no evidence of what this movement was, or that it was related to the fire. The biomedical data indicated that just prior to the fire report the senior pilot was performing essentially no activity until about 23:30:21, when there was a slight increase in pulse and respiratory rate. At 23:30:30, the electrocardiogram indicated muscular activity for several seconds. Similar indications were recorded at 23:30:39. The data showed increased activity but were not indicative of an alarm type of response. By 23:30:45, all the biomedical parameters had reverted to the baseline 'rest' level. Beginning at about 23:30, the command pilot's live microphone transmitted brushing and tapping noises indicative of movement. The noises were similar to those transmitted earlier in the test by the live microphone when the command pilot was known to have been moving. These sounds ended at 23:30:58.6. Any significant crew movement resulted in minor motion of the command module and was detected by the guidance and navigation system. The type of movement, however, could not be determined. Data from this system indicated a slight movement at 23:30:24, with more intense activity starting at 23:30:39 and ending at 23:30:44. More movement began at 23:31:00 and continued until loss of data transmission during the fire. Increases of oxygen flow rate to the crew suits also indicated movement. All suits had some small leakage, the rate of which varied with the position of each crew member in the spacecraft. Earlier in the Plugs-Out Integrated Test, the crew noted that a particular movement, the nature of which was unspecified, provided increased flow rate. This was also confirmed from the flow rate data records. The flow rate showed a gradual rise at 23:30:24 that reached the limit of the sensor at 23:30:59. At 23:30:54.8, a significant voltage transient was recorded. In addition to a surge in the AC Bus 2 voltage, several other parameters being measured displayed anomalous behavior. Beginning at 23:31:04.7, the crew gave the first verbal indication of an emergency by reporting a fire in the command module.

Emergency procedures called for the senior pilot, occupying the center couch, to unlatch and remove the hatch while retaining his harness in a buckled state. A number of witnesses who observed the television picture of the command module hatch window discerned motion that suggested that the senior pilot was reaching for the inner hatch handle. The senior pilot's harness buckle was found unopened after the fire, indicating that he had initiated the standard hatch-opening procedure. Data from the guidance and navigation system indicated considerable activity within the command module after the fire was discovered. This activity was consistent with movement of the crew prompted by the proximity of the fire or with the undertaking of standard emergency egress procedures. Personnel located on adjustable Level A-8 adjacent to the command module responded to the report of the fire. The pad leader ordered the implementation of crew egress procedures and technicians rushed toward the White Room which surrounded the hatch and into which the crew would step upon egress. Then, at 23:31:19, the command module ruptured. All transmission of voice and data from the spacecraft terminated by 23:31:22.4, some 3 sec after rupture. Witnesses monitoring the television showing the hatch window reported that flames spread from the left to the right side of the command module and shortly thereafter covered the entire visible area. Flames and gases flowed rapidly out of the ruptured area, spreading flames into the space between the command module pressure vessel and heat shield through access hatches, and into Levels A-8 and A-7 of the service structure. These flames ignited combustibles, endangered pad personnel, and impeded rescue efforts. The burst of fire, together with the sounds of rupture, caused several pad personnel to believe that the command module had exploded or was about to explode. The immediate reaction of all personnel on Level A-8 was to evacuate that level. This reaction was promptly followed by a return to effect rescue. Several personnel obtained readily accessible fire extinguishers and returned along the swing arm to the White Room to begin rescue efforts. Others obtained fire extinguishers from various other parts of the service structure and rendered assistance in fighting the fires.

Three hatches were installed on the command module. The outermost hatch, called the boost protective cover (BPC) hatch, was part of the cover that shielded the command module during launch and was jettisoned prior to orbital operation. The middle hatch was termed the ablative hatch and became the outer hatch when the BPC was jettisoned after launch. The inner hatch closed the pressure vessel wall of the command module and was the first hatch to be opened by the crew in an unaided crew egress. On the day of the fire, the outer or BPC hatch was in place but not fully latched because of distortion in the BPC caused by wire bundles temporarily installed for the test. The middle hatch and inner hatch were in place and latched after crew ingress. Although the BPC hatch was not fully latched, it was necessary to insert a specially designed tool into the hatch in order to provide a hand-hold for lifting it from the command module. By this time the White Room was filling with dense, dark smoke from the command module interior and from secondary fires throughout Level A-8. While some personnel were able to locate and don operable gas masks, others were not. Some proceeded without masks while others attempted without success to render masks operable. Even operable masks were unable to cope with the dense smoke present because they were designed for use in toxic rather than dense smoke atmospheres. Visibility in the White Room was virtually nonexistent. It was necessary to work essentially by touch since visual observation was limited to a few inches at best. A hatch-removal tool was in the White Room. Once the small fire near the BPC hatch had been extinguished and the tool located, the pad leader and an assistant removed the BPC hatch. Although the hatch was not latched, removal was difficult. The personnel who removed the BPC hatch could not remain in the White Room because of the smoke. They left the White Room and passed the tool required to open each hatch to other individuals. A total of five individuals took part in opening the three hatches. Each was forced to make several trips to and from the White Room in order to reach breathable air. The middle hatch was removed with less effort than was required for the BPC hatch. The inner hatch was unlatched and an attempt was made to raise it from its support and to lower it to the command module floor. The hatch could not be lowered the full distance to the floor and was instead pushed to one side. When the inner hatch was opened, intense heat and a considerable amount of smoke issued from the interior of the command module. When the pad leader ascertained that all hatches were open, he left the White Room, proceeded several feet along the swing arm, donned his headset and reported this fact. From a voice tape it has been determined that this report came approximately 5 min 27 sec after the first report of the fire. The pad leader estimates that his report was made no more than 30 sec after the inner hatch was opened. Therefore, it was concluded that all hatches were opened and the two outer hatches removed approximately 5 min after the report of fire or at about 23:36.

Medical opinion, based upon autopsy reports, concluded that chances of resuscitation decreased rapidly once consciousness was lost (about 15 to 30 sec after the first suit failed) and that resuscitation was impossible by 23:36. Cerebral hypoxia, due to cardiac arrest resulting from myocardial hypoxia, had caused a loss of consciousness. Factors of temperature, pressure, and environmental concentrations of carbon monoxide, carbon dioxide, oxygen, and pulmonary irritants were changing rapidly. The combined effect of these environmental factors dramatically increased the lethal effect of any factor by itself. Because it was impossible to integrate the variables with the dynamic physiological and metabolic conditions they produced, a precise time when consciousness was lost and death supervened could not be conclusively determined. Visibility within the command module was extremely poor. Although the lights remained on, they could be perceived only dimly. No fire was observed. Initially, the crew was not seen. The personnel who had been involved in removing the hatches attempted to locate the crew without success. Throughout this period, other pad personnel were fighting secondary fires on Level A-8. There was considerable fear that the launch escape tower, mounted above the command module, would be ignited by the heat of the fires below and thereby destroy much of the launch complex.

Shortly after the report of the fire, a call was made to the fire department. From log records, it appeared that the fire apparatus and personnel were dispatched at about 23:32. The doctor monitoring the test from the blockhouse near the pad had already proceeded to the base of the umbilical tower. The exact time at which firefighters reached Level A-8 is not known. Personnel who opened the hatches unanimously stated that all hatches were open before any firefighters were seen on the level or in the White Room. The first firefighters who reached Level A-8 stated that all hatches were open, but that the inner hatch was inside the command module when they arrived. This placed the arrival of the firefighters after 23:36. It was estimated on the basis of tests that 7 to 8 min were required to travel from the fire station to the launch complex and to ride the elevator from the ground to Level A-8. Thus, the estimated time the firefighters arrived at Level A-8 was shortly before 23:40. When the firefighters arrived, the positions of the crew couches and crew could be perceived through the smoke but only with difficulty. An unsuccessful attempt was made to remove the senior pilot from the command module.

Initial observations and subsequent inspection indicated several facts. The command pilot's couch (the left couch) was in the '170-degree' position, in which it was essentially horizontal throughout its length. The foot restraints and harness were released and the inlet and outlet oxygen hoses were connected to the suit. The electrical adapter cable was disconnected from the communications cable. The command pilot was lying supine on the aft bulkhead or floor of the command module, with his helmet visor closed and locked and with his head beneath the pilot's head rest and his feet on his own couch. A fragment of his suit material was found outside the command module pressure vessel 5 ft from the point of rupture. This indicated that his suit had failed prior to the time of rupture (23:31:19.4), allowing convection currents to carry the suit fragment through the rupture. The senior pilot's couch (the center couch) was in the '96-degree' position in which the back portion was horizontal and the lower portion was raised. The buckle releasing the shoulder straps and lap belts was not opened. The straps and belts were burned through. The suit oxygen outlet hose was connected but the inlet hose was disconnected. The helmet visor was closed and locked and all electrical connections were intact. The senior pilot was lying transversely across the command module just below the level of the hatchway. The pilot's couch (the couch on the right) was in the '264-degree' position in which the back portion was horizontal and the lower portion dropped toward the floor. All restraints were disconnected, all hoses and electrical connections were intact, and the helmet visor was closed and locked. The pilot was supine on his couch.

From the foregoing, it was determined that the command pilot probably left his couch to avoid the initial fire, the senior pilot remained in his couch as planned for emergency egress, attempting to open the hatch until his restraints burned through. The pilot remained in his couch to maintain communications until the hatch could be opened by the senior pilot as planned. With a slightly higher pressure inside the command module than outside, opening the inner hatch was impossible because of the resulting force on the hatch. Thus the inability of the pressure relief system to cope with the pressure increase due to the fire made opening the inner hatch impossible until after cabin rupture. After cabin rupture, the intense and widespread fire, together with rapidly increasing carbon monoxide concentrations, further precluded egress. Whether the inner hatch handle was moved by the crew cannot be determined, since the opening of the inner hatch from the White Room also sets the handle within the command module in the unlatched position.

Immediately after the firefighters arrived, the pad leader on duty was relieved to allow treatment for smoke inhalation. He had first reported over the headset that he could not describe the situation in the command module. In this manner he attempted to convey the fact that the crew was dead to the test conductor without informing the many people monitoring the communication channels. Upon reaching the ground the pad leader told the doctors that the crew was dead. The three doctors proceeded to the White Room and arrived there shortly after the arrival of the firefighters. The doctors estimate their arrival to have been at 23:45. The three doctors entered the White Room and determined that the crew had not survived the heat, smoke, and thermal burns. The doctors were not equipped with breathing apparatus, and the command module still contained fumes and smoke. It was determined that nothing could be gained by immediate removal of the crew. The firefighters were then directed to stop removal efforts. When the command module had been adequately ventilated, the doctors returned to the White Room with equipment for crew removal. It became apparent that extensive fusion of suit material to melted nylon from the spacecraft would make removal very difficult. For this reason it was decided to discontinue removal efforts in the interest of accident investigation and to photograph the command module with the crew in place before evidence was disarranged. Photographs were taken and the removal efforts resumed at approximately 00:30 GMT, 28 January.

Removal of the crew took approximately 90 min and was completed about 7.5 hours after the accident.

CHRONOLOGY OF THE FIRE

It was most likely that the fire began in the lower forward portion of the left equipment bay, to the left of the command pilot, and considerably below the level of his couch. Once initiated, the fire burned in three stages. The first stage, with its associated rapid temperature rise and increase in cabin pressure, terminated 15 sec after the verbal report of fire. At this time, 23:31:19, the command module cabin ruptured. During this first stage, flames moved rapidly from the point of ignition, traveling along debris traps installed in the command module to prevent items from dropping into equipment areas during ground tests or while in flight. At the same time, Velcro strips positioned near the ignition point also burned.

The fire was not intense until about 23:31:12. The slow rate of buildup of the fire during the early portion of the first stage was consistent with the opinion that ignition occurred in a zone containing little combustible material. The slow rise of pressure could also have resulted from absorption of most of the heat by the aluminum structure of the command module. The initial flames rose vertically and then spread out across the cabin ceiling. The debris traps provided not only combustible material and a path for the spread of the flames, but also firebrands of burning molten nylon. The scattering of these firebrands contributed to the spread of the flames. By 23:31:12, the fire had broken from its point of origin. As a wall of flames extended along the left wall of the module, it prevented the command pilot, occupying the left couch, from reaching the valve that would vent the command module to the outside atmosphere. Although operation of this was the first step in established emergency egress procedures, such action would have been to no avail because the venting capacity was insufficient to prevent the rapid buildup of pressure due to the fire. It was estimated that opening the valve would have delayed command module rupture by no more than 1 sec. The command module was designed to withstand an internal pressure of approximately 13 psi above ambient without rupturing. Data recorded during the fire showed that this design criterion was exceeded late in the first stage of the fire and that rupture occurred at about 23:31:19. The point of rupture was where the floor or aft bulkhead of the command module joined the wall, essentially opposite the point of ignition. About 3 sec prior to rupture, at 23:31:16.8, the final crew communication began. This communication ended shortly after rupture at 23:31:21.8, followed by loss of telemetry at 23:31:22.4. Rupture of the command module marked the beginning of the brief second stage of the fire. This stage was characterized by the period of greatest conflagration due to the forced convection that resulted from the outrush of gases through the rupture in the pressure vessel. The swirling flow scattered firebrands throughout the crew compartment, spreading fire. This stage of the fire ended at approximately 23:31:25. Evidence that the fire spread from the left side of the command module toward the rupture area was found on subsequent examination of the module and crew suits. Evidence of the intensity of the fire includes burst and burned aluminum tubes in the oxygen and coolant systems at floor level. This third stage was characterized by rapid production of high concentrations of carbon monoxide. Following the loss of pressure in the command module and with fire now throughout the crew compartment, the remaining atmosphere quickly became so deficient in oxygen that it could not support continued combustion. Unlike the earlier stages where the flame was relatively smokeless, heavy smoke now formed and large amounts of soot were deposited on most spacecraft interior surfaces as they cooled. The third stage of the fire could not have lasted more than a few seconds because of the rapid depletion of oxygen. It was estimated that the command module atmosphere was lethal by 23:31:30; i.e. 5 sec after the start of the third stage. Although most of the fire inside the command module was quickly extinguished because of a lack of oxygen, a localized, intense fire persisted in the area of the environmental control unit, situated in the left equipment bay, near the point where the fire was believed to have started. Failed oxygen and water/glycol lines in this area continued to supply oxygen and fuel to support the localized fire that melted the aft bulkhead and burned adjacent portions of the inner surface of the command module heat shield.

INVESTIGATION

Immediately after the accident, additional security personnel were positioned at Launch Complex 34 and the facility was impounded. Prior to disturbing any evidence, numerous external and internal photographs were taken of the spacecraft. After crew removal, two experts entered the command module to verify switch positions. Small groups of NASA and North American Aviation management, members of the Apollo 204 Review Board, representatives, and consultants inspected the exterior of CM-012. A series of close-up stereo photographs of the command module were taken to document the as-found condition of the spacecraft systems. After the couches were removed, a special false floor with removable 18-in transparent squares was installed to provide access to the entire inside of the command module without disturbing evidence. A detailed inspection of the spacecraft interior was then performed, followed by the preparation and approval by the Board of a command module disassembly plan. CM-014 was shipped to the Kennedy Space Center (KSC) on 1 February 1967 in order to assist the Board in the investigation. This command module was placed in the Pyrotechnics Installation Building and was used to develop disassembly techniques for selected components prior to their removal from CM-012. By 7 February the disassembly plan was fully operational. After the removal of each component, photographs were taken to record the exposed area. This step-by-step documentation was done throughout the disassembly of the spacecraft. In all, some 5,000 photographs were taken. All interfaces, such as electrical connectors, tubing joints, physical mounting of components, etc., were closely inspected and photographed prior to, during, and following disassembly. Each item removed from the command module was tagged, sealed in a clean plastic container, and transported under the required security to bonded storage. On 17 February the Board decided that removal and wiring tests had progressed sufficiently to enable the command module to be moved without the risk of disturbing further evidence, and it was moved to the Pyrotechnics Installation Building. With the improved working conditions, it was found that a work schedule of two 8-hr shifts per day for 6 days per week was sufficient to maintain pace with the analysis and disassembly planning. The only exception to this was a 3-day period of three 8-hr shifts per day used to remove the aft heat shield, move the command module to a more convenient workstation and remove the crew compartment heat shield. The disassembly of the command module was completed on 27 March 1967.

Apollo 1 fire event	Date (GMT)	Time (h:m:s)
Plugs-Out Integrated Test initiated when power applied to spacecraft.	27 Jan 1967	12:55
Following completion of initial verification tests of system operation, command pilot entered spacecraft, followed by pilot and senior pilot.	27 Jan 1967	18:00
Count held when command pilot noted odor in spacecraft environmental control system suit oxygen. Sample taken.	27 Jan 1967	18:20
Count resumed after hatch installed.	27 Jan 1967	19:42
Cabin purged with oxygen.	27 Jan 1967	19:45
Open microphone first noted by test crew.	27 Jan 1967	22:25
Count held while communication difficulties checked. Various final countdown functions performed during hold as communications permitted.	27 Jan 1967	22:40
From this time until about 23:53 GMT, flight crew interchanged equipment related to communications systems in effort to isolate communications problem. During troubleshooting period, problems developed with ability of various ground stations to communicate with one another and with crew.	27 Jan 1967	22:45
Final countdown functions up to transfer to simulated fuel cell power completed and count held at T–10 min pending resolution of communications problems. For next 10 min, no events related to fire. Major activity was routine troubleshooting of communications problem. All other systems operated normally during this period.	27 Jan 1967	23:20
First indication by either cabin pressure or battery compartment sensors of a pressure increase.	27 Jan 1967	23:21:11
Command pilot live microphone transmitted brushing and tapping noises, indicative of movement. Noises similar to those transmitted earlier in test by live microphone when command pilot was known to be moving.	27 Jan 1967	23:30
No voice transmissions from spacecraft from this time until transmission reporting fire.	27 Jan 1967	23:30:14
Slight increase in pulse and respiratory rate noted from senior pilot.	27 Jan 1967	23:30:21
Data from guidance and navigation system indicated undetermined type of crew movement. Gradual rise in oxygen flow rate to crew suits began, indicating movement. Earlier in Plugs-Out Integrated Test, crew reported that an unspecified movement caused increased flow rate.	27 Jan 1967	23:30:24
Senior pilot's electrocardiogram indicated muscular activity for several seconds.	27 Jan 1967	23:30:30
Additional electrocardiogram indications from senior pilot. Data show increased activity but were not indicative of alarm type of response. More intense crew activity sensed by guidance and navigation system.	27 Jan 1967	23:30:39
Crew movement ended.	27 Jan 1967	23:30:44
All of senior pilot's biomedical parameters reverted to 'rest' level.	27 Jan 1967	23:30:45
Variation in signal output from gas chromatograph.	27 Jan 1967	23:30:50
Significant voltage transient recorded.	27 Jan 1967	23:30:54.8
Command pilot microphone noises ended.	27 Jan 1967	23:30:58.6

Apollo 1 fire event – *continued*	Date (GMT)	Time (h:m:s)
Oxygen flow rate reached limit of sensor.	27 Jan 1967	23:30:59
Additional spacecraft movement noted.	27 Jan 1967	23:31:00
First voice transmission ended.	27 Jan 1967	23:31:10
Fire broke from its point of origin. Evidence suggests a wall of flames extended along left wall of module, preventing command pilot, occupying left couch, from reaching valve that would vent command module to outside atmosphere. Original flames rose vertically and spread out across cabin ceiling. Scattering of firebrands of molten burning nylon contributed to spread of flames. It was estimated that opening valve would have delayed command module rupture by less than 1 sec.	27 Jan 1967	23:31:12
First verbal indication of fire reported by crew.	27 Jan 1967	23:31:04.7
Cabin pressure exceeded range of transducers, 17 psia for cabin and 21 psia for battery compartment transducers. Rupture and resulting jet of hot gases caused extensive damage to exterior.	27 Jan 1967	23:31:16
Beginning of final voice transmission from crew. Entire transmission garbled. Sounded like, "They're fighting a bad fire – let's get out. Open 'er up." Or, "We've got a bad fire – let's get out. We're burning up." Or, "I'm reporting a bad fire. I'm getting out." Transmission ended with cry of pain, perhaps from pilot.	27 Jan 1967	23:31:16.8
Command module ruptured, start of second stage of fire. First stage marked by rapid temperature rise and increase in cabin pressure. Flames had moved rapidly from point of ignition, traveling along net debris traps installed to prevent items from dropping into equipment areas. At same time, Velcro strips positioned near ignition point also burned.	27 Jan 1967	23:31:19.4
End of final voice transmission.	27 Jan 1967	23:31:21.8
All spacecraft transmissions ended. Television monitors showed flames spreading from left to right side of command module and shortly covered entire visible area. Telemetry loss made determination of precise times of subsequent occurrences impossible.	27 Jan 1967	23:31:22.4
Third stage of fire characterized by greatest conflagration due to forced convection from outrush of gases through rupture in pressure vessel. Swirling flow scattered firebrands, spreading fire. Pressure in command module dropped to atmospheric pressure 5 or 6 sec after rupture.	27 Jan 1967	23:31:25
Command module atmosphere reached lethal stage, characterized by rapid production of high concentrations of carbon monoxide. Following loss of pressure, and with fire throughout crew compartment, remaining atmosphere quickly became deficient in oxygen and could not support continued combustion. Heavy smoke formed and large amounts of soot deposited on most spacecraft interior surfaces. Although oxygen leak extinguished most of fire, failed oxygen and water/glycol lines supplied oxygen and fuel to support localized fire that melted aft bulkhead and burned adjacent portions of inner surface of command module heat shield.	27 Jan 1967	23:31:30
Fire apparatus and firefighting personnel dispatched.	27 Jan 1967	23:32
Attempts to remove hatches.	27 Jan 1967	23:32:04
Pad leader reported that attempts had started to remove hatches.	27 Jan 1967	23:32:34

Apollo 1 fire event – *continued*	Date (GMT)	Time (h:m:s)
Hatches opened, outer hatches removed. Resuscitation of crew impossible.	27 Jan 1967	23:36
Pad leader ascertained all hatches open, left White Room, proceeded a few feet along swing arm, donned headset and reported this fact.	27 Jan 1967	23:36:31
Firefighters arrived at Level A-8. Positions of crew couches and crew could be perceived through smoke but only with great difficulty. Unsuccessful attempt to remove senior pilot from command module.	27 Jan 1967	23:40
Doctors arrived.	27 Jan 1967	23:43
Photographs taken, and removal efforts started.	28 Jan 1967	00:30
Removal of crew completed, about 7.5 hours after accident.	28 Jan 1967	07:00
CM-014 shipped to KSC to develop disassembly techniques for selected components prior to their removal from CM-012.	01 Feb 1967	–
Disassembly plan fully operational.	07 Feb 1967	–
CM-012 moved to pyrotechnics installation building at KSC, where better working conditions were available.	17 Feb 1967	–
Disassembly of command module completed.	27 Mar 1967	–

THE CAUSE OF THE FIRE

Although the Board was not able to determine conclusively the specific initiator of the Apollo 204 fire, it identified six of the conditions that led to the disaster.

1. A sealed cabin, pressurized with an oxygen atmosphere.
2. An extensive distribution of combustible materials in the cabin.
3. Vulnerable wiring carrying spacecraft power.
4. Vulnerable plumbing carrying a combustible and corrosive coolant.
5. Inadequate provisions for the crew to escape.
6. Inadequate provisions for rescue or medical assistance.

Having identified these conditions, the Board addressed the question of how such conditions came to exist. Careful consideration of this question led the Board to the conclusion that in its devotion to the many difficult problems of space travel, the Apollo team failed to give adequate attention to certain mundane but equally vital questions of crew safety. The Board's investigation revealed many deficiencies in design and engineering, manufacture, and quality control.

As a result of the investigation, major modifications in design, materials, and procedures were implemented. The two-piece hatch was replaced by a single quick-operating, outward-opening crew hatch made of aluminum and fiberglass. The new hatch could be opened from inside in 7 sec and by a pad safety crew in 10 sec. Ease of opening was enhanced by a gas-powered counterbalance mechanism. The second major modification was the change in the launch pad spacecraft cabin atmosphere for pre-launch testing from 100% oxygen to a mixture of 60% oxygen and 40% nitrogen to reduce support of any combustion. The crew suit loops still carried 100% oxygen. After launch, the 60/40 mix was gradually replaced with pure oxygen until cabin atmosphere reached 100% oxygen at 5 psi. This 'enriched air' mix was selected after extensive flammability tests in various percentages of oxygen at varying pressures. Other changes included: substituting stainless steel for aluminum in high-pressure oxygen tubing; armor-plated water/glycol liquid line solder joints; protective covers over wiring bundles; stowage boxes built of aluminum; replacement of materials to minimize flammability; installation of fireproof storage containers for flammable materials; mechanical fasteners substituted for gripper cloth patches; flameproof coating on wire connections; replacement of plastic switches with metal ones; installation of an emergency oxygen system to isolate the crew from toxic fumes; and the inclusion of a portable fire extinguisher and fire-isolating panels in the cabin. Safety changes were also made at Launch Complex 34. These included structural changes to the White Room for the new quick-opening spacecraft hatch; improved firefighting equipment; emergency egress routes; emergency access to the spacecraft; purging of all electrical equipment in the White Room with nitrogen; installation of a hand-held water hose and a large exhaust fan in the White Room to draw out smoke and fumes; fire-resistant paint; relocation of certain structural members to provide easier access to the spacecraft and faster egress; addition of a water spray system to cool the launch escape system (the solid propellants could be ignited by extreme heat); and the installation of additional water spray systems along the egress route from the spacecraft to ground level.

Gus Grissom (left), Ed White and Roger Chaffee, the Apollo 1 crew, at Pad 34.

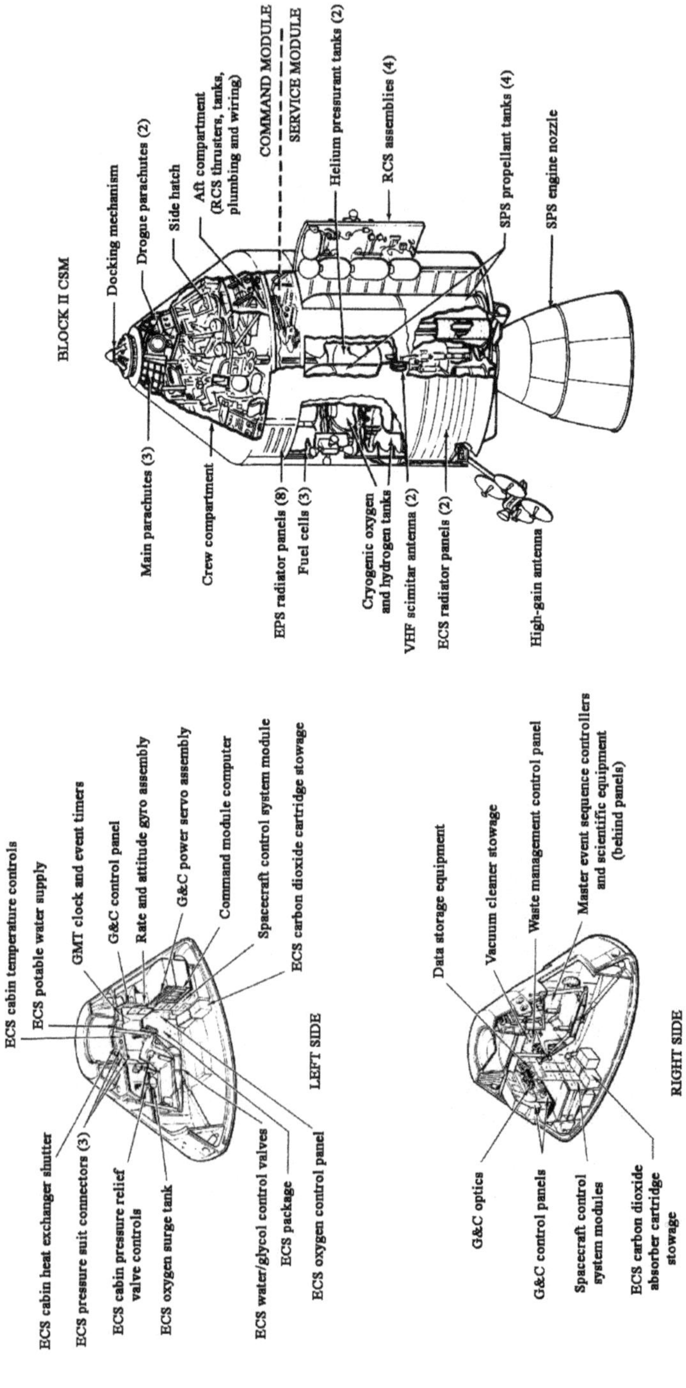

The structure of the Apollo spacecraft

Spacecraft 012 arrives at Pad 34.

The exterior of the fire-damaged Apollo 1 command module in which Grissom, White and Chaffee died (top left); a view through the hatch; the crew positions, with the hatch above the center couch; the vicinity of the environmental control unit, where the ignition source is believed to have been; and its disassembled outer structures. John Glenn, Gordon Cooper and John Young escort Grissom's coffin.

Apollo 4

The first test of the Saturn V

9 November 1967

BACKGROUND

As the first of a planned series of 'A' missions to 'man rate' the Saturn V launch vehicle, Apollo 4 (AS-501) was to be an ambitious 'all up' test in which all three stages (the S-IC, S-II and S-IVB) were 'live'. The payload was a lunar module test article (LTA-10R) and CSM-017, a Block I command/service module with some Block II modifications for certification, including the umbilical running around the rim of the heat shield from the command module to the service module, and a heat shield with a simulated unified crew hatch to be tested at lunar return conditions. In view of the fact that the mission was to be unmanned, certain systems had been deleted from the command module in order to accommodate an electromechanical control sequencer.

Apollo 4 lifted off from Pad A of Launch Complex 39 at the Kennedy Space Center at a Range Zero time of 12:00:01 GMT (07:00:01 EST) on Thursday, 9 November 1967, and the command module was recovered some 8.5 hours later.

LAUNCH PREPARATIONS

The launch countdown for AS-501 was divided into two segments: the 'pre-count' starting with the clock at T–104 hr and running to T–49 hr, followed by the 'terminal count'. The pre-count was started at 12:00 EST on Saturday, 4 November 1967. The pre-count proceeded smoothly, with only two problems on 5 November, neither of which was interruptive: a scratched seal on S-IC helium control valve number 4 needed replacement, and several fluid leaks occurred when the hydraulics of the swing arms of the launch umbilical tower were pressurized. On 6 November a series of early morning alarm reports resulted in the initiation of an emergency evacuation of the tower, but this was cancelled when it was confirmed that the alarms had been accidentally triggered. Later in the day an erratic liquid oxygen pump inlet transducer in the S-IVB had to be replaced, a liquid oxygen pressurization regulator had to be replaced in the S-II, and there were problems involving leaks in pneumatically operated disconnects in the spacecraft. The count resumed at T–49 hr at 22:30 on 6 November, and ran smoothly through to 22:10 on 7 November when the spacecraft reported a potential problem with heat loss in the fuel cell cryogenic hydrogen tank. At 05:32 on 8 November the decision was made to continue the count without reloading the hydrogen tank. Two holds had been scheduled, the first of 6 hours' duration at T–6.5 hr and the second of 1.5 hours at T–4 hr. At 12:31 on 8 November an unscheduled hold was called at T–11 hr. This consumed 1 hr 59 min of the planned 6-hr hold. It was primarily to catch up with the clock after a number of minor problems with the launch vehicle were experienced. One issue involved the pressure of the liquid oxygen in the spacecraft, but analysis indicated a high probability that the pressure would be nominal at liftoff, and would therefore not constrain the count. A second 2-hr unscheduled hold was called at 17:00 on 8 November, at T–8.5 hr, to verify the range safety command receivers. The count was resumed at 19:00. At 21:00 on 8 November the first scheduled hold point was attained, and the clock was held for the remaining 2 hr 1 min of the 6-hr period. The count resumed at 23:01 and ran smoothly to the second scheduled hold point at T–4 hr at 01:30 on 9 November. After picking up at 03:00, the count progressed smoothly to launch at 07:00:01, which was 1 sec later than scheduled. There was 4/10th coverage of stratocumulus cloud with a 4,500-ft base. Visibility was 10 nmi. In view of the unprecedented size of the Saturn V vehicle, the launch site had been cleared of personnel to a radius of 3 nmi.

Apollo 4 preparation event	Date
S-IVB stage arrived at KSC.	14 Aug 1966
S-IVB moved to VAB low bay.	15 Aug 1966
IU arrived.	25 Aug 1966
SLA arrived.	09 Sep 1966
S-IC stage arrived at KSC by barge, and was offloaded and moved into the VAB transfer aisle.	12 Sep 1966
S-IC stage erected on mobile launch platform 1 in high bay 1.	27 Oct 1966
S-II stage 'spool' spacer erected (to stand in for the absent S-II).	31 Oct 1966
S-IVB stage erected.	01 Nov 1966
IU erected.	02 Nov 1966
Initial power applied to S-IC stage.	07 Nov 1966
Initial IU bus power applied.	16 Nov 1966
Launch vehicle electrical mate completed.	21 Nov 1966
Facilities verification model erected (to stand in for absent CSM-017).	28 Nov 1966
Launch vehicle emergency detection system test completed.	12 Dec 1966
Sequential malfunction test completed.	16 Dec 1966
Guidance and control system checks completed.	20 Dec 1966
SM arrived.	21 Dec 1966
CM arrived.	24 Dec 1966
Apollo spacecraft CSM-017 erected.	12 Jan 1967
S-II stage arrived at KSC by barge, and was offloaded and transported to the VAB low bay for checkout.	21 Jan 1967
Spacecraft systems integrated tests with launch vehicle simulator started.	23 Jan 1967
Launch vehicle overall test #1 completed.	24 Jan 1967
Spacecraft de-erected and transported to the Manned Space Operations Building for testing.	13 Feb 1967
IU de-erected.	13 Feb 1967

Apollo 4 preparation event – *continued*	Date
S-IVB de-erected.	14 Feb 1967
S-IVB modifications started.	15 Feb 1967
S-II stage spacer de-erected.	15 Feb 1967
S-II stage erected.	23 Feb 1967
IU and S-IVB staged erected.	24 Feb 1967
Launch vehicle electrical mate completed.	01 Mar 1967
Power transfer test completed.	17 Mar 1967
Launch vehicle electrical support equipment modifications started.	22 Mar 1967
Spacecraft facility verification vehicle erected.	06 Apr 1967
Launch vehicle electrical support equipment modification verification completed.	08 Apr 1967
Launch vehicle overall test #2 completed.	14 Apr 1967
Swing arm compatibility test performed.	24 May 1967
S-II stage LOX tank inspection for presence of structural flaws.	25 May 1967
De-erect spacecraft facility verification vehicle.	26 May 1967
De-erect IU.	27 May 1967
De-erect S-IVB stage.	27 May 1967
S-II liquid oxygen tank dye penetrant inspection.	28 May 1967
S-II stage de-erected.	03 Jun 1967
S-II stage liquid hydrogen tank inspection started.	05 Jun 1967
Spacecraft cabin leak check test accomplished.	12 Jun 1967
S-II stage liquid hydrogen tank inspection completed.	16 Jun 1967
S-II stage erected.	18 Jun 1967
S-IVB stage erected.	19 Jun 1967
IU erected.	19 Jun 1967
Spacecraft erected.	20 Jun 1967
Launch vehicle electrical mate accomplished.	23 Jun 1967
Launch vehicle overall test #2 completed.	14 Jul 1967
Space vehicle electrical mate and emergency detection system test accomplished.	24 Jul 1967
Space vehicle overall test #1 plugs-in, completed.	01 Aug 1967
Space vehicle overall test #2 plugs-out, accomplished.	06 Aug 1967
Spacecraft ordnance installed.	07 Aug 1967
Space vehicle simulated flight test completed.	18 Aug 1967
S-II stage liquid hydrogen insulation modifications completed.	24 Aug 1967
Space vehicle transferred to launch complex 39A.	26 Aug 1967
Spacecraft ground support equipment mobile service structure/mobile launcher interface tests completed.	07 Sep 1967
Liquid oxygen and liquid hydrogen cold flow tests completed.	20 Sep 1967
RP-1 loading of the S-IC stage completed in preparation for the start of the countdown demonstration test (CDDT).	27 Sep 1967
CDDT started with the completion of the pre-count section and continued through 14 October as a result of numerous problems encountered. During this portion of the test it became necessary to initiate the changeout of the fuel cells in the Apollo spacecraft.	29 Sep 1967
Terminal count portion of the CDDT procedure was completed.	14 Oct 1967
Spacecraft fuel cell change out completed.	19 Oct 1967
Inspection of S-II stage liquid oxygen tank anti-vortex baffle completed.	22 Oct 1967
Space vehicle flight readiness test completed.	26 Oct 1967

ASCENT PHASE

During the first 10 sec, the launch vehicle performed a yaw maneuver to 'side step' away from the launch umbilical tower, to preclude colliding with any swing arm that might be tardy in rotating clear. It was launched on a pad azimuth of 90°E of N, but as soon as it was clear of the launch umbilical tower it initiated a roll to align its inertial guidance system with the flight azimuth of 72°E of N, followed by a pitch maneuver. It achieved Mach 1 at T+61.4, and at T+78.4, at an altitude of 37,700 ft and a wind speed of 50 kt, passed through maximum dynamic pressure. The center engine of the S-IC was shut down by a timer at T+135.52, and the outer engines were cut off by liquid oxygen depletion at T+150.769. The separation proceeded in two phases. Firstly, at T+151.43, the S-IC separated from the annular interstage, or 'skirt', that extended down over the engines of the S-II; the S-II ignited at T+152.12; then at T+181.44 the interstage was jettisoned. The sequence was documented by film cameras that were later ejected and recovered. The 'dual plane' separation was designed to ensure that the shed S-IC could not damage the engines of the S-II stage. The telemetry signal from the S-IC ceased at T+410, at about the time it was seen by visual and radar tracking to lose structural integrity and burn up as it re-entered the atmosphere. A small solid rocket motor on the launch escape system fired at T+187.13 to draw the tower clear of the spacecraft. The S-II shut down at T+519.759, and was jettisoned at T+520.53; its re-entry was not observed. As the S-IVB moved off, two solid rocket motors on its periphery (one fewer than on the Saturn IB version of the stage) fired for ullage to settle its propellants prior to igniting the main engine, and were then jettisoned. The vehicle achieved orbit at T+675.6. The command-destruct receivers on the S-IVB were safed by the Bermuda range safety officer shortly thereafter.

The launch phase of the flight was nominal, with all planned events occurring within allowable limits. Through the major portion of the first powered phase, the altitude was greater than nominal and the surface range was less than nominal. The total inertial acceleration was greater than nominal for the S-IC phase, and less for the S-II and S-IVB first burn phases. The total burn time of the S-IC, S-II and S-IVB first burn was 9.6 sec longer than nominal. The S-IC burned 1.1 sec longer than nominal, the S-II burned 4.6 sec longer than nominal, and the S-IVB first burn was 6.1 sec longer than nominal. Nevertheless, the space-fixed velocity at S-IVB first burn cutoff was only 4.07 ft/sec lower than nominal. The longer burn time explained the 23.85 nmi greater surface range at S-IVB first burn cutoff. The altitude of S-IVB first burn cutoff was 0.63 nmi greater than nominal.

The ground support equipment was in good shape, apart from the fact that the engine service platform was extensively damaged and the LUT level platform was completely destroyed.

The COSPAR designation of the spacecraft was 1967-113A and the S-IVB was 1967-113B.

Apollo 4 ascent phase event	GET (h:m:s)	Altitude (nmi)	Range (nmi)	Earth-fixed velocity (ft/sec)	Space-fixed velocity (ft/sec)	Event duration (sec)	Geocentric latitude (°N)	Longitude (°E)	Space-fixed flight path angle (deg)	Space-fixed heading angle (°E of N)
Liftoff	000:00:00.26	0.0	0.0	–	–	–	–	–	–	–
Mach 1 achieved	000:01:01.4	3.97	–	–	–	–	–	–	–	–
Maximum dynamic pressure	000:01:18.4	7.16	–	–	–	–	–	–	–	–
S-IC center engine cutoff	000:02:15.52	26.80	29.31	–	7,241.47	–	28.75	–80.07	23.275	75.952
S-IC outboard engine cutoff	000:02:30.769	34.40	44.62	–	8,831.49	–	28.83	–79.83	20.955	75.293
S-IC/S-II separation	000:02:31.43	34.75	45.36	7,694.62	8,860.60	–	–	–	20.855	75.287
S-II engine cutoff	000:08:39.759	103.86	797.86	–	22,355.61	367.6	31.72	–65.67	0.642	81.485
S-II/S-IVB separation	000:08:40.53	103.89	800.15	21,060.40	22,363.98	–	–	–	0.632	81.510
S-IVB 1st burn cutoff	000:11:05.64	104.00	1,321.95	–	25,556.96	144.9	32.64	–55.53	0.015	87.210
Earth orbit insertion	000:11:15.6	103.96	–	–	25,562.7	–	32.67	–54.67	0.014	87.65

EARTH ORBIT PHASE

The S-IVB stage inserted itself and its payload into a near-circular 'parking orbit' at 100 nmi. The continuous vent system of the S-IVB maintained ullage pressure in the propellant tanks during the coast phase, until being terminated shortly before the engine was to be restarted. At 003:11:26.6, after approximately two revolutions with its longitudinal axis in the orbital plane and parallel to the local horizon, the S-IVB was reignited for a simulated translunar injection burn lasting almost 5 min. This produced an elliptical atmosphere-intersecting 'waiting orbit' with an apogee of 9,292 nmi. At 003:26:28.2 the spacecraft separated from the S-IVB, and at 003:28:06.6 the SPS was ignited for a brief burn which further increased the apogee to 9,769 nmi (9,890 nmi had been planned). This demonstrated the ability of the engine to ignite in the zero-*g* environment without an ullage impulse to settle its propellants. The spacecraft was then aligned to a specific attitude to achieve a thermal gradient across the command module heat shield. This attitude, with the command module hatch window facing toward the Sun and the conical surface of the crew compartment perpendicular to the Sun's rays, was maintained for approximately 4.5 hours in order to induce circumferential thermal stresses and distortions on the command module and its ablator prior to entry. Block II thermal control coating degradation did occur, as indicated by measured data exceeding the nominal coating equilibrium temperatures; this was attributed to a deposit of exhaust material from the solid rocket that had jettisoned the launch escape tower. At 005:46:49.5 the spacecraft made its high apogee. During the 'cold-soak' coast, an automated 70-mm still camera in the spacecraft shot a picture of the Earth once every 10.6 sec, and produced a total of 715 good-quality high-resolution photographs. At 008:10:54.8 the SPS was reignited to accelerate the spacecraft to entry conditions that represented the most severe operational conditions which could possibly result on a lunar return trajectory; an inertial velocity of 34,816 ft/sec was intended but a slight over-burn due to the maneuver being controlled from the ground produced 35,115 ft/sec. Some 2 min 27 sec after SPS shut down, the service module was jettisoned and the command module used its own thrusters to adopt the entry attitude.

Apollo 4 earth orbit phase event	GET (h:m:s)	Space-fixed velocity (ft/sec)	Event duration (sec)	Velocity change (ft/sec)	Apogee (nmi)	Perigee (nmi)	Period (min)	Inclination (deg)
Earth orbit insertion	000:11:15.6	25,563	–	–	101.10	99.14	88.20	32.573
S-IVB 2nd burn ignition	003:11:26.57	25,547	–	–	–	–	–	–
S-IVB 2nd burn cutoff	003:16:26.27	30,882	299.7	–	9,292	–44	303.1	30.31
S-IVB/CSM separation	003:26:28.24	26,233	–	–	–	–	–	–
SPS 1st burn ignition	003:28:06.6	25,504	–	–	–	–	–	–
SPS 1st burn cutoff	003:28:22.6	25,547	16.0	212.50	9,769	–45	316.6	30.31
SPS 2nd burn ignition	008:10:54.8	28,173	–	–	–	–	–	–
SPS 2nd burn cutoff	008:15:35.4	35,115	280.6	4,824.3	–	–	–	–
CS/SM separation	008:18:02.6	36,138	–	–	–	–	–	–
Entry interface	008:19:28.5	36,545	–	–	–	–	–	–

RECOVERY

The entry interface (by definition an altitude 400,000 ft) occurred at 008:29:28.5 while traveling at an inertial velocity of 36,639 ft/sec and a flight path angle of –6.93 deg. As a result of the longer than planned second SPS burn the entry interface conditions were 210 ft/sec greater and 0.20 deg shallower than predicted but still within the desired 'entry corridor'. Due to the change in the entry conditions, the peak load factor of 7.27 *g* was less than the predicted 8.3 *g*. This did not affect the performance of the guidance system in achieving the target, and the entry was well within predicted conditions. The lift-to-drag ratio was 0.365 ($\pm$0.015) compared with the predicted 0.350; a value in the range 0.322 to 0.415 was acceptable. The temperature inside the cabin increased by only 10°F during entry. The drogue parachutes were deployed at 008:31:18.6, and the three main chutes at 008:32:05.8. The command module was sighted by the recovery ship, USS *Bennington*, in the primary recovery zone northwest of Hawaii at 172.52°W and 30.10°N, approximately 6 to 8 nmi from the vessel, and splashed into an 8-ft swell at 008:37:09.2, 10 nmi from the aim point (as determined by post-flight reconstruction of the entry data). Within 20 min, swimmers dropped by helicopters affixed the flotation collar. The apex cover (which had covered the stowed parachutes) was recovered 2.5 hr after splashdown, together with one of the main chutes, marking the first time that an Apollo parachute had been recovered for inspection. The recovery operation lasted about 2 hr in total.

The spacecraft was shipped to Hawaii, where it was deactivated, then flown to North American Aviation in Downey, California, for post-flight testing.

CONCLUSIONS

The principal objectives were to demonstrate the structural and thermal integrity of the launch vehicle and spacecraft, and to verify the adequacy of the Block II heat shield for entry at lunar return conditions. These objectives were satisfactorily accomplished.

The spacecraft subsystems operated satisfactorily in all respects. Performance of the guidance and control subsystems was equal to or better than predictions. All sequencing and computational operations performed by the guidance computer were correct. The environmental control subsystem maintained the cabin pressure and temperature within required limits. The thermal protection subsystem survived the lunar entry environment satisfactorily. The erosion of the heat shield was between 2.5 and 7.6 mm.

MISSION OBJECTIVES

Launch Vehicle Primary Detailed Objectives

1. Demonstrate structural and thermal integrity of launch vehicle throughout powered and coasting flight, and determine in-flight structural loads and dynamic characteristics. *Achieved.*
2. Determine in-flight launch vehicle internal environment. *Achieved.*

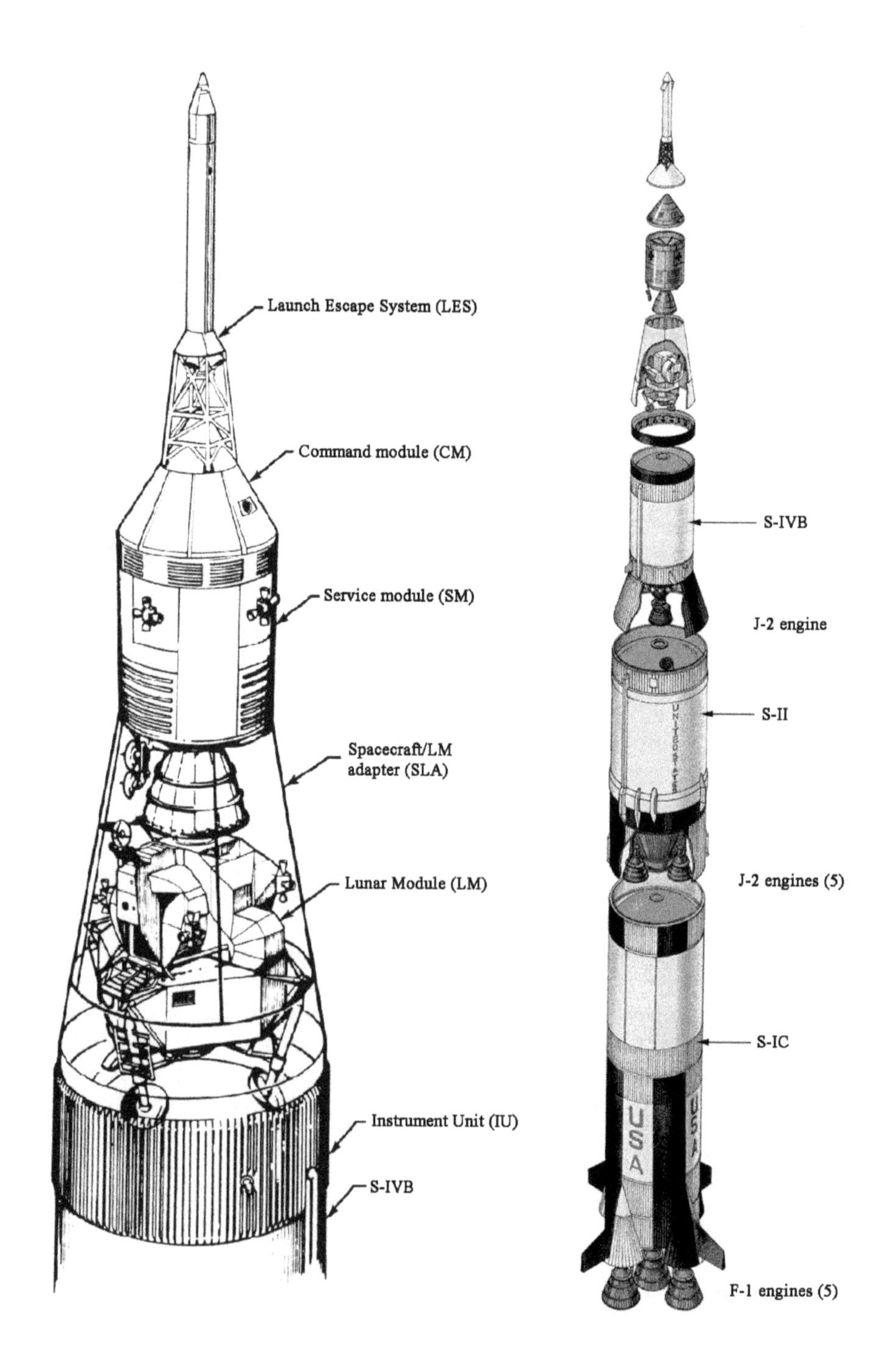

The major structural components of the Apollo–Saturn V space vehicle.

The S-IVB for AS-501 is added to the S-IC/S-II stack in a high bay of the VAB.

With the Instrument Unit in place on the S-IVB, technicians work on its systems.

CSM-017 and LTA-10R (inside the adapter) are added.

The AS-501 space vehicle awaits only the launch escape system.

The first Saturn V lifts off at the start of a perfect flight.

CM-017 awaits recovery.

3. Verify pre-launch and launch support equipment compatibility with launch vehicle and spacecraft systems. *Achieved.*
4. Demonstrate the S-II stage propulsion system and determine in-flight system performance parameters. *Achieved.*
5. Demonstrate the S-II stage propulsion system including programmed mixture ratio shift, propellant management systems, and determine in-flight system performance parameters. *Achieved.*
6. Demonstrate the S-IVB stage propulsion system including the propellant management systems, and determine in-flight system performance parameters. *Achieved.*
7. Demonstrate launch vehicle guidance and control system during S-IC, S-II, and S-IVB powered flight. Achieve guidance cutoff and evaluate system accuracy. *Achieved.*
8. Demonstrate S-IC/S-II dual plane separation. *Achieved.*
9. Demonstrate S-II/S-IVB separation. *Achieved.*
10. Demonstrate launch vehicle sequencing system. *Achieved.*
11. Demonstrate compatibility of the launch vehicle and spacecraft. *Achieved.*
12. Evaluate performance of the emergency detection system in an open-loop configuration, with the automatic abort circuit deactivated in the spacecraft. *Achieved.*
13. Demonstrate the capability of the S-IVB auxiliary propulsion system during S-IVB powered flight and orbital coast periods to maintain attitude control and perform required maneuvers. *Achieved.*
14. Demonstrate the adequacy of the S-IVB continuous vent system while in Earth orbit. *Achieved.*
15. Demonstrate the S-IVB stage restart capability. *Achieved.*
16. Demonstrate the mission support capability required for launch and mission operations to high post-injection altitudes. *Achieved.*

Launch Vehicle Secondary Detailed Test Objectives

1. Determine launch vehicle powered flight external environment. *Achieved.*
2. Determine attenuation effects of exhaust flames on radio-frequency radiating and receiving systems during main engine, retro, and ullage motor firings. *Achieved.*

Spacecraft Primary Objectives

1. Demonstrate the structural and thermal integrity and compatibility of the launch vehicle and spacecraft. Confirm launch loads and dynamic characteristics. *Achieved.*
2. Verify operation of the following subsystems: command module heat shield (adequacy of Block II heat shield design for re-entry at lunar return conditions), service propulsion subsystem (including no-ullage start), and selected subsystems. *Achieved.*
3. Evaluate the performance of the space vehicle emergency detection subsystem in open-loop configuration. *Achieved.*
4. Demonstrate mission support facilities and operations required for launch, mission conduct and command module recovery. *Achieved.*

Spacecraft Mandatory Detailed Test Objectives

1. M1.1: Demonstrate CSM/SLA/LTA/Saturn V structural compatibility and determine spacecraft loads in a Saturn V launch environment. *Achieved.*
2. M1.2: Determine the dynamic and thermal responses of the SLA/CSM structure in the Saturn V launch environment. *Achieved.*
3. M1.4: Determine the force inputs to the simulated LM from the SLA at the spacecraft attachment structure in a Saturn V launch environment. *Only qualitative data were obtained.*
4. M1.5: Obtain data on the acoustic and thermal environment of the SLA/simulated LM interface during a Saturn V launch. *Achieved.*
5. M1.7: Determine vibration response of the LM descent stage engine and propellant tanks in a Saturn V launch environment. *Achieved.*
6. M3.1: Evaluate the thermal and structural performance of the Block II thermal protection system, including effects of cold soak and maximum thermal gradient when subjected to the combination of a high heat load and a high heating rate representative of lunar return entry. *Achieved.*
7. M3.2: Demonstrate an SPS no-ullage restart. *Achieved.*
8. P3.3: Determine performance of the SPS during a long-duration burn. *Achieved.*
9. M3.5: Verify the performance of the SM-RCS thermal control subsystems and engine thermal response in the deep space environment. *Achieved.*
10. M3.6: Verify the thermal design adequacy of the CM-RCS thrusters and extensions during simulated lunar return entry. *Achieved.*
11. M3.8: Evaluate the thermal performance of a gap-and-seal configuration simulating the unified crew hatch for heating conditions anticipated during lunar return entry. *Achieved.*
12. M3.9: Verify operation of the heat rejection system throughout the mission. *Achieved.*
13. M4: Evaluate the performance of the spacecraft emergency detection subsystem in the open-loop configuration. *Achieved.*

Spacecraft Primary Detailed Test Objectives

1. P5.2b: Demonstrate the performance of S-band communications between the spacecraft and MSFN. *Achieved.*
2. P5.6: Measure the integrated skin and depth radiation dose within the command module up to an altitude of at least 2,000 nmi. *Achieved.*

Spacecraft Secondary Detailed Test Objectives

1. S1: Determine the radiation shielding effectiveness of the CM. *Achieved.*
2. S2: Demonstrate satisfactory operation of CSM communication subsystems using the Block II type VHF omnidirectional antennas. *Achieved.*
3. S3.1c: Verify operation of the G&N system after subjection to the Saturn V launch environment. *Achieved.*
4. S3.1d1: Verify operation of the EPS after subjection to the Saturn V launch environment. *Achieved.*
5. S3.1d2: Verify operation of PGS after subjection to the Saturn V launch environment. *Achieved.*

6. S3.2a: Verify operation of the G&N in the space environment after S-IVB separation. *Achieved.*
7. S3.2d1: Verify operation of the EPS in the space environment after S-IVB separation. *Achieved.*
8. S3.2d2: Verify operation of the PGS in the space environment after S-IVB separation. *Achieved.*
9. S3.3a: Verify operation of the CM-RCS during entry and recovery. *Achieved.*
10. S3.3c: Verify operation of the G&N/SCS during entry and recovery. *Achieved.*
11. S3.3d: Verify operation of the EPS during entry and recovery. *Achieved.*
12. S3.3e: Verify operation of the ELS during entry and recovery. *Achieved.*
13. S5: Obtain data via CSM/ARIA communications. *Achieved.*
14. S6: Gather data on the effects of a long-duration SPS burn on spacecraft stability. *Achieved.*
15. S7: Obtain data on the temperature of the simulated LM skin during launch. *Achieved.*

MISSION TIMELINE

Apollo 4 mission event	GET (h:m:s)	Date (GMT)	Time (h:m:s)
Space vehicle hypergolic loading completed in preparation for the start of the launch countdown.	–	03 Nov 1967	–
S-IC stage RP-1 loading accomplished in preparation for the start of the launch countdown.	–	04 Nov 1967	–
Start launch vehicle pre-count at T–104 hr.	–104:00:00	04 Nov 1967	17:00:00
Start launch vehicle terminal count at T–49 hr.	–049:00:00	07 Nov 1967	03:30:00
Start unscheduled hold at T–11 hr.	–011:00:00	08 Nov 1967	17:31:00
Resume count.	–011:00:00	08 Nov 1967	19:30:00
Start unscheduled hold at T–8.5 hr.	–008:30:00	08 Nov 1967	22:00:00
Resume count.	–008:30:00	09 Nov 1967	00:00:00
Start scheduled hold at T–6.5 hr.	–006:30:00	09 Nov 1967	02:00:00
Resume count.	–006:30:00	09 Nov 1967	04:01:00
Start scheduled hold at T–4 hr.	–004:00:00	09 Nov 1967	06:30:00
Resume count.	–004:00:00	09 Nov 1967	08:00:00
Guidance reference release.	–000:00:17.638	09 Nov 1967	11:59:43
S-IC engine start sequence command.	–000:00:09	09 Nov 1967	11:59:52
Launch commit.	–000:00:00.648	09 Nov 1967	12:00:00
First motion (established by ground cameras).	–000:00:00.48	09 Nov 1967	12:00:00
Range zero.	000:00:00.000	09 Nov 1967	12:00:01
Liftoff.	000:00:00.263	09 Nov 1967	12:00:01
Begin yaw maneuver.	000:00:01.263	09 Nov 1967	12:00:02
Complete yaw maneuver.	000:00:10.16	09 Nov 1967	12:00:11
Start pitch (tilt) maneuver.	000:00:11.06	09 Nov 1967	12:00:12
Start roll maneuver.	000:00:11.06	09 Nov 1967	12:00:12

Apollo 4 mission event – *continued*	GET (h:m:s)	Date (GMT)	Time (h:m:s)
Complete roll maneuver.	000:00:31.99	09 Nov 1967	12:00:32
Mach 1.	000:01:01.4	09 Nov 1967	12:01:02
Maximum dynamic pressure.	000:01:18.4	09 Nov 1967	12:01:19
S-IC center engine cutoff.	000:02:15.52	09 Nov 1967	12:02:16
Tilt arrest.	000:02:25.07	09 Nov 1967	12:02:26
S-IC outboard engine cutoff.	000:02:30.769	09 Nov 1967	12:02:31
S-IC/S-II separation.	000:02:31.43	09 Nov 1967	12:02:32
S-II ignition command.	000:02:32.12	09 Nov 1967	12:02:33
S-IC interstage separation.	000:03:01.44	09 Nov 1967	12:03:02
Jettison launch escape system.	000:03:07.13	09 Nov 1967	12:03:08
Initiate iterative guidance mode.	000:03:10.88	09 Nov 1967	12:03:11
S-II engine mixture ratio shift and initiate second phase iterative guidance mode.	000:07:15.69	09 Nov 1967	12:07:16
S-II cutoff.	000:08:39.759	09 Nov 1967	12:08:40
S-II/S-IVB separation.	000:08:40.53	09 Nov 1967	12:08:41
S-IVB 1st ignition command.	000:08:40.72	09 Nov 1967	12:08:41
Third phase iterative guidance mode.	000:08:47.65	09 Nov 1967	12:08:48
Jettison S-IVB ullage motor cases.	000:08:52.53	09 Nov 1967	12:08:53
Initiate Chi-tilde steering.	000:10:32.25	09 Nov 1967	12:10:33
S-IVB 1st cutoff.	000:11:05.64	09 Nov 1967	12:11:06
Initiate coast period.	000:11:07.15	09 Nov 1967	12:11:08
Insertion into parking orbit.	000:11:15.6	09 Nov 1967	12:11:16
Initiate continuous venting of S-IVB for ullage.	000:12:04.8	09 Nov 1967	12:12:05
Terminate continuous venting of S-IVB.	003:05:59.576	09 Nov 1967	15:06:00
S-IVB 2nd ignition.	003:11:26.57	09 Nov 1967	15:11:27
Fourth phase iterative guidance.	003:11:39.99	09 Nov 1967	15:11:40
Iterative guidance mode termination and start Chi-tilde steering.	003:15:58.18	09 Nov 1967	15:15:59
S-IVB 2nd cutoff.	003:16:26.27	09 Nov 1967	15:16:27
Initiate coast period.	003:16.27.66	09 Nov 1967	15:16:28
S-IVB/CSM separation command.	003:26:26.47	09 Nov 1967	15:26:27
CSM free.	003:26:28.244	09 Nov 1967	15:26:29
SPS 1st ignition.	003:28:06.6	09 Nov 1967	15:28:07
SPS 1st cutoff.	003:28:22.6	09 Nov 1967	15:28:23
High apogee.	005:46:49.5	09 Nov 1967	17:46:50
SPS 2nd ignition.	008:10:54.8	09 Nov 1967	20:10:55
SPS 2nd cutoff.	008:15:35.4	09 Nov 1967	20:15:36
CS/SM separation.	008:18:02.6	09 Nov 1967	20:18:03
400,000-ft entry interface.	008:19:28.5	09 Nov 1967	20:19:29
Deployment of drogue parachute.	008:31:18.6	09 Nov 1967	20:31:19
Deployment of main parachutes.	008:32:05.8	09 Nov 1967	20:32:06
Splashdown.	008:37:09.2	09 Nov 1967	20:37:10
CM arrives at Downey, CA.	–	15 Nov 1967	–

Apollo 5

Unmanned test of the LM

22 January 1968

BACKGROUND

The primary objective of Apollo 5 was to test the lunar module (LM-1), which was flown without its landing legs. The mission also tested the Instrument Unit in the configuration for the Saturn V. It lifted off from Pad B of Launch Complex 37 at Cape Kennedy, Florida, at a Range Zero time of 22:48:08 GMT (17:48:08 EST) on Monday, 22 January 1968.

LAUNCH PREPARATIONS

When the AS-206 launch vehicle was erected on Pad B of Launch Complex 37 in January 1967 it was with the intention of launching LM-1 in April, but the delivery date of the payload had already slipped several times, and was unknown. On 20 March NASA announced it had decided to launch LM-1 on AS-204, which had been delivered in August 1966, had been erected on Launch Complex 34, and survived unscathed the fire that destroyed the Apollo 1 spacecraft on 27 January. Accordingly, AS-206 was taken down and returned to storage and, by 11 April had been replaced by AS-204R (as the older vehicle had been redesignated). Unable to ship the payload, Grumman built a plywood mockup of LM-1 on the pad for facilities verification. After the arrival of the ascent and descent stages on 23 June 1967, LM-1 was mated on 27 June, but when it was inspected it was found to require considerable work, and it had to be de-mated in August in order to repair leaks in the ascent stage. After the stages were re-mated, another leak developed in September and they were again de-mated to enable several pieces of hardware to be extracted to be returned to Grumman for attention. After the stages were re-mated in October and testing was completed, LM-1 was mechanically mated to its launch vehicle on 19 November and the aerodynamic nose cone installed on the SLA. The space vehicle flight readiness tests ran through to late December. The LM-1 cabin closeout occurred at 20:50 GMT on 18 January 1968, during the countdown demonstration test. Procedural issues hampered loading hypergolic propellants into LM-1, but the test was finished on 19

January. On 21 January the terminal countdown began at T–10 hr 30 min. When a telemetry computer that was listed as mandatory failed at T–3 hr 30 min, it was deemed safe to continue utilizing a command computer to enable commands to be sent during the ascent and to provide telemetry in orbit. A hold was called at T–2 hr 30 min when ground support equipment failed to supply freon to regulate the temperature of a glycol coolant loop. The LM was transferred to its own power at T–42 min and, after several hours of delays, the vehicle was dispatched shortly before sunset.

Apollo 5 preparation event	Date
S-IB stage delivered by barge to KSC.	15 Aug 1966
S-IVB stage delivered.	16 Aug 1966
S-IB erected on launch complex 37B.	7 Apr 1967
Launch vehicle AS-204R complete.	11 Apr 1967
LM-1 ascent/descent stages delivered.	23 Jun 1967
LM-1 stages mated.	27 Jun 1967
LM-1 mechanically mated with launch vehicle.	19 Nov 1967
Space vehicle flight readiness test completed.	Dec 1967
LM-1 closeout.	18 Jan 1968
Countdown demonstration test concluded.	19 Jan 1968
Launch countdown started.	21 Jan 1968
Launch.	22 Jan 1968

ASCENT PHASE

The vehicle lifted off on a pad azimuth of 90°E of N, but at T + 9.2 sec it began to pitch over and started a roll to align its inertial guidance system with the flight azimuth of 72°E of N. It achieved Mach 1 at T + 59.8 and passed through maximum dynamic pressure at T + 71.5. The pitch maneuver was terminated at T + 135.3 to stabilize the vehicle in readiness for staging. The S-IB cut off at T + 144.3, within 0.1 sec of the predicted time, with the altitude 0.4 nmi higher than nominal, the space-fixed velocity 10.8 ft/sec greater, and the range 0.011 nmi shorter. The guidance system shut down the S-IVB at T + 593.3, some 5.1 sec earlier than predicted, when the velocity was 2.3 ft/sec less than nominal, the altitude was 0.12 nmi higher, and the surface range was 16.3 nmi longer. The orbit had a perigee of 88 nmi, an apogee of 120 nmi, a period of 88.39 min and an inclination of 31.63 deg to the equator.

EARTH ORBIT PHASE

The nose cone was jettisoned 35 seconds after orbital insertion, and at 000:19:53.5 the SLA panels were deployed. The LM separated from the S-IVB stage at 000:53:50. After maneuvering clear of the stage by using its attitude control engines, the LM adopted a

Apollo 5 ascent phase event	GET (h:m:s)	Altitude (nmi)	Range (nmi)	Earth-fixed velocity (ft/sec)	Space-fixed velocity (ft/sec)	Event duration (sec)	Geocentric latitude (°N)	Longitude (°E)	Space-fixed flight path angle (deg)	Space-fixed heading angle (°E of N)
Liftoff	–	0.0	0.0	–	–	–	–	–	–	–
Mach 1 achieved	000:00:59.8	4.1	–	–	–	–	–	–	–	–
Maximum dynamic pressure	000:01:11.5	–	–	–	–	–	–	–	–	–
S-IB center engine cutoff	000:02:19.0	32	–	–	7,563	–	28.68	–80.03	28.12	75.78
S-IB outboard engine cutoff	000:02:22.3	34	–	–	7,760	–	28.70	–79.98	27.62	75.71
S-IB/S-IVB separation	000:02:23.5	–	–	–	–	–	–	–	–	–
S-IVB cutoff	000:09:53.3	88			25,659		31.50	–62.57	–0.01	85.50
Earth orbit insertion	000:10:03.3	88	–	–	25,684	–	31.56	–61.81	0.0	85.92

'cold-soak' orientation which its guidance system then maintained with a minimal engine duty cycle. The plan called for two descent propulsion system (DPS) maneuvers, an abort staging, and two ascent propulsion system (APS) maneuvers. The first DPS burn, on the third revolution, was to last 38 sec. It would run at 10% throttle for the first 26 sec, and then conclude at full throttle. The thrust profile of the second maneuver was to be representative of that of a lunar landing, involving five phases over a total of 734 sec. The abort staging sequence would be initiated while the DPS was at full throttle, and was to include DPS shutdown and a 'fire in the hole' 5-sec APS burn. The second APS burn was to last approximately 445 sec and run to propellant depletion, and mark the end of the primary mission.

At 003:59:42, the guidance system initiated the first DPS firing, but shut down the engine after only 4 sec because the buildup of thrust had not satisfied the programmed velocity–time criteria. Sensing that the vehicle was not accelerating as rapidly as expected, the guidance system had aborted the maneuver. This was a planned feature, because on a manned mission this would allow the crew time to analyze the situation and decide whether to restart the engine in order to continue. In normal circumstances the burn would have started with full tank pressurization, and would have built up thrust within 4 sec, but in this case – by design – the burn had been made at less than normal pressure, and with the tanks only partially pressurized it would have taken 6 sec. The premature cutoff was therefore the result of incomplete coordination between the guidance and propulsion systems. A command was sent to deactivate the guidance system to permit the rest of the mission to be controlled from the ground.

After the premature shutdown, a preplanned alternative sequence designed to address the minimum requirements of the mission was chosen, and at 006:10:00 the automatic sequencer initiated the program for the second and third DPS firings, abort staging, and the first APS firing. Attitude rate control was to be maintained by the backup control system. The descent engine gimbaled properly, and responded smoothly to the commands to full throttle. However, the thermal aspects of the supercritical helium pressurization system could not be adequately evaluated due to the short duration of the three DPS firings. During abort staging, all system operations and vehicle dynamics were satisfactory for manned flight. After the first APS burn, the primary control system was reselected to control the spacecraft attitudes and rates. Unfortunately, since the primary system had been passive during the abort staging sequence, its computer program did not reflect the change of mass resulting from this activity. Consequently, computations of control engine firing times were based on the mass of the two-stage vehicle, and resulted in an extremely high propellant usage by the control engines, eventually causing propellant depletion. As a result of excessive control engine activity, the control engine cluster temperatures exceeded the upper limit, but without detrimental effects. Following propellant depletion, the reaction control system was subjected to abnormal operating conditions because of low manifold pressures. Continued operation resulted in three malfunctions, but without appreciable effect on the mission. The second APS firing, initiated by the automatic sequencer at 007:44:13, ran to thrust decay at 007:50:03. With control system propellant depleted, attitude control was maintained by drawing propellants from the APS tanks until the interconnect valves were automatically closed by the sequencer, at which point, with the APS still firing, the vehicle started to tumble. The rates were soon of such a magnitude as to impede the propellants flowing into the engine, and helium ingestion caused thrust decay prior to propellant depletion. Tracking was lost about 2 min after thrust decay, thereby ending the mission. The ascent stage had been in a retrograde orientation during the controlled portion of its maneuver, and trajectory simulations indicated that soon thereafter it entered the atmosphere over the Pacific Ocean, with the point of impact 400 miles west of the coast of Central America. Tracking predicted that the S-IVB would enter after 15.5 hours.

LM-1 maneuvers*	Ignition GET (h:m:s)	Duration (sec)	X-axis velocity change (ft/sec)	Y-axis velocity change (ft/sec)	Z-axis velocity change (ft/sec)	Total velocity change (ft/sec)
1st DPS	003:59:41.7	4.0	3.05	–0.49	–1.96	3.66
Sequence III	006:10:07.4	186.9	–612.56	–400.66	149.45	747.05
Sequence V	007:44:00.3	210.0	1,531.89	–427.55	–359.41	1,630.54

* The interrupted first DPS burn and the sequence III and V burns, including ullage, as determined by the G&N.

Apollo 5 orbit changes	GET (h:m:s)	Space-fixed velocity (ft/sec)	Apogee (nmi)	Perigee (nmi)	Period (min)	Inclination (deg)
Earth orbit insertion	000:10:03.3	25,684	120	88	88.39	31.63
LM/S-IVB separation	000:53:55.2	25,458	121	90	88.43	31.63
After 1st DPS burn	003:59:45.7	25,490	120	92	88.47	31.64
After sequence III	006:13.14.3	26,319	519	93	96.07	31.48
After sequence V*	007:50:03	–	–	–	–	–

* Note that the ascent stage re-entered the atmosphere shortly after the final burn.

The COSPAR designation of the LM ascent stage was 1968-007A, its descent stage was 1968-007B, and the S-IVB was 1968-007C.

CONCLUSIONS

Despite the switch to an alternative plan following the premature shutdown of the first DPS firing, the overall performance of the LM met all of the requirements for manned orbital flight and the planned second test flight was cancelled.

LM-1 (minus its legs) is lowered into the base of its SLA.

Apollo 5 on Pad 37B ready for launch.

MISSION OBJECTIVES

Launch Vehicle Primary Objectives

1. Evaluate the launch vehicle attitude control system operation and maneuvering capability. *The performance of the guidance and control system was satisfactory, therefore this objective was achieved.*
2. Evaluate the operational adequacy of the launch vehicle systems, including guidance and control, electrical, mechanical, and instrumentation. *Sufficient data were obtained to evaluate these systems, therefore this objective was achieved.*
3. Verify the S-IVB liquid hydrogen and liquid oxygen tank pressure rise rates. *Pressurization control of these systems was normal and within predicted limits, therefore this objective was achieved.*
4. Demonstrate separation of the nose cone from the S-IVB/IU/SLA. *This was accomplished without incident, therefore this objective was achieved.*

Spacecraft Mandatory Detailed Test Objectives

1. M13.1: Determine DPS and APS start, restart and shutdown characteristics in the space environment. *Although the first DPS burn was curtailed and the long DPS burn was not made, the system did perform as commanded for the alternative mission and hence demonstration of these systems was successfully verified.*
2. M13:3: Confirm that no adverse interactions exist between propellant slosh, vehicle stability and DPS/APS performance. *Although there were not enough data for either the DPS or APS burns to determine system performance accurately, telemetered data indicated that engine performance was nominal. This objective, then, was only partially satisfied.*
3. M13.4: Confirm that no vehicle degradation exists that would affect crew safety during APS burn to depletion. *The normal depletion shutdown mode was not accomplished because of propellant slosh caused by the high vehicle attitude rates. A safe depletion shutdown was demonstrated, therefore this objective was partially accomplished.*
4. M13.5: Verify the operation of the DPS propellant feed and pressurization sections. *These test criteria were not completely satisfied.*
5. M13:6: Verify the operation of the APS propellant feed and pressurization sections. *Since the ascent propellant feed and pressurization system operated satisfactorily and had one long burn period, this system can be considered verified.*
6. M13.7: Determine the effects of burn duration on DPS and APS engine performance. *There were not enough steady-state data to determine flow rates, and chamber pressure oscillation occurred intermittently during the APS firings. However, data indicate that the APS engine performance was normal. Long-duration DPS firing was not accomplished. This objective, therefore, was only partially met.*
7. M13.8: Verify APS/RCS operation using the APS propellant interconnect. *Achieved.*
8. M17.1: Demonstrate the structural integrity of the LM ascent and descent stages during Saturn IB launch, RCS, DPS and APS firings in an Earth-orbital environment. *Loads, temperatures, and vibration levels obtained during the LM-1 mission were satisfactory, except for those that required a long DPS burn. As this was not obtained, this objective was only partially satisfied.*

9. M17.2: Determine the interaction loads between the LM and the SLA for a Saturn IB launch environment. *Achieved.*
10. M17.3: Determine the dynamic response of the ascent and descent stage oxidizer tanks during flight-induced environments. *Achieved.*
11. M17.4: Determine vibration levels in critical equipment areas due to launch and to descent and ascent engine firings. *Achieved.*
12. M17.5: Evaluate the thermal and pressure response on the heat shielding, structure, and outriggers when exposed to direct plume impingement and radiant heating from operation of the DPS, APS and RCS engines. *Sufficient data were not obtained to evaluate thermal and pressure response of LM-1. This objective was therefore only partially met.*
13. M20.1: Verify that the descent and ascent stage structure and thermal shield can satisfactorily withstand loads due to pressure and temperature buildup during the 'fire in the hole' abort. *Achieved.*
14. M20.2: Determine staging separation dynamics. *Achieved.*
15. M20.3: Verify attitude control during the 'fire in the hole' abort. *Since the DAP was not in control during Sequence V, no DAP performance data were obtained. Analysis indicates that control would have been adequate with PGNCS/DAP attitude control to accomplish abort staging. This objective was therefore only partially accomplished.*

Spacecraft Primary Detailed Test Objectives

1. P11.1: Verify descent engine gimbaling response to control signals. *The descent engine gimbal trim action satisfactorily reduced the thrust offset during thrusting. The intent of this objective was satisfied.*
2. P11.2: Demonstrate PGNCS thrust vector control and attitude control capability and evaluate the performance of the DAP and IMU in a flight environment. *Verification of the PGNCS performance during DPS and APS engine firing was not obtained. IMU data indicated flight performance was commensurate with pre-flight predictions except for some drift errors. DAP data indicated some performance anomalies. This objective was only partially satisfied.*
3. P.13.2: Verify DPD thrust response to throttling control signals. *DPS throttle control by PGNCS was not accomplished. Throttling capability other than from 10% to FTP was not demonstrated. Therefore, this objective was only partially accomplished.*
4. P13.9: Verify performance and operational characteristics of the RCS. *Achieved.*
5. P15.1: Verify LM electrical power system performance. *Achieved.*
6. P18.1: Demonstrate the operation of the pyrotechnics during LM staging. *Achieved.*

Spacecraft Secondary Detailed Test Objectives

1. S14.1: Verify satisfactory operation of portions of the LM environmental control system equipment. *Achieved.*
2. S15.2: Evaluate the performance of the spacecraft jettison controller and pyrotechnic devices in the execution of nose cone separation, SLA panel deployment and LM/SLA separation functions. *Achieved.*

3. S16.1: Verify performance of portions of the LM S-Band communications subsystem and its compatibility with MSFN. *Achieved.*
4. S16.2: Evaluate the performance of the instrumentation subsystem during boost and LM propulsion subsystem operations. *Achieved.*

MISSION TIMELINE

Apollo 5 mission event	GET (h:m:s)	Date (GMT)	Time (h:m:s)
Automatic sequence start.	–000:02:44	22 Jan 1968	22:45:24
Launch vehicle transfer to internal power.	–000:00:58	22 Jan 1968	22:47:10
Guidance reference release (launch vehicle).	–000:00:04.96	22 Jan 1968	22:48:03
S-IB ignition sequence start.	–000:00:03	22 Jan 1968	22:48:05
Range zero.	000:00:00.00	22 Jan 1968	22:48:08
Liftoff.	000:00:00.4	22 Jan 1968	22:48:08
Guidance reference release (LM).	000:00:00.86	22 Jan 1968	22:48:08
Start tilt (pitch) maneuver.	000:00:09.2	22 Jan 1968	22:48:17
Start roll maneuver.	000:00:09.2	22 Jan 1968	22:48:17
Complete roll maneuver.	000:00:37.9	22 Jan 1968	22:48:45
Mach 1.	000:00:59.8	22 Jan 1968	22:49:07
Maximum dynamic pressure.	000:01:11.5	22 Jan 1968	22:49:19
Tilt arrest.	000:02:15.3	22 Jan 1968	22:50:23
S-IB center engine cutoff.	000:02:19.0	22 Jan 1968	22:50:27
S-IB outer engine cutoff.	000:02:22.3	22 Jan 1968	22:50:30
S-IB/S-IVB separation.	000:02:23.5	22 Jan 1968	22:50:31
S-IB retro ignition.	000:02:24.9	22 Jan 1968	22:50:32
J-2 ignition command.	000:02:24.9	22 Jan 1968	22:50:32
S-IVB cutoff.	000:09:53.3	22 Jan 1968	22:58:01
Orbital insertion.	000:10:03.3	22 Jan 1968	22:58:11
Jettison nose cone.	000:10:38.5	22 Jan 1968	22:58:46
SLA panel deployment.	000:19:53.5	22 Jan 1968	23:08:01
Main solenoid valves opened.	000:50:37	22 Jan 1968	23:38:45
Helium squibs opened.	000:50:50	22 Jan 1968	23:38:58
S-IVB/LM separation.	000:53:55.2	22 Jan 1968	23:42:03
Initiate +X translation.	000:54:05	22 Jan 1968	23:42:13
End +X translation.	000:54:10	22 Jan 1968	23:42:18
Initiate maneuver to cold-soak attitude.	000:54:32.3	22 Jan 1968	23:42:40
Initiate maneuver for DPS firing.	003:55:09.6	23 Jan 1968	02:43:17
Initiate +X translation for ullage.	003:59:33.9	23 Jan 1968	02:47:41
'Engine on' discrete.	003:59:41.4	23 Jan 1968	02:47:49
Initiate 1st DPS firing.*	003:59:41.7	23 Jan 1968	02:47:49
End +X translation.	003:59:42	23 Jan 1968	02:47:50
'Engine off' discrete.	003:59:45.6	23 Jan 1968	02:47:53
Terminate 1st DPS firing.*	003:59:45.7	23 Jan 1968	02:47:53
Begin alternate mission.	004:25:00	23 Jan 1968	03:13:08

Apollo 5 mission event – *continued*	GET (h:m:s)	Date (GMT)	Time (h:m:s)
Select mission programmer sequence III.	004:28:33	23 Jan 1968	03:16:41
Initiate attitude maneuver for mission programmer sequence III.	004:51:31	23 Jan 1968	03:39:39
Select minimum deadband (for attitude control).	004:52:00	23 Jan 1968	03:40:08
Ascent battery to backup.	006:00:28	23 Jan 1968	04:48:36
Master arm on (pyro bus activated).	006:00:46	23 Jan 1968	04:48:54
Select backup control (rate damping only).	006:05:34	23 Jan 1968	04:53:42
Initiate sequence III.	006:10:00	23 Jan 1968	04:58:08
Initiate +X translation.	006:10:07.4	23 Jan 1968	04:58:15
End +X translation.	006:10:17.5	23 Jan 1968	04:58:25
Initiate +X translation.	006:10:22.4	23 Jan 1968	04:58:30
End +X translation.	006:10:27.2	23 Jan 1968	04:58:35
Initiate +X translation for ullage.	006:10:33.4	23 Jan 1968	04:58:41
Initiate 2nd DPS firing.*	006:10:41.7	23 Jan 1968	04:58:49
End +X translation.	006:10:46.3	23 Jan 1968	04:58:54
DPS to full throttle.	006:11:07.8	23 Jan 1968	04:59:15
Terminate 2nd DPS firing.*	006:11:14.7	23 Jan 1968	04:59:22
Initiate +X translation for ullage.	006:11:38.4	23 Jan 1968	04:59:46
Initiate 3rd DPS firing.*	006:11:46.7	23 Jan 1968	04:59:54
End +X translation.	006:11:51.3	23 Jan 1968	04:59:59
DPS to full throttle.	006:12.12.8	23 Jan 1968	05:00:20
Abort staging ('fire in the hole').	006:12:14.3	23 Jan 1968	05:00:22
Terminate 3rd DPD firing.*	006:12:14.7	23 Jan 1968	05:00:22
Initiate 1st APS firing.	006:12:14.7	23 Jan 1968	05:00:22
Terminate 1st APS firing.	006:13.14.3	23 Jan 1968	05:01:22
Select backup control (stops mission programmer sequence III).	006:13:39	23 Jan 1968	05:01:47
Select primary guidance (marking the start of high propellant usage in attitude control).	006:14:05	23 Jan 1968	05:02:13
Control engine system A propellant valves closed.	006:17:11	23 Jan 1968	05:05:19
Control engine system B fuel depleted.	006:21:56	23 Jan 1968	05:10:04
Control engine system B oxidizer depleted.	006:22:20	23 Jan 1968	05:10:28
Control engine system B propellant valves closed.	007:10:14	23 Jan 1968	05:58:22
Control engine system A propellant valves opened.	007:10:53	23 Jan 1968	05:59:01
Control engine propellant cross-feed valves opened.	007:12:24	23 Jan 1968	06:00:32
Ascent engine system A propellant interconnect valves opened.	007:40:59	23 Jan 1968	06:29:07
Ascent engine system B propellant interconnect valves opened.	007:42:17	23 Jan 1968	06:30:25
Control engine system B propellant valves opened.	007:42:59	23 Jan 1968	06:31:07
Initiate sequence V.	007:43:58	23 Jan 1968	06:32:06
Initiate +X translation for ullage.	007:44:00.3	23 Jan 1968	06:32:08
Initiate 2nd APS firing.	007:44:12.7	23 Jan 1968	06:32:20
APS override command (to burn to depletion).	007:44:15	23 Jan 1968	06:32:23
End +X translation.	007:44:17.3	23 Jan 1968	06:32:25
Initiate +X translation.	007:45:11.3	23 Jan 1968	06:33:19
End +X translation.	007:45:27.3	23 Jan 1968	06:33:35

Apollo 5 mission event – *continued*	GET (h:m:s)	Date (GMT)	Time (h:m:s)
Control engine propellant crossfeed valve closed; system A fuel depleted.	007:46:42	23 Jan 1968	06:34:50
Ascent engine system A interconnect valve closed; system A propellant valve opened.	007:46:43	23 Jan 1968	06:34:51
Ascent engine system B interconnect valve closed; system B propellant valve opened.	007:46:53	23 Jan 1968	06:35:01
Control engine system A oxidizer depleted.	007:48:06	23 Jan 1968	06:36:14
APS thrust decay.	007:50:03	23 Jan 1968	06:38:11
Final telemetry from ascent stage.	007:52:10	23 Jan 1968	06:40:18

*Because of parameter sample intervals, these events could have occurred at any time in the second previous to the time shown.

Apollo 6

The second test of the Saturn V

4 April 1968

BACKGROUND

Apollo 6 (AS-502) was the second of a planned series of 'A' missions to 'man rate' the Saturn V launch vehicle. The payload was a lunar module test article (LTA-2R) and CSM-020, a Block I command/service module with some Block II modifications for certification, including a heat shield with a unified crew hatch to be tested at lunar return conditions. In view of the fact that the mission was to be unmanned, certain systems had been deleted from the command module to accommodate an electromechanical control sequencer.

Apollo 6 lifted off from Pad A of Launch Complex 39 at the Kennedy Space Center at a Range Zero time of 12:00:01 GMT (07:00:01 EST) on Thursday, 4 April 1968, and the command module was recovered some 10 hours later.

LAUNCH PREPARATIONS

The countdown picked up at T–24 hr at 01:00:00 EST on Wednesday, 3 April, and proceeded to T–8 hr without imposing unscheduled holds, at which time the scheduled 6-hr hold was started. There were only four significant problems, all of which were resolved during the hold. The problems were, in chronological order: cracks in several liquid hydrogen vent bubble caps exposed the vehicle vent system directly to the atmosphere; a failed transmitter in the S-II required replacement; an electronics package in the S-IC fuel-loading system had to be replaced and recalibrated; and the power supply to the S-IC ground support equipment had to be replaced. On resuming at 23:00:00, the count ran smoothly to launch at 07:00:01 on 4 April.

ASCENT PHASE

At T + 1.9 sec the vehicle began a 1.25 deg/sec yaw maneuver to 'side step' away from the launch umbilical tower in order to preclude colliding with any swing arm that might be

Apollo 6 preparation event	Date
S-IVB stage arrival at KSC.	21 Feb 1967
S-IC stage arrival.	13 Mar 1967
S-IC erected on MLP-2.	17 Mar 1967
IU arrival.	20 Mar 1967
Erection of launch vehicle with S-II spacer.	29 Mar 1967
Launch vehicle electrical interface mate test with S-II spacer.	04 May 1967
Launch vehicle guidance and control tests with S-II spacer.	19 May 1967
S-II stage arrival.	24 May 1967
Launch vehicle propellant dispersion test with S-II spacer.	29 May 1967
Launch vehicle transfer test with S-II spacer.	29 May 1967
Launch vehicle emergency detection system test with S-II spacer.	31 May 1967
Launch vehicle flight sequence and exploding bridge wire functional test with S-II spacer.	01 Jun 1967
Launch vehicle sequence malfunction test with S-II spacer.	08 Jun 1967
Launch vehicle plugs-in overall test #1 with S-II spacer.	13 Jun 1967
De-erection of the launch vehicle through the S-II spacer.	29 Jun 1967
Completed S-II liquid hydrogen tank inspection.	06 Jul 1967
Erection of launch vehicle with S-II flight stage.	13 Jul 1967
Launch vehicle electrical interface mate test.	24 Jul 1967
Launch vehicle switch selector functional test.	24 Jul 1967
Launch vehicle emergency detection system test.	08 Aug 1967
Launch vehicle flight sequence exploding bridge wire functional test.	10 Aug 1967
Launch vehicle power transfer test.	11 Aug 1967
Launch vehicle propellant dispersion test.	11 Aug 1967
Launch vehicle plugs-out overall test #2.	30 Aug 1967
Apollo spacecraft 020 erected.	10 Dec 1967
Swing arm compatibility test.	11 Dec 1967
Launch vehicle plugs-in overall test #1 waived.	21 Dec 1967
Launch vehicle combined guidance and control system tests.	27 Dec 1967
Launch vehicle plugs-out overall test #2.	29 Dec 1967
Launch vehicle VAB/MCC-H interface test.	05 Jan 1968
Space vehicle plugs-in overall test #1.	16 Jan 1968
Space vehicle plugs-out overall test #2.	24 Jan 1968
Swing arm overall test.	29 Jan 1968
Ordnance installed.	02 Feb 1968
Transfer to Launch Complex 39A.	06 Feb 1968
Space vehicle flight readiness test completed.	08 Mar 1968
RP-1 loading of the S-IC stage completed in preparation for the start of the countdown demonstration test (CDDT).	22 Mar 1968
CDDT completed.	31 Mar 1968

tardy in rotating clear, finishing the maneuver at T+9.8. It had been launched on a pad azimuth of 90°E of N, but at T+11.1 initiated a 20-sec roll to align its inertial guidance system with the flight azimuth of 72°E of N, and began to pitch over. At T+60.0 it achieved Mach 1, and at T+75.2 passed through maximum dynamic pressure, enduring moderate winds. As a result of modifications after AS-501, the launch pad and support equipment suffered only minor damage.

From T + 110 to T + 140, the vehicle suffered an oscillation known as 'pogo', as a longitudinal structural mode frequency coupled to the resonant frequency of the oxidizer lines feeding the engines. The greatest disturbance was in the 5.2- to 5.5-hertz range. The oscillations in the engine chamber pressures built up to a peak-to-peak maximum of 8 to 10 psia at T + 125. The low-frequency oscillations of ± 0.6 *g* measured in the command module exceeded the design criteria, and would have been intolerable to a crew. The emergency detection system cast one vote for ending the mission; had it cast a second vote, an abort would have been mandatory. Ground-based and airborne cameras recorded three small pieces and five or six large pieces separating from the vicinity of the SLA between T + 133.31 and T + 133.68, at which time strain, vibration, and acceleration sensors in the S-IVB, IU, SLA, LTA and CSM reported abrupt changes. Subsequent analysis determined that one of the four SLA panels had suffered structural failure, most likely due to a splice plate bond void, and had shed some of its skin. Fortunately, the structural elements were able to sustain the loads for the remainder of the powered flight. The pitch maneuver was terminated at T + 141 to stabilize the vehicle for staging. The cutoff of the center engine at T + 144.72 was 0.11 sec later than nominal, and the cutoff of the outer engines at T + 148.21 occurred 0.85 sec late. The center engine was shut down by oxidizer level sensors and the outer engines by fuel level sensors, demonstrating both of these shutdown modes. At cutoff the space-fixed velocity was 23.89 ft/sec greater than nominal. All S-IC subsystems performed satisfactorily, apart from the fact that at T + 174.25, after the stage had been jettisoned, only one of the film cameras was ejected. The inability to eject the other three cameras was attributed to a lack of pressure in the nitrogen bottle as a result of a failure of the purge system line very near the purge system solenoid valve. Two of the cameras had filmed the S-IC/S-II separation and the others had filmed inside the liquid oxygen tank. The unit that was retrieved had filmed separation. It was decided to supersede the aluminum purge lines with stainless steel, and add structural supports to reduce vibrational stress on the lines. Photographic monitoring indicated that the S-IC broke up at approximately T + 397, some 330 nmi downrange at an altitude of 15.6 nmi.

The five J-2 engines of the S-II ignited at T + 150 and ran satisfactorily for 169 sec, but then at T + 319 the hydrogen flow rate to engine 2 suddenly increased and the thrust decreased by some 23 psi. The engine continued to run at this level until T + 412.92, when the temperature in the engine bay suddenly rose and the engine shut down. Engine 3, which had shown no sign of distress, shut down 1.26 sec later. Post-flight evaluation of telemetered data indicated that the fuel line to the augmented spark igniter (ASI) – a small chamber center-mounted in the injector to produce a small flame for thrust chamber ignition – of engine 2 had failed. The loss of engine 3 was traced to the fact that the control wires for the solenoids of the liquid oxygen prevalves of engines 2 and 3 had been erroneously cross-connected, with the result that when the IU told the prevalve of engine 2 to close, this command went to engine 3, with the result that engine 2 was shut down by cutting off its fuel supply, and engine 3 was shut down by cutting off its oxidizer. Despite the loss of two engines, the S-II was able to gimbal the three remaining engines to maintain stability. However, the controller in the IU had been configured to react only with a single-engine-failure contingency, and was unable to take into account the loss of the second engine. It naively attempted to recover its trajectory as if it had four good engines. When the IU began to adjust the propellant mixture in preparation for S-II shut down, it also initiated the Chi-freeze mode to inhibit attitude changes during the separation phase. Normally this was done 5 sec prior to S-II cutoff, but the loss of two engines reduced the rate of propellant consumption by 40%, and the remaining engines had to burn for much longer than planned to trigger the fuel-depletion cutoff. In fact, at 425.31 sec, the burn was prolonged by 57.81 sec. Not only was the space-fixed velocity at cutoff 335.52 ft/sec less than nominal as a result of diminished acceleration, the extended Chi-freeze steepened the trajectory, which increased the altitude at cutoff by 3.45 nmi and displaced the impact point some 235.86 nmi further downrange than planned.

The inherited trajectory anomaly presented the S-IVB with a serious challenge. The stage ignited at T + 577.28 and burned for 166.52 sec, some 28.95 sec longer than nominal. On finding itself high, slow and short, it had pitched down 50 deg in order to lose altitude, accelerate and gain range. On achieving the desired altitude, it raised its nose above the local horizon to overcome the *negative* radial velocity that it had gained in descending, while simultaneously minimizing further increasing its horizontal velocity. On activating terminal guidance at T + 712.3, the control system set the altitude constraints to zero (referred to as the chi-bar mode) and focused on achieving the desired velocity. Although it ordered the vehicle to pitch up beyond vertical and travel *backwards*, the 1 deg/sec rate meant that the angle was only 65 deg at the eventual cutoff at T + 747.04, as the space-fixed velocity exceeded that intended. The velocity was 160 ft/sec greater than nominal, the altitude was 0.42 nmi lower than nominal, the surface range was 269.15 nmi longer than nominal and the flight path angle was slightly negative. The orbit was significantly different from the 100 nmi circular orbit planned, having an apogee 92.63 nmi above nominal, a perigee 6.57 nmi below nominal, and an eccentricity of 0.0141, but this did not preclude continuing with the mission.[2]

The COSPAR designation of the spacecraft was 1968-025A and the S-IVB was 1968-025B.

EARTH ORBIT PHASE

Once the S-IVB had realigned itself with the horizon, it initiated a sequence of maneuvers, which it completed at the end of the first revolution. First, it rolled 180 deg and pitched down 20 deg, and then it pitched up 20 deg and rolled 180 deg to resume its original attitude. There were no other appreciable effects apart from sloshing of liquid oxygen at the start of each change in pitch, but this was rapidly damped out. This qualified such maneuvers to orient future astronauts for landmark tracking.

The plan was for the S-IVB to reignite for approximately 315 sec at the end of the second revolution, to produce a radius of some 285,110 nmi. Although this simulation of the translunar injection maneuver would achieve lunar distance, the apogee was to be

[2] It is worth noting that the guidance system in the IU performed this recovery entirely on its own.

Apollo 6 ascent phase event	GET (h:m:s)	Altitude (nmi)	Range (nmi)	Earth-fixed velocity (ft/sec)	Space-fixed velocity (ft/sec)	Event duration (sec)	Geocentric latitude (°N)	Longitude (°E)	Space-fixed flight path angle (deg)	Space-fixed heading angle (°E of N)
Liftoff	–	–	–	–	–	–	–	–	–	–
Mach 1 achieved	000:01:00.50	3.86	–	–	–	–	–	–	–	–
Maximum dynamic pressure	000:01:15.20	6.48	–	–	–	–	–	–	–	–
S-IC center engine cutoff	000:02:24.72	30.28	40.70	–	8,598.79	–	28.82	–79.87	20.152	75.131
S-IC outboard engine cutoff	000:02:28.41	32.10	44.90	–	9,080.72	148.4	28.84	–79.78	19.567	75.985
S-IC/S-II separation	000:02:29.08	32.44	45.71	–	9,071.98	–	28.843	–79.780	19.530	74.996
S-II engine 2 cutoff	000:06:52.92	95.71	503.93	–	16,906.63	–	32.14	–62.18	1.611	78.706
S-II engine 3 cutoff	000:06:54.18	–	–	–	–	–	–	–	–	–
S-II engine cutoff	000:09:36.33	105.34	977.66	–	22,065.85	424.31	–	–	1.600	83.388
S-II/S-IVB separation	000:09:37.08	105.43	980.30	–	22,075.62	–	32.144	–62.136	1.597	83.416
S-IVB 1st burn cutoff	000:12:27.04	102.98	1,589.11	–	25,721.29	166.52	32.74	–50.16	–0.40	90.237
Earth orbit insertion	000:12:37.04	102.69	–	–	25,728.64	–	32.730	–	–0.377	90.67

oriented away from the Moon in order not to complicate the task of evaluating the guidance system in deep space. The S-IVB was then to pitch approximately 155 deg and, 3 min after cutoff, separate the CSM in an attitude suitable for a retrograde burn. At 4 min 40 sec after S-IVB cutoff, the SPS was to start a 254-sec retrograde burn to enter an atmosphere-intersecting ellipse with an apogee of approximately 12,000 nmi. The spacecraft was to coast for some 6 hours, oriented to 'cold soak' the heat shield to approximate lunar return thermal conditions. Late in the descending portion of the ellipse, the SPS was to reignite to accelerate to lunar return velocity of 36,500 ft/sec with an inertial flight path angle of –6.5 deg. The service module would then be jettisoned and the command module oriented for entry, with entry interface 4 min after SPS cutoff and the command module splashing down in the primary recovery zone near Hawaii some 9 hr 50 min after launch.

Despite the additional propellant consumed in attaining orbit, the S-IVB was still capable of executing the translunar injection. While coasting in parking orbit, the liquid oxygen measuring side of the propellant utilization system of the S-IVB malfunctioned and thereafter erroneously indicated 100%. This was probably due to an electrical short between the inner and outer elements of the probe caused by metallic debris in the liquid oxygen tank. Fortunately, this did not preclude restarting the engine. However, although the hydraulic system had functioned as expected during the insertion burn, during the restart preparations on the second revolution it failed to produce hydraulic pressure. System temperatures on the first burn indicated the existencc of a cryogenic fuel leak that caused the freezing of the hydraulic fluid and system blockage. During the restart attempt, both the main and auxiliary hydraulic pumps cavitated and gave virtually no system pressure. Ground testing later revealed that on the first burn there had been a leak in the fuel supply to the augmented spark igniter in the engine. On noting the failure of the engine to yield the necessary thrust, the IU cancelled the ignition command at 003:13:50.33 and advanced to the next programmed sequence, just as if it had accomplished the burn. At 003:14:10.33 it initiated the 155-deg pitch maneuver, and 15 sec later was still rotating when the ground commanded the spacecraft to separate. The SLA panels had been designed to hinge open at 40 deg/sec and achieve their maximum divergence within 1.3 sec. As the CSM separated, it suffered a disturbance in pitch of 1.5 deg/sec, implying that with the turn in progress one of the panels nudged the rim of the service module shortly after detaching. With translunar injection ruled out, a (preplanned) alternative mission was selected in which the SPS would put the spacecraft directly into an ellipse with an apogee at an altitude of 12,000 nmi. However, doing this would use so much propellant as to preclude the second SPS burn to accelerate to lunar return velocity and this burn was inhibited by the ground. Meanwhile, as planned, the spacecraft oriented itself to subject its heat shield to a cold soak for 6 hours during the coast phase. While at high altitude, instruments monitored how efficiently the command module was able to block the charged particle radiation in the Van Allen belts. In addition, a 70-mm camera snapped 370 color pictures of Earth in daylight.

Stranded in parking orbit, the S-IVB suffered rapid air drag and re-entered the atmosphere over the Indian Ocean on 26 April 1968.

Apollo 6 earth orbit phase event	GET (h:m:s)	Space-fixed velocity (ft/sec)	Event duration (sec)	Velocity change (ft/sec)	Apogee (nmi)	Perigee (nmi)	Period (min)	Inclination (deg)
Earth orbit insertion	000:12:37.04	25,728.64	–	–	194.44	93.49	89.84	32.57
Orbit at termination of continuous venting by S-IVB	003:08:08.47	–	–	–	200	99	90.01	32.63
S-IVB 2nd reignition (failed)	003:13:34.69	25,724	–	–	–	–	–	–
S-IVB/CSM separation	003:14:27.82	25,743	–	–	–	–	–	–
SPS 1st burn ignition	003:16:06.2	25,774	–	–	–	–	–	–
SPS 1st burn cutoff*	003:23:27.9	31,630	441.7	5,856	12,019.5	18	384.8	32.58
High apogee	006:28:58	7,403	–	–	–	–	–	–
CS/SM separation**	009:36:56.6	32,489	–	–	–	–	–	–
Entry interface	009:38:29	32,830	–	–	–	–	–	–

*Note the 18-nmi perigee of the post-SPS coast indicates an atmosphere-intersecting ellipse.

** The entry speed is lower than on AS-501 because on this occasion the 2nd SPS burn was not made.

RECOVERY

The service module was jettisoned at 009:36:56.6, and then the command module oriented itself for entry. At 009:38:29 the command module made contact with the atmosphere at an inertial velocity of 32,830 (rather less than that of lunar return) at an angle of –5.85 deg. The lift-to-drag ratio at entry time was 0.343. The parachutes were deployed without incident, and at 009:57:19.9 the capsule splashed into the Pacific Ocean north of Hawaii at 158.0°W, 27.7°N, some 49 nmi uprange of the USS *Okinawa* stationed at the recovery point established for the simulated lunar return. For the first time, the capsule adopted the apex-down flotation attitude, but was promptly righted by a set of airbags that inflated on its nose. A fixed-wing aircraft was first on scene, 26 min after splashdown, having located the capsule by its radio beacon. A helicopter arrived 1 hr 46 min after splashdown with swimmers to affix the flotation collar to the capsule, which was bobbing in a 4-ft swell. The capsule was retrieved by the ship when it arrived 6 hours after splashdown. On this occasion none of the chutes was recovered.

CONCLUSIONS

The purpose of the mission was to demonstrate the compatibility and performance of the launch vehicle and spacecraft. It experienced considerably more structural activity than its predecessor, but not of sufficient magnitude to risk the integrity of the vehicle. Although the heat shield was not subjected to the degree of thermal stress planned, this had been achieved on AS-501. The performance of the unified crew hatch was sufficient to enable it to be certified for use on manned missions.

Resonant oscillations had occurred on a variety of rockets. It was acceptable in the Titan II missile, but had required to be 'designed out' in adapting that vehicle for the manned Gemini missions, with an upper limit of 0.25 *g* being set. However, an error in preparing the vehicle resulted in Gemini 5 enduring 0.38-*g* oscillations at a frequency of 10 hertz superimposed on the 3.3-*g* acceleration. On AS-501 longitudinal oscillations had not exceeded 0.1 *g*. Engines in the S-IC were supposed to be tuned to different frequencies to prevent any two or more engines unbalancing the vehicle, but it was found that two of the engines of AS-502 had been inadvertently tuned to the *same* frequency, and this had aggravated the problem. To determine the cause of the pogo, an S-IC was erected on the test stand at Huntsville. Prevalves in the liquid oxygen ducts just above the firing chambers of the five engines were present to block the flow of oxygen until the fluid was passed to the main liquid oxygen valves late in the countdown, in readiness for ignition. With pogo, the flow of propellant became irregular, inducing thrust fluctuations that increased the effect. When the prevalves were modified to allow the injection of helium into the cavity in the final minutes of the countdown, the gas acted as a shock absorber against liquid oxygen pressure surges in flight and eliminated the pogo. An analysis of the structural failure of one of the SLA panels concluded that this had occurred independently of the pogo: the heating caused by passing through the atmosphere had increased the pressure of moisture in the honeycomb material, so weakening the structural bonding that a section of panel blew out. It was decided to apply a layer of cork to the exterior of the adapter to absorb moisture, and holes were drilled in order to prevent the build up of pressure within the underlying honeycomb.

After the success of AS-501, the problems with the J-2 engines in the S-II and S-IVB of AS-502 was a surprise. Determining the cause required careful analysis. It was clear that there had been a leak in the fuel pipes that fed the ASI of engine 2 of the S-II, and that after some time this had failed completely. Furthermore, the telemetry indicated that the engine on the S-IVB had suffered the same problem, although to a lesser degree, and had failed only after attaining orbit, preventing it from restarting. The ASI fuel lines were subjected to a variety of tests in an effort to replicate the failure. In order to allow the lines to absorb expansion and contraction, they incorporated short sections resembling bellows, and one idea was that these might have resonated and sprung a leak. When firing tests failed to show such resonation, it was decided to improve the fidelity of the test by installing a rig inside a vacuum chamber and pumping liquid hydrogen through the pipes at rates appropriate to operational use, and in each test a section of bellows failed within 100 sec. It was realized that in firing the engines at sea level water vapor had settled on the cryogenically chilled bellows and served to damp out vibrations, whereas in flight the stage was above most of the atmosphere prior to engine-start and the bellows were free to resonate. The remedy was to delete the bellows, and use pipes that incorporated bends designed to absorb the expansion and contraction for which the bellows had been designed.

The problems encountered by the AS-502 mission initially gave rise to concern for the pace of the program, but rapid corrective action obviated a major delay and, indeed, it was decided to cancel the third unmanned test flight.

Preparing LM Test Article-2R.

AS-502 emerges from the VAB ready for the Apollo 6 mission.

AS-502 on its crawler passes the Mobile Service Structure.

Apollo 6 lifts off.

After the S-IC stage had been jettisoned, the S-II discarded its protective 'skirt'.

CM-020 awaits recovery.

MISSION OBJECTIVES

Launch Vehicle Primary Detailed Objectives

1. Demonstrate structural and thermal integrity of launch vehicle throughout powered and coasting flight, and determine in-flight structural loads and dynamic characteristics. *Partially achieved. The vibrations in the S-IC tail fins exceeded the range of the accelerometers near maximum dynamic pressure, and oscillatory coupling occurred between the launch vehicle structure and the S-IC engines. The thermal and pressure environments in the forward skirt of the S-IC during S-IC/S-II separation were greater than designed.*
2. Determine in-flight launch vehicle internal environment. *Achieved.*
3. Verify pre-launch and launch support equipment compatibility with the launch vehicle and spacecraft systems. *Achieved.*
4. Demonstrate the S-IC propulsion system and determine in-flight system performance parameters. *Achieved, although oscillatory coupling occurred between the launch vehicle structure and the S-IC engines.*
5. Demonstrate the S-II propulsion system, including the programmed mixture ratio shift and the propellant management systems, and determine in-flight system performance parameters. *Engines number 2 and 3 suffered premature shutdown at T+412.92 sec and T+414.18 sec respectively, instead of the nominal cutoff time of T+517.69 sec. The propellant management system performed satisfactorily. The programmed mixture ratio shift was successfully demonstrated, but it occurred substantially later than nominal because the premature shutdown of two of the engines resulted in attitude shifts that disturbed the normal propellant level relationships.*
6. Demonstrate the launch vehicle guidance and control system during S-IC, S-II, and S-IVB powered flight; achieve guidance cutoff and evaluate system accuracy. *The performance of the guidance and navigation system was as predicted from liftoff to T+412.92 sec when the first of two of the S-II engines prematurely shutdown. Guidance computations responded to variations in altitude and velocity caused by the decrease in thrust during the S-II burn period. Due to the two-engine-out perturbation, flight path angle and velocity were not optimal at the time that guidance commanded S-IVB engine cutoff. All orbital maneuvers were satisfactorily performed. The IU commands were properly executed for S-IVB restart, but the engine did not reignite. Since acceleration test conditions were not met, the engine cutoff command was issued. The flight control computer, thrust vector controller, and auxiliary propulsion system satisfied all of the requirements for attitude control and stability in both the boost and orbital coast phases of the mission.*
7. Demonstrate S-IC/S-II dual plane separation. *Achieved.*
8. Demonstrate S-II/S-IVB separation. *Achieved.*
9. Demonstrate the launch vehicle sequencing system. *Achieved.*
10. Demonstrate the compatibility of the launch vehicle and the spacecraft. *This was partially achieved. There was oscillatory coupling between the launch vehicle structure and the S-IC engines between T+110 and T+140 sec, as evidenced by the buildup and decay of longitudinal acceleration. Also, vibrational coupling occurred between longitudinal and lateral modes.*

11. Evaluate the performance of the emergency detection system in a closed-loop configuration. *This was satisfactorily achieved for the launch vehicle.*
12. Demonstrate the capability of the S-IVB auxiliary propulsion system during S-IVB powered flight and orbital coast phases to maintain attitude control and perform required maneuvers. *Achieved.*
13. Demonstrate the adequacy of the S-IVB continuous vent system while in orbit. *Achieved.*
14. Demonstrate the S-IVB restart capability. *This was not achieved. Conditions for a restart were nominal apart from those a hot-start of the gas generator and lack of main chamber ignition. The cause of the restart failure was a failed fuel line to the augmented spark igniter.*
15. Demonstrate the mission support capability required for launch and mission operations to high post-injection altitudes. *Due to the failure of the S-IVB to restart, high post-insertion altitudes were not achieved.*
16. Verify the S-IVB propulsion system, including the propellant management system, and determine in-flight performance parameters. *The propulsion system met all operational requirements until the failure of the engine to restart. The liquid oxygen mass bridge erroneously indicated a 100% level from T+ 11,091 sec. If a restart had been possible, this would have caused the engine to run at a 5.5:1 mixture ratio.*

Launch Vehicle Secondary Detailed Test Objectives

1. Determine launch vehicle powered flight external environment. *Achieved.*
2. Determine attenuation effects of exhaust flames on radio-frequency radiating and receiving systems during main engine, retro, and ullage motor firings. *Achieved.*

Spacecraft Primary Objectives

1. Demonstrate the emergency detection subsystem in the closed-loop configuration. *Achieved.*

Spacecraft Primary Detailed Test Objectives

1. P3.1: Evaluate the thermal and structural performance of the Block II heat shield, including effects of cold-soak and maximum thermal gradient when subjected to the combination of a high heat load and a high heating rate representative of lunar return entry. *This was successfully demonstrated. The temperature data were within design limits for the flight, although the ablator temperature rises were higher than for the spacecraft of the Apollo 4 mission.*
2. P3.8: Evaluate the thermal performance of the unified crew hatch for heating conditions anticipated during lunar return entry. *Achieved.*
3. P1.1a: Demonstrate CSM/SLA/LTA/Saturn V structural compatibility and determine spacecraft loads in a Saturn V launch environment. *This was only partially achieved. At T+ 133 sec pieces of the SLA separated from the vehicle.*
4. P1.2: Determine the dynamic and thermal responses of the SLA/CSM structure in the Saturn V launch environment. *Achieved, but a separation transient occurred indicating that the motion of the two vehicles was momentarily coupled.*
5. P1.4: Determine the force inputs to the simulated LM from the SLA at the spacecraft

attachment structures in a Saturn V launch environment. *Achieved. The loads at liftoff, maximum dynamic pressure, and end of S-IC boost were less than LM design conditions.*
6. P4: Demonstrate the spacecraft emergency detection subsystem in the closed-loop configuration. *At approximately 00:02:13 one of the three hot-wire automatic abort voting circuits became de-energized as a result of a break in the wire between the distributor in the IU and the CM umbilical.*
7. P1.5: Obtain data on the acoustic and thermal environment of the SLA/simulated LM interface during a Saturn V launch. *Loads at liftoff, maximum dynamic pressure and the end of S-IC boost were less than LM design conditions. However, at approximately 00:01:50 (between the time of maximum dynamic pressure and the end of S-IC boost) axial and lateral accelerations of approximately 5 hertz began in the LTA-2R, lasting until 00:02:13, when a major change in character occurred.*
8. P1.7: Determine vibration response of LM descent stage engine and propellant tanks in a Saturn V launch environment. *Vibration levels of the LTA-2R oxidizer tank exceeded the expected mission levels in narrow frequency bands.*
9. P3.2: Demonstrate an SPS no-ullage start. *Achieved.*
10. P3.5: Verify the performance of the thermal control subsystem of the reaction control system of the service module, and the thermal response of the engines in the deep space environment. *The thermal control system for the SM-RCS maintained the engine mounting structures and injector head temperatures at satisfactory levels for Quads A, B, and D, but Quad C displayed anomalous temperatures during the early portion of the cold-soak phase of the mission.*
11. P3.6: Verify the thermal design adequacy of the command module reaction control system thrusters and extensions during simulated lunar return entry. *The system adequately withstood the effects of a high heating load entry after having been subjected to an extended cold-soak period.*
12. P3.9: Verify operation of the heat rejection system throughout the mission. *The environmental control system performed satisfactorily throughout the mission but due to loss of accurate data at 00:01:28 it could not be determined when active cooling was initiated.*
13. P5.6: Measure the integrated skin and depth radiation dose within the command module up to an altitude of at least 2,000 nmi. *The two dose-rate measurements of the Van Allen belt dosimeter randomly switched between low range and high range.*
14. P3.3: Determine performance of the SPS during a long-duration burn. *Achieved.*
15. P5.2b: Demonstrate the performance of the CSM/MSFN S-band communications. *The S-band telemetry performance analysis indicated an intermittent problem from 00:01:28 to 00:08:20 and during the coast ellipse phase.*

Spacecraft Secondary Detailed Test Objectives

1. S4: Determine and display, in real-time, Van Allen belt radiation dose rate and integrated dose date at MCC in Houston, Texas. *An interference problem during descent from apogee prevented real-time observation of the data, but the information was recovered by post-flight data reduction.*
2. S3.2d: Verify operation of the primary guidance system (PGS) in the space environment after S-IVB separation. *Achieved.*

3. S2: Demonstrate satisfactory operation of the CSM communication subsystem using the Block II type VHF omnidirectional antennas. *Achieved.*
4. S3.3c: Verify operation of the G&N/SCS during entry and recovery. *Achieved.*
5. S3.1d: Verify operation of the PGS after being subjected to the Saturn V launch environment. *Achieved.*
6. S6: Gather data on the effects of a long-duration SPS burn on spacecraft stability. *The performance of the SPS thrust vector control loop was as predicted.*
7. S3.3a: Verify operation of the CM-RCS during entry and recovery. *Achieved.*
8. S3.3e: Verify operation of the Earth landing system during entry and recovery. *Achieved.*
9. S3.2d: Verify operation of the electrical power system (EPS) in the space environment after S-IVB separation. *This functioned normally throughout the mission. Circuit breaker 100 tripped causing AC Bus transfer of essential loads.*
10. S3.1c: Verify operation of the G&N system after subjection to the Saturn V launch environment. *No data were available from 01:24:00 to 08:16:00. The performance of the inertial system was excellent and well within pre-flight predictions.*
11. S3.3d: Verify operation of the electrical power system during entry and recovery. *During entry, Bus 2 phase B voltage increased over normal 117.5 volts prior to SM/CM separation for 15 min.*
12. S3.2a: Verify operation of the G&N system in the space environment after S-IVB separation. *Achieved.*
13. S3.1d: Verify operation of the EPS after being subjected to the Saturn V launch environment. *Achieved.*
14. S1: Determine the radiation shielding effects of the CM. *All instrumentation operated properly.*
15. S7: Obtain data on the temperature of the simulated LM skin during launch. *Achieved.*
16. S5: Obtain data via CSM/ARIA communications. *The ARIA at Bermuda supported the mission in excellent fashion.*

MISSION TIMELINE

Apollo 6 mission event	GET (h:m:s)	Date (GMT)	Time (h:m:s)
Start countdown clock at T–24 hr.	–	03 Apr 1968	06:00:00
Move mobile service structure.	–	03 Apr 1968	13:00:00
Spacecraft closeout complete.	–	03 Apr 1968	21:30:00
Retract Apollo access arm.	–	03 Apr 1968	22:00:00
Start planned 6-hr hold.	–	03 Apr 1968	22:00:00
Resume count.	–	04 Apr 1968	04:00:00
Launch vehicle cryogenics load completed.	-000:55:00	04 Apr 1968	11:05:01
Begin terminal count phase (SC).	-000:45:00	04 Apr 1968	11:15:01
Start to retract swing arm #3 (S-II aft).	-000:26.06.74	04 Apr 1968	11:33:54

Apollo 6 mission event – *continued*	GET (h:m:s)	Date (GMT)	Time (h:m:s)
Start to retract swing arm #9 (Egress).	–000:26:06.40	04 Apr 1968	11:33:54
Spacecraft transfer to internal power.	–000:15:00	04 Apr 1968	11:45:01
Arm launch escape system.	–000:10:00	04 Apr 1968	11:50:01
Remove Q-ball cover.	–000:04:46.57	04 Apr 1968	11:55:14
Start automatic sequence.	–000:03:07	04 Apr 1968	11:56:54
Launch vehicle transfer to internal power.	–000:00:50	04 Apr 1968	11:59:11
Start to retract swing arm #1 (S-IC intertank).	–000:00:26.27	04 Apr 1968	11:59:34
Start to retract swing arm #2 (S-IC forward).	–000:00:21.42	04 Apr 1968	11:59:39
Guidance reference release.	–000:00:16.85	04 Apr 1968	11:59:44
S-IC engine start sequence command.	–000:00:08.77	04 Apr 1968	11:59:52
Launch commit.	–000:00:00.12	04 Apr 1968	12:00:00
Range zero.	000:00:00	04 Apr 1968	12:00:01
Arm liftoff switches.	000:00:00.18	04 Apr 1968	12:00:01
Hold-down arm release.	000:00:00.36	04 Apr 1968	12:00:01
First motion.	000:00:00.38	04 Apr 1968	12:00:01
Liftoff.	000:00:00.6	04 Apr 1968	12:00:01
Start to retract swing arm #8 (CM/SM).	000:00:00.60	04 Apr 1968	12:00:01
Start to retract swing arm #7 (IU/S-IVB forward).	000:00:00.69	04 Apr 1968	12:00:01
Tail services masts.	000:00:00.76	04 Apr 1968	12:00:01
Start to retract swing arm #4 (S-II intermediate).	000:00:00.83	04 Apr 1968	12:00:01
Start to retract swing arm #5 (S-II forward).	000:00:00.91	04 Apr 1968	12:00:01
Start to retract swing arm #6 (S-IVB aft).	000:00:00.92	04 Apr 1968	12:00:01
Tail plug disconnect.	000:00:01.74	04 Apr 1968	12:00:02
Begin yaw maneuver.	000:00:01.9	04 Apr 1968	12:00:02
Complete yaw maneuver.	000:00:09.8	04 Apr 1968	12:00:10
Start pitch maneuver.	000:00:11.1	04 Apr 1968	12:00:12
Start roll maneuver.	000:00:11.1	04 Apr 1968	12:00:12
Complete roll maneuver.	000:00:31.1	04 Apr 1968	12:00:32
Mach 1.	000:01:00.50	04 Apr 1968	12:01:01
Maximum dynamic pressure.	000:01:15.20	04 Apr 1968	12:01:16
Tilt arrest.	000:02:20.9	04 Apr 1968	12:02:21
S-IC center engine cutoff.	000:02:24.72	04 Apr 1968	12:02:25
S-IC outboard engine cutoff.	000:02:28.41	04 Apr 1968	12:02:29
S-II ullage ignition.	000:02:28.9	04 Apr 1968	12:02:29
S-IC/S-II separation command.	000:02:29.08	04 Apr 1968	12:02:30
S-IC retro ignition.	000:02:29.10	04 Apr 1968	12:02:30
S-IC/S-II separation.	000:02:29.14	04 Apr 1968	12:02:30
S-II ignition command.	000:02:29.76	04 Apr 1968	12:02:30
S-IC camera ejected.	000:02:54.25	04 Apr 1968	12:02:55
S-IC interstage separation.	000:02:59.06	04 Apr 1968	12:03:00
Jettison launch escape system.	000:03:04.77	04 Apr 1968	12:03:05
Jettison S-II cameras.	000:03:06.4	04 Apr 1968	12:03:07
Initiate iterative guidance mode.	000:03:10.85	04 Apr 1968	12:03:11
S-II engine 2 premature cutoff.	000:06:52.92	04 Apr 1968	12:06:53
S-II engine 3 premature cutoff.	000:06:54.18	04 Apr 1968	12:06:55

Apollo 6 mission event – *continued*	GET (h:m:s)	Date (GMT)	Time (h:m:s)
Terminate iterative guidance mode.	000:06:55.4	04 Apr 1968	12:06:56
S-II engine mixture ratio shift and initiate second phase iterative guidance mode.	000:08:10.76	04 Apr 1968	12:08:11
Terminate iterative guidance mode and initiate Chi-freeze.	000:08:37.7	04 Apr 1968	12:08:38
S-II cutoff.	000:09:36.33	04 Apr 1968	12:09:37
S-IVB ullage ignition.	000:09:36.98	04 Apr 1968	12:09:37
S-II/S-IVB separation command.	000:09:37.08	04 Apr 1968	12:09:38
S-II retro ignition.	000:09:37.08	04 Apr 1968	12:09:38
S-II/S-IVB separation starts.	000:09:37.13	04 Apr 1968	12:09:38
S-IVB 1st ignition command.	000:09:37.28	04 Apr 1968	12:09:38
S-II/S-IVB separation complete.	000:09:38.07	04 Apr 1968	12:09:39
Terminate Chi-freeze.	000:09:42.9	04 Apr 1968	12:09:43
Third phase iterative guidance mode.	000:09:44.78	04 Apr 1968	12:09:45
Jettison S-IVB ullage motor cases.	000:09:49.08	04 Apr 1968	12:09:50
Pitch command nose-up attitude.	000:10:44.02	04 Apr 1968	12:10:45
Initiate chi-bar steering.	000:11:52.3	04 Apr 1968	12:11:53
Terminate iterative guidance mode and initiate Chi-freeze.	000:12:25.4	04 Apr 1968	12:12:26
S-IVB 1st cutoff.	000:12:27.04	04 Apr 1968	12:12:28
Initiate coast period.	000:12:28.30	04 Apr 1968	12:12:29
Insertion into parking orbit.	000:12:37.04	04 Apr 1968	12:12:38
Start to maneuver to local horizontal.	000:12:42.30	04 Apr 1968	12:12:43
Initiate continuous venting of S-IVB for ullage.	000:13:26.25	04 Apr 1968	12:13:27
Initiate 180-deg roll.	000:13:57.30	04 Apr 1968	12:13:58
Initiate 20-deg pitch down.	000:53:27.30	04 Apr 1968	12:53:28
Initiate 20-deg pitch up.	001:31:27.30	04 Apr 1968	13:31:28
Initiate 180-deg roll.	001:36:27.30	04 Apr 1968	13:36:28
Initiate S-IVB restart sequence.	003:07:58.73	04 Apr 1968	15:07:59
Terminate continuous venting of S-IVB .	003:08:08.47	04 Apr 1968	15:08:09
S-IVB 2nd ignition command.	003:13:34.69	04 Apr 1968	15:13:35
S-IVB 2nd ignition cancelled by IU.	003:13:50.33	04 Apr 1968	15:13:51
Initiate maneuver to separation attitude.	003:14:10.33	04 Apr 1968	15:14:11
S-IVB/CSM command (from ground).	003:14:26.02	04 Apr 1968	15:14:27
CSM free.	003:14:27.82	04 Apr 1968	15:14:28
SPS 1st ignition.	003:16:06.2	04 Apr 1968	15:16:07
SPS 1st cutoff.	003:23:27.9	04 Apr 1968	15:23:28
High apogee.	006:28:58	04 Apr 1968	18:28:59
CS/SM separation.	009:36:56.6	04 Apr 1968	21:36:57
400,000-ft entry interface.	009:38:29	04 Apr 1968	21:38:30
Begin radio blackout.	009:38:53.2	04 Apr 1968	21:38:54
End radio blackout (estimated from trajectory).	009:48:18	04 Apr 1968	21:48:19
Deployment of drogue parachutes.	009:51:27.4	04 Apr 1968	21:51:28
Deployment of main parachutes.	009:52:13.4	04 Apr 1968	21:52:14
Splashdown.	009:57:19.9	04 Apr 1968	21:57:20

Apollo 7

The first manned mission: testing the CSM in Earth orbit

11–22 October 1968

BACKGROUND

Twenty-one months after the Apollo 1 fire, the US was ready to begin the manned phase of the program with a Type 'C' mission, with the primary objectives of demonstrating:

- CSM and crew performance;
- crew, space vehicle, and mission support facilities performance; and
- CSM rendezvous capability.

The crew members were Captain Walter Marty 'Wally' Schirra Jr [shi-RAH] (USN), commander; Major Donn Fulton Eisele [EYES-lee] (USAF), command module pilot; and Ronnie Walter 'Walt' Cunningham, lunar module pilot. Selected in the original astronaut group in 1959, Schirra had been pilot of the fifth (third orbital) Mercury mission (MA-8) and command pilot of Gemini 6-A. With Apollo 7, Schirra would become the first person to make three flights into space. Born on 12 March 1923 in Hackensack, New Jersey, Schirra was 45 years old at the time of the Apollo 7 mission. Schirra received a BS degree from the US Naval Academy in 1945. His backup for the mission was Colonel Thomas Patten Stafford (USAF). Eisele and Cunningham were each making their first spaceflight. Born on 23 June 1930 in Columbus, Ohio, Eisele was 38 years old at the time of the Apollo 7 mission. He received a BS in astronautics in 1952 from the US Naval Academy, an MS in astronautics in 1960 from the US Air Force Institute of Technology, and was selected as an astronaut in 1963.[3] His backup was Commander John Watts Young (USN). Born on 16 March 1932 in Creston, Iowa, Cunningham was 36 years old at the time of the Apollo 7 mission. He received a BA in physics in 1960 and an MA in physics in 1961 from the University of California at Los Angeles. He was selected as an astronaut in 1963. His backup was Commander Eugene Andrew 'Gene' Cernan (USN). The capsule communicators (CAPCOMs) for the mission were Stafford, Lieutenant Commander Ronald

[3] Eisele died of a heart attack 1 December 1987 in Tokyo, Japan (*Houston Chronicle*, 3 December 1987, p. 8).

Ellwin Evans (USN), Major William Reid Pogue (USAF),[4] John Leonard 'Jack' Swigert Jr [SWY-girt], Young, and Cernan. The support crew were Swigert, Evans, and Pogue. The flight directors were Glynn S. Lunney (first shift), Eugene F. Kranz (second shift), and Gerald D. Griffin (third shift).

The Saturn IB launch vehicle assigned to Apollo 7 was designated AS-205. The mission also carried the designation Eastern Test Range #66. The spacecraft was CSM-101. This was the first Block II configuration to be flown; i.e., with the capability to accommodate the LM and other systems advancements. Schirra wanted to name his ship 'Phoenix', but NASA refused.

LAUNCH PREPARATIONS

The countdown began at 19:00 GMT on 6 October 1968. There were three planned holds. The first two, at T–72 hr for 6 hours and at T–33 hr for 3 hours, allowed sufficient time to fix any spacecraft problems. The final hold, at T–6 hr, provided a rest period for the launch crew. The clock resumed 6 hours later at 09:00 GMT, 11 October, and proceeded smoothly until T–10 min when the thrust chamber jacket chilldown was initiated for the launch vehicle S-IVB stage. This progressed more slowly than expected, and would have required a recycling of the clock to T–15 min if the proper temperature was not achieved in time for initiation of the automatic countdown sequence. As a result, a hold was called at T–6 min 15 sec, and lasted for 2 min 45 sec. Post-launch analysis determined that chilldown would have occurred without the hold, but the hold was advisable in real-time to meet revised temperature requirements. At 14:56:30 GMT, the countdown resumed and continued to liftoff without further problems. A large high-pressure system centered over Nova Scotia produced high easterly surface winds at launch time. The upper winds, above 30,000 ft, were light from the west. Surface wind speeds were the highest observed for any Saturn vehicle to date. A few scattered clouds were in the area. Cumulonimbus clouds covered 30% of the sky with a base at 2,100 ft, visibility was 10 statute miles, the temperature was 82.9°F, the relative humidity was 65%, the dew point was 70.0°F, the barometric pressure was 14.765 psi, and there were winds of 19.8 kt at 90° from true north as measured by the anemometer on the light pole 59.4 ft above ground at the launch site.

ASCENT PHASE

Apollo 7 lifted off from Launch Complex 34 at Cape Kennedy, Florida, at a Range Zero time of 15:02:45 GMT (11:02:45 EDT) on 11 October 1968, well within the planned launch window of 15:00:00 to 19:00:00 GMT. The ascent phase was nominal. Moments after liftoff, the vehicle rolled from a pad azimuth of 100°E of N to a flight azimuth of 72°E of N.

[4] Pogue replaced Major Edward Galen Givens Jr (USAF), who died in an automobile accident in Pearland, TX, on 6 June 1967. Givens had been selected in the astronaut class of 1966 (*Houston Chronicle*, 8 June 1967).

Apollo 7 preparation event	Date
Individual and combined CM and SM systems test completed at factory.	18 Mar 1968
Saturn S-IB stage delivered to KSC.	28 Mar 1968
Saturn S-IVB stage delivered to KSC.	07 Apr 1968
Saturn IB instrument unit delivered to KSC.	11 Apr 1968
Integrated CM and SM systems test completed at factory.	29 Apr 1968
CM-101 and SM-101 ready to ship from factory to KSC.	29 May 1968
CM-101 and SM-101 delivered to KSC.	30 May 1968
CM-101 and SM-101 mated.	11 Jun 1968
CSM-101 combined systems test completed.	19 Jun 1968
CSM-101 altitude tests completed.	29 Jul 1968
Space vehicle moved to Launch Complex 34.	09 Aug 1968
CSM-101 integrated systems test completed.	27 Aug 1968
CSM-101 electrically mated to launch vehicle.	20 Aug 1968
Space vehicle overall test completed.	04 Sep 1968
Space vehicle countdown demonstration test completed.	17 Sep 1968
Space vehicle flight readiness test completed.	25 Sep 1968

The maximum wind conditions encountered during ascent were 81 kt at 172,000 ft. Wind shear in the high dynamic pressure region reached 0.0113/sec in the pitch plane at 48,100 ft. The maximum wind speed in the high dynamic pressure region was 30.3 kt from 309 deg at 44,500 ft. The S-IB provided continuous thrust until center engine cutoff at T + 140.65 sec. The outboard engine shut down 3.67 sec later at an Earth-fixed velocity of 6,479.1 ft/sec. Cutoff conditions were very close to prediction. The S-IB was separated from the upper stage at T + 145.59, followed by S-IVB ignition at T + 146.97. Cutoff occurred at T + 616.76, with deviations from the planned trajectory of only 2.3 ft/sec in velocity and 0.054 nmi in altitude. The S-IVB burn time of 469.79 sec was within 1 sec of prediction, and all structural loadings remained well within design tolerances during ascent. The probable impact of the spent S-IB was determined from a theoretical, tumbling, free flight trajectory. Assuming the booster remained intact during entry, the impact occurred in the Atlantic Ocean at 29.76°N and 75.72°W, 265.01 nmi from the launch site. At insertion, at T + 626.76 (i.e. S-IVB cutoff plus 10 sec to account for engine tailoff and other transient effects), the parking orbit showed an apogee and perigee of 152.34 × 123.03 nmi, an inclination of 31.608 deg, a period of 89.55 min, and a velocity of 25,532.2 ft/sec. The apogee and perigee were based upon a spherical Earth with a radius of 3,442.934 nmi.

The COSPAR designation for the spacecraft upon achieving orbit was 1968-089A and the S-IVB was designated 1968-089B.

EARTH ORBIT PHASE

The crew adapted quickly and completely to the weightless environment. There were no disorientation problems associated with movement inside the CM or looking out the windows at Earth. In fact, an attempt by the lunar module pilot to induce vertigo or

Apollo 7 ascent phase event	GET (h:m:s)	Altitude (nmi)	Range (nmi)	Earth-fixed velocity (ft/sec)	Space-fixed velocity (ft/sec)	Event duration (sec)	Geocentric latitude (°N)	Longitude (°E)	Space-fixed flight path angle (deg)	Space-fixed heading angle (°E of N)
Liftoff	000:00:00.36	0.019	0.000	0.0	1,341.7	–	28.3608	–80.5611	0.06	90.01
Mach 1 achieved	000:01:02.15	4.120	0.753	1,039.1	1,960.1	–	28.3649	–80.5477	29.63	86.70
Maximum dynamic pressure	000:01:15.5	6.567	1.933	1,459.4	2,408.8	–	28.3708	–80.5264	31.64	83.65
S-IB center engine cutoff	000:02:20.65	30.626	29.184	6,264.7	7,394.5	123.64	28.5090	–80.0349	27.09	75.87
S-IB outboard engine cutoff	000:02:24.32	32.678	32.418	6,479.1	7,616.8	147.31	28.5252	–79.9765	26.55	75.78
S-IB/S-IVB separation	000:02:25.59	33.389	33.561	6,472.1	7,612.6	–	28.5310	–79.9558	26.32	75.79
S-IVB engine cutoff	000:10:16.76	123.167	983.290	24,181.2	25,525.9	469.79	31.3633	–61.9777	0.00	85.91
Earth orbit insertion	000:10:26.76	123.177	1,121.743	24,208.5	25,553.2	–	31.4091	–61.2293	0.005	86.32

motion sickness by movement of the head in all directions at rapid rates met with negative results. Early in the mission, however, the crew reported some soreness of their back muscles in the kidney area. The soreness was relieved by exercise and hyperextension of the back.

Prior to separation from the S-IVB, a 2-min 56-sec manual takeover of attitude control from the launch vehicle stage was performed at 002:30:48.80. The crew exercised the manual S-IVB/IU orbital attitude control capability. This consisted of a test of the closed loop spacecraft/launch vehicle control system by performing manual pitch, roll, and yaw maneuvers. The control system responded properly. After completion of the test, the crew switched attitude control back to the automatic launch vehicle system, which resumed the normal attitude timeline. By the time the CSM/S-IVB separated at 002:55:02.40, venting of S-IVB propellants had raised the orbit to 170.21 × 123.01 nmi. One key objective was to perform a 'safing' of the S-IVB stage by reducing the pressures in its propellant tanks and high-pressure bottles to a level that would permit safe rendezvous and simulated docking maneuvers. This was to occur over several stages. First, the LH_2 tank safing was to be performed by three preprogrammed ventings; however, four additional ventings were required because the preprogrammed ones did not adequately safe the tank under the orbital conditions experienced. The first venting occurred at 000:10:17, and the final one ended at 005:11:15. The seven ventings totaled 3,274.1 sec. Second, a liquid oxygen dump was initiated at 001:34:28 and lasted 721.00 sec. Third, a cold helium dump was performed at 001:42:28 and again at 004:30:16, lasting 2,868.00 and 1,199.99 sec, respectively. Finally, a stage control sphere helium dump occurred at 003:17:33, but was terminated by ground command after 2,967 sec to save the remaining helium for control of the LH_2 tank vent-and-relief valve. Eventually, the safing was adequately accomplished. During the second revolution the crew noted that one of the spacecraft/LM adapter panels on the S-IVB was deployed only 25 deg instead of the desired 45 deg. It had opened fully, but a retention cable designed to prevent the panel from closing had become stuck and caused the panel to partially close. It was decided to jettison the panels on future missions. By the 19th revolution, the panel had moved to the full-open position.

In order to establish the conditions required for rendezvous with the S-IVB, a 16.3-sec phasing maneuver was performed at 003:20:09.9 using the service module reaction control system. This resulted in an orbit of 165.2 × 124.8 nmi. The phasing burn was intended to place the spacecraft 76.5 nmi ahead of the S-IVB, but because the orbit of the S-IVB decayed more rapidly than expected during the six next revolutions an additional phasing maneuver of 17.6 sec was made at 015:52:00.9 to reestablish the desired conditions. The resulting orbit was 164.7 × 120.8 nmi.

At 014:46, it was reported that the commander had developed a bad head cold, which had begun about one hour after liftoff, and that he had taken two aspirins. The next day, the other two crew members also experienced head cold symptoms. This condition, which continued throughout the mission, caused extreme discomfort because it was very difficult to clear the ears, nose, and sinuses in 'zero-*g*' conditions. Medication was taken, but the symptoms persisted. At 023:33, Schirra cancelled the first television transmission, scheduled to begin in 20 min. Irked that Mission Control had added two burns and a urine dump to the crew's workload while they were testing a new vehicle, he announced that, "TV will be delayed without further discussion".

Two service propulsion system firings were required for rendezvous with the S-IVB. The first firing, a 9.36-sec corrective combination maneuver at 026:24:55.66, was necessary to achieve the desired 1.32-deg angular and 8.0-nmi altitude offset so that the second firing would produce an orbit coelliptic with that of the S-IVB. The result was an orbit of 194.1 × 123.0 nmi. During this period, the sextant was used to track the S-IVB, which was visible in reflected sunlight. The 7.76-sec firing at 028:00:56.47 occurred when the spacecraft was 80 nmi behind and 7.8 nmi below the S-IVB, and created a more circularized orbit of 153.6 × 113.9 nmi. These two firings achieved the desired conditions for the 46-sec rendezvous terminal phase initiation, which occurred at 029:16:33, about 4.5 min earlier than planned because of a minor differential in the orbit. A small midcourse correction was made at 029:37:48, followed by a 708-sec braking maneuver at 029:43:55, and final closure to within 70 ft of the tumbling S-IVB. Stationkeeping was performed for 25 min starting at 029:55:43 in an orbit of 161.0 × 122.1 nmi, after which a 5.4-sec service module reaction control system posigrade maneuver removed the CSM from the vicinity of the S-IVB stage. The crew maneuvered the CSM around the S-IVB in order to inspect and photograph it. This demonstrated the ability of the spacecraft to rendezvous with the LM (represented by the S-IVB) if the ascent stage became disabled after leaving the lunar surface. However, the crew reported that the manually controlled braking maneuver was frustrating because no reliable backup ranging information was available, as would be the case during an actual rendezvous with the LM. The next 24-hr period was devoted to a sextant calibration test at 041:00, two attitude control tests at 049:00 and 050:40, and two primary evaporator tests at 049:50 and 050:30. In addition, the crew performed a rendezvous navigation test using the sextant to track the S-IVB visually to a distance of 160 nmi at 044:40 and to 320 nmi at 053:20. The crew later reported sighting the S-IVB at a range of nearly 1,000 nmi.

To ensure maximum return from Apollo 7, it was planned to complete as many primary and secondary objectives as possible early in the flight, and, by the end of the second day, more than 90% had been accomplished. Three tests of the rendezvous radar transponder were performed. This system would be essential for docking the LM ascent stage to the CM after liftoff from the lunar surface. The first two tests occurred at 061:00 and 071:40. The third was performed during revolution 48 at 076:27, when the ground radar at White Sands Missile Range, New Mexico, acquired and locked onto the spacecraft transponder at a range of 390 nmi and tracked it to 415 nmi. At 071:43 the first of seven television transmissions began. It was the first live television transmission from a manned American spacecraft. The crew opened the telecast with a sign that read "From the lovely Apollo room high atop everything", then pointed the camera out the window as the spacecraft passed over New Orleans and then over the Florida peninsula. The orbital motion of the spacecraft was evident. The transmission lasted 7 min.

The service propulsion system was fired six additional times during the mission. The third firing, at 075:48:00.27 (advanced 16 hours from the original plan), was a 9.10-sec burn controlled by the stabilization and control system. This maneuver was performed early to increase the backup deorbit capability of the service module reaction control system by lowering the perigee to 90 nmi and placing it in the northern hemisphere. The resulting orbit was 159.7 × 89.5 nmi. After the third firing, a 3-hr cold soak of the service propulsion thermal control system was performed. The cold soak stabilized the spacecraft and exposed one side away from the Sun for a time to lower the temperature and monitor the effects of the cold space environment. The thermal characteristics of the system were better than anticipated for random, drifting flight, since the temperature decrease was less than predicted. A test to determine whether the environmental control system radiator surface coating had degraded was conducted between 092:37 and 097:00, with the results indicating that the solar absorptivity of the radiator panel tested was within the predicted limits, thereby validating the system for lunar flight.

The second television transmission started at 095:25, and lasted about 11 min. The program included a tour of the CM, featuring some of the controls, a demonstration of the exercise device, and an attempt to show water condensation inside the spacecraft. In fact, condensation was a major issue associated with the cabin and suit circuits. This had been anticipated in the cabin, since the cold coolant lines from the radiator to the environment control unit and from the environment control unit to the inertial measurement unit were not insulated. Each time excessive condensation was observed on the coolant lines or in a 'puddle' on the aft bulkhead after a service propulsion system maneuver, the crew would vacuum the water overboard. Experiment S005 (Synoptic Terrain Photography) began at 098:40, using a hand-held modified Hasselblad model 500C 70-mm still camera. The aim was to study the origin of the Carolina bays in the USA, wind erosion in desert regions, coastal morphology, and the origin of the African rift valley. Near-vertical, high-sun-angle photographs of Baja California, other parts of Mexico, and parts of the Middle East were useful for geologic studies. Photographs of New Orleans and Houston were generally better for geographic urban studies than those from previous programs. Areas of oceanographic interest, particularly islands in the Pacific Ocean, were photographed for the first time. In addition, the mission obtained the first extensive photographic coverage of northern Chile, Australia, and other areas. Of the 500 photographs taken of land and ocean areas, approximately 200 were usable, and, in general, the color and exposure were excellent. The need to change the film magazines, filters, and exposure settings hurriedly when a target came into view, and to hold the camera steady, accounted for the improper exposure of many frames. The purpose of Experiment S006 (Synoptic Weather Photography) was to photograph as many as possible of 27 basic categories of weather phenomena. This began at 099:10. The camera was the same used for Experiment S005. Of the 500 photographs taken, approximately 300 showed clouds or other items of meteorological interest, and approximately 80 contained features of interest in oceanography. Categories considered worthy of additional interest included weather systems, winds and their effects on clouds, ocean surfaces, underwater zones of Australian reefs, the Pacific atolls, the Bahamas and Cuba, landform effects, climactic zones, and hydrology. Oceanographic surface features were revealed more clearly than in any of the preceding manned flights. The pictures of Hurricane Gladys and Typhoon Gloria, taken on 17 October and 20 October 1968, respectively, were the best views of tropical storms to date. Image sharpness of photographs for this experiment ranged from fair to excellent, again affected by the difficulty in holding the camera steady. Regardless, ocean swells could be resolved from altitudes near 100 nmi.

The third television transmission began at 119:08 and lasted about 10 min. It featured a demonstration of how to prepare food in space, in particular a package of dried fruit juice reconstituted with water. The telecast also demonstrated the process of vacuuming water that had accumulated on the cold glycol lines. Various controls at the commander's workstation were also viewed. The fourth service propulsion system firing was performed at 120:43:00.44 to evaluate the minimum-impulse capability of the engine. It lasted only 0.48 sec and produced an orbit of 156.7 × 89.1 nmi. The fourth television transmission began at 141:11. It showed deposits on window 1 and on optical site markings utilized to measure pitch angles on window 2. As the camera panned around the spacecraft, it gave the viewers a look at sleep stations, stowage areas, helmet bags and pressure suit hoses. Schirra also demonstrated weightlessness by blowing on a floating pen in order to control its motion. By 141:27, the crew had signed off and the transmission signal had faded. At 09:30 GMT on 18 October, during its 108th revolution of the Earth, the S-IVB re-entered the atmosphere and impacted the Indian Ocean at an estimated position of 8.9°S and 81.6°E. A fifth service propulsion system firing was performed to position the spacecraft for an optimum deorbit maneuver at the end of the planned orbital phase by facilitating at least 2 min of tracking by the ground station in Hawaii in the event that another orbit was required. This occurred at 165:00:00.42. To ensure verification of the propellant gauging system, the duration of the burn was increased from the original plan. At 1,691.3 ft/sec, this 66.95-sec firing gave the largest velocity change of the mission. It also incorporated a manual thrust-vector-control takeover halfway through the maneuver. The resulting orbit was 244.2 × 89.1 nmi. During translunar and transearth coasting on future missions, it would be necessary to put the spacecraft into a slow 'barbecue' roll, informally dubbed the 'smorgasbord mode', to maintain an even external temperature. This maneuver, called passive thermal control, was tested twice on Apollo 7, first at 167:00 and next at 212:00. The fifth television transmission, starting at 189:04, featured another spacecraft tour. The program began with a view of the instrument panel including attitude thruster switches and the display keyboard, and cryogenic controls, and ended with the crew performing a military 'close order drill'. An attempt to show scenes of Earth below was unsuccessful. The sixth service propulsion system maneuver was performed during the eighth day, at 210:07:59.99, and was the second minimum-impulse maneuver. At the time, the apogee was 234.6 nmi and the perigee was 88.4 nmi. This firing lasted 0.50 sec and was directed out-of-plane since no altitude change was desired. For the sixth television transmission, which started at 213:10, the crew aimed the camera out of the window and gave ground controllers a view of the Florida peninsula, then turned the camera inside the spacecraft to show off the beards they had grown during the mission. At 231:08, the Solar Particle Alert Network facility at Carnarvon, Australia, detected a Class 1B solar flare. Although the flare would have no effect on the spacecraft or crew, this was a timely checkout of the systems and procedures that would be used in the event of a solar flare during a lunar mission. This event was followed by the seventh service propulsion system firing, a 7.70-sec maneuver at 239:06:11.97 that placed the spacecraft perigee at the proper longitude for entry and recovery, and lowered the orbit to 229.8 × 88.5 nmi. For the final television transmission, starting at 236:18 and lasting for about 11 min, the crew once again showed off their beards, and reported seeing several jet contrails over the Gulf Coast.

The midcourse navigation program, using the Earth horizon and a star, could not be accomplished because the Earth horizon was indistinct and variable. The air glow was about 3 deg wide and had no distinct boundaries or lines when viewed through the sextant. This problem seemed to be associated with the spacecraft being in a low Earth orbit. Using this same program on lunar landmarks and a star, however, the task was very easy to perform. Lunar landmarks showed up nearly as well as Earth landmarks. Stars could be seen at 10 and 15 deg, and greater, from the Moon. Sextant/star counts and star checks and star/horizon sightings were made throughout the mission; lunar landmark/star sightings were attempted at 147:00.

Apollo 7 earth orbit phase event	GET (h:m:s)	Space-fixed velocity (ft/sec)	Event duration (sec)	Velocity change (ft/sec)	Apogee (nmi)	Perigee (nmi)	Period (min)	Inclination (deg)
Earth orbit insertion	000:10:26.76	25,553.2	–	–	152.34	123.03	89.55	31.608
Separation of CSM from S-IVB	002:55:02.40	25,499.5	–	–	170.21	123.01	89.94	31.640
1st rendezvous phasing ignition	003:20:09.9	25,531.7	–	–	167.0	125.3	89.99	31.61
1st rendezvous phasing cutoff	003:20:26.2	25,525.0	16.3	5.7	165.2	124.8	89.95	31.62
2nd rendezvous phasing ignition	015:52:00.9	25,283.1	–	–	165.1	124.7	89.95	31.62
2nd rendezvous phasing cutoff	015:52:18.5	25,277.4	17.6	7.0	164.7	120.8	89.86	31.62
1st SPS ignition	026:24:55.66	25,289.9	–	–	164.6	120.6	89.86	31.62
1st SPS cutoff	026:25:05.02	25,354.0	9.36	204.1	194.1	123.0	90.57	31.62
2nd SPS ignition	028:00:56.47	25,446.5	–	–	194.1	123.0	90.57	31.62
2nd SPS cutoff	028:01:04.23	25,357.2	7.76	173.8	153.6	113.9	89.52	31.63
Terminal phase initiation ignition	029:16:33	25,327.1	–	–	153.6	113.9	89.52	31.63
Terminal phase initiation cutoff	029:17:19	–	46	17.7	–	–	–	–
Terminal phase finalize (braking)	029:43:55	–	–	–	154.1	121.6	89.68	31.61
Terminal phase end	029:55:43	25,546.1	708	49.1	161.0	122.1	89.82	31.61
Separation ignition	030:20:00.0	25,514.1	–	–	161.0	122.1	89.82	31.61
Separation cutoff	030:20:05.4	25,515.1	5.4	2.0	161.0	122.2	89.82	31.61
3rd SPS ignition	075:48:00.27	25,326.1	–	–	159.4	121.3	89.77	31.61
3rd SPS cutoff	075:48:09.37	25,273.9	9.10	209.7	159.7	89.5	89.17	31.23
4th SPS ignition	120:43:00.44	25,661.2	–	–	149.4	87.5	88.94	31.25
4th SPS cutoff	120:43:00.92	25,670.6	0.48	12.3	156.7	89.1	89.11	31.24
5th SPS ignition	165:00:00.42	25,519.3	–	–	146.5	87.1	88.88	31.25
5th SPS cutoff	165:01:07.37	25,714.9	66.95	1,691.3	244.2	89.1	90.77	30.08
6th SPS ignition	210:07:59.99	25,354.7	–	–	234.8	88.5	90.59	30.08
6th SPS cutoff	210:08:00.49	25,354.6	0.50	14.2	234.6	88.4	90.58	30.07
7th SPS ignition	239:06:11.97	25,864.6	–	–	228.3	88.4	90.24	30.07
7th SPS cutoff	239:06:19.67	25,866.4	7.70	220.1	229.8	88.5	90.48	29.87
8th SPS ignition (deorbit)	259:39:16.36	25,155.3	–	–	225.3	88.2	90.39	29.88
8th SPS cutoff	259:39:28.15	24,966.5	11.79	343.6	–	–	–	–

RECOVERY

The final day was devoted primarily to preparations for the deorbit maneuver, which was accomplished by the eighth SPS firing, an 11.79-sec burn at 259:39:16.36 while passing over Hawaii during the 163rd orbit. During the final orbit, the apogee was 225.3 nmi, the perigee was 88.2 nmi, the period was 90.39 min, and the inclination was 29.88 deg. Due to their cold symptoms, there was a considerable amount of discussion about whether the crew should wear their helmets and gloves during entry. With helmets on, it might be impossible to properly clear the throat and ears as increasing gravity drew mucus down from the head area, where it remained during zero-*g* conditions. At the crew's insistence, it was decided 48 hours prior to entry that they should not wear helmets and gloves. The service module was jettisoned at 259:43:33, and the command module's entry followed both automatic and manually guided profiles. The CM re-entered the Earth's atmosphere (at the 400,000-ft altitude of the 'entry interface') at 259:53:26 at a velocity of 25,846.4 ft/sec. Trajectory reconstruction indicated that the service module impacted the Atlantic Ocean at 260:03 at a point estimated to be 29°N and 72°W. During entry, three objects – the CM, the service module, and the 12-ft insulation disk between the two – were tracked simultaneously and sighted visually. The parachute system effected a soft splashdown of the CM in the Atlantic Ocean southeast of Bermuda at 11:11:48 GMT on 22 October 1968. Mission duration was 260:09:03. The impact point was 1.9 nmi from the target point and 7 nmi from the recovery ship USS *Essex*. The splashdown site was estimated to be 27.63°N and 64.15°W. The CM assumed an apex-down (Stable II) flotation attitude, but was turned apex-up (Stable I) within 13 min by the inflatable bag uprighting system. While this was in progress, the recovery beacon was not visible and voice communication with the crew was interrupted. The three men were retrieved by helicopter and stepped onto the recovery ship 56 min after splashdown. The CM was recovered 55 min later. The estimated CM weight at splashdown was 11,409 lb, and the estimated distance traveled for the mission was 3,953,842 nmi. At CM retrieval, the weather recorded by the *Essex* showed light rain showers, a 600-ft ceiling, a visibility of 2 nmi, a wind speed of 16 kt at 260° from true north, an air temperature of 74°F, a water temperature of 81°F, and waves of up to 3 ft at 260° from true north. The CM was offloaded from the *Essex* on 24 October at the Norfolk Naval Air Station, Norfolk, Virginia, and the Landing Safing Team began the evaluation and deactivation procedures at 14:00 GMT. The deactivation was completed at 01:30 GMT on 27 October 1968. The CM was then flown to Long Beach, California, and trucked to the North American Rockwell Space Division facility at Downey, California, for post-flight analysis.

CONCLUSIONS

The Apollo 7 mission was successful in every respect. All spacecraft systems operated satisfactorily, and all but one of the detailed test objectives were met. As an engineering test flight, Apollo 7 demonstrated the performance of the S-IVB orbital safing process, the adequacy of attitude control in both the manual and automatic modes, and that the vehicle systems could perform for extended periods in orbit. For the first time, a mixed cabin atmosphere consisting of 65% oxygen and 35% nitrogen was used on board an American manned spacecraft. All previous flights had used 100% oxygen – a procedure changed as a result of recommendations made by the Apollo 1 fire investigation board. Another 'first' was the availability of hot and cold drinking water for the crew as a byproduct of the service module fuel cells, an important advance for lunar excursions. Consumables usage was maintained at safe levels, and permitted the introduction of additional flight activities toward the conclusion of the mission. The most significant aerodynamic effect encountered was the unexpected phenomenon noted as 'perigee torquing', a rotation of the CSM most noticeable when the perigee was at its lowest.

The following conclusions were made from an analysis of post-mission data:

1. The results of the Apollo 7 mission, when combined with results of previous missions and ground tests, demonstrated that the CSM was qualified for operation in the Earth orbital environment and was ready for tests in the cislunar and lunar orbital environments.
2. The concepts and operational functioning of the crew/spacecraft interfaces, including procedures, provisioning, accommodations, and displays and controls, were acceptable.
3. The overall thermal balance of the spacecraft, for both active and passive elements, was more favorable than predicted for the near-Earth environment.
4. The endurance required for systems operation on a lunar mission was demonstrated.
5. The capability of performing rendezvous using the CSM, with only optical and onboard data, was demonstrated; however, it was determined that ranging information would be extremely desirable for the terminal phase.
6. Navigation techniques in general were demonstrated to be adequate for lunar missions. Specifically:
   (a) Onboard navigation using the landmark tracking technique proved feasible in Earth orbit.
   (b) The Earth horizon was not usable for optics measurements in low Earth orbit with the available optics design and techniques.
   (c) Although a debris cloud of frozen liquid particles following venting obscured star visibility with the scanning telescope, it could be expected to dissipate rapidly in Earth orbit without significantly contaminating the optical surfaces.
   (d) Star visibility data with the scanning telescope indicated that in cislunar space, with no venting and with proper spacecraft orientation to shield the optics from the Sun and Earth or Moon light, constellation recognition would be adequate for platform inertial orientation.
   (e) Sextant star visibility was adequate for platform realignments in daylight using Apollo navigation stars as close as 30 deg from the Sun line-of-sight.
7. The rendezvous radar acquisition and tracking test demonstrated the capability of performance at ranges required for rendezvous between the CSM and the LM.
8. Mission support facilities, including the Manned Space Flight Network and the recovery forces were satisfactory for an Earth orbital mission.

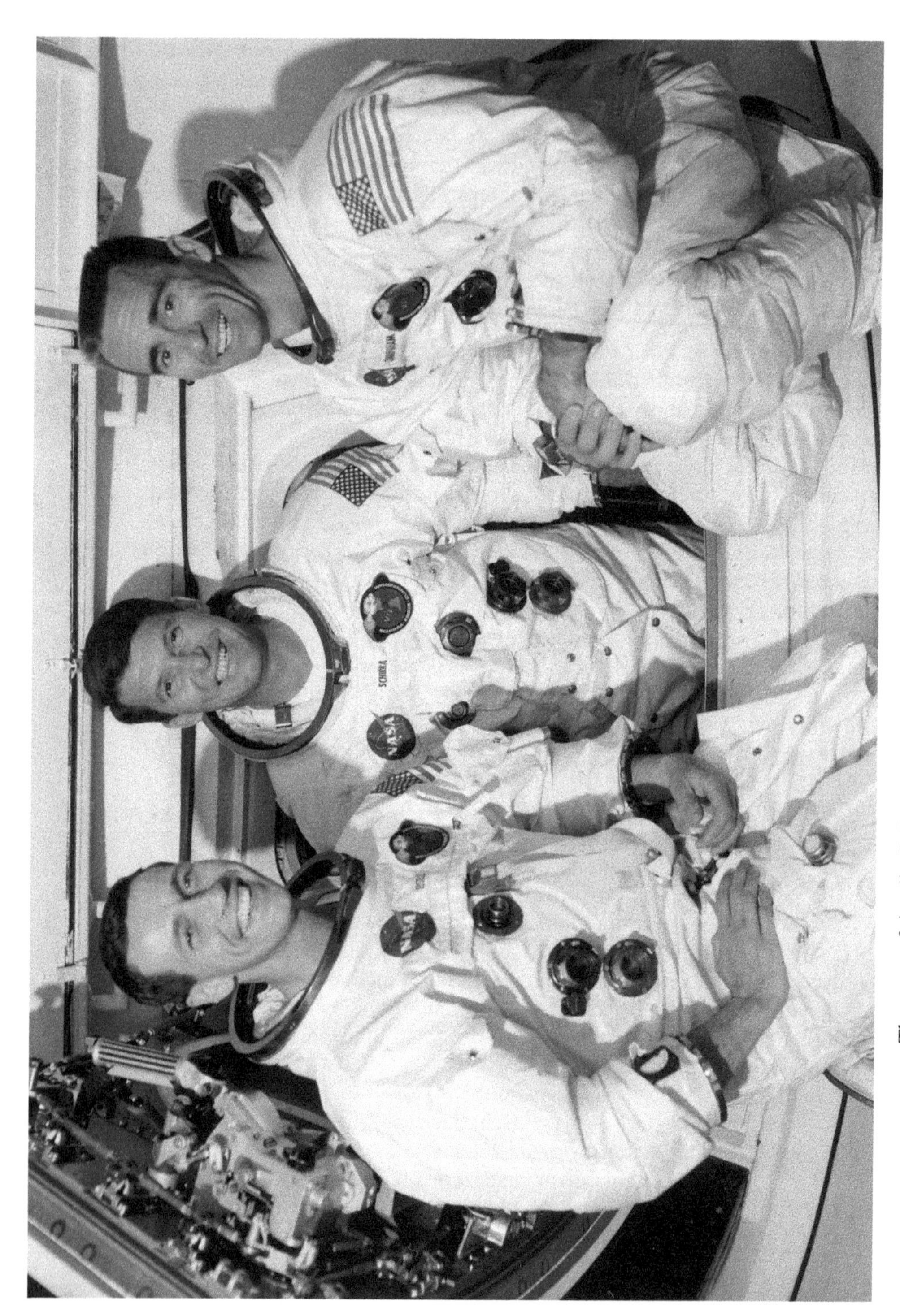

The crew of Apollo 7: Donn Eisele (left), Wally Schirra and Walt Cunningham.

Preparing AS-205 on Pad 34 for Apollo 7.

Apollo 7 lifts off.

Apollo 7's spent S-IVB stage in space, with its SLA panels fully deployed.

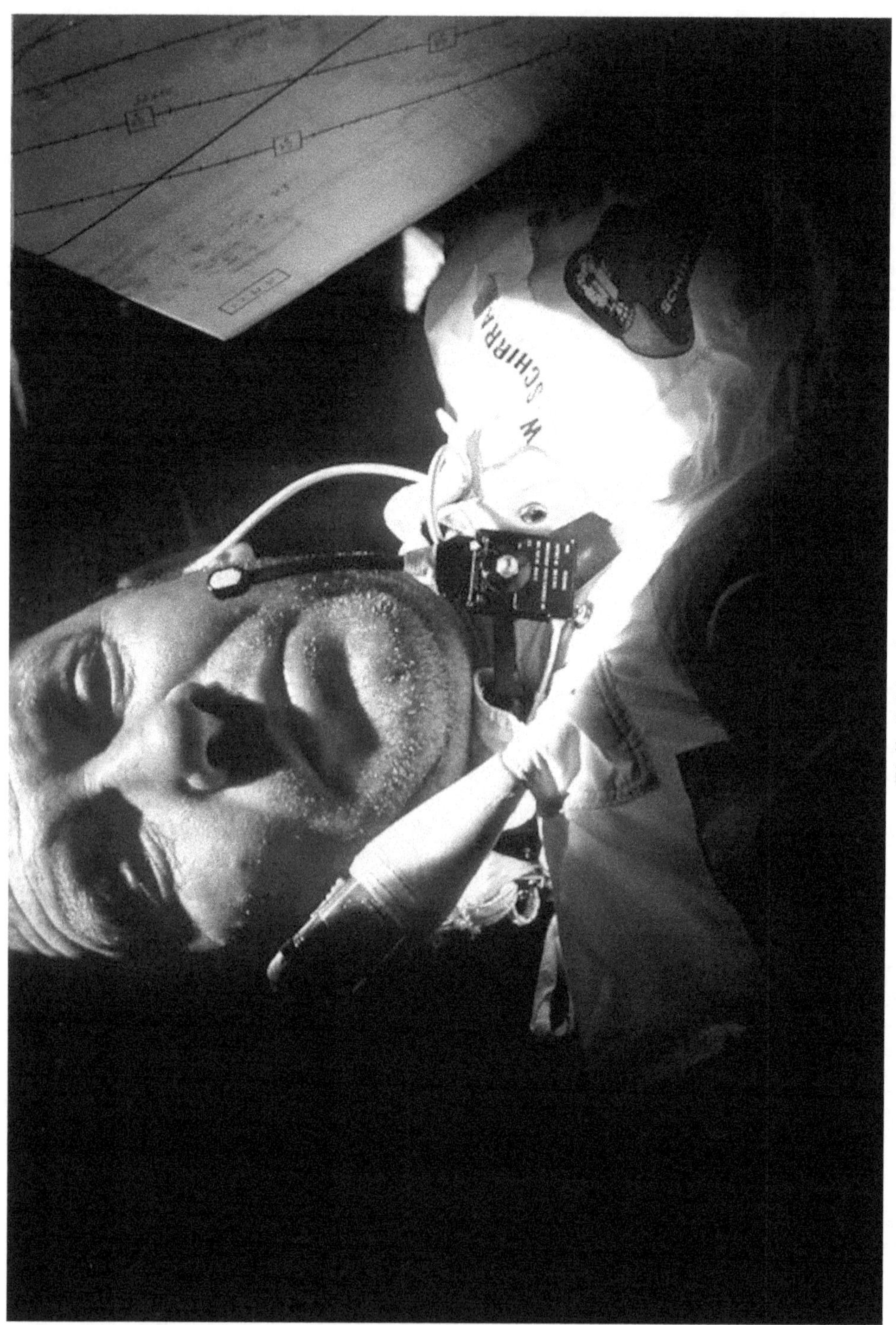

Wally Schirra in space.

MISSION OBJECTIVES

Launch Vehicle Primary Detailed Objectives

1. To demonstrate the adequacy of the launch vehicle attitude control system for orbital operation. *Achieved.*
2. To demonstrate S-IVB orbital safing capability. *Achieved.*
3. To evaluate S-IVB J-2 engine augmented spark igniter line modifications. *Achieved.*

Launch Vehicle Secondary Detailed Test Objectives

1. To evaluate the S-IVB/instrument unit orbital coast lifetime capability. *Achieved.*
2. To demonstrate command and service module manned launch vehicle orbital attitude control. *Achieved.*

Spacecraft Primary Objectives

1. To demonstrate command and service module and crew performance. *Achieved.*
2. To demonstrate crew, space vehicle, and mission support facilities performance. *Achieved.*
3. To demonstrate command and service module rendezvous capability. *Achieved.*

Spacecraft Primary Detailed Test Objectives

1. P1.6: To perform inertial measurement unit alignments using the sextant. *Achieved.*
2. P1.7: To perform an inertial measurement unit orientation determination and a star pattern daylight visibility check. *Achieved.*
3. P1.8: To perform onboard navigation using the technique of the scanning telescope landmark tracking. *Achieved.*
4. P1.10: To perform optical tracking of a target vehicle using the sextant. *Achieved, during rendezvous.*
5. P1.12: To demonstrate guidance navigation control system automatic and manual attitude-controlled reaction control system maneuvers. *Partially achieved, by the automatic mode prior to the service propulsion system burns and the manual mode. Although all required modes were demonstrated, all rates were not checked.*
6. P1.13: To perform guidance navigation control system controlled service propulsion system and reaction control system velocity maneuvers. *Achieved, at various times during the mission.*
7. P1.14: To evaluate the ability of the guidance navigation control system to guide the entry from Earth orbit. *Achieved, during entry.*
8. P1.15: To perform star and Earth horizon sightings to establish an Earth horizon model. *Not achieved. On the two occasions attempted, the Earth horizon was indistinct and variable, with no defined boundaries or lines, thus precluding obtaining the necessary data.*
9. P1.16: To obtain inertial measurement unit performance data in the flight environment. *Achieved, in conjunction with the inertial measurement unit alignment checks. Two pulse integrating pendulous accelerometer bias tests were also performed.*
10. P2.3: To monitor the entry monitoring system during service propulsion velocity changes and entry. *Achieved, during the first service propulsion service burn and entry.*

11. P2.4: To demonstrate the stabilization control system automatic and manual attitude controlled reaction control system maneuvers. *Achieved, except for testing the high and auto rate modes.*
12. P2.5: To demonstrate the command and service module stabilization control system velocity control capability. *Achieved.*
13. P2.6: To perform a manual thrust vector control takeover. *Achieved.*
14. P2.7: To obtain data on the stabilization control systems capability to provide a suitable inertial reference in a flight environment. *Achieved, during the zero-g phase of the mission prior to the fourth service propulsion system burn and prior to the S-IVB separation. Desired data during the boost phase were not obtained.*
15. P2.10: To accomplish the backup mode of the gyro display coupler-flight director attitude indicator alignment using the scanning telescope in preparation for an increment velocity maneuver. *Achieved, although there was a problem with the flight director attitude indicator in the later part of the mission.*
16. P3.14: To demonstrate the service propulsion system minimum impulse burns in a space environment. *Achieved, during the fourth and sixth service propulsion burns.*
17. P3.15: To perform a service propulsion system performance burn in the space environment. *Achieved, during the fifth service propulsion burn.*
18. P3.16: To monitor the primary and auxiliary gauging system. *Achieved, during the fifth service propulsion burn.*
19. P3.20: To verify the adequacy of the propellant feed line thermal control system. *Achieved, by the demonstration of normal operation and the cold soak test.*
20. P4.4: To verify the life support functions of the environmental control system. *Achieved.*
21. P4.6: To obtain data on operation of the waste management system in the flight environment. *Achieved.*
22. P4.8: To operate the secondary coolant loop. *Achieved, and included daily redundant component tests.*
23. P4.9: To demonstrate the water management subsystems operation in the flight environment. *Achieved, throughout the mission, despite a problem with the chlorination procedure and some hardware problems.*
24. P4.10: To demonstrate the post-landing ventilation circuit operation. *Achieved.*
25. P5.8: To obtain data on thermal stratification with and without the cryogenic fans of the cryogenic gas storage system. *Achieved. Although only two of the three stratification tests were successful and part of the third test was accomplished (the rest was deleted), sufficient data were obtained.*
26. P5.9: To verify automatic pressure control of the cryogenic tank systems in a zero-g environment. *Achieved.*
27. P5.10: To demonstrate fuel cell water operations in a zero-g environment. *Achieved.*
28. P6.7: To demonstrate S-band data uplink capability. *Achieved.*
29. P6.8: To demonstrate a simulated command and service module overpass of the lunar module rendezvous radar during the lunar stay. *Achieved, during the 48th revolution.*
30. P7.19: To obtain data on the environmental control system primary radiator thermal coating degradation. *Achieved, from 092:37 to 097:00 GET.*
31. P7.20: To obtain data on the Block II forward heat shield thermal protection system. *Achieved, during entry.*

32. P20.8: To perform a command and service module/S-IVB separation, transposition and simulated docking. *Achieved.*
33. P20.10: To demonstrate the performance of the command and service module/ Manned Space Flight Network S-band communication system. *Achieved.*
34. P20.11: To obtain data on all command and service module consumables. *Achieved.*
35. P20.13: To perform a command and service module active rendezvous with the S-IVB. *Achieved.*
36. P20.15: To obtain crew evaluation of intravehicular activity in general. *Achieved.*

Spacecraft Secondary Detailed Test Objectives

1. S1.11: To monitor the guidance navigation control systems and displays during launch. *Achieved.*
2. S3.17: To obtain data on the service module reaction control subsystem pulse and steady-state performance. *Achieved.*
3. S7.24: To obtain data on initial coning angles when in the spin mode as used during transearth flight. *Partially achieved. The first of three tests was accomplished. A pitch control mode was also accomplished but was not planned prior to launch. The third test was deleted (the crew objected because they expected excessive cross-coupling).*
4. S7.28: To obtain command and service module vibration data. *Achieved, during boost, powered flight, and deorbit.*
5. S20.9: To perform manual out-of-window command and service module attitude orientation for retrofire. *Achieved, by two tests.*
6. S20.12: To perform crew controlled manual S-IVB attitude maneuvers in three axes. *Achieved.*
7. S20.14: To verify that the launch vehicle propellant pressure displays are adequate to warn of a common bulkhead reversal. *Achieved.*
8. S20.16: To obtain photographs of the command module rendezvous windows during discrete phases of the mission. *Achieved, although the second and third of four scheduled tests were deleted.*
9. S20.17: To obtain data on propellant slosh damping following service propulsion system cutoff and following reaction control subsystem burns. *Achieved, by three tests.*
10. S20.18: To obtain data via the command and service module/Apollo Range Instrumentation Aircraft communication subsystems. *Achieved.*
11. S20.19: To demonstrate command and service module VHF voice communications with the Manned Space Flight Network. *Achieved, throughout the mission and during recovery.*
12. S20.20: To evaluate the crew optical alignment sight for docking, rendezvous, and proper attitude verification. *Achieved, throughout the mission and in conjunction with deorbit attitude.*
13. S7.21: To obtain data on the spacecraft/LM adapter deployment system operation. *Achieved.*

Experiments

1. S005 (Synoptic Terrain Photography): To obtain selective, high quality photographs

with color and panchromatic film of selected land and ocean areas. *Achieved. Of the more than 500 photographs obtained, approximately 200 were usable for the purposes of the experiment. The objective of comparing color with black-and-white photography of the same areas was not successful because of problems with focus, exposure, and filters.*

2. S006 (Synoptic Weather Photography): To obtain selective, high quality color cloud photographs to study the fine structure of the Earth's weather system. *Achieved. In particular, excellent views of Hurricane Gladys and Typhoon Gloria were obtained. The color photographs enabled meteorologists to ascertain much more accurately the types of clouds involved than with black-and-white satellite photographs. Oceanographic surface features were also revealed more clearly than in any of the preceding manned flights.*
3. M006: To establish the occurrence and degree of bone demineralization during long spaceflights. *Achieved, by pre-flight and post-flight X-Ray studies of selected bones of crew members.*
4. M011: To determine if the space environment fosters any cellular changes in human blood. *Achieved, by comparison of pre-flight and post-flight crew blood samples.*
5. M023: To measure changes in lower body negative pressure as evidence of cardiovascular deconditioning resulting from prolonged weightlessness. *Achieved, by pre-flight and post-flight medical examinations.*

Test Objectives Added In-Flight

1. Pitch about Y axis. *Achieved.*
2. Optics degradation evaluation. *Achieved.*
3. Sextant/horizon sightings. *Not achieved. Erroneous procedures were given to the crew.*
4. Three additional S-band communication modes. *Achieved.*

MISSION TIMELINE

Apollo 7 mission event	GET (h:m:s)	Date (GMT)	Time (h:m:s)
Countdown started at T–101 hr.	–101:00:00	06 Oct 1968	19:00:00
Scheduled 6-hr hold at T–72 hr.	–072:00:00	08 Oct 1968	00:00:00
Countdown resumed at T–72 hr.	–072:00:00	08 Oct 1968	06:00:00
Scheduled 3-hr hold at T–33 hr.	–033:00:00	09 Oct 1968	21:00:00
Countdown resumed at T–33 hr.	–033:00:00	10 Oct 1968	00:00:00
Terminal countdown started.	–018:00:00	10 Oct 1968	14:30:00
Scheduled 6-hr hold at T–6 hr.	–006:00:00	11 Oct 1968	03:00:00
Terminal countdown started.	–006:00:00	11 Oct 1968	09:00:00
Crew ingress.	–002:27	11 Oct 1968	12:35
Unscheduled 2-min 45-sec hold to complete propellant chilldown.	–000:06:15	11 Oct 1968	14:53:45
Countdown resumed at T–6 min 15 sec.	–000:06:15	11 Oct 1968	14:56:30
Guidance reference release.	–000:00:04.972	11 Oct 1968	15:02:40
S-IB engine start command.	–000:00:02.988	11 Oct 1968	15:02:42

Apollo 7 mission event – *continued*	GET (h:m:s)	Date (GMT)	Time (h:m:s)
Range zero.	000:00:00.00	11 Oct 1968	15:02:45
All holddown arms released (1st motion) (1.21 *g*).	000:00:00.17	11 Oct 1968	15:02:45
Liftoff (umbilical disconnected).	000:00:00.36	11 Oct 1968	15:02:45
Pitch and roll maneuver started.	000:00:10.31	11 Oct 1968	15:02:55
Roll maneuver ended.	000:00:38.46	11 Oct 1968	15:03:23
Mach 1 achieved.	000:01:02.15	11 Oct 1968	15:03:47
Maximum bending moment achieved (7,546,000 lbf-in).	000:01:13.1	11 Oct 1968	15:03:58
Maximum dynamic pressure (665.60 lb/ft^2).	000:01:15.5	11 Oct 1968	15:04:00
Pitch maneuver ended.	000:02:14.26	11 Oct 1968	15:04:59
S-IB maximum total inertial acceleration (4.28 *g*).	000:02:20.10	11 Oct 1968	15:05:05
S-IB center engine cutoff.	000:02:20.65	11 Oct 1968	15:05:05
S-IB outboard engine cutoff.	000:02:24.32	11 Oct 1968	15:05:09
S-IB maximum Earth-fixed velocity.	000:02:24.6	11 Oct 1968	15:05:09
S-IB/S-IVB separation command.	000:02:25.59	11 Oct 1968	15:05:10
S-IVB engine ignition command.	000:02:26.97	11 Oct 1968	15:05:12
S-IVB ullage case jettisoned.	000:02:37.58	11 Oct 1968	15:05:22
Launch escape tower jettisoned.	000:02:46.54	11 Oct 1968	15:05:31
Iterative guidance mode initiated.	000:02:49.76	11 Oct 1968	15:04:54
S-IB apex.	000:04:19.4	11 Oct 1968	15:06:54
S-IB impact in the Atlantic Ocean (theoretical).	000:09:20.2	11 Oct 1968	15:12:05
S-IVB engine cutoff.	000:10:16.76	11 Oct 1968	15:13:01
S-IVB maximum total inertial acceleration (2.56 *g*).	000:10:16.9	11 Oct 1968	15:12:45
S-IVB safing experiment – Start 1st LH_2 tank vent.	000:10:17.37	11 Oct 1968	15:13:02
S-IVB safing experiment – Tank passivization valve open.	000:10:17.56	11 Oct 1968	15:13:02
S-IVB maximum Earth-fixed velocity.	000:10:19.3	11 Oct 1968	15:12:54
Earth orbit insertion.	000:10:26.76	11 Oct 1968	15:13:11
Orbital navigation started.	000:10:32.2	11 Oct 1968	15:13:17
S-IVB safing experiment – Start LOX tank vent.	000:10:47.17	11 Oct 1968	15:13:32
S-IVB safing experiment – End LOX tank vent.	000:11:17.17	11 Oct 1968	15:14:02
S-IVB safing experiment – End 1st LH_2 tank vent (approximate due to data dropout).	000:31:17:36	11 Oct 1968	15:34:02
S-IVB safing experiment – Start 2nd LH_2 tank vent.	000:54:06.95	11 Oct 1968	15:56:52
S-IVB safing experiment – End 2nd LH_2 tank vent.	000:59:06.95	11 Oct 1968	16:01:52
Start of 2-min power failure in Mission Control Center started. No loss of communications.	001:18:34	11 Oct 1968	16:21:19
S-IVB safing experiment – LOX dump started.	001:34:28.96	11 Oct 1968	16:37:14
S-IVB safing experiment – LOX tank non-propulsive vent valve open (until end of mission).	001:34:38.95	11 Oct 1968	16:37:24
S-IVB safing experiment – Start 3rd LH_2 tank vent.	001:34:42.95	11 Oct 1968	16:37:28
S-IVB safing experiment – Start 1st cold helium dump.	001:42:28.95	11 Oct 1968	16:45:14
S-IVB safing experiment – End 3rd LH_2 tank vent.	001:44:42.95	11 Oct 1968	16:47:28
S-IVB safing experiment – LOX dump ended.	001:46:29.96	11 Oct 1968	16:49:15
S-IVB safing experiment – End 1st cold helium dump.	002:30:16.95	11 Oct 1968	17:33:02
Manual takeover of S-IVB attitude control started.	002:30:48.80	11 Oct 1968	17:32:45
Manual takeover – Pitch maneuver started.	002:31:22	11 Oct 1968	17:34:07

Apollo 7 mission event – *continued*	GET (h:m:s)	Date (GMT)	Time (h:m:s)
Manual takeover – Pitch maneuver ended.	002:32:15	11 Oct 1968	17:35:00
Manual takeover – Roll maneuver started.	002:32:22	11 Oct 1968	17:35:07
Manual takeover – Roll maneuver ended.	002:32:51	11 Oct 1968	17:35:36
Manual takeover – Yaw maneuver started.	002:33:01	11 Oct 1968	17:35:46
Manual takeover – Yaw maneuver ended.	002:33:31	11 Oct 1968	17:36:16
Manual takeover of S-IVB attitude control ended.	002:33:44.80	11 Oct 1968	17:35:45
Window photography.	002:45	11 Oct 1968	12:17
Separation of CSM from S-IVB.	002:55:02.40	11 Oct 1968	17:57:47
S-IVB safing experiment – Start 4th LH_2 tank vent.	003:09:14.48	11 Oct 1968	18:11:59
S-IVB safing experiment – End 4th LH_2 tank vent.	003:15:56.11	11 Oct 1968	18:18:41
S-IVB safing experiment – Start stage control sphere helium dump.	003:17:33.95	11 Oct 1968	18:20:19
1st rendezvous phasing maneuver ignition.	003:20:09.9	11 Oct 1968	18:22:54
1st rendezvous phasing maneuver cutoff.	003:20:26.2	11 Oct 1968	18:23:11
S-IVB safing experiment – Start 5th LH_2 tank vent.	004:05:47.27	11 Oct 1968	19:08:32
S-IVB safing experiment – End stage control sphere helium dump.	004:07:01.27	11 Oct 1968	19:09:46
S-IVB safing experiment – End 5th LH_2 tank vent.	004:10:08.43	11 Oct 1968	19:12:53
S-IVB safing experiment – Start 2nd cold helium dump.	004:30:16.96	11 Oct 1968	19:33:02
S-IVB safing experiment – Start 6th LH_2 tank vent.	004:43:55.85	11 Oct 1968	19:46:40
S-IVB safing experiment – End 6th LH_2 tank vent.	004:49:01.73	11 Oct 1968	19:51:46
S-IVB safing experiment – End 2nd cold helium dump.	004:50:16.95	11 Oct 1968	19:53:02
S-IVB safing experiment – Start 7th LH_2 tank vent.	005:08:58.99	11 Oct 1968	20:11:44
S-IVB safing experiment – End 7th LH_2 tank vent.	005:11:15.43	11 Oct 1968	20:14:00
Hydrogen stratification test.	013:28:00	12 Oct 1968	04:30:45
2nd rendezvous phasing maneuver ignition.	015:52:00.9	12 Oct 1968	06:54:45
2nd rendezvous phasing maneuver cutoff.	015:52:18.5	12 Oct 1968	06:55:03
Y-Pulse Integrating Pendulum Accelerometer test.	022:30	12 Oct 1968	13:32
S-IVB optical tracking.	025:10	12 Oct 1968	16:12
Oxygen stratification test.	025:14:00	12 Oct 1968	16:16:45
1st SPS ignition (N_{CC}/corrective combination maneuver – initiation of rendezvous sequence).	026:24:55.66	12 Oct 1968	17:27:40
1st SPS cutoff.	026:25:05.02	12 Oct 1968	17:27:50
2nd SPS ignition (N_{SR}/coelliptic maneuver).	028:00:56.47	12 Oct 1968	19:03:41
2nd SPS cutoff.	028:01:04.23	12 Oct 1968	19:03:49
S-IVB optical tracking.	028:20	12 Oct 1968	19:22
Terminal phase initiation ignition.	029:16:33	12 Oct 1968	20:19:18
Terminal phase initiation cutoff.	029:17:19	12 Oct 1968	20:20:04
Midcourse correction.	029:30:42	12 Oct 1968	20:33:27
Terminal phase finalize (braking).	029:43:55	12 Oct 1968	20:46:40
Terminal phase end/start station-keeping.	029:55:43	12 Oct 1968	20:58:28
Separation maneuver ignition.	030:20:00.0	12 Oct 1968	21:22:45
Separation maneuver cutoff.	030:20:05.4	12 Oct 1968	21:22:50
Sextant calibration test.	041:00	13 Oct 1968	08:02
Sextant tracking of S-IVB started.	044:40	13 Oct 1968	11:42

Apollo 7 mission event – *continued*	GET (h:m:s)	Date (GMT)	Time (h:m:s)
Sextant tracking of S-IVB ended at 160 nmi.	045:30	13 Oct 1968	12:32
Attitude hold test.	049:00	13 Oct 1968	16:02
Primary evaporator test.	049:50	13 Oct 1968	16:52
Primary evaporator test.	050:30	13 Oct 1968	17:32
Attitude hold test.	050:40	13 Oct 1968	17:42
Sextant tracking of S-IVB started.	052:10	13 Oct 1968	19:12
Sextant tracking of S-IVB ended at 320 nmi.	053:20	13 Oct 1968	20:22
Rendezvous radar transponder test.	061:00	14 Oct 1968	04:02
Rendezvous radar transponder test.	071:40	14 Oct 1968	14:42
1st television transmission started.	071:43	14 Oct 1968	14:45
1st television transmission ended.	071:50	14 Oct 1968	14:52
3rd SPS ignition (to position and size orbital ellipse).	075:48:00.27	14 Oct 1968	18:50:45
3rd SPS cutoff.	075:48:09.37	14 Oct 1968	18:50:54
Rendezvous radar transponder test.	076:27	14 Oct 1968	19:29
Radiator degradation test started.	092:37	15 Oct 1968	11:39
2nd television transmission started.	095:25	15 Oct 1968	14:27
2nd television transmission ended.	095:36	15 Oct 1968	14:38
Radiator surface coating degradation test ended.	097:00	15 Oct 1968	16:02
Hydrogen stratification test.	098:11	15 Oct 1968	17:13
Experiment S005 photography.	098:40	15 Oct 1968	17:42
Experiment S006 photography.	099:10	15 Oct 1968	18:12
Window photography.	101:10	15 Oct 1968	20:12
3rd television transmission started.	119:08	16 Oct 1968	14:10
3rd television transmission ended.	119:18	16 Oct 1968	14:20
4th SPS ignition (minimum impulse burn).	120:43:00.44	16 Oct 1968	15:45:45
4th SPS cutoff.	120:43:00.92	16 Oct 1968	15:45:45
Star/horizon sightings.	124:00	16 Oct 1968	19:02
Oxygen stratification test.	131:52	16 Oct 1968	02:54
4th television transmission started.	141:11	17 Oct 1968	12:13
4th television transmission ended.	141:27	17 Oct 1968	12:29
Lunar landmark star sightings.	147:00	17 Oct 1968	18:02
S-IVB impact (theoretical).	162:27:15	18 Oct 1968	09:30:00
5th SPS ignition (to position and size orbital ellipse).	165:00:00.42	18 Oct 1968	12:02:45
5th SPS cutoff.	165:01:07.37	18 Oct 1968	12:03:52
Passive thermal control test started.	167:00	18 Oct 1968	14:02
Passive thermal control test ended.	167:50	18 Oct 1968	14:52
Service propulsion cold soak test started.	168:00	18 Oct 1968	15:02
Service propulsion cold soak test ended,	171:10	18 Oct 1968	18:12
5th television transmission.	189:04	19 Oct 1968	12:06
Morse code emergency keying test started.	190:36:06	19 Oct 1968	13:38:51
Morse code emergency keying test ended.	190:43:01	19 Oct 1968	13:45:46
Oxygen stratification test.	198:27:00	19 Oct 1968	21:29:45
6th SPS ignition (minimum impulse burn).	210:07:59.99	20 Oct 1968	09:10:45
6th SPS cutoff.	210:08:00.49	20 Oct 1968	09:10:45
Passive thermal control test (pitch procedure) started.	212:00	20 Oct 1968	11:02

Apollo 7 mission event – *continued*	GET (h:m:s)	Date (GMT)	Time (h:m:s)
Passive thermal control test ended.	212:50	20 Oct 1968	11:52
6th television transmission.	213:10	20 Oct 1968	12:12
Star/horizon sightings.	213:30	20 Oct 1968	12:32
Hydrogen stratification test.	227:12	21 Oct 1968	02:14
Optics degradation test started.	228:30	21 Oct 1968	03:32
Solar Particle Alert Network Facility at Carnarvon reported class 1B solar flare.	231:08	21 Oct 1968	06:10
7th television transmission started.	236:18	21 Oct 1968	11:20
7th television transmission ended.	236:29	21 Oct 1968	11:31
7th SPS ignition (time anomaly adjust for deorbit burn).	239:06:11.97	21 Oct 1968	14:08:57
7th SPS cutoff.	239:06:19.67	21 Oct 1968	14:09:04
Window photography.	242:30	21 Oct 1968	17:32
8th SPS ignition (deorbit burn).	259:39:16.36	22 Oct 1968	10:42:01
8th SPS cutoff.	259:39:28.15	22 Oct 1968	10:42:13
CM/SM separation.	259:43:33.8	22 Oct 1968	10:46:18
Entry.	259:53:26	22 Oct 1968	10:56:11
Communication blackout started.	259:54:58	22 Oct 1968	10:57:43
Communication blackout ended.	259:59:46	22 Oct 1968	11:02:31
Maximum entry g force (3.33 g).	260:01:09	22 Oct 1968	11:03:54
SM impact in the Atlantic Ocean. S-band contact with CM by recovery aircraft.	260:03	22 Oct 1968	11:05
Drogue parachute deployment.	260:03:23	22 Oct 1968	11:06:08
Main parachute deployment. VHF voice contact with CM established by recovery forces.	260:04:13	22 Oct 1968	11:06:58
Splashdown (went to apex-down).	260:09:03	22 Oct 1968	11:11:48
Inflation of flotation bags started.	260:18	22 Oct 1968	11:20
CM returned to apex-up position.	260:22	22 Oct 1968	11:24
VHF recovery beacon signal received by recovery aircraft.	260:23	22 Oct 1968	11:25
VHF voice communication with CM re-established.	260:24	22 Oct 1968	11:26
CM sighted by recovery helicopter.	260:30	22 Oct 1968	11:32
Swimmers and flotation collar deployed.	260:32	22 Oct 1968	11:34
Flotation collar inflated.	260:41	22 Oct 1968	11:43
CM hatch opened.	260:45	22 Oct 1968	11:47
Crew on board recovery helicopter.	260:58	22 Oct 1968	12:00
Recovery ship at CM. Crew on board recovery ship.	261:06	22 Oct 1968	12:08
CM on board recovery ship.	262:01	22 Oct 1968	13:03
Crew departed recovery ship.	285:54	23 Oct 1968	12:56
Crew arrived at Cape Kennedy, FL.	288:43	23 Oct 1968	15:45
CM offloaded at Norfolk Naval Air Station, VA.	310:58	24 Oct 1968	14:00
Deactivation of CM completed.	370:28	27 Oct 1968	01:30

Apollo 8

The second manned mission: testing the CSM in lunar orbit

21–27 December 1968

BACKGROUND

Apollo 8 was a Type 'C-prime' mission, a CSM manned flight demonstration in lunar orbit instead of Earth orbit, as Apollo 7. It was the first mission on which men ventured to the vicinity of the Moon – a bold step forward in the development of a lunar landing capability.

The mission was originally designated AS-503, an unmanned Earth orbital mission to be launched in May 1968 with BP-30 and LM test article LTA-B as the third 'A' mission, but after Apollo 6 it was decided on 27 April that providing CSM-101 was successful on Apollo 7, AS-503 would fly a manned CSM and LM-3 on the 'D' mission. The change to a manned flight required that the S-II stage be returned to the Mississippi Test Facility for 'man rating'. Additional tests for a manned flight continued at the Kennedy Space Center. After the Mississippi tests were completed on 30 May the stage was returned to the Cape on 27 June. After two months of testing, which started on 11 June, it was determined that the LM would not be ready for the projected launch in early December. Therefore, the decision was made on 19 August that a 19,900-lb LM test article would be installed in the spacecraft/LM adapter for mass loading purposes, as a substitute for the LM. It was also on this date that the crew was instructed to train for a mission to the Moon, officially designated 'Apollo 8'. The possibility of conducting a lunar mission was first discussed with the crew on 10 August, and the results of Apollo 7, to be launched in October, would determine whether the mission would be lunar orbital, circumlunar, or Earth orbital. All training immediately focused on the lunar orbital mission, the most difficult of the three, and ground support preparations were accelerated. The first simulation exercise was conducted on 9 September, and the space vehicle was transferred to the launch site on 9 October. Following the successful completion of Apollo 7 on 22 October, the official decision to conduct a lunar orbit mission was made on 12 November, just five weeks before the scheduled launch. The decision was made after a thorough evaluation of spacecraft performance during Apollo 7's 11 days in Earth orbit and an assessment of the risks involved in a lunar orbit mission. These risks included the total dependency upon the service propulsion engine for propelling the spacecraft from lunar orbit, and a lunar orbit return time of 3 days, compared to an Earth orbit return of just 30 min to 3 hr. Also considered was the value of the flight in furthering the program goal of landing on the Moon before the end of 1969. The principal gains from a lunar mission would include: experience in deep space navigation, communications, and tracking; greater knowledge of spacecraft thermal response to deep space; and crew operational experience – all directly applicable to lunar landing missions.

Apollo 8 was the first manned mission launched with the three-stage Saturn V vehicle. The spacecraft was a Block II CSM, and the spacecraft/LM adapter was the first to incorporate a mechanism to jettison the panels that would enclose the LM on future missions.

The primary objectives of Apollo 8 were:

- to demonstrate the combined performance of the crew, space vehicle, and mission support team during a manned Saturn V mission with the CSM; and
- to demonstrate the performance of nominal and selected backup lunar orbit rendezvous procedures.

The crew members were Colonel Frank Frederick Borman II (USAF), commander; Captain James Arthur Lovell Jr (USN), command module pilot; and Major William Alison Anders (USAF), lunar module pilot. Selected in the astronaut group of 1962, Borman had been command pilot of Gemini 7. Born on 14 March 1928 in Gary, Indiana, he was 40 years old at the time of the Apollo 8 mission. Borman received a BS from the US Military Academy in 1950 and an MS in aeronautical engineering in 1957 from the California Institute of Technology. His backup for the mission was Neil Alden Armstrong. Lovell had been pilot of Gemini 7 and command pilot of Gemini 12. Born on 25 March 1928 in Cleveland, Ohio, he was 40 years old at the time of the Apollo 8 mission. Lovell received a BS in 1952 from the US Naval Academy, and was selected as an astronaut in 1962. His backup was Colonel Edwin Eugene 'Buzz' Aldrin Jr (USAF). Born on 17 October 1933 in Hong Kong, Anders was 35 years old at the time of the Apollo 8 mission, which was his first spaceflight. He received a BS in electrical engineering in 1955 from the US Naval Academy and an MS in nuclear engineering in 1962 from the US Air Force Institute of Technology, and was selected as an astronaut in 1963. His backup was Fred Wallace Haise Jr. The capsule communicators (CAPCOMs) for the mission were Lieutenant Colonel Michael Collins (USAF), Lieutenant Commander Thomas Kenneth 'Ken' Mattingly II (USN), Major Gerald Paul Carr (USMC), Armstrong, Aldrin, Vance DeVoe Brand, and Haise. The support crew were Brand, Mattingly, and Carr. The flight directors were Clifford E. Charlesworth (first shift), Glynn S. Lunney (second shift), and Milton L. Windler (third shift).

The Apollo 8 launch vehicle was a Saturn V, designated AS-503. The mission also carried the designation Eastern Test Range #170. The CSM combination was designated CSM-103. The lunar module test article was designated LTA-B. Because this was a lunar mission, it was necessary for the vehicle to be launched within a particular daily 'launch window', or time period, within a monthly launch window. Part of the constraints were dictated by the desire to pass over selected lunar sites with lighting conditions similar to those planned for the first landing mission. Lunar orbit inclination, inclination of the free-return trajectory, and spacecraft propellant reserves were other primary factors considered in the mission planning. The first monthly window was in December 1968, with launch dates of 20 to 27 December, and January 1969 as a backup. It was decided to make the first attempt on 21 December to have the total available daily window during daylight. Targeting for this day would allow the crew to observe the ALS-1 candidate for the first landing site at the ideal Sun elevation angle of 6.74 deg. The window for 21 December lasted from 12:50:22 to 17:31:40 GMT, with liftoff scheduled for 12:51:00 GMT.

LAUNCH PREPARATIONS

The countdown for Apollo 8 began at 00:00 GMT on 16 December 1968. The terminal countdown sequence began at 01:51 GMT on 19 December (T–28 hr). At that time, space vehicle operations were functionally ahead of the clock. Later in the count, it was discovered that the onboard liquid oxygen supply for the spacecraft environmental control system and fuel cell systems was contaminated with nitrogen and preparations were made to replace the liquid oxygen. The reservicing operations were completed and the tanks were pressurized at T–10 hr. During the planned 6-hr hold period at T–9 hr, virtually all of the countdown tasks, delayed by the liquid oxygen detanking and retanking operations, were brought back in line. When the count was resumed at T–9 hr, space vehicle operations were essentially on the timeline. At T–8 hr, S-IVB liquid oxygen loading operations began. The cryogenic loading operations were completed at 08:29 GMT on 21 December, 8 min into the 1-hr planned hold. The countdown was resumed at 09:21 GMT, with the clock at T–3 hr 30 min, and the crew entered the spacecraft at T–2 hr 53 min. A cold front passed through the launch area the afternoon before launch and became a stationary front about the time for launch, lying through the Miami area. At launch time, surface winds were from the north but changed to westerly at 4,900 ft and remained generally from the west above that region. Cirrus clouds covered 40% of the sky (cloud base not recorded), the visibility was 10 statute miles, the temperature was 59.0°F, the relative humidity was 88%, the dew point was 56%, the barometric pressure was 14.804 psi and the winds were 18.7 ft/sec at 348° from true north as measured by the anemometer on the light pole 60.0 ft above ground at the launch site.

Apollo 8 preparation event	Date
S-II-3 stage delivered to KSC.	26 Dec 1967
S-IC-3 stage delivered to KSC.	27 Dec 1967
S-IC-3 stage erected on MLP-1.	30 Dec 1967
S-IVB-503 stage delivered to KSC.	30 Dec 1967
Saturn V instrument unit IU-503 delivered to KSC.	04 Jan 1968
Boilerplate payload (BP-30) delivered to KSC.	06 Jan 1968
LM Test Article LTA-B delivered to KSC.	09 Jan 1968
LTA-B mated to spacecraft/LM adapter SLA-10.	19 Jan 1968
S-II-3 stage erected.	31 Jan 1968
S-IVB-503 stage erected.	01 Feb 1968
IU-503 erected.	01 Feb 1968

Apollo 8 preparation event – *continued*	Date
BP-30 and summary launch escape system erected.	05 Feb 1968
Launch vehicle electrically mated.	12 Feb 1968
Space vehicle overall test #1 completed (for unmanned mission).	11 Mar 1968
Space vehicle pull test completed (for unmanned mission).	25 Mar 1968
Space vehicle overall test #2 completed (for unmanned mission).	08 Apr 1968
Decision made to de-erect BP-30 for service propulsion system skirt modifications.	10 Apr 1968
'C' mission changed to 'C-prime' mission.	27 Apr 1968
LTA-B/SLA-10, IU-503 and S-IVB-503 stage de-erected.	28 Apr 1968
S-II-3 stage de-erected.	29 Apr 1968
S-II-3 stage departed for Mississippi Test Facility for 'man rating' tests.	01 May 1968
Individual and combined CM and SM systems test completed at factory.	02 Jun 1968
Descent stage of LM-3 delivered to KSC.	09 Jun 1968
Ascent stage of LM-3 delivered to KSC.	14 Jun 1968
S-II-3 stage delivered to KSC from Mississippi Test Facility.	27 Jun 1968
Integrated CM and SM systems test completed at factory.	21 Jul 1968
S-II-3 stage re-erected.	24 Jul 1968
SM-103 quads delivered to KSC.	06 Aug 1968
CM-103 and SM-103 ready to ship from factory to KSC.	11 Aug 1968
SM-103 delivered to KSC.	11 Aug 1968
CM-103 delivered to KSC.	12 Aug 1968
S-IVB-503 stage erected.	14 Aug 1968
IU-503 erected.	15 Aug 1968
Facility verification vehicle erected.	16 Aug 1968
AS-503 designated Apollo 8. Decision to replace LM-3 with LTA-B/SLA-11A.	19 Aug 1968
CM-103 and SM-103 mated.	22 Aug 1968
Launch vehicle electrical systems test completed.	23 Aug 1968
CSM-103 combined systems test completed.	05 Sep 1968
Facility verification vehicle de-erected.	14 Sep 1968
BP-30 erected for service arm checkout.	15 Sep 1968
SLA-11A delivered to KSC.	18 Sep 1968
CSM-103 altitude tests completed.	22 Sep 1968
LTA-B mated with SLA-11A.	29 Sep 1968
Service arm overall test completed.	02 Oct 1968
BP-30 de-erected.	04 Oct 1968
CSM-103 moved to VAB.	07 Oct 1968
Space vehicle and MLP-1 transferred to Launch Complex 39A.	09 Oct 1968
Mobile service structure transferred to Launch Complex 39A.	12 Oct 1968
Space vehicle cutoff and malfunction test completed.	22 Oct 1968
CSM-103/Mission Control Center Houston test completed.	29 Oct 1968
CSM-103 integrated systems test completed.	02 Nov 1968
CSM-103 electrically mated to launch vehicle.	04 Nov 1968
Space vehicle electrically mated.	05 Nov 1968
Space vehicle overall test completed.	06 Nov 1968
Space vehicle overall test #1 (plugs in) completed.	07 Nov 1968
Launch vehicle/Mission Control Center Houston test completed.	11 Nov 1968
Launch umbilical tower/pad water system test completed.	12 Nov 1968

Apollo 8 preparation event – *continued*	Date
Space vehicle flight readiness test completed.	19 Nov 1968
Space vehicle hypergolic fuel loading completed.	30 Nov 1968
S-IC-3 stage RP-1 fuel loading completed.	02 Dec 1968
Space vehicle countdown demonstration test (wet) completed.	10 Dec 1968
Space vehicle countdown demonstration test (dry) completed.	11 Dec 1968

ASCENT PHASE

Apollo 8 was launched from Pad A of Launch Complex 39 at the Kennedy Space Center at a Range Zero time of 12:51:00 GMT (07:51:00 EST) on 21 December 1968, well within the launch window. The ascent phase was nominal. Moments after liftoff, the vehicle rolled from a pad azimuth of 90°E of N to a flight azimuth of 72.124°E of N. The maximum wind conditions encountered during ascent were 114.1 ft/sec at 284° from true north at 49,900 ft (high dynamic pressure region). Component wind shears were of low magnitude at all altitudes. The maximum was a pitch plane shear of 0.0103/sec at 52,500 ft. The S-IC shut down at T + 153.82 sec, followed by S-IC/S-II separation, and S-II ignition. The S-II shut down at T + 524.04 followed by separation from the S-IVB, which ignited at T + 528.29. The first S-IVB cutoff occurred at T + 684.98, with deviations from the planned trajectory of only +1.44 ft/sec in velocity and only –0.01 nmi in altitude. The S-IC impacted at T + 540.410 in the Atlantic Ocean at 30.2040°N, 74.1090°W, 353.462 nmi from the launch site. The S-II impacted at T + 1,165.106 in the Atlantic at 31.8338°N, 37.2774°W, 2,245.913 nmi from the launch site. Four recoverable film camera capsules were carried on board the S-IC. Two were located in the forward interstage looking forward to view S-IC/S-II separation and S-II engine start. The other two were mounted on top of the S-IC stage LOX tank and contained pulse cameras which viewed aft into the LOX tank through fiber-optic bundles. One of the LOX tank capsules was recovered by helicopter at T + 1,170 at 30.22°N and 73.97°W. Despite damage caused by sea water and dye marker that had leaked into the camera compartment, the film provided usable data. It was not known if the other three capsules were ejected. There were also two television cameras on the S-IC to view propulsion and control system components. Both of these cameras provided good-quality data. At insertion, at T + 694.98 (i.e. S-IVB cutoff plus 10 sec to account for engine tailoff and other transient effects), the parking orbit showed an apogee and perigee of 99.99 × 99.57 nmi, an inclination of 32.509 deg, a period of 89.19 min, and a velocity of 25,567.06 ft/sec. The apogee and perigee were based upon a spherical Earth with a radius of 3,443.934 nmi.

The COSPAR designation for the spacecraft upon achieving orbit was 1968-118A and the S-IVB was designated 1968-118B.

Apollo 8 ascent phase event	GET (h:m:s)	Altitude (nmi)	Range (nmi)	Earth-fixed velocity (ft/sec)	Space-fixed velocity (ft/sec)	Event duration (sec)	Geocentric latitude (°N)	Longitude (°E)	Space-fixed flight path angle (deg)	Space-fixed heading angle (°E of N)
Liftoff	000:00:00.67	0.032	0.000	2.2	1,340.7	–	28.4470	–80.6041	0.00	90.00
Mach 1 achieved	000:01:01.45	3.971	1.297	1,076.3	2,078.4	–	28.4526	–80.5805	26.79	85.21
Maximum dynamic pressure	000:01:18.9	7.252	3.545	1,735.4	2,754.7	–	28.4645	–80.5398	29.56	82.43
S-IC center engine cutoff*	000:02:05.93	22.398	22.704	5,060.1	6,213.78	132.52	28.5581	–80.1934	24.527	76.572
S-IC outboard engine cutoff	000:02:33.82	35.503	48.306	7,698.0	8,899.77	160.41	28.6856	–79.7302	20.699	75.387
S-IC/S-II separation*	000:02:34.47	35.838	49.048	7,727.36	8,930.15	–	28.6893	–79.7168	20.605	75.384
S-II engine cutoff	000:08:44.04	103.424	812.267	21,055.6	22,379.1	367.85	31.5492	–65.3897	0.646	81.777
S-II/S-IVB separation*	000:08:44.90	103.460	815.159	21,068.14	22,391.60	–	31.5565	–65.3338	0.636	81.807
S-IVB 1st burn cutoff	000:11:24.98	103.324	1,391.631	24,238.3	25,562.43	156.69	32.4541	–54.0565	–0.001	88.098
Earth orbit insertion	000:11:34.98	103.326	1,430.363	24,242.9	25,567.06	–	32.4741	–53.2923	0.0006	88.532

* Only the commanded time is available for this event.

EARTH ORBIT PHASE

At 000:42:05, the optics cover was jettisoned and the crew performed star checks over the tracking station at Carnarvon in Australia to verify the platform alignment. During the second revolution, at 001:56:00, all spacecraft systems were approved for translunar injection. Because of the risks involved, the mission had been structured with three commit points: launch, Earth parking orbit, and translunar coast preceding the point where the CSM was to brake into lunar orbit. Had any problems been detected at these points, the plan was to shift to alternative missions, which provided for maximum crew safety and maximum scientific and engineering benefit. Had there been reason for not to commit to the third point, the CSM would have continued on its 'free-return' trajectory, looping behind the Moon and returning directly to Earth. After in-flight systems checks, it was determined that liquid oxygen venting through the J-2 engine had increased the apogee by 6.4 nmi; a condition that was only 0.7 nmi greater than predicted. The 317.72-sec translunar injection maneuver (second S-IVB firing) was performed at 002:50:37.79. The S-IVB engine shut down at 002:55:55.51 and translunar injection occurred 10 sec later, at a velocity of 35,505.41 ft/sec, after 1.5 Earth revolutions lasting 2 hr 44 min 30.53 sec.

Apollo 8 earth orbit phase event	GET (h:m:s)	Space-fixed velocity (ft/sec)	Event duration (sec)	Velocity change (ft/sec)	Apogee (nmi)	Perigee (nmi)	Period (min)	Inclination (deg)
Earth orbit insertion	000:11:34.98	25,567.06	-	-	99.99	99.57	88.19	32.509
S-IVB 2nd burn ignition	002:50:37.79	25,558.6	-	-	-	-	-	-
S-IVB 2nd burn cutoff	002:55:55.51	35,532.41	317.72	9,973.81	-	-	-	30.639

TRANSLUNAR PHASE

The spacecraft was separated from the S-IVB at 003:20:59.3 by a small maneuver of the service module reaction control system, and the high-gain antenna was deployed (later used for the first time at 006:33:04). After spacecraft turnaround, the crew observed and photographed the S-IVB and tested station-keeping. At 003:40:01, a 1.1-ft/sec maneuver was performed using the service module reaction control system to increase the distance between the spacecraft and the S-IVB. The distance did not increase as rapidly as desired, and a second, 7.7-ft/sec maneuver was performed at 004:45:01. One objective of the mission was to 'slingshot' the S-IVB into solar orbit. The maneuver to accomplish this objective included a continuous LH_2 vent, a LOX dump, and an auxiliary propulsion system ullage burn. At 004:55:56.02, the LH_2 vent valve was opened. The liquid oxygen was expelled through the J-2 engine for 5 min, beginning at 005:07:55.82. The auxiliary propulsion motors were ignited at 005:25:55.85 and fired to depletion at 005:38:34.00. The resulting velocity increment targeted the S-IVB to go past the trailing limb of the Moon. At the moment of closest approach, 069:58:55.2, the lunar radius was 1,682 nmi and the S-IVB was at an altitude of 681 nmi above selenographic latitude 19.2°N and longitude 88.0°E. The velocity increase relative to Earth from this lunar flyby was 0.79 nmi/sec. The orbital parameters after passing from the lunar sphere of influence resulted in a solar orbit with an aphelion and perihelion of 79,770,000 × 74,490,000 nmi, a semi-major axis of 77,130,000 nmi, an inclination of 23.47 deg, and a period of 340.80 days. The translunar injection maneuver was so accurate that only one small midcourse correction would have been sufficient to achieve the desired lunar orbit insertion altitude of 65 nmi. However, the second of the two maneuvers that separated the spacecraft from the S-IVB altered the trajectory so that a 2.4-sec midcourse correction of 20.4 ft/sec at 010:59:59.2 was required to achieve the desired trajectory.[5] For this midcourse correction, the service propulsion system was used to reduce the altitude of closest approach to the Moon from 458.1 to 66.3 nmi. An additional 11.9-sec midcourse correction of only 1.4 ft/sec was performed at 060:59:55.9 to refine the lunar insertion conditions further. On the translunar coast, the crew made systems checks and navigation sightings, and tested the spacecraft high-gain antenna, a four-dish unified S-band antenna that swung out from the service module after separation from the S-IVB.

Apollo 8 was the first manned US mission in which the crew members experienced symptoms of a mild motion sickness, identical to incipient mild sea sickness. Soon after leaving their couches, all three experienced nausea as a result of rapid body movements. The duration of their symptoms varied between 2 and 24 hours but did not interfere with operational effectiveness. After waking from a fitful rest period at 016:00:00, Borman experienced a headache, nausea, vomiting, and diarrhea, and was diagnosed in-flight as a possible viral gastroenteritis, an epidemic of which had been noted in the Cape Kennedy area prior to the mission. During the post-mission medical debriefing, he reported that the symptoms may have been a side effect of a sleeping tablet taken at 011:00:00, which had produced similar symptoms during pre-mission testing of the drug (Seconal). Two of the six live television transmissions were also made during the translunar flight. The first was a 23-min 37-sec transmission at 031:10:36. The wide-angle lens was used to obtain excellent pictures of the inside of the spacecraft and Lovell preparing a meal; however the telephoto lens passed too much light and pictures of Earth were very poor. A procedure for taping filters from the still camera onto the television camera improved later transmissions. A 25-min 38-sec transmission at 55:02:45 provided scenes of Earth's western hemisphere.

At 055:38:40 the crew were notified that they had become the first humans to travel to a place where the pull of Earth's gravity was less than that of another body. The spacecraft was 176,250 nmi from Earth, 33,800 nmi from the Moon, and its velocity had slowed to

[5] The maneuver at 010:59:59.2 was targeted for a velocity change of 24.8 ft/sec. Only 20.4 ft/sec was achieved because thrust was less than expected. The firing time of 2.4 sec was correct for the constants loaded into the computer, but approximately 0.4 sec too short for the actual engine performance.

3,261 ft/sec. As it moved farther into the Moon's gravitational field, the spacecraft picked up speed. Ignition for lunar orbit insertion was performed with the service propulsion system at 069:08:20.4, at an altitude of 75.6 nmi above the Moon. The 246.9-sec burn resulted in an orbit of 168.5 × 60.0 nmi and a velocity of 5,458 ft/sec. The translunar coast had lasted 66 hr 16 min 21.79 sec.

Apollo 8 translunar phase event	GET (h:m:s)	Altitude (nmi)	Space-fixed velocity (ft/sec)	Event duration (sec)	Velocity change (ft/sec)	Space-fixed flight path angle (deg)	Space-fixed heading angle (°E of N)
Translunar injection	002:56:05.51	187.221	35,505.41	–	–	7.897	67.494
CSM separated from S-IVB	003:20:59.3	3,797.775	24,974.90	–	–	45.110	107.122
Midcourse correction ignition	010:59:59.2	52,768.4	8,187	–	–	73.82	120.65
Midcourse correction cutoff	011:00:01.6	52,771.7	8,172	2.4	20.4	73.75	120.54
Midcourse correction ignition	060:59:55.9	21,064.5	4,101	–	–	–84.41	–86.90
Midcourse correction cutoff	061:00:07.8	21,059.2	4,103	11.9	1.4	–84.41	–87.01

LUNAR ORBIT PHASE

As the spacecraft passed behind the Moon for the first time, and communications were interrupted, the Apollo 8 crew became the first humans to directly observe the far side of the Moon. After 4 hours of navigation checks, ground-based determination of the orbital parameters, and a 12-min television transmission of the lunar surface at 071:40:52, a 9.6-sec lunar orbit circularization maneuver was performed at 073:35:06.6 which resulted in an orbit of 60.7 × 59.7 nmi. The next 12 hours of crew activity in lunar orbit involved photography of both the near and far sides of the Moon and landing-area sightings. The principal photographic objectives were to obtain vertical and oblique overlapping (stereo strip) photographs during at least two revolutions, photographs of specified targets of opportunity, and photographs through the spacecraft sextant of a potential landing site. The purpose of the overlapping photography was to determine elevation and geographical position of lunar far-side features. The targets of opportunity were areas recommended for photography if time and circumstances permitted. These were selected to yield either detailed coverage of specific features or broad coverage of areas not adequately covered by satellite photography. Most were proposed to improve knowledge of areas on the near side. The sextant photography was included to provide image comparisons for landmark evaluation and navigation training purposes. A secondary objective was to photograph one of the certified Apollo landing sites.

The Apollo 8 photography afforded the first opportunity to analyze the intensity and spectral distribution of lunar surface illumination free from the atmospheric modulation present in Earth telescopic photography and without the electronic processing losses present in satellite photography. The 70-mm photography was completed in an excellent manner, with in excess of 800 pictures being obtained. Of these, 600 were good-quality reproductions of lunar surface features, and the remainder were of the S-IVB during separation and venting, and long-distance Earth and lunar photography. Over 700 ft of 16-mm film was also exposed during the S-IVB separation, lunar landmark photography through the sextant, lunar surface sequence photography, and documentation of intravehicular activity. The still photography contributed significantly to knowledge of the lunar environment. In addition, many valuable observations were made by the crew. Their initial comments during the lunar orbit phase included descriptions of the color of the lunar surface as "black-and-white", "absolutely no color" or "whitish gray, like dirty beach sand". As expected, the crew could recognize surface features in shadow zones and extremely bright areas of the lunar surface, but these features were not well delineated in the photographs. This recognition combined with the photographic information enabled new interpretations of lunar surface features and phenomena. As a result, lunar-surface lighting constraints for the lunar landing missions were widened. Specifically, the lower limit for lunar lighting had been set at 6 deg. The Apollo 8 crew saw surface detail at Sun angles in the vicinity of 2 or 3 deg and stated that these low angles should present no problem for a lunar landing, but landing sites in long shadow areas, however, were to be avoided. At the higher limit, an upper bound of 16 deg would still provide very good definition of surface features for most of the critical landing phase near touchdown. Between 16 and 20 deg, lighting was judged to be acceptable for viewing during final descent. A Sun angle above 20 deg was considered unsatisfactory for a manual landing maneuver. The crew report of the absence of sharp color boundaries was significant. The lack of visible contrast from an altitude of 60 nmi reduced the probability that a crew would be able to use color to distinguish geologic units while operating near or on the lunar surface. Just prior to sunrise on one of the early lunar orbit revolutions, the command module pilot observed what was believed to be zodiacal light and solar corona through the telescope. The lunar module pilot observed a cloud or bright area in the sky during lunar darkness on two successive revolutions; if correct, this indicated that one of the Magellanic clouds had been observed. Long-distance Earth photography of general interest highlighted global weather and terrain features. Lunar photography had not been accomplished during the translunar coast because of rigid spacecraft attitude constraints, but good-quality photography of most of the Moon's illuminated disk was subsequently accomplished during the transearth coast.

The crew initially followed the lunar orbit mission plan and performed all scheduled tasks. However, because of crew fatigue, the commander made the decision at 084:30 to cancel all activities during the final 4 hours in lunar orbit to allow the crew to rest. The only activities during this period were a required platform alignment and preparation for transearth injection. A planned 26-min 43-sec television transmission of the Moon and Earth was made at 085:43:03, on Christmas eve. It was during this transmission that the crew read from the Bible the first 10 verses of Genesis, and then wished viewers "Good night, good luck, a Merry Christmas, and God bless all of you, all of you on the good Earth". An estimated one billion people in 64 countries heard or viewed the live reading and greeting; delayed broadcasts reached an additional 30 countries that same day.

Orbit analysis indicated that previously unknown mass concentrations, or 'mascons', were perturbing the orbit. As a result, the final lunar orbit had an apogee and perigee of 63.6 × 58.6 nmi. The 203.7-sec transearth injection maneuver was performed with the service propulsion system at an altitude of 60.2 nmi at 089:19:16.6 after 10 revolutions and 20 hr 10 min 13.0 sec in lunar orbit. The velocity at transearth injection was 8,842 ft/sec. During the mission, the spacecraft reached a maximum distance from Earth of 203,752.37 nmi.

Apollo 8 lunar orbit phase event	GET (h:m:s)	Altitude (nmi)	Space-fixed velocity (ft/sec)	Event duration (sec)	Velocity change (ft/sec)	Apolune (nmi)	Perilune (nmi)
Lunar orbit insertion ignition	069:08:20.4	75.6	8,391	–	–	–	–
Lunar orbit insertion cutoff	069:12:27.3	62.0	5,458	246.9	2,997	168.5	60.0
Lunar orbit circularization ignition	073:35:06.6	59.3	5,479	–	–	–	–
Lunar orbit circularization cutoff	073:35:16.2	60.7	5,345	9.6	134.8	60.7	59.7

TRANSEARTH PHASE

After emerging from lunar occlusion following transearth injection, Apollo 8 experienced the only significant communications difficulty during the mission. Although two-way phaselock was established at 089:28:47, two-way voice contact and telemetry synchronization were not achieved until 089:33:28 and 089:43:00, respectively. Data indicated that high-gain antenna acquisition may have been attempted while line-of-sight was within the service module reflection region and that the reflections may have caused the antenna to track on a side lobe. In addition, the spacecraft was erroneously configured for high-bit-rate transmission; therefore the command at 089:29:29 that configured the spacecraft for normal voice and subsequent playback of the data storage equipment selected an S-band signal combination that was not compatible with the received carrier power.

The transearth coast activities included star/horizon navigation sightings using both Moon and Earth horizons. Passive thermal control using a roll rate of one revolution per hour was employed during most of the translunar and transearth coast phases to maintain nearly stable onboard temperatures. Only one small transearth midcourse correction, a 15.0-sec maneuver using the service module reaction control system, was required at 104:00:00, and changed the velocity by 4.8 ft/sec. A crew procedural error caused the loss of the onboard state vector and platform alignment at 106:26; realignment was performed at 106:45. A special test of the automatic acquisition mode of the high-gain antenna was performed at 110:16:55. Results indicated that the antenna performed as predicted. The final two television transmissions were made during transearth coast. The fifth was a 9-min 31-sec transmission of the spacecraft interior at 104:24:04. The sixth transmission was for 19 min 54 sec at 127:45:33 and featured views of Earth, particularly of the western hemisphere.

Apollo 8 transearth phase event	GET (h:m:s)	Altitude (nmi)	Space-fixed velocity (ft/sec)	Event duration (sec)	Velocity change (ft/sec)	Space-fixed flight path angle (deg)	Space-fixed heading angle (°E of N)
Transearth injection ignition	089:19:16.6	60.2	5,342	–	–	−0.16	−110.59
Transearth injection cutoff	089:22:40.3	66.1	8,842	203.7	3,519.0	5.10	−115.00
Midcourse correction ignition	104:00:00.00	165,561.5	4,299	–	–	−80.59	52.65
Midcourse correction cutoff	104:00:15.00	167,552.0	4,298	15.0	4.8	−80.60	52.65

RECOVERY

The service module was jettisoned at 146:28:48.0. No radar tracking data were available during entry, but photographic coverage information correlated well with the predicted trajectory in altitude, latitude, longitude, and time. The CM followed an automatically guided entry profile, and re-entered Earth's atmosphere (at the 400,000-ft altitude of the 'entry interface') at 146:46:12.8 at a velocity of 36,221.1 ft/sec, after a transearth coast of 57 hr 23 min 32.5 sec. The ionization became so bright during entry that the CM interior was bathed in a cold blue light as bright as daylight. At 180,000 ft, as expected, the lift of the CM deflected it to 210,000 ft, where it then resumed its downward course. The parachute system effected splashdown of the CM in the Pacific Ocean at 15:51:42 GMT on 27 December. Mission duration was 147:00:42.0. The impact point was 1.4 nmi from the target point and 2.6 nmi from the recovery ship USS *Yorktown*. The splashdown site was estimated to be 8.10°N and 165.00°W. As a result of the splashdown impact, the CM initially adopted an apex-down flotation attitude, but the inflatable-bag-uprighting system returned it to the normal flotation position 6 min and 3 sec later. As planned, helicopters and aircraft hovered over the spacecraft and pararescue personnel were not deployed until local sunrise, 43 min after splashdown. At dawn, the crew was retrieved by helicopter and were on the recovery ship 88 min after splashdown. The spacecraft was recovered 60 min later. The estimated CM weight at splashdown was 10,977 lb and the estimated distance traveled for the mission was 504,006 nmi When the recovery swimmers were deployed, the weather recorded on the *Yorktown* showed scattered clouds at 2,000 ft and overcast at 9,000 ft, a visibility of 10 nmi, a wind speed of 19 kt at 70° from true north, a water temperature of 82°F, and waves of up to 6 ft at 110° from true north. The CM was removed from the *Yorktown* on 29 December at Ford Island, Hawaii. The Landing Safing Team began the evaluation and deactivation procedures at 21:00 GMT, and completed this work on 1 January 1969. The CM was then flown to Long Beach, California, and trucked to the North American Rockwell Space Division facility at Downey, California, for post-flight analysis. It arrived on 2 January 1969 at 21:00 GMT.

CONCLUSIONS

With only minor problems, all Apollo 8 spacecraft systems operated as intended, and all the primary mission objectives were successfully accomplished. Crew performance was admirable throughout the mission. Approximately 90% of the photographic objectives were accomplished and 60% of the additional lunar photographs requested as 'targets of opportunity' were also taken, despite fogging of three of the spacecraft windows due to exposure of the window sealant to the space environment and early curtailment of crew activities due to fatigue. Many smaller lunar features, previously unknown, were also photographed. These features were located principally on the far side of the Moon in areas that had been photographed only at much greater distances and lower resolution by automated spacecraft. In addition, the heat shield system was not adversely affected by exposure to cislunar space or to the lunar environment, and performed as expected. The following conclusions were made from an analysis of post-mission data:

1. The CSM systems were operational for a manned lunar mission.
2. All system parameters and consumable quantities were maintained well within their design operating limits during both cislunar and lunar orbit flight.
3. Passive thermal control, a slow rolling maneuver perpendicular to the Sun line, was a satisfactory means of maintaining critical spacecraft temperatures near the middle of the acceptable response ranges.
4. The navigation techniques developed for translunar and lunar orbit flight were proved to be more than adequate to maintain required lunar orbit insertion and transearth injection guidance accuracies.
5. Non-simultaneous sleep periods adversely affected the normal circadian cycle of each crew member and provided a poor environment for undisturbed rest. Mission activity scheduling for the lunar orbit coast phase also did not provide adequate time for required crew rest periods.
6. Communications and tracking at lunar distances were excellent in all modes. The high-gain antenna, flown for the first time, performed exceptionally well and withstood dynamic structural loads and vibrations that exceeded anticipated operating levels.
7. Crew observations of the lunar surface showed the 'washout' effect (surface detail being obscured by backscatter) to be much less severe than anticipated. In addition, smaller surface details were visible in shadow areas at low Sun angles, indicating that lighting for lunar landing should be photometrically acceptable.
8. To accommodate the change in Apollo 8 from an Earth orbital to a lunar mission, pre-mission planning, crew training, and ground support reconfigurations were completed in a time period significantly shorter than usual. The required response was particularly demanding on the crew and, although not desirable on a long-term basis, exhibited a capability that had never before been demonstrated.

The Apollo 8 crew: Bill Anders (left), Jim Lovell and Frank Borman.

AS-503 before the roll back of the Mobile Service Structure.

Apollo 8 lifts off.

Apollo 8's spent S-IVB with its SLA panels jettisoned to expose the rudimentary LTA-B.

After entering lunar orbit, the Apollo 8 crew witnessed their first 'Earthrise'.

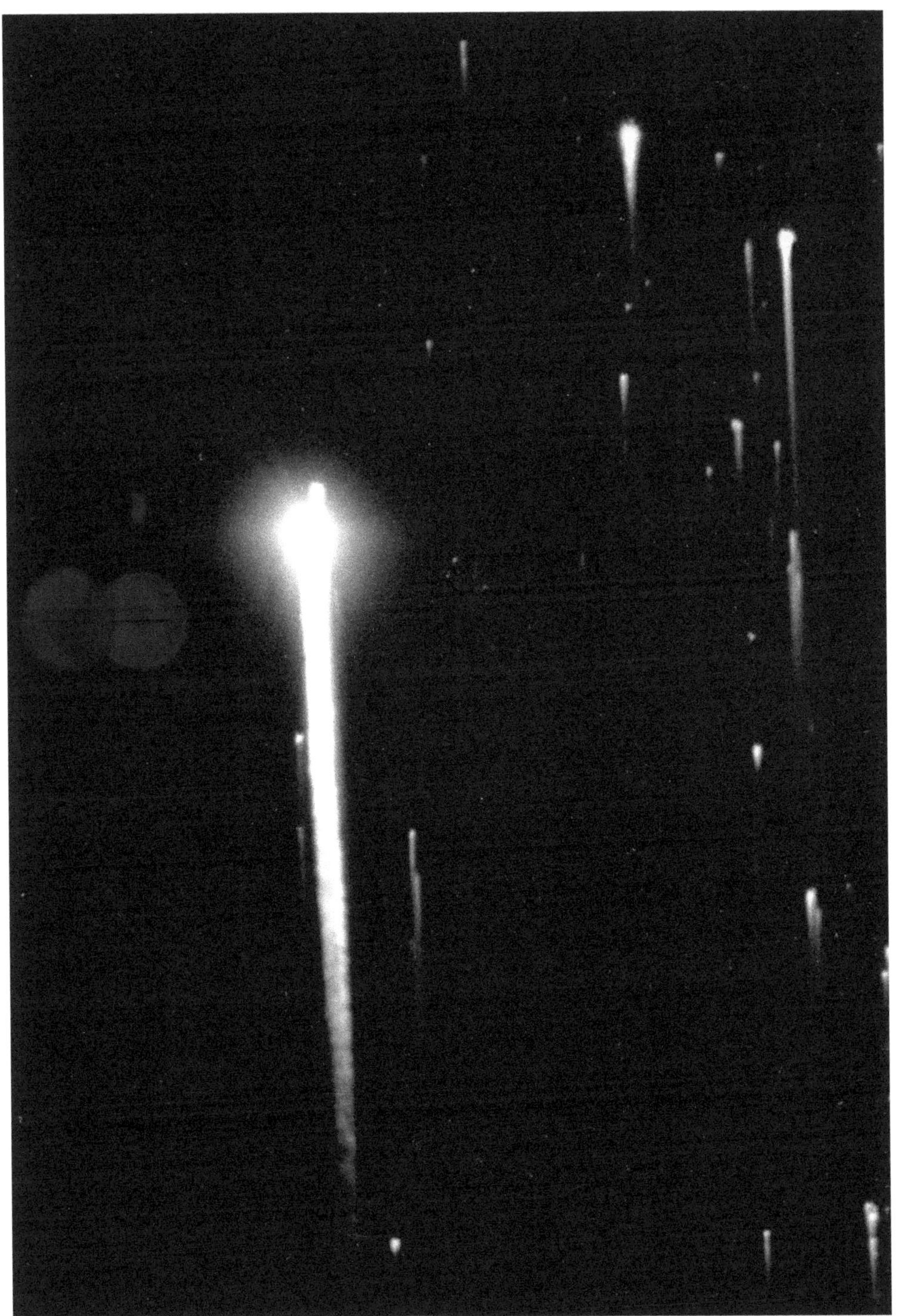

Apollo 8's high-speed re-entry was photographed by an aircraft with an optical tracking system.

The Apollo 8 crew safely on the recovery ship.

MISSION OBJECTIVES

Launch Vehicle Primary Detailed Test Objectives

1. To verify that modifications incorporated in the S-IC stage since the Apollo 6 flight suppress low-frequency longitudinal oscillations (known as 'pogo'). *Achieved.*
2. To confirm the launch vehicle longitudinal oscillation environment during the S-IC stage burn. *Achieved.*
3. To verify the modifications made to the J-2 engine since the Apollo 6 flight. *Achieved.*
4. To confirm the J-2 engine environment in the S-II and S-IVB stages. *Achieved.*
5. To demonstrate the capability of the S-IVB to restart in Earth orbit. *Achieved.*
6. To demonstrate the operation of the S-IVB helium heater repressurization system. *Achieved.*
7. To demonstrate the capability to safe the S-IVB stage in orbit. *Achieved.*
8. To verify the capability to inject the S-IVB/instrument unit/lunar module test article into a lunar 'slingshot' trajectory. *Achieved.*
9. To verify the capability of the launch vehicle to perform a free-return translunar injection. *Achieved.*

Launch Vehicle Secondary Detailed Test Objective

1. To verify the onboard command and communications system and ground system interface and the operation of the command and communications system in the deep space environment. *Achieved.*

Spacecraft Primary Objectives

1. To demonstrate crew/space vehicle/mission support facilities performance during a manned Saturn V mission with the command and service module. *Achieved.*
2. To demonstrate the performance of nominal and selected backup lunar orbit rendezvous mission activities, including:
   (a) Saturn targeting for translunar injection. *Achieved.*
   (b) Long-duration service propulsion burns and midcourse corrections. *Achieved.*
   (c) Pre-translunar injection procedures. *Achieved.*
   (d) Translunar injection. *Achieved.*
   (e) Command and service module orbital navigation. *Achieved.*

Spacecraft Primary Detailed Test Objectives

1. P1.31: To perform a guidance and navigation control system controlled entry from a lunar return. *Achieved.*
2. P1.33: To perform star–lunar horizon sightings during the translunar and transearth phases. *Achieved, although the field of view in the scanning telescope was obscured by what appeared to be particles whenever the telescope optics were repositioned.*
3. P1.34: To perform star–Earth horizon sightings during translunar and transearth phases. *Achieved, although the field of view in the scanning telescope was obscured by what appeared to be particles whenever the telescope optics were repositioned.*
4. P6.11: To perform manual and automatic acquisition, tracking, and communication

with the Manned Space Flight Network using the high-gain command and service module S-band antenna during a lunar mission. *Achieved.*

5. P7.31: To obtain data on the passive thermal control system during a lunar orbit mission. *Achieved.*
6. P7.32: To obtain data on the spacecraft dynamic response. *Achieved.*
7. P7.33: To demonstrate spacecraft/LM adapter panel jettison in a zero-*g* environment. *Achieved.*
8. P20.105: To perform lunar orbit insertion service propulsion system guidance and navigation control system controlled burns with a fully loaded command and service module. *Achieved.*
9. P20.106: To perform a transearth insertion guidance and navigation control system controlled service propulsion system burn. *Achieved.*
10. P20.107: To obtain data on the command module crew procedures and timeline for lunar orbit mission activities. *Achieved.*
11. P20.109: To demonstrate command service module passive thermal control modes and related communication procedures during a lunar orbit mission. *Achieved.*
12. P20.110: To demonstrate ground operational support for a command and service module lunar orbit mission. *Achieved.*
13. P20.111: To perform lunar landmark tracking in lunar orbit from the command and service module. (The intent of this objective was to establish that an onboard capability existed to compute relative position data for the lunar landing mission. This mode was to be used in conjunction with the Manned Space Flight Network state-vector update.) *Partially achieved. All portions of the objective were satisfied except for the functional test, which required the use of onboard data to determine the error uncertainties in the landing site location. A procedural error caused the time intervals between the mark designations to be too short; thus, the data may have been correct but may not have been representative. The accuracy of the onboard capability was not determined because the data analysis was not complete at the time the mission report was published. Sufficient data were obtained to determine that no constraint existed for subsequent missions. A demonstration of this technique was planned for the next lunar mission.*
14. P20.112: To prepare for translunar injection and monitor the guidance and navigation control system and launch vehicle tank pressure displays during the translunar injection burn. *Achieved.*
15. P20.114: To perform translunar and transearth midcourse corrections. *Achieved, although the service propulsion system engine experienced a momentary drop in chamber pressure from 94 to 50 psi during the SPS burn for the midcourse correction, and the entry monitoring system velocity counter counted through zero at the termination of the transearth midcourse correction.*

Spacecraft Secondary Detailed Test Objectives

1. S1.27: To monitor the guidance and navigation control system and displays during launch. *Achieved.*
2. S1.30: To obtain inertial measurement unit performance data in the flight environment. *Achieved.*

3. S1.32: To perform star–Earth landmark sighting navigation during translunar and transearth phases. *Partially achieved. The three sets of sightings required at less than 50,000 nmi altitude were not obtained. The accuracy of other navigation modes was sufficient to preclude the necessity of using star–Earth landmarks for midcourse navigation. No constraint on subsequent missions resulted from this problem.*
4. S1.35: To perform an inertial measurement unit alignment and a star pattern visibility check in daylight. *Achieved.*
5. S3.21: To perform service propulsion system lunar orbit injection and transearth injection burns and monitor the primary and auxiliary gauging systems. *Achieved.*
6. S4.5: To obtain data on the Block II environmental control system performance during manned lunar return entry conditions. *Achieved, although the #2 cabin fan was noisy.*
7. S6.10: To communicate with the Manned Space Flight Network using the command and service module S-band omnidirectional antennas at lunar distance. *Achieved.*
8. S7.30: To demonstrate the performance of the Block II thermal protection system during a manned lunar return entry. *Achieved.*
9. S20.104: To perform a command and service module/S-IVB separation and a command and service module transposition on a lunar mission timeline. *Achieved.*
10. S20.108: To obtain data on command and service module consumables for a command and service module lunar orbit mission. *Achieved.*
11. S20.115: To obtain photographs during the transearth, translunar and lunar orbit phases for operational and scientific purposes. *Achieved, although the hatch and side windows were obscured by fog or frost throughout the mission.*
12. S20.116: To obtain data to determine the effect of the tower jettison motor, S-II retro and service module reaction control system exhausts and other sources of contamination on the command module windows. *Achieved. The hatch and side windows were obscured by fog or frost throughout the mission.*

Functional Tests Added In-Flight

1. P1.34: Star–Earth horizon photography through the sextant. *Achieved.*
2. P1.34: Midcourse navigation with helmets on. *Achieved.*
3. P1.34: Navigation with long eyepiece. *Achieved.*
4. P6.11: High-gain antenna, automatic reacquisition. *Achieved.*
5. P20.109: Passive thermal control, roll rate of 0.3 deg/sec. *Achieved.*

MISSION TIMELINE

Apollo 8 mission event	GET (h:m:s)	Date (GMT)	Time (h:m:s)
Countdown started.	–103:00:00	16 Dec 1968	00:00:00
Terminal countdown started.	–028:00:00	20 Dec 1968	01:51:00
Scheduled 6-hr hold at T–9 hr.	–009:00:00	20 Dec 1968	20:51:00
Countdown resumed at T–9 hr.	–009:00:00	21 Dec 1968	02:51:00
Scheduled 1-hr hold at T–3 hr 30 min.	–003:30:00	21 Dec 1968	08:21:00
Countdown resumed at T–3 hr 30 min.	–003:30:00	21 Dec 1968	09:21:00
Crew ingress.	–002:53	21 Dec 1968	09:58
Guidance reference release.	–000:00:16.970	21 Dec 1968	12:50:43
S-IC engine start command.	–000:00:08.89	21 Dec 1968	12:50:51
S-IC engine ignition (#5).	–000:00:06.585	21 Dec 1968	12:50:53
All S-IC engines thrust OK.	–000:00:01.387	21 Dec 1968	12:50:58
Range zero.	000:00:00.00	21 Dec 1968	12:51:00
All holddown arms released.	000:00:00.27	21 Dec 1968	12:51:00
1st motion (1.16 *g*).	000:00:00.33	21 Dec 1968	12:51:00
Liftoff (umbilical disconnected).	000:00:00.67	21 Dec 1968	12:51:00
Tower clearance yaw maneuver started.	000:00:01.76	21 Dec 1968	12:51:01
Yaw maneuver ended.	000:00:09.72	21 Dec 1968	12:51:09
Pitch and roll maneuver started.	000:00:12.11	21 Dec 1968	12:51:12
Roll maneuver ended.	000:00:31.52	21 Dec 1968	12:51:31
Mach 1 achieved.	000:01:01.48	21 Dec 1968	12:52:01
Maximum bending moment achieved (60,000,000 lbf-in).	000:01:14.7	21 Dec 1968	12:52:14
Maximum dynamic pressure (776.938 lb/ft^2).	000:01:18.9	21 Dec 1968	12:52:18
S-IC center engine cutoff command.	000:02:05.93	21 Dec 1968	12:53:05
Pitch maneuver ended.	000:02:25.50	21 Dec 1968	12:53:25
S-IC outboard engine cutoff.	000:02:33.82	21 Dec 1968	12:53:33
S-IC maximum total inertial acceleration (3.96 *g*).	000:02:33.92	21 Dec 1968	12:53:33
S-IC maximum Earth-fixed velocity; S-IC/S-II separation command.	000:02:34.47	21 Dec 1968	12:53:34
S-II engine start command.	000:02:35.19	21 Dec 1968	12:53:35
S-II ignition.	000:02:36.19	21 Dec 1968	12:53:36
S-II aft interstage jettisoned.	000:03:04.47	21 Dec 1968	12:54:04
Launch escape tower jettisoned.	000:03:08.6	21 Dec 1968	12:54:08
Iterative guidance mode initiated.	000:03:16.22	21 Dec 1968	12:54:16
S-IC apex.	000:04:26.54	21 Dec 1968	12:55:26
S-II engine cutoff.	000:08:44.04	21 Dec 1968	12:59:44
S-II maximum total inertial acceleration (1.86 *g*).	000:08:44.14	21 Dec 1968	12:59:44
S-II maximum Earth-fixed velocity; S-II/S-IVB separation command.	000:08:44.90	21 Dec 1968	12:59:44
S-IVB 1st burn start command.	000:08:45.00	21 Dec 1968	12:59:45
S-IVB 1st burn ignition.	000:08:48.29	21 Dec 1968	12:59:48
S-IVB ullage case jettisoned.	000:08:56.8	21 Dec 1968	12:59:56
S-IC impact (theoretical).	000:09:00.41	21 Dec 1968	13:00:00
S-II apex.	000:09:20.34	21 Dec 1968	13:00:20

Apollo 8 mission event – *continued*	GET (h:m:s)	Date (GMT)	Time (h:m:s)
S-IVB 1st burn cutoff.	000:11:24.98	21 Dec 1968	13:02:25
S-IVB 1st burn maximum total inertial acceleration (0.72 *g*).	000:11:25.08	21 Dec 1968	13:02:25
S-IVB 1st burn maximum Earth-fixed velocity.	000:11:25.50	21 Dec 1968	13:02:25
Earth orbit insertion.	000:11:34.98	21 Dec 1968	13:02:35
Maneuver to local horizontal attitude started.	000:11:45.19	21 Dec 1968	13:02:45
Orbital navigation started.	000:13:05.19	21 Dec 1968	13:04:05
S-II impact (theoretical).	000:19:25.106	21 Dec 1968	13:10:25
Optics cover jettisoned.	000:42:05	21 Dec 1968	13:33:05
All spacecraft systems approved for translunar injection.	001:56:00	21 Dec 1968	14:47:00
CAPCOM (Mike Collins): "All right, Apollo 8. You are go for TLI."	002:27:22	21 Dec 1968	15:18:22
S-IVB 2nd burn restart preparation.	002:40:59.54	21 Dec 1968	15:31:59
S-IVB 2nd burn restart command.	002:50:29.51	21 Dec 1968	15:41:29
S-IVB 2nd burn ignition.	002:50:37.79	21 Dec 1968	15:41:37
S-IVB 2nd burn cutoff.	002:55:55.51	21 Dec 1968	15:46:55
S-IVB 2nd burn maximum total inertial acceleration (1.55 *g*).	002:55:55.61	21 Dec 1968	15:46:55
S-IVB LH2 tank latch relief valve open.	002:55:55.91	21 Dec 1968	15:46:55
S-IVB 2nd burn maximum Earth-fixed velocity.	002:55:56.00	21 Dec 1968	15:46:56
S-IVB LH_2 tank CVS valve open/ S-IVB safing procedures started.	002:55:56.19	21 Dec 1968	15:46:56
S-IVB LOX tank non-propulsive vent valve open.	002:55:56.42	21 Dec 1968	15:46:56
Translunar injection.	002:56:05.51	21 Dec 1968	15:47:05
Maneuver to local horizontal attitude and orbital navigation started.	002:56:15.77	21 Dec 1968	15:47:15
S-IVB LOX tank non-propulsive vent valve closed.	002:58:26.39	21 Dec 1968	15:49:26
S-IVB LH_2 tank CVS valve and tank relief valve closed.	003:10:55.71	21 Dec 1968	16:01:55
Maneuver to transposition and docking attitude started.	003:10:58.40	21 Dec 1968	16:01:58
Sequence to separate CSM from S-IVB/LTA started. High-gain antenna deployed.	003:20:56.3	21 Dec 1968	16:11:56
CSM separated from S-IVB.	003:20:59.3	21 Dec 1968	16:11:59
1st CSM evasive maneuver from S-IVB (RCS).	003:40:01	21 Dec 1968	16:31:01
S-IVB LH_2 tank latch relief valve open.	003:55:56.16	21 Dec 1968	16:46:56
S-IVB LH_2 tank latch relief valve closed.	004:10:55.77	21 Dec 1968	17:01:55
Last reported VHF uplink reception.	004:39:54	21 Dec 1968	17:30:54
S-IVB lunar slingshot attitude maneuver initiated.	004:44:56.63	21 Dec 1968	17:35:56
2nd CSM evasive maneuver from S-IVB (RCS).	004:45:01	21 Dec 1968	17:36:01
Last reported VHF downlink reception.	004:48	21 Dec 1968	17:39
S-IVB lunar slingshot maneuver – LH_2 tank CVS vent valve open command.	004:55:56.02	21 Dec 1968	17:46:56
S-IVB lunar slingshot maneuver – LOX dump started.	005:07:55.82	21 Dec 1968	17:58:55
S-IVB lunar slingshot maneuver – Apply velocity change.	005:07:56.03	21 Dec 1968	17:58:56
S-IVB start bottle vent dump start.	005:08:25.82	21 Dec 1968	17:59:25
S-IVB start bottle vent dump end.	005:10:55.83	21 Dec 1968	18:01:55
S-IVB pneumatic sphere dump start.	005:12:25.83	21 Dec 1968	18:03:25
S-IVB LOX dump end.	005:12:55.82	21 Dec 1968	18:03:55

Apollo 8 mission event – *continued*	GET (h:m:s)	Date (GMT)	Time (h:m:s)
S-IVB lunar slingshot maneuver – LOX dump ended.	005:12:56.03	21 Dec 1968	18:03:56
S-IVB LOX tank non-propulsive vent valve open.	005:12:59.0	21 Dec 1968	18:03:59
S-IVB LH_2 tank latch relief valve open.	005:13:01.23	21 Dec 1968	18:04:01
S-IVB cold helium dump start.	005:13:03.6	21 Dec 1968	18:04:03
S-IVB lunar slingshot maneuver – APS ignition.	005:25:55.85	21 Dec 1968	18:16:55
S-IVB lunar slingshot maneuver – APS cutoff.	005:38:08.56	21 Dec 1968	18:29:08
S-IVB lunar slingshot maneuver – APS depletion.	005:38:34.00	21 Dec 1968	18:29:34
S-IVB cold helium dump end.	006:03:03.5	21 Dec 1968	18:54:03
S-IVB pneumatic sphere dump end.	006:11:05.88	21 Dec 1968	19:02:05
1st use of high gain antenna.	006:33:04	21 Dec 1968	19:24:04
Midcourse correction ignition.	010:59:59.2	21 Dec 1968	23:50:59
Midcourse correction cutoff.	011:00:01.6	21 Dec 1968	23:51:01
Data processing by missions operations computer and backup computer lost for 10 min due to undesirable instruction sequence.	011:51:00	22 Dec 1968	00:42:00
S-band mode testing started.	012:03:01	22 Dec 1968	00:54:01
1st television transmission started.	031:10:36	22 Dec 1968	20:01:36
1st television transmission ended.	031:24:13	22 Dec 1968	20:15:13
2nd television transmission started.	055:02:45	23 Dec 1968	19:53:45
2nd television transmission ended.	055:28:23	23 Dec 1968	20:19:23
Equigravisphere.	055:38	23 Dec 1968	20:29
Midcourse correction ignition.	060:59:55.9	24 Dec 1968	01:50:55
Midcourse correction cutoff.	061:00:07.8	24 Dec 1968	01:51:07
CAPCOM (Jerry Carr): "Apollo 8, this is Houston. At 68:04, you are go for LOI."	068:04:07	24 Dec 1968	08:55:07
CAPCOM: "Apollo 8, Houston. One minute to LOS. All systems go."	068:57:06	24 Dec 1968	09:48:00
CAPCOM: "Safe journey, guys."	068:57:19	24 Dec 1968	09:48:19
LMP (Anders): "Thanks a lot, troops."	068:57:24	24 Dec 1968	09:48:24
CMP (Lovell): "We'll see you on the other side."	068:57:26	24 Dec 1968	09:48:26
CAPCOM: "Apollo 8, 10 sec to go. You're go all the way."	068:57:54	24 Dec 1968	09:48:54
CDR (Borman): "Roger."	068:58:00	24 Dec 1968	09:49:00
Lunar orbit insertion ignition.	069:08:20.4	24 Dec 1968	09:59:20
Lunar orbit insertion cutoff.	069:12:27.3	24 Dec 1968	10:03:27
CAPCOM: "Apollo 8, Houston. Over."	069:33:44	24 Dec 1968	10:24:44
CMP: "Go ahead, Houston. This is Apollo 8. Burn complete..."	069:33:52	24 Dec 1968	10:24:52
CAPCOM: "Apollo 8, this is Houston. Roger...good to hear your voice."	069:34:07	24 Dec 1968	10:25:07
S-IVB closest approach to lunar surface.	069:58:55.2	24 Dec 1968	10:49:55
Control point sightings.	071:00	24 Dec 1968	11:51
16-mm camera photography started.	071:10	24 Dec 1968	12:01
3rd television transmission started.	071:40:52	24 Dec 1968	12:31:52
3rd television transmission ended.	071:52:52	24 Dec 1968	12:43:52
Pseudo-landing site sightings.	071:55	24 Dec 1968	12:46
16-mm photography stopped.	072:20	24 Dec 1968	13:11

Apollo 8 mission event – *continued*	GET (h:m:s)	Date (GMT)	Time (h:m:s)
Lunar orbit circularization ignition.	073:35:06.6	24 Dec 1968	14:26:06
Lunar orbit circularization cutoff.	073:35:16.2	24 Dec 1968	14:26:16
Training photography.	074:00	24 Dec 1968	14:51
CSM landmark tracking and photography.	074:15	24 Dec 1968	15:06
Stereo photography started.	075:20	24 Dec 1968	16:11
Stereo photography ended.	076:00	24 Dec 1968	16:51
Landmark lighting evaluation.	076:15	24 Dec 1968	17:06
Control point sightings.	077:20	24 Dec 1968	18:11
Pseudo-landing site sightings.	078:00	24 Dec 1968	18:51
Control point sightings.	079:20	24 Dec 1968	20:11
Pseudo-landing site sightings.	080:00	24 Dec 1968	20:51
Control point sightings.	081:20	24 Dec 1968	22:11
Pseudo-landing site sightings.	082:00	24 Dec 1968	22:51
4th television transmission started.	085:43:03	25 Dec 1968	02:34:03
LMP: "We are now approaching the lunar sunrise, and for all the people back on Earth, the crew of Apollo 8 has a message that we would like to send to you. In the beginning. . ." (reading from the Bible).	086:06:56	25 Dec 1968	02:57:56
CMP: "And God called the light Day. . ."	086:07:29	25 Dec 1968	02:58:29
CDR: "And from the crew of Apollo 8, we close with good night, good luck, a merry Christmas, and God bless all of you, all of you on the good Earth."	086:08:36	25 Dec 1968	02:59:36
4th television transmission ended.	086:09:46	25 Dec 1968	03:00:46
Maneuver to transearth injection attitude.	087:15	25 Dec 1968	04:06
CAPCOM: (Ken Mattingly) "Okay, Apollo 8. . .you have a go for TEI."	088:03:36	24 Dec 1968	04:54:36
Transearth injection ignition (SPS).	089:19:16.6	25 Dec 1968	06:10:16
Transearth injection cutoff.	089:22:40.3	25 Dec 1968	06:13:40
Two-way communication phaselock established, but no voice or telemetry.	089:28:47	25 Dec 1968	06:19:47
Two-way voice synchronization established.	089:33:28	25 Dec 1968	06:24:28
CMP: "Houston, Apollo 8, over."	089:34:16	25 Dec 1968	06:25:16
CAPCOM: "Hello, Apollo 8. Loud and clear."	089:34:19	25 Dec 1968	06:25:19
CMP: "Roger. Please be informed there IS a Santa Claus."	089:34:25	25 Dec 1968	06:25:25
CAPCOM: "That's affirmative. You are the best ones to know."	089:34:31	25 Dec 1968	06:25:31
Two-way telemetry synchronization established.	089:43:00	25 Dec 1968	06:34:00
Midcourse correction ignition.	104:00:00	25 Dec 1968	20:51:00
Midcourse correction cutoff.	104:00:15	25 Dec 1968	20:51:15
5th television transmission started.	104:24:04	25 Dec 1968	21:15:04
5th television transmission ended.	104:33:35	25 Dec 1968	21:24:35
Onboard state vector and platform alignment data corrupted due to crew error.	106:26	25 Dec 1968	23:17
State vector and platform alignment data corrected.	106:45	25 Dec 1968	23:36
Test of high-gain antenna automatic acquisition.	110:16:55	26 Dec 1968	03:07:55
6th television transmission started.	127:45:33	26 Dec 1968	20:36:33

Apollo 8 mission event – *continued*	GET (h:m:s)	Date (GMT)	Time (h:m:s)
6th television transmission ended.	128:05:27	26 Dec 1968	20:56:27
1st reception of ground VHF during transearth coast.	142:16:00	27 Dec 1968	11:07:00
CM/SM separation.	146:28:48.0	27 Dec 1968	15:19:48
Entry.	146:46:12.8	27 Dec 1968	15:37:12
Communication blackout started.	146:46:37	27 Dec 1968	15:37:37
Maximum entry *g* force (6.84 *g*).	146:47:38.4	27 Dec 1968	15:38:38
Recovery aircraft received direction-finding signals from CM and established visual contact.	146:49	27 Dec 1968	15:40
Radar contact with CM established by recovery ship at 270 nmi.	146:50	27 Dec 1968	15:41
Radar contact with CM established by recovery ship at 109 nmi.	146:51	27 Dec 1968	15:42
Communication blackout ended.	146:51:42.0	27 Dec 1968	15:42:42
Radar contact with CM established by recovery ship at 60 nmi.	146:52	27 Dec 1968	15:43
Drogue parachute deployed	146:54:47.8	27 Dec 1968	15:45:47
Main parachute deployed.	146:55:38.9	27 Dec 1968	15:46:38
Voice contact established with CM by recovery helicopter. Recovery beacon signal contact established with CM by recovery aircraft.	146:56:01	27 Dec 1968	15:47
Flashing light on CM visible on recovery ship.	146:57:05	27 Dec 1968	15:48:05
Splashdown.	147:00:42.0	27 Dec 1968	15:51:42
CM went to apex down position. Voice contact lost.	147:00:50	27 Dec 1968	15:51:50
CM returned to apex-up position.	147:07:45	27 Dec 1968	15:58:45
Swimmers deployed to CM.	147:44	27 Dec 1968	16:35
Flotation collar inflated.	148:07	27 Dec 1968	16:58
CM hatch opened.	148:12	27 Dec 1968	17:03
Crew in life raft.	148:15	27 Dec 1968	17:06
Crew on board recovery helicopter.	148:23	27 Dec 1968	17:14
Crew on board recovery ship.	148:29	27 Dec 1968	17:20
Recovery ship arrived at CM.	149:22	27 Dec 1968	18:13
CM on board recovery ship.	149:29	27 Dec 1968	18:20
Deactivation of CM started at Ford Island, Hawaii.	200:09	29 Dec 1968	21:00
CM deactivation completed.	–	01 Jan 1969	–
CM arrived at contractor's facility in Downey, CA.	296:09	02 Jan 1969	21:00

Apollo 9

The third manned mission: testing the LM in Earth orbit

3–13 March 1969

BACKGROUND

Apollo 9 was a Type 'D' mission, a lunar module manned flight demonstration in Earth orbit. It was the first manned test of the 'lunar ferry' that would put astronauts on the Moon. A lunar module had first flown without a crew during Apollo 5 on 22 January 1968. Many of the LM tests on Apollo 9 would exceed conditions expected in a lunar landing. To ensure the major objectives would be accomplished if Apollo 9 ended early, the schedule for the first half of the mission also included more work for the crew than the schedule of either Apollo 7 or Apollo 8.

The primary objectives were:

- to demonstrate crew, space vehicle, and mission support facilities performance during a manned Saturn V mission with command and service modules and lunar module;
- to demonstrate lunar module crew performance;
- to demonstrate performance of nominal and selected backup lunar orbit rendezvous mission activities; and
- to assess command and service module and lunar module consumables.

To meet these objectives, the lunar module was evaluated during three separate piloting periods that required multiple activation and deactivation of systems, a situation unique to this mission.

The crew members were Colonel James Alton McDivitt (USAF), commander; Colonel David Randolph Scott (USAF), command module pilot; and Russell Louis 'Rusty' Schweickart, lunar module pilot. Selected in the astronaut group of 1962, McDivitt had been command pilot of Gemini 4. Born on 10 June 1929 in Chicago, Illinois, he was 39 years old at the time of the Apollo 9 mission. McDivitt received a BS in aeronautical engineering from the University of Michigan in 1959. His backup for the mission was Commander Charles 'Pete' Conrad Jr (USN). Scott had been pilot of Gemini 8. Born on 6 June 1932 in San Antonio, Texas, he was 36 years old at the time of the Apollo 9 mission. Scott received a BS from the US Military Academy in 1954 and an MS in aeronautics and astronautics from the Massachusetts Institute of Technology in 1962. He was selected as an astronaut in 1963. His backup was Commander Richard Francis 'Dick' Gordon Jr (USN). Schweickart, a civilian, was making his first spaceflight. Born on 25 October 1935 in Neptune, New Jersey, he was 33 years old at the time of the Apollo 9 mission. Schweickart received a BS in aeronautical engineering in 1956 and an MS in aeronautics and astronautics in 1963 from the Massachusetts Institute of Technology. His backup was Commander Alan LaVern Bean (USN). The capsule communicators (CAPCOMs) for the mission were Major Stuart Allen Roosa (USAF), Lieutenant Commander Ronald Ellwin Evans (USN), Major Alfred Merrill Worden (USAF), Conrad, Gordon, and Bean. The support crew were Major Jack Robert Lousma (USMC), Lieutenant Commander Edgar Dean Mitchell (USN/Sc.D.), and Worden. The flight directors were Eugene F. Kranz (first shift), Gerald D. Griffin (second shift), and M.P. 'Pete' Frank (third shift).

The Apollo 9 launch vehicle was a Saturn V, designated AS-504. The mission also carried the designation Eastern Test Range #9025. The CSM was designated CSM-104 and had the call-sign 'Gumdrop', derived from the appearance of the command module when it was transported on Earth: during shipment, it was covered in blue wrappings that gave it the appearance of a wrapped gumdrop. The lunar module was designated LM-3 and had the call-sign 'Spider', derived from its arachnid-like configuration.

LAUNCH PREPARATIONS

The launch was originally scheduled for 28 February 1969, and the terminal countdown had begun for that launch at 03:00:00 GMT on 27 February at T–28 hr. However, one-half hour into the scheduled 3-hr hold at T–16 hr, the countdown was recycled to T–42 hr to allow the crew to recover from a mild viral respiratory illness. The count was resumed at 07:30:00 GMT on 1 March. A low-pressure disturbance southwest of Cape Kennedy in the Gulf of Mexico was the principal cause of overcast conditions. At launch time, stratocumulus clouds covered 70% of the sky (base 3,500 ft) and altostratus clouds covered 100% (base 9,000 ft), the temperature was 67.3°F, the relative humidity was 61%, and the barometric pressure was 14.642 lb/in^2. The winds, as measured by the anemometer on the light pole 60.0 ft above ground at the launch site, were 13.4 kt at 160° from true north.

ASCENT PHASE

Apollo 9 was launched from Pad A of Launch Complex 39 at the Kennedy Space Center at a Range Zero time of 16:00:00 GMT (11:00:00 EST) on 3 March 1969, well within the launch window which remained open to 19:15:00 GMT. Between T+13.3 and T+33.0 sec, the vehicle rolled from a pad azimuth of 90°E of N to a flight azimuth of 72°E of N. The maximum wind conditions encountered during ascent were 148.1 kt at 264° from true north at 38,480 ft, with a maximum wind shear of 0.0254/sec at 48,160 ft. The S-IC shut down at T+162.76, followed by S-IC/S-II separation and S-II ignition. The S-II shut down at T+536.22, followed by separation from the S-IVB, which ignited at T+540.82.

Apollo 9 preparation event	Date
LM-3 integrated test at factory.	31 Jan 1968
S-II-4 stage delivered to KSC.	15 May 1968
LM-3 final engineering evaluation acceptance test at factory.	17 May 1968
Descent stage of LM-3 ready to ship from factory to KSC.	04 Jun 1968
Descent stage of LM-3 delivered to KSC.	09 Jun 1968
Ascent stage of LM-3 ready to ship from factory to KSC.	12 Jun 1968
Ascent stage of LM-3 delivered to KSC.	14 Jun 1968
LM-3 stages mated.	30 Jun 1968
LM-3 combined systems test completed.	01 Jul 1968
Individual and combined CM and SM systems test completed at factory.	20 Jul 1968
LM-3 reassigned to Apollo 9.	19 Aug 1968
Integrated CM and SM systems test completed at factory.	31 Aug 1968
S-IVB-504 stage delivered to KSC.	12 Sep 1968
LM-3 altitude tests completed.	27 Sep 1968
S-IC-4 stage delivered to KSC.	30 Sep 1968
Saturn V instrument unit IU-504 delivered to KSC.	30 Sep 1968
CM-104 and SM-104 ready to ship from factory to KSC.	05 Oct 1968
CM-104 and SM-104 delivered to KSC.	05 Oct 1968
CM-104 and SM-104 mated.	08 Oct 1968
CSM-104 combined systems test completed.	24 Oct 1968
CSM-104 altitude tests completed.	18 Nov 1968
CSM-104 mated to space vehicle.	03 Dec 1968
CSM-104 moved to VAB.	03 Dec 1968
LM-3 combined systems test completed.	07 Dec 1968
CSM-104 integrated systems test completed.	11 Dec 1968
CSM-104 electrically mated to launch vehicle.	26 Dec 1968
Space vehicle overall test completed.	27 Dec 1968
Space vehicle and MLP-2 transferred to launch complex 39A.	03 Jan 1969
Space vehicle flight readiness test completed.	18 Jan 1969
LM-3 flight readiness test completed.	19 Jan 1969
Space vehicle countdown demonstration test (wet) completed.	11 Feb 1969
Space vehicle countdown demonstration test (dry) completed.	12 Feb 1969
Terminal countdown initiated.	26 Feb 1969
Terminal countdown interrupted due to illness of crew.	27 Feb 1969
Terminal countdown reinitiated following crew medical clearance.	01 Mar 1969

The first S-IVB cutoff occurred at T + 664.66, with deviations from the planned trajectory of +2.86 ft/sec in velocity and –0.17 nmi in altitude. The S-IC impacted at T + 536.436 in the Atlantic Ocean at 30.183°N and 74.238°W, 346.64 nmi from the launch site. The S-II impacted at T + 1,225.346 in the Atlantic Ocean at 31.462°N and 34.041°W, 2,413.2 nmi from the launch site. At insertion, at T + 674.66 (i.e. S-IVB cutoff plus 10 sec to account for engine tailoff and other transient effects), the parking orbit showed an apogee and perigee of 100.74 × 99.68 nmi, an inclination of 32.552 deg, a period of 88.20 min, and a velocity of 25,569.78 ft/sec. The apogee and perigee were based upon a spherical Earth with a radius of 3,443.934 nmi.

Apollo 9 ascent phase event	GET (h:m:s)	Altitude (nmi)	Range (nmi)	Earth-fixed velocity (ft/sec)	Space-fixed velocity (ft/sec)	Event duration (sec)	Geocentric latitude (°N)	Longitude (°E)	Space-fixed flight path angle (deg)	Space-fixed heading angle (°E of N)
Liftoff	000:00:00.67	0.032	000.0	1.8	1,340.7	–	28.4470	–80.6041	0.08	90.00
Mach 1 achieved	000:01:08.2	4.243	1.383	1,088.4	2,100.7	–	28.4545	–80.5794	26.35	84.50
Maximum dynamic pressure	000:01:25.5	7.429	3.789	1,737.7	2,783.2	–	28.4666	–80.5369	28.08	81.87
S-IC center engine cutoff*	000:02:14.34	22.459	24.602	5,154.1	6,329.49	140.64	28.5720	–80.1602	22.5766	76.420
S-IC outboard engine cutoff	000:02:42.76	34.808	51.596	7,793.3	9,013.71	169.06	28.7071	–79.6718	18.5394	75.335
S-IC/S-II separation*	000:02:43.45	35.144	52.410	7,837.89	9,059.28	–	28.7111	–79.6571	18.449	75.337
S-II engine cutoff	000:08:56.22	100.735	830.505	21,431.9	22,753.54	371.06	31.6261	–65.0422	0.9177	81.872
S-II/S-IVB separation*	000:08:57.18	100.794	833.794	21,440.5	22,762.27	–	31.6343	–64.9786	0.906	81.907
S-IVB 1st burn cutoff	000:11:04.66	103.156	1,296.775	24,240.6	25,563.98	123.84	32.4266	–55.9293	–0.0066	86.979
Earth orbit insertion	000:11:14.66	103.154	1,335.515	24,246.39	25,569.78	–	32.4599	–55.1658	–0.0058	87.412

* Only the commanded time is available for this event.

The COSPAR designation for the CSM upon achieving orbit was 1969-018A; the S-IVB was designated 1969-018B. After undocking, the LM ascent stage would be designated 1969-018C and the descent stage 1969-018D.

EARTH ORBIT PHASE

After the post-insertion checkout, the CSM was separated from the S-IVB stage at 002:41:16.0. The SLA panels that housed the LM and shielded it from the rigors of launch were then jettisoned. The CM was turned so its apex, holding the docking probe, faced the LM. Docking with the LM was completed at 003:01:59.3. The commander and lunar module pilot then initiated preparations for their eventual entry into the LM. They pressurized the tunnel between the two spacecraft, and with the aid of the CMP, removed the CM hatch and checked the latches on the docking ring to verify the seal. Then they connected the electrical umbilical lines that would provide power to the LM while docked to the CM. The hatch was then replaced.

At 004:08:09, an ejection mechanism, being used for the first time, ejected the docked spacecraft from the S-IVB. Following a separation maneuver, the S-IVB was restarted at 004:45:55.54, burned for 62.06 sec, and 10 sec later the stage entered a 1,671.58 × 105.75 nmi intermediate coasting orbit that would enable the engine to cool down prior to being restarted within one revolution. The period of the orbit was 119.22 min, the inclination was 32.302 deg, and the velocity at insertion was 27,753.61 ft/sec. At 005:59:01.07, the crew performed the first of eight service propulsion system firings, a 5.23-sec maneuver that raised the CSM/LM orbit to 127.6 × 111.3 nmi to improve ground tracking lighting conditions for the upcoming rendezvous. The third and final S-IVB ignition at 006:07:19.26 was a 242.06-sec maneuver to demonstrate restart capability after the 80-min coast and to test the engine performance under 'out-of-specification' conditions. The escape orbit was achieved 10 sec after shutdown, and the velocity was 31,619.85 ft/sec. S-IVB performance was not as predicted due to various anomalies, including the failure of an LH_2 and LOX dump. The LH_2 dump through the engine could not be accomplished due to loss of pneumatic control of the engine valves. The LOX dump was not performed due loss of engine pneumatic control during the third burn. However, the LOX tank was satisfactorily safed by utilizing the LOX non-propulsive venting system. The third ignition also served to place the S-IVB into a solar orbit with an aphelion and perihelion of 80,280,052 × 69,417,732 nmi, an inclination of 24.390 deg, an eccentricity of 0.07256, and a period of 325.8 days.

Crew activity on the second day was devoted to systems checks, pitch, roll and yaw maneuvers, and the second, third, and fourth service propulsion system burns while docked to the LM. The second burn, a 110.29-sec maneuver at 022:12:04.07, raised the orbit to 192.5 × 110.7 nmi. The third burn, at 025:17:39.27, lasted 279.88 sec. It raised the orbit to 274.9 × 112.6 nmi and lightened the spacecraft so that it could be controlled by the reaction control system engines later in the mission and be in a better rescue position for rendezvous activities. During these two burns, tests were made to measure the oscillatory response of a docked spacecraft to provide data to improve the autopilot response for this configuration. The fourth burn, at 028:24:41.37, was a 27.87-sec phasing maneuver to shift the node east and put the spacecraft in a better position later for lighting, braking, and docking.

On the third day, at 043:15, the lunar module pilot transferred to the LM to activate and check out the systems. The commander followed at 044:05. The LM landing gear was deployed at 045:00. At about 045:52, the commander reported that the lunar module pilot had been sick on two occasions and that the crew was behind in the timeline. For these reasons, the extravehicular activity was restricted to one daylight pass and would include only the opening of the hatches of the CM and LM. It was also decided to keep the lunar module pilot connected to the environmental control system hoses. Following communication checks for both vehicles, a 5-min television transmission was broadcast at 046:28 from inside the LM. The camera was trained on the instrument displays, other features of the LM interior, and the crew. The picture was good, but the sound was unsatisfactory. The LM descent engine was fired for 371.51 sec at 49:41:34.46 with the vehicles still docked. Attitude control with the digital autopilot and manual throttling of the descent engine to full thrust were also demonstrated. Crew transfer back to the CM began at 050:15, and the LM was deactivated at 051:00. The fifth service propulsion system firing, 43.26 sec in duration, occurred at 054:26:12.27 to circularize the orbit for the LM active rendezvous. The resulting orbit was 131.0 × 125.9 nmi, compared to a desired circular orbit of 130.0 nmi, but it was considered acceptable for the rendezvous sequence.

Extravehicular operations were demonstrated on the fourth day of the mission. The plan was for the lunar module pilot to exit the LM, transfer to the open hatch of the CM, and then return. This plan was abbreviated from 2 hr 15 min to 39 min because of several bouts of nausea experienced by the lunar module pilot on the preceding day and because of the many activities required for rendezvous preparation. The LM was depressurized at 072:45 and the forward hatch opened at 072:46. The lunar module pilot began his egress to the outer platform at 72:59:02, feet first and face up, and completed egress at 073:07. He was wearing the extravehicular mobility unit backpack, which provided communications and oxygen; it also circulated water through the suit to keep him cool. His only connection to the LM was a 25-ft nylon rope to prevent him drifting into space. He secured his feet in the 'golden slippers', the gold-painted restraints affixed to the surface outside the hatch, called the 'front porch' by the astronauts, where he remained while outside the LM. During this same period, the command module pilot, dependent on CSM systems for life support, depressurized the CM and opened the side hatch at 073:02:00. He partially exited the hatch for observation, photography, and retrieval of thermal samples from the side of the CM. The samples were missing, so he retrieved the service module thermal samples at 073:26. The lunar module pilot retrieved the LM thermal samples at 073:39. About 3 min later, he began an abbreviated evaluation of translation and body-attitude-control capability using the extravehicular transfer handrails. The initially planned hand-over-hand trip from the LM to the CM was not made. During this period, the lunar module pilot also completed 16-mm and 70-mm photography of the command module pilot's activities and the exterior of both spacecraft. The lunar module pilot began his ingress at 073:45 and completed it at 073:46:03. The LM forward hatch was then closed, locked, and the LM was repressurized by 073:53. The CMP had already closed and locked the CM hatch, completing this task by 073:49:23. The CM was repressurized by 074:02:00. The first images of the television transmission from inside the LM were received at 074:58:03

and the transmission was concluded with loss of signal at 075:13:13. Voice and pictures were both good, an improvement over the previous day's transmission. The commander returned to the CM at 075:15, followed by the lunar module pilot at 076:55.

On the fifth day, the lunar module pilot transferred to the LM at 088:05, followed by the commander at 088:55, to prepare for the first LM free flight and active rendezvous. The CSM was maneuvered to the inertial undocking attitude at 092:22. Undocking was attempted at 092:38 but the capture latches did not release immediately. Undocking occurred at 092:39:36, and the LM was rolled on its axis so that the CMP could make a visual inspection. A small separation maneuver at 093:02:54, using the service module reaction control system, placed the LM 2.0 nmi behind the CSM 45 min later. The maximum range between the LM and CSM was 98 nmi, achieved about halfway between the coelliptical sequence initiation and constant differential height maneuver. During this maneuver, the LM engine ran smoothly until throttled to 20%, at which time it chugged noisily. The commander stopped throttling and waited. Within seconds, the chugging stopped. He accelerated to 40% before shutting down and had no more problems. The LM crew then checked their systems and fired the descent engine again to 10%. It ran evenly. The first LM rendezvous phasing maneuver was executed at 093:47:35.4 with the descent propulsion system under abort guidance control. This maneuver placed the LM in a near equiperiod orbit with apogee and perigee altitudes 12.2 nmi above and below the CSM. The second maneuver was not applied; it was a computation to be used only in case of a contingency requiring a LM abort. The solution time was 094:57:53. The third rendezvous maneuver was executed at 095:39:08.06 and resulted in a LM orbit of 138.9 × 133.9 nmi. Coelliptic sequence initiation was performed at 096:16:06.54, and the descent stage was jettisoned immediately after the start of reaction control system thrusting. The maneuver left the LM 10 nmi below and 82 nmi behind the CSM. The descent stage remained in Earth orbit until 03:45 GMT on 22 March, when it impacted the Indian Ocean off the coast of eastern Africa. The resulting ascent stage orbit was 116 × 111 nmi. After coelliptic sequence initiation using the CSM reaction control system, rendezvous radar tracking was reestablished, but the CM was unable to acquire the ascent stage tracking light, which had failed. The constant differential height maneuver was performed at 096:58:15.0, using the ascent stage engine for the first time. The onboard solution for terminal phase initiation was executed at 097:57:59, creating an ascent stage orbit of about 126 × 113 nmi. Two small midcourse corrections were performed at 10 and 22 min after terminal phase initiation. Terminal phase braking began at 098:30:03, followed by stationkeeping, formation flying, photography, and docking at 099:02:26. The ascent stage had been separated from the CSM for 6 hr 22 min 50 sec. After docking, the crew transferred back to the CSM by 101:00. The ascent stage was jettisoned at 101:22:45.0, and the ascent engine fired for 362.3 sec at 101:53:15.4 until oxidizer depletion. The final orbit for the ascent stage was 3,760.9 × 126.6 nmi, with an expected orbital lifetime of five years; however, entry occurred on 23 October 1981. The sixth service propulsion system maneuver, a 1.43-sec burn at 123:25:06.97, had been postponed for one revolution because the reaction control translation required prior to ignition for propellant settling was improperly programmed. The maneuver, originally scheduled for 121:48:00, was an orbit-shaping retrograde maneuver to lower the perigee so that the reaction control system deorbit capability would be enhanced in the event of a contingency.

During the final four days of the mission, the crew conducted Earth resources and multispectral terrain photography experiments of the southern USA, Mexico, Brazil, and Africa. One objective, designated experiment S065, was to determine the extent to which multi-band photography in the visible and near-infrared regions from orbit could be effectively applied to the Earth resources disciplines. The other objective was to obtain simultaneous photographs with four different film/filter combinations from orbit to assist in defining future multispectral photographic systems. The results were excellent. The quality and subject material exceeded that of any previous orbital mission and would aid Skylab program planning. The reasons for the excellent results were the amount of time available (four days so the crew could wait for cloud cover to pass); the orbital inclination of 33.6°, which permitted vertical and near-vertical coverage of areas never photographed before; sufficient reaction control system propellants to allow the crew to orient the spacecraft whenever necessary; the lack of contamination on the spacecraft windows; and the continuous assistance and evaluation of the science support room at the Manned Spacecraft Center. The crew also made an inertial measurement unit alignment with a sighting of the planet Jupiter (the first time a planet had been used) and performed a number of daylight star sightings, landmark sightings, and star sextant sightings. During two successive revolutions, at 192:43 and 194:13, the crew successfully tracked the Pegasus 3 satellite, launched on 30 July 1965, at a range of 1,000 nmi. While over Hawaii, the crew made a sighting of the LM ascent stage from 222:38:40 to 222:45:40.

The service propulsion system had been fired for the seventh time at 169:30:00.36, a 24.90-sec maneuver that raised the orbit to 253.2 × 100.7 nmi and established the desired conditions for the nominal deorbit point. If the service propulsion system had failed at deorbit, the reaction control system could have conducted a deorbit maneuver from this apogee condition and still landed near the primary recovery area. The deorbit maneuver was accomplished after 151 orbits with the eighth service propulsion firing, an 11.74-sec maneuver at 240:31:14.84. It was performed one revolution later than planned because of unfavorable weather in the planned recovery area.

Apollo 9 earth orbit phase event	GET (h:m:s)	Space-fixed velocity (ft/sec)	Event duration (sec)	Velocity change (ft/sec)	Apogee (nmi)	Perigee (nmi)	Period (min)	Inclination (deg)
Earth orbit insertion	000:11:14.66	25,569.78	–	–	100.74	99.68	88.20	32.552
CSM separated from S-IVB	002:41:16.0	25,553	–	–	–	–	–	–
CSM/LM ejected from S-IVB	004:08:09	25,565.3	–	–	–	–	–	–
S-IVB 2nd burn restart*	004:45:55.54	25,556.1	–	–	–	–	–	–
S-IVB 2nd burn cutoff	004:46:57.60	27,742.03	62.06	–	–	–	–	32.303
S-IVB intermediate orbit insertion	004:47:07.60	27,753.61	–	–	1,671.58	105.75	119.22	32.302
1st SPS ignition	005:59:01.07	25,549.8	–	–	–	–	–	–
1st SPS cutoff	005:59:06.30	25,583.8	5.23	36.6	127.6	111.3	88.8	32.56

Apollo 9 earth orbit phase event – *continued*	GET (h:m:s)	Space-fixed velocity (ft/sec)	Event duration (sec)	Velocity change (ft/sec)	Apogee (nmi)	Perigee (nmi)	Period (min)	Inclination (deg)
S-IVB 3rd burn restart*	006:07:19.26	20,766.0	–	–	–	–	–	–
S-IVB 3rd burn cutoff	006:11:21.32	31,589.17	242.06	–	–	–	–	33.824
S-IVB escape orbit insertion	006:11:31.32	31,619.85	–	–	–	–	–	33.825
2nd SPS ignition	022:12:04.07	25,588.2	–	31.8	–	–	–	–
2nd SPS cutoff	022:13:54.36	25,701.7	110.29	850.5	192.5	110.7	90.0	33.46
3rd SPS ignition	025:17:39.27	25,692.4	–	–	–	–	–	–
3rd SPS cutoff	025:22:19.15	25,794.3	279.88	2567.9	274.9	112.6	91.6	33.82
4th SPS ignition	028:24:41.37	25,807.7	–	–	–	–	–	–
4th SPS cutoff	028:25:09.24	25,798.9	27.87	300.5	275.0	112.4	91.6	33.82
DPS docked ignition	049:41:34.46	25,832.7	–	–	–	–	–	–
DPS docked cutoff	049:47:45.97	25,783.0	371.51	1737.5	274.6	112.1	91.5	33.97
5th SPS ignition	054:26:12.27	25,700.8	–	–	–	–	–	–
5th SPS cutoff	054:26:55.53	25,473.2	43.26	572.5	131.0	125.9	89.2	33.61
CSM/LM separation ignition	093:02:54	25,480.5	–	–	–	–	–	–
CSM/LM separation cutoff	093:03:03.5	25,480.5	9.5	–	127	122	–	–
LM descent phasing ignition	093:47:35.4	25,518.9	–	–	–	–	–	–
LM descent phasing cutoff	093:47:54.4	25,518.2	19.0	–	137	112	–	–
LM descent insertion ignition	095:39:08.06	25,412.6	–	–	–	–	–	–
LM descent insertion cutoff	095:39:30.43	25,453.0	22.37	–	138.9	133.9	–	–
LM coelliptic sequence ignition	096:16:06.54	25,452.0	–	–	–	–	–	–
LM coelliptic sequence cutoff	096:16:38.25	25,412.0	31.71	–	138	113	–	–
LM constant differential height ignition	096:58:15.0	25,592.0	–	–	–	–	–	–
LM constant differential height cutoff	096:58:17.9	25,550.6	2.9	–	116	111	–	–
LM terminal phase initiation ignition	097:57:59	25,540.8	–	–	–	–	–	–
LM terminal phase initiation cutoff	097:58:36.6	25,560.5	37.6	–	126	113	–	–
LM ascent engine depletion ignition	101:53:15.4	25,480.3	–	–	–	–	–	–
LM ascent engine depletion	101:59:17.7	29,415.4	362.3	5,373.4	3,760.9	126.6	165.3	28.95
6th SPS ignition	123:25:06.97	25,522.2	–	–	–	–	–	–

Apollo 9 earth orbit phase event – *continued*	GET (h:m:s)	Space-fixed velocity (ft/sec)	Event duration (sec)	Velocity change (ft/sec)	Apogee (nmi)	Perigee (nmi)	Period (min)	Inclination (deg)
6th SPS cutoff	123:25:08.40	25,489.0	1.43	33.7	123.1	108.5	88.7	33.62
7th SPS ignition	169:39:00.36	25,589.6	–	–	–	–	–	–
7th SPS cutoff	169:39:25.26	25,825.9	24.90	650.1	253.2	100.7	90.9	33.51
8th SPS ignition	240:31:14.84	25,318.4	–	–	–	–	–	–
8th SPS cutoff	240:31:26.58	25,142.8	11.74	322.7	240.0	–4.2	88.8	33.52

* Only the commanded time is available for this event.

RECOVERY

The service module was jettisoned at 240:36:03.8, and the CM entry followed a primary guidance system profile. The command module re-entered Earth's atmosphere (at the 400,000-ft altitude of the 'entry interface') at 240:44:10.2 at a velocity of 25,894 ft/sec. The parachute system effected splashdown of the CM in the Atlantic Ocean at 17:00:54 GMT on 13 March. The mission duration was 241:00:54. The impact point, estimated to be 23.22°N and 67.98°W, was about 2.7 nmi from the target point and 3 nmi from the recovery ship USS *Guadalcanal.* Although the service module could not survive entry intact, radar tracking data predicted impact in the Atlantic Ocean at a point estimated to be 22.0°N and 65.3°W, some 175 nmi downrange from the CM. After splashdown, the CM assumed an apex-up flotation attitude. The crew was retrieved by helicopter and was aboard the recovery ship 49 min after splashdown. The CM was recovered 83 min later. The estimated CM weight at splashdown was 11,094 lb, and the estimated distance traveled for the mission was 3,664,820 nmi. At CM retrieval, the weather recorded on the *Guadalcanal* showed scattered clouds at 2,000 ft and broken clouds at 9,000, a visibility of 10 nmi, a wind speed of 9 kt at 200° from true north, an air temperature of 79°F, and a water temperature of 76°F, with waves of up to 7 ft at 340° from true north. The crew departed the *Guadalcanal* by helicopter at 15:00 GMT on 14 March and arrived at Eleuthera, Bahamas, at 16:30 GMT, where they transferred to an aircraft to fly to Houston. The CM was removed from the *Guadalcanal* on 16 March at the Norfolk Naval Air Station, Norfolk, Virginia, where the Landing Safing Team began the evaluation and deactivation procedures at 16:00 GMT, completing this work on 19 March. The CM was then flown to Long Beach, California, and trucked to the North American Rockwell Space Division facility at Downey, California, for post-flight analysis, where it arrived on 21 March.

CONCLUSIONS

The following conclusions were made from an analysis of post-mission data:

1. The onboard rendezvous equipment and procedures in both spacecraft provided the required precision for rendezvous operations to be conducted during a lunar landing mission. The CSM computations and preparations for mirror-image maneuvers were completed on time by the command module pilot.
2. The functional operation of the docking process of the two spacecraft was demonstrated. However, the necessity for proper lighting conditions for the docking alignment aids was illustrated.
3. The performance of all systems in the extravehicular mobility unit was excellent throughout the entire extravehicular operation. The results of this mission, plus satisfactory results from additional qualification tests of minor design changes, provided verification of the operation of the extravehicular mobility unit on the lunar surface.
4. The extent of the extravehicular activity indicated the practicality of extravehicular crew transfer in the event of a contingency. Cabin depressurization and normal repressurization were demonstrated in both spacecraft.
5. Performance of the lunar module systems demonstrated the operational capability to conduct a lunar mission, except for the steerable antenna, which was not operated, and the landing radar, which could not be fully evaluated in Earth orbit. None of the anomalies adversely affected the mission. The concepts and operational functioning of the crew/spacecraft interfaces, including procedures, provisioning, restraints, displays, and controls, were satisfactory for manned lunar module functions. The interfaces between the two spacecraft, while both docked and undocked, were also verified.
6. The lunar module consumable expenditures were well within predicted values, thus demonstrating adequate margins to perform the lunar mission.
7. Gas in the CM potable water supply interfered with proper food rehydration and therefore had some effect on food taste and palatability. Lunar module water was acceptable.
8. Orbital navigation of the CSM, using the yaw-control technique for landmark tracking, was demonstrated and reported to be adequate. The star visibility threshold of the CM scanning telescope was not definitely established for the docked configuration; therefore, platform orientation using the Sun, the Moon, and planets may be required if inertial reference is inadvertently lost during translunar flight.
9. Mission support, including the Manned Space Flight Network, adequately provided simultaneous ground control of two manned spacecraft.

MISSION OBJECTIVES

Launch Vehicle Primary Objectives

1. To demonstrate S-IVB/instrument unit control capability during transposition, docking and lunar module ejection maneuver. *Achieved.*

Launch Vehicle Secondary Objectives

1. To demonstrate S-IVB restart capability. *Achieved.*
2. To verify J-2 engine modifications. *Achieved.*

The Apollo 9 crew: Jim McDivitt (left), Dave Scott and Rusty Schweickart.

Apollo 9 preparations: CSM-104 (top) and LM-3.

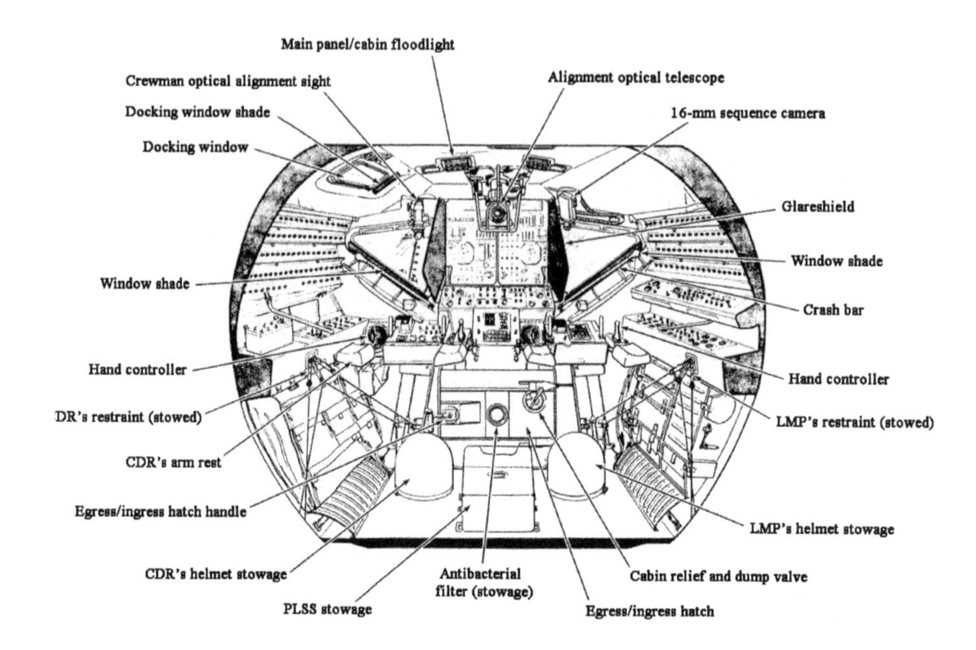

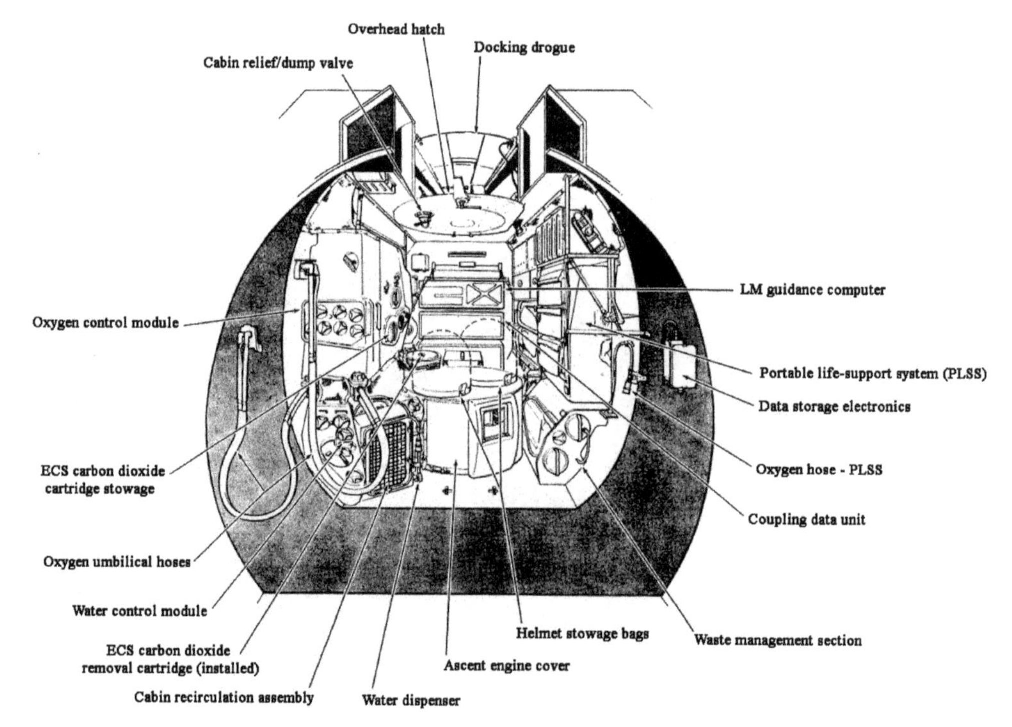

The interior layout of the LM.

Rusty Schweickart and Jim McDivitt (foreground) in a LM simulator.

The launch of Apollo 9.

Apollo 9's spent S-IVB with its SLA panels jettisoned to expose LM-3.

Rusty Schweickart standing on the 'porch' of LM-3 to test the lunar suit.

Dave Scott stands in the hatch of CSM 'Gumdrop'.

LM-3 'Spider' after undocking, with its legs deployed.

On returning from its independent mission, 'Spider' exposed its ascent engine to inspection.

3. To confirm J-2 environment in S-II stage. *Achieved.*
4. To confirm launch vehicle longitudinal oscillation environment during S-IC stage burn period. *Achieved.*
5. To demonstrate helium heater repressurization system operation. *Achieved.*
6. To demonstrate S-IVB propellant dump and safing with a large quantity of residual S-IVB propellants. *Partially achieved. The S-IVB stage was adequately safed, however the propellant dump was not achieved due to loss of engine helium control regulator discharge pressure.*
7. To verify that modifications incorporated in the S-IC stage suppress low frequency longitudinal oscillations. *Achieved.*
8. To demonstrate 80-min restart capability. *Partially achieved. The experimental start was achieved and accomplished the planned S-IVB third burn. However, rough combustion, a gas generator spike at ignition, and control oscillations resulted in a low performance at start, performance loss during the burn, and loss of engine helium control regulator discharge pressure.*
9. To demonstrate dual repressurization capability. *Achieved.*
10. To demonstrate helium heater restart capability. *Achieved.*
11. To verify the onboard command and communications system/ground system interface and operation in the deep space environment. *Achieved.*

Spacecraft Primary Objectives

1. To demonstrate crew/space vehicle/mission support facilities performance during a manned Saturn V mission with command and service module and lunar module. *Achieved.*
2. To demonstrate lunar module/crew performance. *Achieved.*
3. To demonstrate performance of nominal and selected backup lunar orbit rendezvous mission activities, including:
   a. Transposition, docking, and lunar module withdrawal. *Achieved.*
   b. Intravehicular and extravehicular crew transfer. *Achieved.*
   c. Extravehicular capability. *Achieved.*
   d. Service propulsion system and descent propulsion system burns. *Achieved.*
   e. Lunar module active rendezvous and docking. *Achieved.*
4. To assess command and service module and lunar module consumables. *Achieved.*

Spacecraft Mandatory Detailed Test Objectives

1. M11.6: To perform a medium duration descent propulsion system firing to include manual throttling with command and service module and lunar module docked, and a short duration descent propulsion system firing with an undocked lunar module and approximately half full descent propulsion system propellant tanks. *Achieved, the primary guidance and navigation control system/digital auto-pitch performance was monitored and found acceptable during the first and second descent propulsion system burns.*
2. M13.11: To perform a long duration ascent propulsion system burn. *Achieved, a burn to depletion was performed by the ascent propulsion system for an extended period.*
3. M13.12: To perform a long duration descent propulsion system burn and obtain data

to determine that no adverse interactions existed between propellant slosh, vehicle engine vibration, and descent propulsion system performance during a burn. *Achieved, data were collected during the docked descent propulsion system burn and the rendezvous.*

4. M14: To demonstrate the performance of the environmental control system during lunar module activity periods. *Achieved, although minor problems occurred in the system.*
5. M15.3: To determine the performance of the lunar module electrical power subsystem in the primary and backup modes. *Achieved, despite some problems with the batteries.*
6. M16.7: To operate the landing radar during the descent propulsion system burns. *Achieved.*
7. M17.9: To deploy the lunar module landing gear and obtain data on landing gear temperatures resulting from descent propulsion system operation. *Achieved.*
8. M17.17: To verify the performance of the passive thermal subsystems (thermal blanket, plume protection, ascent and descent stage base heat shields, and thermal control coatings) to provide adequate thermal control when the spacecraft is exposed to the natural and propulsion induced thermal environments. *Achieved, lunar module environmental and thermal effect data were collected during the docked descent propulsion system burn, extravehicular activity, and the post-rendezvous inspection.*
9. M17.18: To demonstrate the structural integrity of the lunar module during Saturn V launch and during descent propulsion system and ascent propulsion system burn in an orbital environment. *Achieved.*

Spacecraft Primary Detailed Test Objectives

1. P1.23: To demonstrate Block II command and service module attitude control during service propulsion system thrusting with the command and service module and lunar module docked. *Achieved, during the first, second, and third service propulsion burns.*
2. P1.24: To perform inertial measurement unit alignments using the sextant while docked. *Achieved.*
3. P1.25: To perform an inertial measurement unit and a star pattern visibility check in daylight while docked. *Achieved, many daytime sightings were made with visible star patterns although reflective light hindered some tests.*
4. P2.9: To perform manual thrust vector control takeover of a guidance navigation control system initiated service propulsion docked burn. *Achieved, during the third service propulsion system burn.*
5. P7.29: To obtain data on the effects of the tower jettison motor, S-II retrorockets, and service module reaction control system exhaust on the command and service module. *Achieved. Spacecraft exhaust effects data were collected following Earth orbital insertion, lunar module/command and service module ejection, during the revised extravehicular period, and during the post-rendezvous inspection; however, the revised extravehicular activity permitted recovery of only part of the thermal samples.*
6. P11.5: To perform lunar module inertial measurement unit alignments using the alignment optical telescope and calibrate the coarse optical alignment sight. *Achieved, lunar module in-flight inertial measurement unit alignment data were collected at various times during lunar module activity periods.*

7. P11.7: To demonstrate reaction control system translation and attitude control of the staged lunar module using automatic and manual primary guidance and navigation control system controls. *Achieved.*
8. P11.10: To obtain data to verify inertial measurement unit performance in the flight environment. *Achieved, lunar module primary guidance and navigation control system and command and service module guidance navigation control system inertial measurement unit performance data were collected throughout the mission.*
9. P11.14: To perform a primary guidance and navigation control system/digital autopilot controlled long duration ascent propulsion burn. *Achieved.*
10. P12.2: To demonstrate an abort guidance system calibration and obtain abort guidance system performance data in the flight environment. *Achieved, during docked descent propulsion system burn and the rendezvous phasing burn.*
11. P12.3: To demonstrate reaction control system translation and attitude control of unstaged lunar module using automatic and manual abort guidance system/control electric section control modes. *Achieved.*
12. P12.4 To perform an abort guidance system/control electric section controlled descent propulsion system burn with a heavy descent stage. *Achieved.*
13. P16.4: To demonstrate tracking of command and service module rendezvous radar transponder at various ranges between the command and service module and the lunar module. *Achieved.*
14. P16.6: To perform a landing radar self-test. *Achieved.*
15. P16.19: To obtain data on rendezvous radar corona susceptibility during lunar module -X translation reaction control system engine firings while undocked and during -X reaction control system engine firings while docked. *Partially achieved. Data were obtained but the rendezvous radar failed to lock.*
16. P20.21: To demonstrate the lunar module/Manned Space Flight Network operational S-band communication subsystem capability. *Achieved, despite intermittent discrepancies.*
17. P20.22: To demonstrate lunar module/command and service module/Manned Space Flight Network/extravehicular activity operational S-band and VHF communication compatibility. *Achieved, despite sporadic failures.*
18. P20.24: To demonstrate command and service module docking with the S-IVB/SLA/LM. *Achieved.*
19. P20.25: To demonstrate lunar module separation and ejection of the command and service module/lunar module from the SLA. *Achieved.*
20. P20.26: To demonstrate the technique to be employed for the undocking of the lunar module from the command and service module prior to lunar descent. *Achieved.*
21. P20.27: To perform a lunar module active rendezvous with a passive command and service module. *Achieved.*
22. P20.28: To demonstrate lunar module active docking capability with the passive command and service module. *Achieved.*
23. P20.29: To perform a pyrotechnic separation of the lunar module and command and service module in flight. *Achieved.*
24. P20.31: To demonstrate mission support facilities performance during an Earth orbital mission. *Achieved.*

25. P20.33: To perform procedures required to prepare for a command and service module active rendezvous with the lunar module. *Achieved, the command and service modules were maintained in a recovery mode during the lunar module simulated descent.*
26. P20.34: To demonstrate crew capability to transfer themselves and equipment from the command and service module to the lunar module and return. *Achieved, the crew was successful in making the transfer in the time allotted.*
27. P20.35: To demonstrate extravehicular transfer and obtain extravehicular activity data. *Achieved, although the program was modified during the mission.*

Spacecraft Secondary Detailed Test Objectives

1. S1.26: To perform onboard navigation using the technique of scanning telescope landmark tracking. *Achieved.*
2. S13.10: To perform an unmanned ascent propulsion system burn to depletion. *Achieved.*
3. S20.32: To evaluate one-man lunar module operation capability, and obtain data on crew maneuverability, crew compartmentation, and propulsive venting. *Achieved.*
4. S20.37: To obtain data on descent propulsion plume effects on astronauts' visibility. *Achieved, the descent propulsion system did not affect the crew's visibility during the two burns.*
5. S20.120: To obtain data on the electromagnetic compatibility of the command and service module, lunar module, and portable life support system. *Achieved, the command and service module, lunar module, and portable life support system were electromagnetically compatible with respect to any conducted or radiated electromagnetic interference.*

Functional Tests Added In-Flight

1. Command and lunar module intravehicular transfer, unsuited. *Achieved.*
2. Tunnel clearing, unsuited. *Achieved.*
3. Command module platform alignment in daylight. *Achieved.*
4. Command module platform alignment using a planet (Jupiter). *Achieved.*
5. Digital autopilot orbital rate, pitch and roll. *Achieved.*
6. Backup gyro display coupler alignment of stabilization and control system. *Achieved.*
7. Window degradation photography. *Achieved.*
8. Satellite tracking, ground inputs. *Achieved.*
9. Command and service module high-gain S-band antenna reacquisition test. *Achieved.*
10. Passive thermal control cycling at 0.1 deg/sec at three deadbands: ± 10, ± 20, and ± 25 deg. *Achieved.*

Experiments

1. S-065: To obtain selective, simultaneous multispectral photographs, with four different film/filter combinations, of selected land and ocean areas. *Achieved.*

MISSION TIMELINE

Apollo 9 mission event	GET (h:m:s)	Date (GMT)	Time (h:m:s)
Terminal countdown started.	–028:00:00	27 Feb 1969	03:00:00
Scheduled 3-hr hold at T–16 hr.	–016:00:00	27 Feb 1969	15:00:00
Decision made to recycle countdown to T–42 hr due to health of crew.	–016:00:00	27 Feb 1969	15:30:00
Countdown resumed at T–42 hr.	–042:00:00	01 Mar 1969	07:30:00
Scheduled 5-hr 30-min hold at T–28 hr.	–028:00:00	01 Mar 1969	21:30:00
Countdown resumed at T–28 hr.	–028:00:00	02 Mar 1969	03:00:00
Scheduled 3-hr hold at T–16 hr.	–016:00:00	02 Mar 1969	15:00:00
Countdown resumed at T–16 hr.	–016:00:00	02 Mar 1969	18:00:00
Scheduled 6-hr hold at T–9 hr.	–009:00:00	03 Mar 1969	01:00:00
Countdown resumed at T–9 hr.	–009:00:00	03 Mar 1969	07:00:00
Guidance reference release.	–000:00:16.97	03 Mar 1969	15:59:43
S-IC engine start command.	–000:00:08.9	03 Mar 1969	15:59:51
S-IC engine ignition (#5).	–000:00:06.3	03 Mar 1969	15:59:53
All S-IC engines thrust OK.	–000:00:01.3	03 Mar 1969	15:59:58
Range zero.	000:00:00.00	03 Mar 1969	16:00:00
All holddown arms released (1st motion) (1.10 *g*).	000:00:00.26	03 Mar 1969	16:00:00
Liftoff (umbilical disconnected).	000:00:00.67	03 Mar 1969	16:00:00
Tower clearance yaw maneuver started.	000:00:01.7	03 Mar 1969	16:00:01
Yaw maneuver ended.	000:00:09.7	03 Mar 1969	16:00:09
Pitch and roll maneuver started.	000:00:13.3	03 Mar 1969	16:00:13
Roll maneuver ended.	000:00:33.0	03 Mar 1969	16:00:33
Mach 1 achieved.	000:01:08.2	03 Mar 1969	16:01:08
Maximum bending moment (86,000,000 lbf-in).	000:01:19.4	03 Mar 1969	16:01:19
Maximum dynamic pressure (630.73 lb/ft^2).	000:01:25.5	03 Mar 1969	16:01:25
S-IC center engine cutoff command.	000:02:14.34	03 Mar 1969	16:02:14
Pitch maneuver ended.	000:02:38.0	03 Mar 1969	16:02:38
S-IC outboard engine cutoff.	000:02:42.76	03 Mar 1969	16:02:42
S-IC maximum total inertial acceleration (3.85 *g*).	000:02:42.84	03 Mar 1969	16:02:42
S-IC/S-II separation command and S-IC maximum Earth-fixed velocity.	000:02:43.45	03 Mar 1969	16:02:43
S-II engine start command.	000:02:44.17	03 Mar 1969	16:02:44
S-II ignition.	000:02:45.16	03 Mar 1969	16:02:45
S-II aft interstage jettisoned.	000:03:13.5	03 Mar 1969	16:03:13
Launch escape tower jettisoned.	000:03:18.3	03 Mar 1969	16:03:18
Iterative guidance mode initiated.	000:03:24.6	03 Mar 1969	16:03:24
S-IC apex.	000:04:26.03	03 Mar 1969	16:04:26
S-II engine cutoff.	000:08:56.22	03 Mar 1969	16:08:56
S-II maximum total inertial acceleration (2.00 *g*).	000:08:56.31	03 Mar 1969	16:08:56
S-IC impact (theoretical).	000:08:56.436	03 Mar 1969	16:08:56
S-II maximum Earth-fixed velocity.	000:08:56.45	03 Mar 1969	16:08:56
S-II/S-IVB separation command.	000:08:57.18	03 Mar 1969	16:08:57
S-IVB 1st burn start command.	000:08:57.28	03 Mar 1969	16:08:57

Apollo 9 mission event – *continued*	GET (h:m:s)	Date (GMT)	Time (h:m:s)
S-IVB 1st burn ignition.	000:09:00.82	03 Mar 1969	16:09:00
S-IVB ullage case jettisoned.	000:09:09.0	03 Mar 1969	16:09:09
S-II apex.	000:09:53.58	03 Mar 1969	16:09:53
S-IVB 1st burn cutoff.	000:11:04.66	03 Mar 1969	16:11:04
S-IVB 1st burn maximum total inertial acceleration (0.80 g).	000:11:04.74	03 Mar 1969	16:11:04
Earth orbit insertion. S-IVB 1st burn maximum Earth-fixed velocity.	000:11:14.66	03 Mar 1969	16:11:14
Maneuver to local horizontal attitude started.	000:11:24.9	03 Mar 1969	16:11:24
Orbital navigation started.	000:12:47.7	03 Mar 1969	16:12:47
S-II impact (theoretical).	000:20:25.346	03 Mar 1969	16:20:25
Maneuver to transposition and docking attitude.	002:34:01.0	03 Mar 1969	18:34:01
CSM separated from S-IVB (command).	002:41:16.0	03 Mar 1969	18:41:16
Formation flying. CSM docked with LM/S-IVB.	003:01:59.3	03 Mar 1969	19:01:59
CSM/LM ejected from S-IVB.	004:08:09	03 Mar 1969	20:08:09
Maneuver to local horizontal attitude started.	004:25:05.1	03 Mar 1969	20:25:05
S-IVB 2nd burn restart preparation.	004:36:17.24	03 Mar 1969	20:36:17
S-IVB 2nd burn restart command.	004:45:47.20	03 Mar 1969	20:45:47
S-IVB 2nd burn ignition (for intermediate orbit insertion).	004:45:55.54	03 Mar 1969	20:45:55
S-IVB 2nd burn cutoff.	004:46:57.60	03 Mar 1969	20:46:57
S-IVB 2nd burn maximum total inertial acceleration (1.24 g).	004:46:57.68	03 Mar 1969	20:46:57
S-IVB 2nd burn maximum Earth-fixed velocity.	004:46:58.20	03 Mar 1969	20:46:58
S-IVB intermediate orbit insertion.	004:47:07.60	03 Mar 1969	20:47:07
Orbital navigation started.	004:47:14.2	03 Mar 1969	20:47:14
Maneuver to local horizontal attitude started.	004:47:18.6	03 Mar 1969	20:47:18
1st SPS ignition.	005:59:01.07	03 Mar 1969	21:59:01
1st SPS cutoff.	005:59:06.30	03 Mar 1969	21:59:06
Powered flight navigation started.	005:59:39.0	03 Mar 1969	21:59:39
S-IVB 3rd burn restart preparation.	005:59:40.98	03 Mar 1969	21:59:41
S-IVB 3rd burn restart command.	006:06:27.35	03 Mar 1969	22:06:27
S-IVB 3rd burn ignition (Earth escape trajectory).	006:07:19.26	03 Mar 1969	22:07:19
S-IVB 3rd burn maximum total inertial acceleration (1.69 g).	006:08:53.00	03 Mar 1969	22:08:53
S-IVB 3rd burn cutoff.	006:11:21.32	03 Mar 1969	22:11:21
S-IVB safing procedures started.	006:11:21.92	03 Mar 1969	22:11:21
S-IVB 3rd burn maximum Earth-fixed velocity.	006:11:23.50	03 Mar 1969	22:11:23
S-IVB escape orbit insertion.	006:11:31.32	03 Mar 1969	22:11:31
Orbital navigation started.	006:11:38.0	03 Mar 1969	22:11:38
Maneuver to local horizontal attitude started.	006:11:42.0	03 Mar 1969	22:11:42
S-IVB safing – LOX dump started (failed due to loss of engine pneumatic control during 3rd burn).	006:12:15.5	03 Mar 1969	22:12:15
S-IVB safing – LOX NPV valve latched open to safe LOX tank	006:24:02	03 Mar 1969	22:24:02
S-IVB safing – LH_2 dump started (failed due to loss of pneumatic control of engine valves).	006:24:11.3	03 Mar 1969	22:24:11
S-IVB safing – APS depletion firing ignition.	007:34:04.6	03 Mar 1969	23:34:04
S-IVB safing – APS depletion firing cutoff.	007:41:53	03 Mar 1969	23:41:53
2nd SPS ignition.	022:12:04.07	04 Mar 1969	14:12:04

Apollo 9 mission event – *continued*	GET (h:m:s)	Date (GMT)	Time (h:m:s)
2nd SPS cutoff.	022:13:54.36	04 Mar 1969	14:13:54
3rd SPS ignition.	025:17:39.27	04 Mar 1969	17:17:39
3rd SPS cutoff.	025:22:19.15	04 Mar 1969	17:22:19
4th SPS ignition.	028:24:41.37	04 Mar 1969	20:24:41
4th SPS cutoff.	028:25:09.24	04 Mar 1969	20:25:09
Pressure suits donned.	041:00	05 Mar 1969	09:00
LMP entered LM.	043:15	05 Mar 1969	11:15
LM transferred to internal power.	043:40	05 Mar 1969	11:40
LM systems activated.	043:45	05 Mar 1969	11:45
CDR entered LM.	044:04	05 Mar 1969	12:04
Landing gear deployed.	045:00	05 Mar 1969	13:00
Portable life support systems prepared.	045:05	05 Mar 1969	13:05
CDR requested private communication regarding LMP illness.	045:39:05	05 Mar 1969	13:39:05
CAPCOM replies that he is ready to receive CDR's private communication.	045:51:56	05 Mar 1969	13:51:56
TV transmission (approximately 5 min).	046:28	05 Mar 1969	14:28
Self test of landing radar and rendezvous radar.	048:15	05 Mar 1969	16:15
DPS docked ignition.	049:41:34.46	05 Mar 1969	17:41:34
DPS docked cutoff.	049:47:45.97	05 Mar 1969	17:47:46
Landing radar self test.	050:00	05 Mar 1969	18:00
Transfer to CM started.	050:15	05 Mar 1969	18:15
LM deactivated.	051:00	05 Mar 1969	19:00
5th SPS ignition.	054:26:12.27	05 Mar 1969	22:26:12
5th SPS cutoff.	054:26:55.53	05 Mar 1969	22:26:55
Pressure suits removed.	055:00	05 Mar 1969	23:00
Pressure suits donned.	068:15	06 Mar 1969	12:15
Transfer to LM started.	069:45	06 Mar 1969	13:45
LM systems activated.	070:00	06 Mar 1969	14:00
CDR assessed LMP condition as excellent.	071:53	06 Mar 1969	15:53
LM depressurized.	072:45	06 Mar 1969	16:45
LM forward hatch open.	072:46	06 Mar 1969	16:46
CM depressurized.	072:59	06 Mar 1969	16:59
LMP started egress.	072:59:02	06 Mar 1969	16:59:02
CM side hatch open.	073:02:00	06 Mar 1969	17:02:00
CDR reported LMP's foot extending through LM forward hatch. LMP lowered EVA visor.	073:04	06 Mar 1969	17:04
LMP egress completed. Entered foot restraints. CDR photographed LMP activities.	073:07	06 Mar 1969	17:07
CDR passed 70-mm camera to LMP. LMP started photography.	073:10	06 Mar 1969	17:10
LMP ended 70-mm photography and handed camera to CDR.CMP photographed LM with 16-mm camera.	073:20	06 Mar 1969	17:20
CDR passed 16-mm camera to LMP. CMP activities photographed by LMP.	073:23	06 Mar 1969	17:23
CMP retrieved SM thermal samples.	073:26	06 Mar 1969	17:26
LMP passed 16-mm camera to CDR.	073:34	06 Mar 1969	17:34

Apollo 9 mission event – *continued*	GET (h:m:s)	Date (GMT)	Time (h:m:s)
LMP 16-mm camera failed. LMP evaluated handrail, retrieved LM thermal sample, and passed to CDR.	073:39	06 Mar 1969	17:39
LMP started handrail evaluation.	073:42	06 Mar 1969	17:42
LMP ingress started.	073:45	06 Mar 1969	17:45
LMP ingress completed.	073:46:03	06 Mar 1969	17:46:03
LM hatch closed.	073:48	06 Mar 1969	17:48
CM side hatch reported closed and locked.	073:49:23	06 Mar 1969	17:49:23
LM hatch reported locked.	073:49:56	06 Mar 1969	17:49:56
LM repressurized at 3.0 psi.	073:53	06 Mar 1969	17:53
CM repressurization started.	073:55	06 Mar 1969	17:55
CM repressurized at 2.7 psi.	074:02:00	06 Mar 1969	18:02:00
TV transmission started.	074:58:03	06 Mar 1969	18:58:03
TV transmission ended.	075:13:13	06 Mar 1969	19:13:13
CDR entered CM.	075:15	06 Mar 1969	19:15
LMP entered CM.	076:55	06 Mar 1969	20:55
Pressure suits removed.	077:15	06 Mar 1969	21:15
LMP entered LM to open translunar bus tie circuit breakers.	078:09	06 Mar 1969	22:09
Pressure suits donned.	086:00	07 Mar 1969	06:00
LMP entered LM.	088:05	07 Mar 1969	08:05
LM systems activated.	088:15	07 Mar 1969	08:15
CDR entered LM.	088:55	07 Mar 1969	08:55
Check LM systems.	089:05	07 Mar 1969	09:05
Rendezvous radar transponder test.	091:00	07 Mar 1969	11:00
Landing radar self test.	091:55	07 Mar 1969	11:55
Rendezvous radar transponder test.	092:05	07 Mar 1969	12:05
Maneuver to undocking attitude.	092:22	07 Mar 1969	12:22
Unsuccessful undocking attempt. Capture latches failed to release.	092:38	07 Mar 1969	12:38
CSM/LM reported undocked.	092:39:36	07 Mar 1969	12:39:36
Formation flying and photography.	092:45	07 Mar 1969	12:45
CSM/LM separation maneuver ignition.	093:02:54	07 Mar 1969	13:02:54
CSM/LM separation maneuver cutoff.	093:03:03.5	07 Mar 1969	13:03:03
LM descent propulsion phasing maneuver ignition.	093:47:35.4	07 Mar 1969	13:47:35
LM descent propulsion phasing maneuver cutoff.	093:47:54.4	07 Mar 1969	13:47:54
Landing radar self test.	094:15	07 Mar 1969	14:15
Rendezvous radar on.	095:10	07 Mar 1969	15:10
LM descent propulsion insertion maneuver ignition.	095:39:08.06	07 Mar 1969	15:39:08
LM descent propulsion insertion maneuver cutoff.	095:39:30.43	07 Mar 1969	15:39:30
CAPCOM reported "everything looks good for staging".	095:58:15	07 Mar 1969	15:58:15
LM coelliptic sequence initiation maneuver ignition. Approximate time of LM descent stage jettison.	096:16:06.54	07 Mar 1969	16:16:06
LM coelliptic sequence initiation maneuver cutoff.	096:16:38.25	07 Mar 1969	16:16:38
CDR reports LM "staging went okay".	096:33:11	07 Mar 1969	16:33:11
LM constant differential height ignition.	096:58:15.0	07 Mar 1969	16:58:15
LM constant differential height cutoff.	096:58:17.9	07 Mar 1969	16:58:17

Apollo 9 mission event – *continued*	GET (h:m:s)	Date (GMT)	Time (h:m:s)
LM terminal phase initiation ignition.	097:57:59	07 Mar 1969	17:57:59
LM terminal phase initiation cutoff.	097:58:36.6	07 Mar 1969	17:58:36
1st RCS midcourse correction ignition.	098:25:19.66	07 Mar 1969	18:25:19
1st RCS midcourse correction cutoff.	098:25:23.57	07 Mar 1969	18:25:23
Terminal phase braking.	098:30:03	07 Mar 1969	18:30:03
Stationkeeping.	098:30:51.2	07 Mar 1969	18:30:51
Formation flying and photography.	098:40	07 Mar 1969	18:40
CSM/LM docked.	099:02:26	07 Mar 1969	19:02:26
CDR entered CM.	100:35	07 Mar 1969	20:35
LM prepared for jettison.	100:40	07 Mar 1969	20:40
LMP entered CM.	101:00	07 Mar 1969	21:00
LM ascent stage jettisoned.	101:22:45.0	07 Mar 1969	21:22:45
Post-jettison CSM separation maneuver.	101:32:44	07 Mar 1969	21:32:44
LM ascent engine depletion ignition.	101:53:15.4	07 Mar 1969	21:53:15
LM ascent engine depletion.	101:59:17.7	07 Mar 1969	21:59:17
Pressure suits removed.	102:00	07 Mar 1969	22:00
6th SPS ignition.	123:25:06.97	08 Mar 1969	19:25:07
6th SPS cutoff.	123:25:08.40	08 Mar 1969	19:25:08
Experiment S065 photography.	124:10	08 Mar 1969	20:10
CSM landmark tracking.	125:30	08 Mar 1969	21:30
CSM landmark tracking.	143:00	09 Mar 1969	15:00
Experiment S065 photography.	146:00	09 Mar 1969	18:00
Experiment S065 photography.	147:30	09 Mar 1969	19:30
Target of opportunity photography.	149:00	09 Mar 1969	21:00
Target of opportunity photography.	150:10	09 Mar 1969	22:10
7th SPS ignition.	169:39:00.36	10 Mar 1969	17:39:00
7th SPS cutoff.	169:39:25.26	10 Mar 1969	17:39:25
16-mm photography.	171:10	10 Mar 1969	19:10
Experiment S065 photography.	171:20	10 Mar 1969	19:20
Experiment S065 photography.	171:50	10 Mar 1969	19:50
Target of opportunity photography.	173:10	10 Mar 1969	21:10
Experiment S065 photography.	190:40	11 Mar 1969	14:40
Experiment S065 photography.	192:10	11 Mar 1969	16:10
Tracking of Pegasus 3 satellite started.	192:43	11 Mar 1969	16:43
Tracking of Pegasus 3 satellite ended.	192:44	11 Mar 1969	16:44
High-gain antenna test.	193:10	11 Mar 1969	17:10
High-gain antenna test.	193:40	11 Mar 1969	17:40
Target of opportunity photography.	193:50	11 Mar 1969	17:50
Tracking of Pegasus 3 satellite started.	194:13	11 Mar 1969	18:13
Tracking of Pegasus 3 satellite ended.	194:15	11 Mar 1969	18:15
CSM landmark tracking.	195:10	11 Mar 1969	19:10
Target of opportunity photography.	195:30	11 Mar 1969	19:30
Target of opportunity photography.	213:25	12 Mar 1969	13:25
Observation of descent stage attempted.	213:50	12 Mar 1969	13:50
Target of opportunity photography.	215:00	12 Mar 1969	15:00

Apollo 9 mission event – *continued*	GET (h:m:s)	Date (GMT)	Time (h:m:s)
Experiment S065 photography.	215:10	12 Mar 1969	15:10
Target of opportunity photography.	215:30	12 Mar 1969	15:30
Experiment S065 photography.	216:10	12 Mar 1969	16:10
Target of opportunity photography.	216:20	12 Mar 1969	16:20
Experiment S065 photography.	216:40	12 Mar 1969	16:40
Target of opportunity photography.	217:00	12 Mar 1969	17:00
CSM landmark tracking.	217:50	12 Mar 1969	17:50
Passive thermal control evaluated.	218:30	12 Mar 1969	18:30
Passive thermal control evaluated.	222:00	12 Mar 1969	22:00
Tracking of ascent stage with optics started.	222:38:40	12 Mar 1969	22:38:40
Tracking of ascent stage with optics ended.	222:45:40	12 Mar 1969	22:45:40
8th SPS ignition (deorbit).	240:31:14.84	13 Mar 1969	16:31:14
8th SPS cutoff.	240:31:26.58	13 Mar 1969	16:31:26
CM/SM separation.	240:36:03.8	13 Mar 1969	16:36:03
Entry.	240:44:10.2	13 Mar 1969	16:44:10
Communication blackout started.	240:47:01	13 Mar 1969	16:47:01
Communication blackout ended.	240:50:43	13 Mar 1969	16:50:43
Radar contact with CM established by recovery aircraft.	240:51	13 Mar 1969	16:51
Drogue parachute deployed	240:55:07.8	13 Mar 1969	16:55:07
Main parachute deployed.	240:55:59.0	13 Mar 1969	16:55:59
Recovery beacon contact with CM established by recovery aircraft. VHF voice contact with CM established by recovery helicopter.	240:57	13 Mar 1969	16:57
Visual contact with CM established by recovery helicopter.	240:58	13 Mar 1969	16:58
Splashdown (went to apex-up).	241:00:54	13 Mar 1969	17:00:54
Swimmers and flotation collar deployed.	241:07	13 Mar 1969	17:07
Flotation collar inflated.	241:14	13 Mar 1969	17:14
CM hatch opened.	241:27	13 Mar 1969	17:27
Crew aboard recovery helicopter.	241:45	13 Mar 1969	17:45
Crew aboard recovery ship.	241:49:33	13 Mar 1969	17:49:33
CM aboard recovery ship.	243:13	13 Mar 1969	19:13
Crew departed recovery ship.	263:00	14 Mar 1969	15:00
Crew arrived in Eleuthera in the Bahamas.	264:30	14 Mar 1969	16:30
Deactivation of CM started at Norfolk Naval Air Station, VA.	312:00	16 Mar 1969	16:00
CM arrived at contractor's facility in Downey, CA.	–	21 Mar 1969	–
LM descent stage entry.	443:45	22 Mar 1969	03:45
LM ascent stage entry.	–	23 Oct 1981	–

Apollo 10

The fourth manned mission: testing the LM in lunar orbit

18–26 May 1969

BACKGROUND

Apollo 10 was a Type 'F' mission, a manned lunar module demonstration in lunar orbit, the dress rehearsal for the first manned landing on the Moon.

The primary objectives were:

- to demonstrate crew, space vehicle, and mission support facilities performance during a manned lunar mission with command and service modules and lunar module; and
- to evaluate lunar module performance in the cislunar and lunar environment.

In addition, visual observations and stereoscopic strip photography of ALS-2, the prime candidate for the first lunar landing, would be attempted.

This was also the first time all members of a three-person crew had previously flown in space. The crew members were Colonel Thomas Patten Stafford (USAF), commander; Commander John Watts Young (USN), command module pilot; and Commander Eugene Andrew 'Gene' Cernan (USN), lunar module pilot. Selected as an astronaut in 1962, Stafford was making his third spaceflight. He had been pilot of Gemini 6-A and command pilot of Gemini 9-A. Born on 17 September 1930 in Weatherford, Oklahoma, Stafford was 38 years old at the time of the Apollo 10 mission. He received a BS from the US Naval Academy in 1952. His backup was Colonel Leroy Gordon Cooper Jr (USAF). Young was also making his third spaceflight, having been pilot on Gemini 3 and command pilot of Gemini 10. Born on 24 September 1930 in San Francisco, California, Young was 38 years old at the time of the Apollo 10 mission. Young received a BS in aeronautical engineering from the Georgia Institute of Technology in 1952, and was selected as an astronaut in 1962. His backup was Lieutenant Colonel Donn Fulton Eisele [EYES-lee] (USAF). Cernan had been pilot of Gemini 9-A. Born on 14 March 1934 in Chicago, Illinois, he was 35 years old at the time of the Apollo 10 mission. Cernan received a BS in electrical engineering from Purdue University in 1956 and an MS in aeronautical engineering from the US Naval Postgraduate School in 1963, and was selected as an astronaut in 1963. His backup was Commander Edgar Dean Mitchell (USN). The capsule communicators (CAPCOMs) were

Major Charles Moss Duke Jr (USAF), Major Joe Henry Engle (USAF), Major Jack Robert Lousma (USMC), and Lieutenant Commander Bruce McCandless II (USN). The support crew consisted of Engle, Lieutenant Colonel James Benson Irwin (USAF), and Duke. The flight directors were Glynn S. Lunney and Gerald D. Griffin (first shift), Milton L. Windler (second shift), and M.P. 'Pete' Frank (third shift).

The Apollo 10 launch vehicle was a Saturn V, designated AS-505. The mission also carried the designation Eastern Test Range #920. The CSM was designated CSM-106, and had the call-sign 'Charlie Brown'. The lunar module was designated LM-4, and had the call-sign 'Snoopy'. The call-signs were taken from the popular comic strip Peanuts©[1] by Charles L. Schultz. For this mission, Snoopy the beagle exchanged his traditional World War I flying ace goggles and scarf for a space helmet. At the Manned Spacecraft Center, Snoopy was the symbol of quality performance.

LAUNCH PREPARATIONS

The terminal countdown was picked up at 01:00:00 GMT on 17 May 1969 and proceeded with no unscheduled holds. The primary LOX replenish pump failed to start at T–8 hr due to a blown fuse in the pump motor starter circuit. Troubleshooting and fuse replacement delayed LOX loading by 50 min but it was completed by T–4 hr 22 min. The lost time was made up during the scheduled 1-hr hold at T–3 hr 30 min. A high pressure cell in the Atlantic Ocean off the New England coast caused southeasterly surface winds and brought moisture into the Cape Canaveral area, which contributed to overcast conditions. At launch time, cumulus clouds covered 40% of the sky (base 2,200 ft), altocumulus covered 20% (base 11,000 ft), and cirrus covered 100% (base not recorded), the temperature was 80.1°F, the relative humidity was 75%, and the barometric pressure was 14.779 psi. The winds, as measured by the anemometer on the light pole 60.0 ft above ground at the launch site, were 19.0 kt at 142° from true north.

ASCENT PHASE

Apollo 10 was launched from Pad B of Launch Complex 39 at the Kennedy Space Center at a Range Zero time of 16:49:00 GMT (12:49:00 EDT) on 18 May 1969, marking the first use of this pad. The launch window extended to 21:09 GMT in order to take advantage of a Sun elevation angle at the ALS-2 site of 11 deg. Between T + 13.05 and T + 32.3 sec, the vehicle rolled from a pad azimuth of 90°E of N to a flight azimuth of 72.028°E of N. The maximum wind conditions encountered during ascent were 82.6 kt at 270° from true north at 46,520 ft, with a maximum wind shear of 0.0203/sec at 50,200 ft. The S-IC shut down at T + 161.63, followed by S-IC/S-II separation, and S-II ignition. The S-II shut down at T + 552.64 followed by separation from the S-IVB, which ignited at T + 556.81.

[1] Copyright United Features Syndicate.

Apollo 10 preparation event	Date
LM-4 integrated test at factory.	25 May 1968
Individual and combined CM and SM systems test completed at factory.	08 Sep 1968
LM-4 final engineering evaluation acceptance test at factory.	02 Oct 1968
Descent stage of LM-4 ready to ship from factory to KSC.	09 Oct 1968
Descent stage of LM-4 delivered to KSC.	11 Oct 1968
Ascent stage of LM-4 ready to ship from factory to KSC.	12 Oct 1968
Ascent stage of LM-4 delivered to KSC.	16 Oct 1968
Integrated CM and SM systems test completed at factory.	19 Oct 1968
LM-4 stages mated.	02 Nov 1968
LM-4 combined systems test completed.	06 Nov 1968
CM-106 and SM-106 delivered to KSC.	23 Nov 1968
CM-106 and SM-106 ready to ship from factory to KSC.	24 Nov 1968
CM-106 and SM-106 mated.	26 Nov 1968
S-IC-5 stage delivered to KSC.	27 Nov 1968
S-II-5 stage delivered to KSC.	03 Dec 1968
S-IVB-505 stage delivered to KSC.	03 Dec 1968
LM-4 altitude tests completed.	06 Dec 1968
Saturn V instrument unit IU-505 delivered to KSC.	15 Dec 1968
CSM-106 combined systems test completed.	16 Dec 1968
Launch vehicle erected.	30 Dec 1968
CSM-106 altitude tests completed.	17 Jan 1969
Launch vehicle propellant dispersion/malfunction overall test completed.	03 Feb 1969
CSM-106 moved to VAB.	06 Feb 1969
Spacecraft erected.	06 Feb 1969
LM-4 combined systems test completed.	10 Feb 1969
CSM-106 integrated systems test completed.	13 Feb 1969
CSM-106 electrically mated to launch vehicle.	27 Feb 1969
Space vehicle overall test completed.	03 Mar 1969
Space vehicle overall test #1 (plugs in) completed.	05 Apr 1969
Space vehicle and MLP-3 transferred to Launch Complex 39B.	11 May 1969
LM-4 flight readiness test completed.	27 Jun 1969
Emergency egress test completed.	28 Jul 1969
Space vehicle flight readiness test completed.	19 Apr 1969
Space vehicle hypergolic fuel loading completed.	25 Apr 1969
S-IC-5 stage RP-1 fuel loading completed.	02 May 1969
Space vehicle countdown demonstration test (wet) completed.	05 May 1969
Space vehicle countdown demonstration test (dry) completed.	06 May 1969

The first S-IVB cutoff occurred at T + 703.76, with deviations from the planned trajectory of only –0.23 ft/sec in velocity and only –0.08 nmi in altitude. The S-IC impacted the Atlantic Ocean at T + 539.12 at 30.188°N and 74.207°W, 348.80 nmi from the launch site. The S-II impacted the Atlantic Ocean at T + 1,217.89 at 31.522°N and 34.512°W, 2,389.29 nmi from the launch site. At insertion, at T + 713.76 (i.e. S-IVB cutoff plus 10 sec to account for engine tailoff and other transient effects), the parking orbit showed an apogee

Apollo 10 ascent phase event	GET (h:m:s)	Altitude (nmi)	Range (nmi)	Earth-fixed velocity (ft/sec)	Space-fixed velocity (ft/sec)	Event duration (sec)	Geocentric latitude (°N)	Longitude (°E)	Space-fixed flight path angle (deg)	Space-fixed heading angle (°E of N)
Liftoff	000:00:00.58	0.035	0.000	1.3	1,340.4	–	28.4658	–80.6209	0.06	90.00
Mach 1 achieved	000:01:06.8	4.244	1.037	1,057.9	2,028.6	–	28.4714	–80.6023	27.82	85.03
Maximum dynamic pressure	000:01:22.6	7.137	2.893	1,623.4	2,645.8	–	28.4813	–80.5690	28.83	82.23
S-IC center engine cutoff*	000:02:15.16	23.430	25.009	5,299.0	6,473.20	141.56	28.5967	–80.1577	22.807	76.461
S-IC outboard engine cutoff	000:02:41.63	35.247	50.419	7,810.2	9,028.58	168.03	28.7182	–79.7090	18.946	75.538
S-IC/S-II separation*	000:02:42.31	35.580	51.223	7,833.4	9,052.79	–	28.7222	–79.6943	18.848	75.538
S-II center engine cutoff	000:07:40.61	96.710	599.079	17,310.1	18,630.15	296.56	30.9579	–69.4941	1.029	79.585
S-II outboard engine cutoff	000:09:12.64	101.204	883.670	21,309.9	22,632.02	388.59	31.7505	–64.0222	0.741	82.458
S-II/S-IVB separation	000:09:13.50	101.247	886.634	21,317.8	22,639.93	–	31.7574	–63.9647	0.730	82.490
S-IVB 1st burn cutoff*	000:11:43.76	103.385	1,430.977	24,238.8	25,562.40	146.95	32.5150	–53.2920	–0.0064	88.497
Earth orbit insertion	000:11:53.76	103.334	1,469.790	24,244.3	25,567.88	–	32.5303	–52.5260	–0.0049	89.933

*Only the commanded time is available for this event.

and perigee of 100.32 × 99.71 nmi, an inclination of 32.546 deg, a period of 88.20 min, and a velocity of 25,567.88 ft/sec. The apogee and perigee were based upon a spherical Earth with a radius of 3,443.934 nmi.

The COSPAR designation for the CSM upon achieving orbit was 1969-043A and the S-IVB was designated 1969-043B. After undocking at the Moon, the LM would be designated 1969-043C.

EARTH ORBIT PHASE

After in-flight systems checks, the 343.08-sec translunar injection maneuver (second S-IVB firing) was performed at 002:33:27.52. The S-IVB engine shut down at 2:39:10.58 and translunar injection occurred 10 sec later, after 1.5 Earth orbits lasting 2 hr 27 min 16.82 sec, at a velocity of 35,562.96 ft/sec.

Apollo 10 earth orbit phase event	GET (h:m:s)	Space-fixed velocity (ft/sec)	Event duration (sec)	Velocity change (ft/sec)	Apogee (nmi)	Perigee (nmi)	Period (min)	Inclination (deg)
Earth orbit insertion	000:11:53.76	25,567.88	–	–	100.32	99.71	88.20	32.546
S-IVB 2nd burn ignition	002:33:27.52	25,561.4	–	–	–	–	–	–
S-IVB 2nd burn cutoff	002:39:10.58	35,585.83	343.06	–	–	–	–	31.701

TRANSLUNAR PHASE

At 003:02:42.4, the CSM was separated from the S-IVB stage. It was transposed and then docked with the LM at 003:17:36.0. The docked spacecraft were ejected at 003:56:25.7 and a separation maneuver was performed at 004:39:09.8. The sequence was televised to Earth starting at 003:06:00 for 22 min and from 003:56:00 for 13 min 25 sec. Additional television broadcasts during translunar coast included:

Apollo 10 translunar television GET (h:m:s)	Duration (m:s)	Subject
005:06:34	13:15	View of Earth and spacecraft interior
007:11:27	24:00	View of Earth and spacecraft interior
027:00:48	27:43	View of Earth and spacecraft interior
048:00:51	14:39	View of Earth and spacecraft interior (recorded)
048:24:00	03:51	View of Earth and spacecraft interior (recorded)
049:54:00	04:49	View of Earth
053:35:30	25:00	View of Earth and spacecraft interior
072:37:26	17:16	View of Earth and spacecraft interior

A ground command for propulsive venting of residual propellants targeted the S-IVB to go past the Moon. The closest approach of the S-IVB to the Moon was 1,680 nmi, at 078:51:03.6 on 21 May at 23:40 GMT. The trajectory after passing from the lunar sphere of influence resulted in a solar orbit with an aphelion and perihelion of 82,160,000 × 73,330,000 nmi, an inclination of 23.46 deg, and a period of 344.88 days. As regards the spacecraft, a pre-planned, 7.1-sec, midcourse correction of 49.2 ft/sec was executed at 026:32:56.8 and adjusted the trajectory to coincide with a July lunar landing trajectory. The maneuver was so accurate that two additional planned midcourse corrections were cancelled. The passive thermal control technique was employed to maintain desired spacecraft temperatures throughout the translunar coast except when a specific attitude was required. At 075:55:54.0, at an altitude of 95.1 nmi above the Moon, the service propulsion system engine was fired for 356.1 sec to insert the spacecraft into a lunar orbit of 170.0 × 60.2 nmi. The translunar coast had lasted 73 hr 22 min 29.5 sec.

Apollo 10 translunar phase event	GET (h:m:s)	Altitude (nmi)	Space-fixed velocity (ft/sec)	Event duration (sec)	Velocity change (ft/sec)	Space-fixed flight path angle (deg)	Space-fixed heading angle (°E of N)
Translunar injection	002:39:20.58	179.920	35,562.96	–	–	7.379	61.065
CSM separated from S-IVB (ignition)	003:02:42.4	3,502.626	25,548.72	–	–	43.928	67.467
CSM SPS evasive maneuver ignition	004:39:09.8	17,938.5	14,220.2	–	–	65.150	91.21
CSM SPS evasive maneuver cutoff	004:39:12.7	17,944.7	14,203.7	2.9	18.8	65.100	91.22
Midcourse correction ignition	026:32:56.8	110,150.2	5,094.4	–	–	77.300	108.36
Midcourse correction cutoff	026:33:03.9	110,155.9	5,110.0	7.1	49.2	77.800	108.92

LUNAR ORBIT PHASE

After two revolutions of tracking and ground updates, a 13.9-sec maneuver was initiated at 080:25:08.1 to circularize the orbit at 61.0 × 59.2 nmi. A 29-min 9-sec scheduled color television transmission of the lunar surface was conducted at 080:44:40, with the crew describing the lunar features below them. The picture quality of lunar scenes was excellent. The lunar module pilot entered the LM at 081:55 for 2 hours of 'housekeeping' activities and some LM communications tests. The tests were terminated following the LM relay communications tests due to time limitations. Results were excellent, and the remaining tests were conducted later in the mission. At 095:02, the commander and lunar module pilot entered to activate LM systems and discovered that the LM had moved 3.5 deg out of line with the CM. The crew feared that separating the two spacecraft might shear off some of the latching pins, possibly preventing redocking. But Mission Control reported that as long as the misalignment was less than 6 deg, there would be no problem. Undocking occurred at 098:29:20 and was televised for 20 min 10 sec starting at 098:13:00. During this period, the LM landing gear were deployed and all LM systems checked out.

A 8.3-sec CSM reaction control system maneuver at 098:47:17.4 separated the CSM to about 30 ft from the LM. The CSM was in an orbit of 62.9 × 57.7 nmi at the time. Stationkeeping was initiated at this point while the command module pilot visually inspected the LM. The CSM reaction control system was then used to perform the separation maneuver directed radially downward toward the Moon's center. This maneuver provided a separation at descent orbit insertion of about 2 nmi from the LM. A 27.4-sec LM descent propulsion system burn at 099:46:01.6 placed the LM into a descent orbit of 60.9 × 8.5 nmi with its lowest point 15 deg east of the ALS-2 landing site. Numerous photographs of the lunar surface were taken. Some camera malfunctions were reported, and despite communications difficulties, the crew managed to provide a continuous commentary of their observations. An hour later, the LM made a low-level pass over the landing site. The pass was highlighted by a test of the landing radar, visual observation of lunar lighting, stereoscopic strip photography, and execution of the phasing maneuver using the descent engine. The lowest measured point in the trajectory was 47,400 ft (7.8 nmi) over the lunar surface at 100:41:43. The second LM maneuver, a 39.95-sec descent propulsion system phasing burn at 100:58:25.93, established a lead angle equivalent to that which would occur at powered ascent cutoff during a lunar landing, and put the LM into an orbit of 190.1 × 12.1 nmi. At 102:44:49, during preparations for rendezvous with the CSM, the LM started to wallow off slowly in yaw, and then stopped. At 102:45:12, it started a rapid roll accompanied by small pitch and yaw rates. The ascent stage was then separated from the descent stage at 102:45:16.9 at an altitude of 31.4 nmi and the motion was terminated 8 sec later. A 15.55-sec firing of the ascent engine at 102:55:02.13 placed the ascent stage into an orbit of 46.5 × 11.0 nmi. The descent stage was left in the low orbit, but perturbations by 'mascons' would have caused this to decay, sending the stage to crash onto the lunar surface. Analysis revealed that the cause of the anomalous motion was human error. Inadvertently, the control mode of the LM abort guidance system was returned to AUTO rather than being left in the ATTITUDE HOLD mode for staging. In AUTO, the abort guidance system drove the LM to acquire the CSM, which was not in accordance with the planned attitude timeline. The commander took over manual control to re-establish the proper attitude. At the orbital low point, the insertion maneuver was performed on time using the LM ascent propulsion system. This burn established the equivalent of the standard LM insertion orbit of a lunar landing mission (45 × 11.2 nmi). The LM coasted in that orbit for about 1 hour. The terminal maneuver occurred at about the midpoint of darkness, and braking for the terminal phase finalization was performed manually as planned. The rendezvous simulated one that would follow a nominal ascent from the lunar surface. It started with a 27.3-sec coelliptic sequence initiation maneuver at 103:45:55.3, which placed the spacecraft into an orbit of 48.7 × 40.7 nmi. This was followed by a 1.65-sec constant differential height maneuver at 104:43:53.28 to raise the perilune to 42.1 nmi. The 16.50-sec terminal phase initiation maneuver at 105:22:55.58 then raised the orbit to 58.3 × 46.8 nmi. Docking was complete at 106:22:02 at an altitude of 54.7 nmi after 8 hr 10 min 5 sec of lunar flight. Once docked, the LM crew transferred the exposed film packets to the CM. The LM ascent stage was jettisoned at 108:24:36. A 6.5-sec separation maneuver at 108:43:23.3 raised the LM orbit to 64.0 × 56.3 nmi. This was followed at 108:52:05.5 (about one revolution after jettison) by a 249.0-sec remote control firing to depletion of the ascent engine. This burn, commanded as planned, utilized the LM ascent engine arming assembly and was targeted to place the LM into a solar orbit. Communications were maintained until LM ascent stage battery depletion at about 120:00. The ascent stage batteries lasted about 12 hours after LM jettison.

Prior to transearth injection, views of the lunar surface and spacecraft interior were transmitted to Earth for 24 min 12 sec starting at 132:07:12. After a rest period, the crew conducted landmark tracking and photography exercises. During the remaining lunar orbital period of operation, 18 landmark sightings, and extensive stereo and oblique photographs were taken. Two scheduled TV periods were deleted because of crew fatigue. Transearth injection was achieved at 137:39:13.7 at a velocity of 8,987.2 ft/sec, following a 164.8-sec engine firing at 56.5 nmi altitude. The spacecraft had made 31 lunar orbits, lasting 61 hr 37 min 23.6 sec.

Apollo 10 lunar orbit phase event	GET (h:m:s)	Altitude (nmi)	Space-fixed velocity (ft/sec)	Event duration (sec)	Velocity change (ft/sec)	Apolune (nmi)	Perilune (nmi)
Lunar orbit insertion ignition	075:55:54.0	95.1	8,232.3	–	–	–	–
Lunar orbit insertion cutoff	076:01:50.1	61.2	5,471.9	356.1	2,982.4	170.0	60.2
Lunar orbit circularization ignition	080:25:08.1	60.4	5,484.7	–	–	–	–
Lunar orbit circularization cutoff	080:25:22.0	59.3	5,348.9	13.9	139.0	61.0	59.2
CSM/LM undocked	098:11:57	58.1	5,357.8	–	–	–	–
CSM/LM separation ignition	098:47:17.4	59.2	5,352.2	–	–	–	–
CSM/LM separation cutoff	098:47:25.7	59.2	5,352.1	8.3	2.5	62.9	57.7
LM descent orbit insertion ignition	099:46:01.6	61.6	5,339.6	–	–	–	–
LM descent orbit insertion cutoff	099:46:29.0	61.2	5,271.2	27.4	71.3	60.9	8.5
LM closest approach to lunar surface	100:41:43	7.8	–	–	–	–	–
LM phasing ignition	100:58:25.93	17.7	5,212.4	–	–	–	–
LM phasing cutoff	100:59:05.88	19.0	5,672.9	39.95	176.0	190.1	12.1
LM ascent/descent stage separation	102:45:16.9	31.4	5,605.6	–	–	–	–
LM ascent orbit insertion ignition	102:55:02.13	11.6	5,705.2	–	–	–	–
LM ascent orbit insertion cutoff	102:55:17.68	11.7	5,520.6	15.55	220.9	46.5	11.0
LM coelliptic sequence initiation ignition	103:45:55.3	44.7	5,335.5	–	–	–	–
LM coelliptic sequence initiation cutoff	103:46:22.6	44.6	5,381.7	27.3	45.3	48.7	40.7
LM constant differential height ignition	104:43:53.28	44.3	5,394.7	–	–	–	–

Apollo 10 lunar orbit phase event – *continued*	GET (h:m:s)	Altitude (nmi)	Space-fixed velocity (ft/sec)	Event duration (sec)	Velocity change (ft/sec)	Apolune (nmi)	Perilune (nmi)
LM constant differential height cutoff	104:43:54.93	43.8	5,394.9	1.65	3.0	48.8	42.1
LM terminal phase initiation ignition	105:22:55.58	48.4	5,369.2	–	–	–	–
LM terminal phase initiation cutoff	105:23:12.08	47.0	5,396.7	16.50	24.1	58.3	46.8
LM 1st midcourse correction	105:37:56	–	–	–	1.27	–	–
LM 2nd midcourse correction	105:52:56	–	–	–	1.84	–	–
LM braking	106:05:49	–	–	–	31.6	63.3	56.4
CSM/LM docked	106:22:02	54.7	5,365.9	–	–	–	–
LM separation ignition	108:43:23.3	57.3	5,352.3	–	–	–	–
LM separation cutoff	108:43:29.8	57.6	5,352.1	6.5	2.1	64.0	56.3
LM ascent propulsion system ignition	108:52:05.5	59.1	5,343.0	–	–	–	–
LM ascent propulsion system depletion	108:56.14.5	89.7	9,056.4	249.0	4,600.0	2,211.6	56.2

TRANSEARTH PHASE

Transearth activities included a number of star–Earth horizon navigation sightings and the CSM S-band high-gain reflectivity test which was conducted at 168:00. Six television transmissions were made on the return trip.

Apollo 10 transearth television GET (h:m:s)	Duration (m:s)	Subject
137:50:51	43:03	View of Moon after transearth injection
139:30:16	06:55	View of Moon after transearth injection
147:23:00	11:25	View of receding Moon and spacecraft interior
152:29:19	29:05	View of Earth, Moon, and spacecraft interior
173:27:17	10:22	View of Earth and spacecraft interior
186:51:49	11:53	View of Earth and spacecraft interior

The passive thermal control technique and the navigation procedures employed on the translunar coast were also used on the return trip. The only midcourse correction required was a 6.7-sec, 2.2-ft/sec, maneuver at 188:49:58.0, 3 hours prior to CM/SM separation.

Apollo 10 transearth phase event	GET (h:m:s)	Altitude (nmi)	Space-fixed velocity (ft/sec)	Event duration (sec)	Velocity change (ft/sec)	Space-fixed flight path angle (deg)	Space-fixed heading angle (°E of N)
Transearth injection ignition	137:36:28.9	56.0	5,362.7	–	–	–0.44	–73.60
Transearth injection cutoff	137:39:13.7	56.5	8,987.2	164.8	3,680.3	2.53	–76.68
Midcourse correction ignition	188:49:58.0	25,570.4	12,540.0	–	–	–69.65	119.34
Midcourse correction cutoff	188:50:04.7	25,557.4	12,543.5	6.7	2.2	–69.64	119.34

RECOVERY

The service module was jettisoned at 191:33:26, and the CM entry followed a normal profile. The command module re-entered Earth's atmosphere (at the 400,000-ft altitude of the 'entry interface') at 191:48:54.5 traveling at a velocity of 36,314 ft/sec, following a transearth coast of 54 hr 3 min 40.9 sec.[2] The service module impacted the Pacific Ocean at a point estimated to be 19.4°S and 173.37°W. The parachute system effected a soft splashdown of the CM in the Pacific Ocean at 16:52:23 GMT on 26 May. The mission duration was 192:03:23. The impact point, estimated to be 15.07°S and 164.65°W, was about 1.3 nmi from the target point and 2.9 nmi from the recovery ship USS *Princeton.* After splashdown, the CM assumed an apex-up flotation attitude. The crew was retrieved by helicopter and was on board the recovery ship 39 min after splashdown. The CM was recovered 57 min later. The estimated CM weight at splashdown was 10,901 lb, and the estimated distance traveled for the mission was 721,250 nmi. At CM retrieval, the weather recorded on the *Princeton* showed 10% cloud cover at 2,000 ft and 20% at 7,000 ft, a visibility of 10 nmi, a wind speed of 5 kt at 100° from true north, a water temperature of 85°F, with waves to 3 ft; the air temperature was not recorded. The CM was offloaded from the *Princeton* on 31 May at Ford Island, Hawaii, and the Landing Safing Team began the evaluation and deactivation procedures at 18:00 GMT, completing this work at 05:56 GMT on 3 June. The CM was flown to Long Beach, California, where it arrived at 10:15 GMT on 4 June. It was trucked the same day to the North American Rockwell

[2] The *Guinness Book of World Records* states that Apollo 10 holds the record for the fastest a human has ever traveled: 24,791 statute miles per hour at 400,000 ft altitude (entry) on 26 May 1969, but the Apollo 10 mission report states the maximum speed at entry was 36,397 ft per second, or 24,816 statute miles per hour.

Space Division facility in Downey, California, for post-flight analysis. All systems in the CSM and the LM were managed very well. Although some problems occurred, most were minor and none caused a constraint to completion of mission objectives. Valuable data concerning lunar gravitation were obtained during the 61 hr in lunar orbit. Spacecraft systems performance was satisfactory, and all mission objectives were accomplished. All detailed test objectives were satisfied with the exception of the LM steerable antenna and relay modes for voice and telemetry communications.

CONCLUSIONS

The Apollo 10 mission provided the concluding data and final environmental evaluation to proceed with a lunar landing. The following conclusions were made from an analysis of post-mission data:

1. The systems in the command and service modules and lunar module were operational for a manned lunar landing.
2. The crew activity timeline, in those areas consistent with the lunar landing profile, demonstrated that critical crew tasks associated with lunar module checkout, initial descent, and rendezvous were both feasible and practical without unreasonable crew workload.
3. The lunar module S-band communications capability using either the steerable or the omnidirectional antenna was satisfactory at lunar distances.
4. The operating capability of the landing radar in the lunar environment during a descent propulsion firing was satisfactorily demonstrated for the altitudes experienced.
5. The range capability of the lunar module rendezvous radar was demonstrated in the lunar environment with excellent results. Used for the first time, VHF ranging information from the CM provided consistent correlation with radar range and range-rate data.
6. The lunar module abort guidance system capability to control an ascent propulsion system maneuver and to guide the spacecraft during rendezvous was demonstrated.
7. The capability of the Mission Control Center and the Manned Space Flight Network to control and monitor two vehicles at lunar distance during both descent and rendezvous operations was proven adequate for a lunar landing.
8. The lunar potential model was significantly improved over that of Apollo 8, and the orbit determination and prediction procedures proved remarkably more precise for both spacecraft in lunar orbit. After a combined analysis of Apollo 8 and 10 trajectory reconstructions, the lunar potential model was expected to be entirely adequate for support of a lunar descent and ascent.

The Apollo 10 crew: Gene Cernan (left), John Young and Tom Stafford.

A bird's-eye view of the AS-505 roll out.

The launch of Apollo 10.

The Apollo 10 CSM ‘Charlie Brown’ in lunar orbit.

The landmark crater Maskelyne to the right of the LM's track during its low pass over ALS-2.

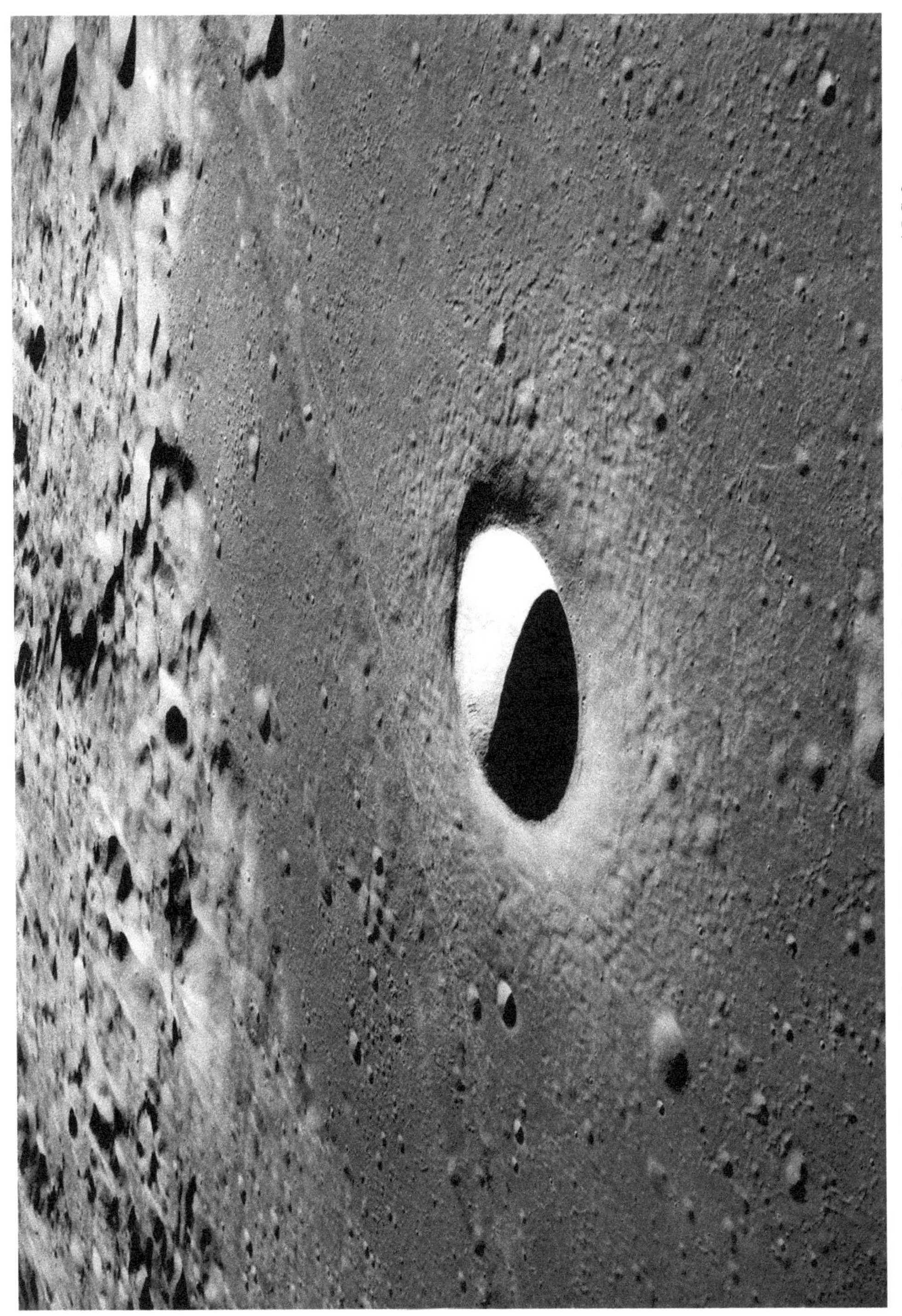

The crater Moltke and US Highway 1 to the left of the LM's track during its low pass over ALS-2.

MISSION OBJECTIVES

Launch Vehicle Objectives

1. To demonstrate launch vehicle capability to inject the spacecraft into the specified translunar trajectory. *Achieved.*
2. To demonstrate launch vehicle capability to maintain a specified attitude for transposition, docking, and spacecraft ejection maneuver. *Achieved.*
3. To demonstrate S-IVB propellant dump and safing. *Achieved.*
4. To verify J-2 engine modifications. *Achieved.*
5. To confirm J-2 engine environment in S-II and S-IVB stages. *Achieved.*
6. To confirm launch vehicle longitudinal oscillations environment during S-IC stage burn period. *Achieved.*
7. To verify that modifications incorporated in the S-IC stage suppressed low-frequency longitudinal oscillations. *Achieved.*
8. To confirm launch vehicle longitudinal oscillation environment during S-II stage burn period. *Achieved.*
9. To demonstrate that early center engine cutoff for S-II stage suppressed low-frequency longitudinal oscillations. *Achieved.*

Spacecraft Primary Objectives

1. To demonstrate crew/space vehicle/mission support facilities performance during a manned lunar mission with a command and service module and lunar module. *Achieved.*
2. To evaluate lunar module performance in the cislunar and lunar environment. *Achieved.*

Spacecraft Primary Detailed Objectives

1. P11.15: To perform primary guidance and navigation control system/descent propulsion system undocked descent orbit insertion and a high-thrust maneuver. *Achieved.*
2. P16.10: To perform manual and automatic acquisition, tracking, and communications with the Manned Space Flight Network using the steerable S-band antenna at lunar distance. *Achieved, despite some problems during the 13th lunar revolution.*
3. P16.14: To operate the landing radar at the closest approach to the Moon and during descent propulsion system burns. *Achieved.*
4. P20.66: To obtain data on the command module and lunar module crew procedures and timeline for the lunar orbit phase of a lunar landing mission. *Achieved.*
5. P20.78: To perform a lunar module active simulated lunar landing mission rendezvous. *Achieved.*
6. P20.91: To perform lunar landmark tracking in lunar orbit from the command and service module with the lunar module attached. *Achieved.*
7. P20.121: To perform lunar landmark tracking from the command and service module while in lunar orbit. *Achieved.*

Spacecraft Secondary Detailed Objectives

1. S1.39: To perform star–lunar landmark sightings during the transearth phase. *Achieved.*
2. S6.9: To perform a reflectivity test using the command and service module S-band high-gain antenna while docked. *Not achieved, cancelled while docked. S-band communications lost because steerable antenna track mode not switched properly; however, operation of steerable antenna during abnormal staging excursions demonstrated ability of antenna to track under very high rates.*
3. 7.26: To obtain data on the passive thermal control system during a lunar orbit mission. *Achieved.*
4. S11.17: To obtain data to verify inertial measurement unit performance in the flight environment. *Achieved.*
5. S12.6: To obtain abort guidance system performance data in the flight environment. *Achieved.*
6. S12.8: To demonstrate reaction control system translation and attitude control of the staged lunar module using automatic and manual abort guidance system/control electronics system control. *Achieved.*
7. S12.9: To perform an unmanned abort guidance system controlled ascent propulsion system burn. *Achieved.*
8. S12.10: To evaluate the ability of the abort guidance system to perform a lunar module active rendezvous. *Achieved.*
9. S13.13: To perform a long-duration unmanned ascent propulsion system burn. *Achieved.*
10. S13.14: To obtain supercritical helium system pressure data while in standby conditions and during all descent propulsion system engine firings. *Achieved.*
11. S16.12: To communicate with the Manned Space Flight Network using the lunar module S-band omnidirectional antennas at lunar distance. *Achieved, despite some problems during the 13th lunar revolution.*
12. S16.15: To obtain data on the rendezvous radar performance and capability near maximum range. *Achieved.*
13. S16.17: To demonstrate lunar module, command and service module/Manned Space Flight Network communications at lunar distance. *Achieved, despite some problems due to procedural errors.*
14. S20.46: To perform command and service module transposition, docking, and command and service module/lunar module ejection after the S-IVB translunar injection burn. *Achieved.*
15. S20.77: To obtain data on the operational capability of VHF ranging during a lunar module active rendezvous. *Achieved.*
16. S20.79: To demonstrate command and service module/lunar module passive thermal control modes during a lunar orbit mission. *Achieved.*
17. S20.80: To demonstrate operational support for a command and service module/lunar module orbit mission. *Achieved despite some communication problems.*
18. S20.82: To monitor primary guidance and navigation control system/abort guidance system performance during lunar orbit operations. *Achieved.*
19. S20.83: To obtain data on lunar module consumables for a simulated lunar landing

mission in lunar orbit to determine required lunar landing mission consumables. *Achieved.*

20. S20.86: To obtain data on the effects of lunar illumination and contrast conditions on crew visual perception while in lunar orbit. *Achieved.*
21. S20.95: To perform translunar midcourse corrections. *Achieved. Only one of four possible midcourse corrections was required.*
22. S20.117: To perform lunar orbit insertion using service propulsion system/guidance and navigation control system controlled burns with a docked command and service module/lunar module. *Achieved.*

MISSION TIMELINE

Apollo 10 mission event	GET (h:m:s)	Date (GMT)	Time (h:m:s)
Terminal countdown started.	–028:00:00	17 May 1969	01:00:00
Scheduled 1-hr hold at T–3 hr 30 min.	–003:30:00	18 May 1969	12:19:00
Countdown resumed at T–3 hr 30 min.	–003:30:00	18 May 1969	13:19:00
Guidance reference release.	–000:00:16.978	18 May 1969	16:48:43
S-IC engine start command.	–000:00:08.9	18 May 1969	16:48:51
S-IC engine ignition (#5).	–000:00:06.4	18 May 1969	16:48:53
All S-IC engines thrust OK.	–000:00:01.6	18 May 1969	16:48:58
Range zero.	000:00:00.00	18 May 1969	16:49:00
All holddown arms released (1st motion) (1.06 *g*).	000:00:00.25	18 May 1969	16:49:00
Liftoff (umbilical disconnected).	000:00:00.58	18 May 1969	16:49:00
Tower clearance yaw maneuver started.	000:00:01.6	18 May 1969	16:49:01
Yaw maneuver ended.	000:00:10.0	18 May 1969	16:49:10
Pitch and roll maneuver started.	000:00:13.05	18 May 1969	16:49:13
Roll maneuver ended.	000:00:32.3	18 May 1969	16:49:32
Mach 1 achieved.	000:01:06.8	18 May 1969	16:50:06
Maximum dynamic pressure (694.232 lb/ft^2).	000:01:22.6	18 May 1969	16:50:22
Maximum bending moment (88,000,000 lbf-in).	000:01:24.6	18 May 1969	16:50:24
S-IC center engine cutoff command.	000:02:15.16	18 May 1969	16:51:15
Pitch maneuver ended.	000:02:38.7	18 May 1969	16:51:38
S-IC outboard engine cutoff.	000:02:41.63	18 May 1969	16:51:41
S-IC maximum total inertial acceleration (3.92 *g*).	000:02:41.71	18 May 1969	16:51:41
S-IC maximum Earth-fixed velocity.	000:02:41.96	18 May 1969	16:51:41
S-IC/S-II separation command.	000:02:42.31	18 May 1969	16:51:42
S-II engine start command.	000:02:43.05	18 May 1969	16:51:43
S-II ignition.	000:02:44.05	18 May 1969	16:51:44
S-II aft interstage jettisoned.	000:03:12.3	18 May 1969	16:52:12
Launch escape tower jettisoned.	000:03:17.8	18 May 1969	16:52:17
Iterative guidance mode initiated.	000:03:22.9	18 May 1969	16:52:22
S-IC apex.	000:04:26.87	18 May 1969	16:53:26
S-II center engine cutoff.	000:07:40.61	18 May 1969	16:56:40
S-II maximum total inertial acceleration (1.82 *g*).	000:07:40.69	18 May 1969	16:56:40

Apollo 10 mission event – *continued*	GET (h:m:s)	Date (GMT)	Time (h:m:s)
S-IC impact (theoretical).	000:08:59.12	18 May 1969	16:57:59
S-II outboard engine cutoff.	000:09:12.64	18 May 1969	16:58:12
S-II maximum Earth-fixed velocity. S-II/S-IVB separation command.	000:09:13.50	18 May 1969	16:58:13
S-IVB 1st burn start command.	000:09:13.60	18 May 1969	16:58:13
S-IVB 1st burn ignition.	000:09:16.81	18 May 1969	16:58:16
S-IVB ullage case jettisoned.	000:09:25.4	18 May 1969	16:58:25
S-II apex.	000:09:57.21	18 May 1969	16:58:57
S-IVB 1st burn cutoff command.	000:11:43.76	18 May 1969	17:00:43
S-IVB 1st burn maximum Earth-fixed velocity and maximum total inertial acceleration (0.70 g).	000:11:43.84	18 May 1969	17:00:43
Earth orbit insertion.	000:11:53.76	18 May 1969	17:00:53
Maneuver to local horizontal attitude and orbital navigation started.	000:12:04.1	18 May 1969	17:01:04
S-II impact (theoretical).	000:20:17.89	18 May 1969	17:09:17
S-IVB 2nd burn restart preparation.	002:23:49.26	18 May 1969	19:12:49
S-IVB 2nd burn restart command.	002:33:19.20	18 May 1969	19:22:19
S-IVB 2nd burn ignition.	002:33:27.52	18 May 1969	19:22:27
S-IVB 2nd burn cutoff.	002:39:10.58	18 May 1969	19:28:10
S-IVB 2nd burn maximum total inertial acceleration (1.49 g).	002:39:10.66	18 May 1969	19:28:10
S-IVB 2nd burn maximum Earth-fixed velocity. S-IVB safing procedures started.	002:39:11.30	18 May 1969	19:28:11
Translunar injection.	002:39:20.58	18 May 1969	19:28:20
Orbital navigation started.	002:39:29.6	18 May 1969	19:28:29
CSM separated from S-IVB (ignition).	003:02:42.4	18 May 1969	19:51:42
CSM separated from S-IVB (cutoff).	003:02:45.7	18 May 1969	19:51:45
TV transmission started.	003:06:00	18 May 1969	19:55:00
CSM docked with LM/S-IVB.	003:17:36.0	18 May 1969	20:06:36
TV transmission ended.	003:28:00	18 May 1969	20:17:00
TV transmission started.	003:56:00	18 May 1969	20:45:00
CSM/LM ejected from S-IVB.	003:56:25.7	18 May 1969	20:45:25
TV transmission ended.	004:09:25	18 May 1969	20:58:25
CSM SPS evasive maneuver ignition.	004:39:09.8	18 May 1969	21:28:09
CSM SPS evasive maneuver cutoff.	004:39:12.7	18 May 1969	21:28:12
Maneuver to S-IVB slingshot attitude initiated.	004:42:15.8	18 May 1969	21:31:15
S-IVB lunar slingshot maneuver – APS ignition.	004:45:36.4	18 May 1969	21:34:36
S-IVB lead experiment – LOX lead started.	004:48:21.3	18 May 1969	21:37:21
S-IVB lead experiment – LOX lead ended.	004:48:30.3	18 May 1969	21:37:30
S-IVB lead experiment – LH_2 lead started.	004:50:09.9	18 May 1969	21:39:09
S-IVB lunar slingshot maneuver – APS cutoff.	004:50:17.0	18 May 1969	21:39:17
S-IVB lead experiment – LH_2 lead ended.	004:50:58.8	18 May 1969	21:39:58
S-IVB safing – LH_2 tank CVS open.	004:51:36.1	18 May 1969	21:40:36
S-IVB lunar slingshot maneuver – LOX dump started.	004:54:15.79	18 May 1969	21:43:15
S-IVB lunar slingshot maneuver – LOX dump ended.	004:59:16.00	18 May 1969	21:48:16
TV transmission started.	005:06:34	18 May 1969	21:55:34

Apollo 10 mission event – *continued*	GET (h:m:s)	Date (GMT)	Time (h:m:s)
S-IVB safing – LH_2 tank NPV valve open.	005:16:09.8	18 May 1969	22:05:09
TV transmission ended.	005:19:49	18 May 1969	22:08:49
S-IVB lunar slingshot maneuver – APS ignition.	005:28:55.8	18 May 1969	22:17:55
S-IVB lunar slingshot maneuver – APS cutoff.	005:29:04.9	18 May 1969	22:18:04
TV transmission started.	007:11:27	19 May 1969	00:00:27
TV transmission ended.	007:35:27	19 May 1969	00:24:27
Midcourse correction ignition.	026:32:56.8	19 May 1969	19:21:56
Midcourse correction cutoff.	026:33:03.9	19 May 1969	19:22:03
TV transmission started.	027:00:48	19 May 1969	19:49:48
TV transmission ended.	027:28:31	19 May 1969	20:17:31
High-gain antenna reacquisition test.	028:50	19 May 1969	21:39
TV transmission started (recorded).	048:00:51	20 May 1969	16:49:51
TV transmission ended.	048:15:30	20 May 1969	17:04:30
TV transmission started (recorded).	048:24:00	20 May 1969	17:13:00
TV transmission ended.	048:27:51	20 May 1969	17:16:51
TV transmission started.	049:54:00	20 May 1969	18:43:00
TV transmission ended.	049:58:49	20 May 1969	18:47:49
TV transmission started.	053:35:30	20 May 1969	22:24:30
TV transmission ended.	054:00:30	20 May 1969	22:49:30
Equigravisphere.	061:50:50	21 May 1969	06:39:50
TV transmission started.	072:37:26	21 May 1969	17:26:26
TV transmission ended.	072:54:42	21 May 1969	17:43:42
Lunar orbit insertion ignition.	075:55:54.0	21 May 1969	20:44:54
Lunar orbit insertion cutoff.	076:01:50.1	21 May 1969	20:50:50
Lunar surface observations.	076:30	21 May 1969	21:19
S-IVB closest approach to lunar surface.	078:51:03.6	21 May 1969	23:40:03
Lunar orbit circularization ignition.	080:25:08.1	22 May 1969	01:14:08
Lunar orbit circularization cutoff.	080:25:22.0	22 May 1969	01:14:22
TV transmission started.	080:44:40	22 May 1969	01:33:40
Lunar surface observations.	080:50	22 May 1969	01:39
TV transmission ended.	081:13:49	22 May 1969	02:02:49
LM cabin pressurized.	081:30	22 May 1969	02:19
Transfer to LM power and systems checked.	081:55	22 May 1969	02:44
Transfer to LM power. Systems tested.	082:40	22 May 1969	03:29
Transfer to CM, hatch and tunnel closed.	084:30	22 May 1969	05:19
CDR and LMP entered LM to activate systems.	095:02	22 May 1969	15:51
Landing gear deployed.	098:00	22 May 1969	18:49
CSM/LM undocked.	098:11:57	22 May 1969	19:00:57
TV transmission started.	098:29:20	22 May 1969	19:18:20
CSM/LM separation maneuver ignition.	098:47:17.4	22 May 1969	19:36:17
CSM/LM separation maneuver cutoff.	098:47:25.7	22 May 1969	19:36:25
TV transmission ended.	098:49:30	22 May 1969	19:38:30
CSM rendezvous radar transponder anomaly.	098:51:54	22 May 1969	19:40:54
LM system checks.	099:00	22 May 1969	19:49
Descent orbit insertion ignition (SPS).	099:46:01.6	22 May 1969	20:35:01

Apollo 10 mission event – *continued*	GET (h:m:s)	Date (GMT)	Time (h:m:s)
Descent orbit insertion cutoff.	099:46:29.0	22 May 1969	20:35:29
LM oriented for radar overpass test.	100:32:00	22 May 1969	21:21:00
LM acquisition of radar beam.	100:32:22	22 May 1969	21:21:22
LM near-lunar-surface activity.	100:40	22 May 1969	21:29
LM closest approach to lunar surface.	100:41:43	22 May 1969	21:30:43
Phasing maneuver ignition.	100:58:25.93	22 May 1969	21:47:25
Phasing maneuver cutoff.	100:59:05.88	22 May 1969	21:48:05
Anomalous motion of LM started.	102:44:49	22 May 1969	23:33:49
LM in rapid roll.	102:45:12	22 May 1969	23:34:12
LM ascent stage/descent stage separated.	102:45:16.9	22 May 1969	23:34:16
LM ascent stage anomalous motion under control.	102:45:25	22 May 1969	23:34:25
Ascent orbit insertion ignition	102:55:02.13	22 May 1969	23:44:02
Ascent orbit insertion cutoff.	102:55:17.68	22 May 1969	23:44:17
Coelliptic sequence initiation ignition.	103:45:55.3	23 May 1969	00:34:55
Coelliptic sequence initiation cutoff.	103:46:22.6	23 May 1969	00:35:22
Constant differential height maneuver ignition.	104:43:53.28	23 May 1969	01:32:53
Constant differential height maneuver cutoff.	104:43:54.93	23 May 1969	01:32:54
Terminal phase initiation ignition.	105:22:55.58	23 May 1969	02:11:55
Terminal phase initiation cutoff.	105:23:12.08	23 May 1969	02:12:12
Midcourse correction (lunar orbit).	105:37:56	23 May 1969	02:26:56
Midcourse correction (lunar orbit).	105:52:56	23 May 1969	02:41:56
Braking maneuver.	106:05:49	23 May 1969	02:54:49
CSM/LM docked.	106:22:02	23 May 1969	03:11:02
CDR and LMP entered CM.	106:42	23 May 1969	03:31
LM closeout activities started.	107:20	23 May 1969	04:09
LM ascent stage jettisoned.	108:24:36	23 May 1969	05:13:36
LM separation maneuver ignition.	108:43:23.3	23 May 1969	05:32:23
LM separation maneuver cutoff.	108:43:29.8	23 May 1969	05:32:29
LM ascent propulsion system ignition.	108:52:05.5	23 May 1969	05:41:05
LM ascent propulsion system depletion.	108:56.14.5	23 May 1969	05:45:14
Terminator-to-terminator strip photographs.	119:20	23 May 1969	16:09
Orbital navigation and landmark tracking.	124:30	23 May 1969	21:19
Orbital navigation and landmark tracking.	128:00	24 May 1969	00:49
TV transmission started.	132:07:12	24 May 1969	04:56:12
TV transmission ended.	132:31:24	24 May 1969	05:20:24
Target of opportunity photography.	133:00	24 May 1969	05:49
Target of opportunity and strip photography.	134:40	24 May 1969	07:29
Transearth injection ignition (SPS).	137:36:28.9	24 May 1969	10:25:28
Transearth injection cutoff.	137:39:13.7	24 May 1969	10:28:13
TV transmission started.	137:50:51	24 May 1969	10:39:51
TV transmission ended.	138:33:54	24 May 1969	11:22:54
TV transmission started.	139:30:16	24 May 1969	12:19:16
TV transmission ended.	139:37:11	24 May 1969	12:26:11
TV transmission started.	147:23:00	24 May 1969	20:12:00
TV transmission ended.	147:34:25	24 May 1969	20:23:25

Apollo 10 mission event – *continued*	GET (h:m:s)	Date (GMT)	Time (h:m:s)
TV transmission started.	152:29:19	25 May 1969	01:18:19
TV transmission ended.	152:58:24	25 May 1969	01:47:24
CSM S-band high-gain reflectivity test.	168:00	25 May 1969	16:49
TV transmission started.	173:27:17	25 May 1969	22:16:17
TV transmission ended.	173:37:39	25 May 1969	22:26:39
TV transmission started.	186:51:49	26 May 1969	11:40:49
TV transmission ended.	187:03:42	26 May 1969	11:52:42
Midcourse correction ignition.	188:49:58.0	26 May 1969	13:38:58
Midcourse correction cutoff.	188:50:04.7	26 May 1969	13:39:04
Maneuver to entry attitude.	189:40	26 May 1969	14:29
CM/SM separation.	191:33:26	26 May 1969	16:22:26
Entry.	191:48:54.5	26 May 1969	16:37:54
Communication blackout started.	191:49:12	26 May 1969	16:38:12
Maximum entry g force (6.78 g).	191:50:14	26 May 1969	16:39:14
Visual contact with CM established by recovery forces.	191:51	26 May 1969	16:40
Radar contact with CM established by recovery ship.	191:52	26 May 1969	16:41
Communication blackout ended.	191:53:40	26 May 1969	16:42:40
Drogue parachute deployed	191:57:18.0	26 May 1969	16:46:18
Main parachute deployed.	191:58:05	26 May 1969	16:47:05
Splashdown (went to apex-up).	192:03:23	26 May 1969	16:52:23
Flotation collar inflated.	192:21	26 May 1969	17:10
CM hatch opened.	192:28	26 May 1969	17:17
Crew in life raft.	192:31	26 May 1969	17:20
Crew on board recovery helicopter.	192:37	26 May 1969	17:26
Crew on board recovery ship.	192:42	26 May 1969	17:31
CM on board recovery ship.	193:39	26 May 1969	18:28
CM at Ford Island, Hawaii.	–	31 May 1969	–
CM deactivation started.	313:11	31 May 1969	18:00
CM deactivation completed	353:07	03 Jun 1969	05:56
CM arrived in Long Beach, CA	381:26	04 Jun 1969	10:15
CM arrived at contractor's facility in Downey, CA.	–	04 Jun 1969	–

Apollo 11

The fifth manned mission: the first lunar landing

16–24 July 1969

BACKGROUND

As the Type 'G' mission, Apollo 11 was a manned lunar landing demonstration. The primary objective of the Apollo program was to perform a manned lunar landing and return safely to Earth.

This was only the second time that an all-experienced crew had flown an American mission (the first was Apollo 10).[3] The crew members for this historic mission were Neil Alden Armstrong, commander; Lieutenant Colonel Michael Collins (USAF), command module pilot; and Colonel Edwin Eugene 'Buzz' Aldrin Jr (USAF), lunar module pilot. Selected as an astronaut in 1962, Armstrong had been the first civilian ever to command an American space mission when he was command pilot of Gemini 8, which featured the first-ever docking of two vehicles in space. Apollo 11 made him the first civilian to command two missions. Armstrong was born on 5 August 1930 in Wapakoneta, Ohio, and was 38 years old at the time of the Apollo 11 mission. He received a BS in aeronautical engineering from Purdue University in 1955 and an MS in aerospace engineering from the University of Southern California in 1970, following the Apollo mission. His backup was Captain James Arthur Lovell Jr (USN). Collins had been pilot of Gemini 10. He was born on 31 October 1930 in Rome, Italy, and was 38 years old at the time of the Apollo 11 mission. Collins received a BS from the US Military Academy in 1952 and was selected as an astronaut in 1963. His backup was Lieutenant Colonel William Alison Anders (USAF). Aldrin had been pilot of Gemini 12. He was born on 20 January 1930 in Montclair, New Jersey, and was 39 years old at the time of the Apollo 11 mission. Aldrin received a BS in mechanical engineering from the US Military Academy in 1951 and an ScD in astronautics from the Massachusetts Institute of Technology in 1963. Also in 1963, he was selected as an astronaut. Aldrin has the distinction of being the first astronaut with a doctorate to fly in space. His backup was Fred Wallace Haise Jr. The capsule communicators (CAPCOMs) for the mission were Major Charles Moss Duke Jr (USAF),

[3] It would be the last such crew until space shuttle mission STS-26 nearly two decades later.

Lieutenant Commander Ronald Ellwin Evans (USN), Lieutenant Commander Bruce McCandless II (USN), Lovell, Anders, Lieutenant Commander Thomas Kenneth 'Ken' Mattingly II (USAF), Haise, Don Leslie Lind PhD, Owen Kay Garriott Jr PhD, and Harrison Hagan 'Jack' Schmitt PhD. The support crew were Mattingly, Evans, Major William Reid Pogue (USAF), and John Leonard 'Jack' Swigert Jr. The flight directors were Clifford E. Charlesworth and Gerald D. Griffin (first shift), Eugene F. Kranz (second shift), and Glynn S. Lunney (third shift).

The Apollo 11 launch vehicle was a Saturn V, designated AS-506. The mission also carried the designation Eastern Test Range #5307. The CSM was designated CSM-107 and had the call-sign 'Columbia'. The lunar module was designated LM-5 and had the call-sign 'Eagle'.

LANDING SITE

NASA began the Ranger project in December 1959 as the flagship for its reconnaissance of the Moon. The first two flights in August and November 1961 were intended to test the spacecraft's basic systems in deep space, but the Agena rocket stages failed and stranded their payloads in 'parking orbit'. Despite these losses, the Jet Propulsion Laboratory (JPL) decided to proceed with the second batch of spacecraft, whose plunging dive to the Moon was to be documented by a TV camera and, a split second before the spacecraft hit, it was to eject a spherical shock-resistant 'hard landing' capsule containing a seismometer. Unfortunately, Ranger 3's Agena over-performed and the spacecraft missed the Moon by 20,000 nmi. The next Agena was so accurate that Ranger 4 hit the Moon, but by then an electrical fault had crippled the spacecraft. Ranger 5, which missed the Moon by 420 nmi, was also disabled by a power failure. In December 1962, with its best result being a dead spacecraft hitting the Moon, the project was in danger of cancellation. After a review of spacecraft assembly procedures, NASA redefined the project's goals: the next vehicles would have only the TV package, and their sole objective would be to gain close-up pictures of the lunar surface to assess whether it was likely to support a spacecraft.

The location of the target was constrained by flight dynamics considerations. The initial TV viewpoint would match the best telescopic pictures. The spacecraft was to make a near-vertical dive in order that successive images would overlap. Watching the final moments of the spacecraft's descent in 'real time' would be stunning, but the analysis would be pain-staking. And not only would the film be replayed to identify the manner in which the character of the surface altered on different scales, 'zooming out' by running the film backward would place fine detail into a wider context. A target in the western hemisphere was required for a plunging dive, but Apollo's dynamical constraints favored eastern sites. Ranger 6's television system was disabled by an electrical arc at launch, but this did not become evident until the system failed to start as the spacecraft neared its target in Mare Tranquilitatis (Sea of Tranquility). The project's luck changed spectacularly on 31 July 1964, when Ranger 7 struck a patch of mare in the general vicinity of Mare Nubium (Sea of Clouds). The final image showed detail only a few feet across, which was a factor of a *thousand* improvement in resolution over the best telescope. The terrain was fairly soft and rolling, with none of the jagged features of science fiction.

Although an automated spacecraft might well come to grief by setting down on a rock or in a crater, there were evidently open spaces too, and an Apollo crew should be able to maneuver to a safe spot. A set of shallow ridges indicated that the mare was a lava flow. The presence of boulders implied that the surface was likely to support a spacecraft. On 31 August 1964, the International Astronomical Union marked Ranger 7's success by renaming the site Mare Cognitum (Known Sea). In February 1965, Ranger 8 followed a shallow trajectory from the west that took it over the central highlands to Tranquilitatis. While this trajectory increased surface coverage, it resulted in substantial smearing in the final frames. Having inspected two mare sites, NASA released the final mission to the scientists who, after considering the craters Copernicus, Kepler and Aristarchus, chose Alphonsus, a 60-nmi crater with a central peak and a flat floor full of interesting rilles and 'dark halo' craters that looked to be volcanoes. On 24 March 1965, Ranger 9 fell into the crater and, for the first time, the TV was fed to the commercial networks, which broadcast it with the banner 'Live From The Moon'.

Ranger answered the most urgent question concerning the nature of the lunar surface: it looked as if it would support the weight of a spacecraft. JPL had hoped to reinstate the 'hard landing' seismometer and mount a series of follow-on flights, but funding had been denied in December 1963. Originally intended to serve as the primary means of studying the Moon, Ranger had been overtaken by the incredible pace of events.

With Ranger, JPL had seized the initiative in the development of spacecraft for missions in deep space. In May 1960 it took on a far more adventurous project to develop two related spacecraft, one to enter lunar orbit to conduct mapping and the other to land. Unfortunately, the development of the powerful Centaur stage which was to dispatch the new probes to the Moon proved protracted. In coming to terms with Kennedy's timescale for Apollo, NASA cancelled JPL's mapper, and ordered the Langley Research Center to develop a lightweight orbiter that would be able to ride the Atlas–Agena. This new spacecraft was not to be a global mapper, it was to chart predetermined sites as potential Apollo landing sites. This unimaginatively named Lunar Orbiter project was initiated in August 1963. Ranger had yet to prove itself, but it was obvious that developing an orbiter would not just be a matter of fitting a motor to insert Ranger into lunar orbit. While JPL's TV camera package was ideal for documenting a 20-min plunging dive that would result in the destruction of the spacecraft, it was capable of providing the required high surface resolution only during its final few seconds, by which time its field of view was extremely constrained. To survey wide areas with such resolution from an altitude of 35 nmi, Lunar Orbiter would expose film and would develop and scan this for transmission to Earth. Furthermore, because the orbiter had to be lightweight, the camera could not be shielded from radiation in space, and very 'slow' film was required, which in turn meant that the camera had to be able to compensate for its motion. A two-lens system was used, with a wide lens providing the context for a narrow lens. Clearly, film stock would be the limiting factor – in fact, the camera could accommodate 212 frames for each lens – but luckily NASA was able to use a camera built by Kodak for a military reconnaissance satellite. In December 1963, at the same time as it cancelled the follow-on Rangers, NASA issued the Lunar Orbiter contract to Boeing; in effect, the budget was transferred. As with Ranger, Lunar Orbiter would employ 3-axis stabilization, but a spacecraft's configuration is intimately related to its payload, and although it was possible to use many off-the-shelf systems the orbiter turned out to be very different. The budget allowed for five operational spacecraft, plus a spare for engineering trials. It was expected that three successful flights would be sufficient to survey all the sites on the list for the first Apollo landing, which was as far into the future as NASA was considering at the time. To accomplish this, Langley devised three interleaved flight plans, which were designated 'A', 'B' and 'C'.

The spacecraft was to fly an elliptical orbit with a 33-nmi perilune on the near side, in a plane selected to enable the spacecraft to expose its pictures at a low Sun angle to highlight the surface relief. With the plane of the orbit fixed in space, the perilune point would migrate to more western longitudes as the Moon rotated, and after 10 days the perilune point would have traveled the length of the equatorial zone in which the candidate landing sites were located, documenting each in ideal illumination. Furthermore, since the sites were distributed across a strip 160 nmi wide, it was necessary to tilt the trajectory with respect to the lunar equator to pass over the sites near the periphery of this zone. In fact, it was decided to adopt an inclination of 11 deg, with the perilune of the first orbiter south of the equator, the second to the north, and the third inclined as necessary to fill in gaps and make follow-up studies of the most interesting sites. Lunar Orbiter 1 was launched on 10 August 1966 and entered lunar orbit on 14 August. In its 'mapping' mode, it was to snap four narrow frames for each wide-field shot in order that both provided contiguous coverage of the surface, but a malfunction in the motion compensator smeared the close-up pictures. Although the flight controllers considered raising the perilune in order to reduce the smearing and map the *entire* Apollo zone with a resolution of about 80 ft, it was decided to remain at low level and document the planned sites at medium resolution. On 29 August, Lunar Orbiter 1 imaged the 9th potential landing site on its target list, and thus completed its primary mission. It transmitted telemetry for two months more to enable the degradation of its systems to be assessed, then was deorbited to clear the way for its successor, which started work on 18 November. In addition to inspecting the remaining 11 candidate sites, Lunar Orbiter 2 was able to snap a number of secondary sites which, while of no immediate interest to the Apollo selectors, were of 'scientific' interest. It finished its photography on 26 November. In addition to its own targets, it had taken high-resolution images of its predecessor's best sites: Site 'A3' (now called 2P-6) was very promising; of the new sites, 2P-2 was deemed to be suitable and 2P-4 was rejected as too rough. With all of the candidate sites documented, the US Geological Survey (USGS) made terrain maps for the Apollo planners. In addition to photographing the most promising sites from different angles to facilitate stereoscopic studies for detailed analysis of the topography, Lunar Orbiter 3 charted approach routes. With the goal of the project achieved using the first three spacecraft, NASA released the remaining spacecraft to the scientists, who opted to fly them in near-polar orbits at higher altitudes to conduct more general mapping.

In the period of only a year, the Lunar Orbiters not only satisfied the objective of surveying likely sites for the first Apollo landing, they returned the first clear views of the far side, tremendously advanced our knowledge of near-side regional geology, and identified more 'feature' sites for later missions than could possibly be visited. Lunar Orbiter had proven to be remarkably successful. Even after they had finished imaging, the spacecraft provided insight into the lunar *interior*. Although the first was deorbited before the arrival of its successor, it was noted that the spacecraft's orbit was being perturbed, which meant that the gravitational field was uneven. To follow up, its successors were not deorbited until their attitude-control propellant was almost exhausted, and with vehicles in both equatorial and polar orbits it was possible to chart the field in sufficient detail to reveal that the mare-flooded basins were the source of the strongest gravity. The discovery of the 'mascons' (the excess *mass* that was evidently *con*centrated in these basins) was a serendipitous bonus from the project.

As the mechanical properties of the lunar surface would influence the design of the Apollo lander, in October 1962 the Apollo planners said that the development of JPL's 'soft landing' spacecraft should receive a higher priority than the orbital mapper, whose results would only be of operational value, but the development of the Centaur stage was so prolonged that the missions could not start until 1966. The planners faced the same dilemma as their Apollo counterparts: where should they target their first mission? Safety considerations effectively obliged them to select a mare site. In fact, this was consistent with the goal of characterizing the surface in the Apollo zone. When Surveyor 1 was launched on 30 May 1966, the 'old hands' at JPL might well have wondered whether they were in for a rerun of the teething troubles that had plagued Ranger, but the spacecraft landed safely in Flamsteed, a crater whose rim had been breached by Oceanus Procellarum (Ocean of Storms). As with Ranger, the solitary instrument was a television camera. The first picture showed the vehicle's footpad resting on the surface, which was barely indented. The camera then proceeded to take a multitude of individual frames from which a panoramic mosaic was created. There was a profusion of small craters and rocks, but the site was generally flat, with a featureless horizon. It gave the appearance of being a flow of very fluid lava. The spacecraft continued to transmit panoramas to document the appearance of the surface under different illumination, and, as the Sun set, went into hibernation for the fortnight-long lunar 'night'. To everyone's surprise, not only did it awaken with the return of the Sun, it did so each 'morning' for the rest of the year. Having succeeded at the first attempt, it came as a shock when Surveyor 2 started to tumble during a trajectory correction en route to the Moon and was lost. On 20 April 1967 Surveyor 3 set down in a 660-ft-diameter crater in Procellarum, bouncing a number of times before coming to a halt. The inner wall was pocked by smaller craters, one of which had excavated large blocks of rock, indicating that the 'regolith' that coated the surface was not hundreds of feet thick. In addition to a camera, this lander had an arm with a scoop to enable it to determine the mechanical properties of the loose surface material, dig trenches to inspect the subsurface, and roll rocks over to determine the extent to which their state of erosion was selective. In contrast to its hardy predecessor, Surveyor 3 survived only one lunar night. Contact was lost with Surveyor 4 several minutes before it was scheduled to land. Having sampled two western mare and failed twice to reach a site on the meridian, JPL dispatched Surveyor 5 to sample an eastern mare. It set down in Tranquilitatis, some 14 nmi from the 'A3' site that had passed its Lunar Orbiter inspection and graduated to the short-list for the initial Apollo landing. It, too, landed in a crater, this time on a 20-deg slope. Instead of a scoop arm, it had an instrument to study the chemical composition of the regolith. After taking one reading, the spacecraft 'pulsed' its thrusters to 'hop' a little further down slope to sample a second patch of regolith. The results indicated calcium, silicon, oxygen, aluminum and magnesium. This implied basalt, but the high ratio of iron and titanium hinted that it was subtly different to its terrestrial counterpart. Surveyor 6 was sent to Sinus Medii (Meridian Sea) to fill in for its lost forerunners, and landed without incident on 10 November 1967. The results of the chemical analysis suggested an iron-rich basalt. Since the mare in the Apollo landing zone had proved to be remarkably similar, NASA released the final spacecraft to the scientists, who decided to send it to Tycho, a bright 'ray' crater in the southern highlands, where it landed on 10 January 1968. By cutting margins, JPL was able to fit both the robotic arm and the chemical analyzer, which proved fortunate because the boxy instrument became hung up, and if it were not for the arm nudging it free the scientific study would have been undermined. In addition, rather than make the spacecraft hop to sample different patches of regolith, the arm was used to place the instrument on a patch of excavated soil to check that this was the same as the surface layer, and later to balance it on a rock. No one knew how old Tycho was, but its prominent system of rays indicated that it was 'recent'. On seeing the elemental abundance data, some argued that the lunar highlands were an alumina-rich basalt, but E.M. Shoemaker of the astrogeology branch of the USGS concluded the dominant rock in the Tycho ejecta was anorthositic gabbro.

The goal of the Lunar Orbiter series of spacecraft was to reconnoiter possible landing sites for Apollo. As they did not have enough film to *search* for sites, they concentrated on sites that appeared suitable on the basis of telescopic studies. Possible sites for the first lunar landing were under study by the Apollo Site Selection Board for more than 2 years. The 30 original candidate sites located on the near side of the Moon within 45 deg of the meridian and 5 deg of the equator were short-listed to three by the following *operational* factors:

- *Smoothness:* Relatively few craters and boulders.
- *Approach:* No large hills, high cliffs, or deep craters that could result in incorrect altitude signals to the lunar module landing radar.
- *Propellant requirements:* Least potential expenditure of spacecraft propellants.
- *Recycling:* Effective launch preparation recycling if the countdown were delayed.
- *Free return:* Sites within reach of the spacecraft launched on a free-return translunar trajectory.
- *Slope:* Less than 2-deg slope in the approach path and landing area.

Firstly, the flight dynamics team insisted that the site be generally east of the lunar meridian in order to allow room further west for one or two suitably lit backup sites in the event of the launch having to be postponed by several days: launch 'windows' for a given site occurred only once per month, and it was thought better to go for a secondary site a few days late than wait a month for the optimum site to present itself again. This required the prime target to be in the eastern hemisphere. The time of landing had to be just after local sunrise at the site because the Sun was required to be low on the horizon in order to cast sufficient shadow to show surface topography. Because the Moon's rotation is 'locked' to the period of its orbit of the Earth, it spins once a month and the Sun traverses the lunar sky at a rate of 12 deg in 24 hr, which required the backup sites to be spaced 12 deg apart in lunar longitude so that the illumination would be right for each day's delay in launch. On the other hand, the primary site could not be too far east, as this would not allow sufficient time after the final limb crossing to make the navigational checks prior to initiating the powered descent. Secondly, the landing site had to be within a narrow strip within 5 deg of latitude of the lunar equator. A higher latitude site would involve a less propellant-efficient trajectory, and propellant economy was a priority for the first landing.

Furthermore, not only did all the sites have to be flat in order to minimize the need to maneuver to avoid obstacles during the final phase of the descent, but the terrain on the approach had also to be flat in order not to complicate the task of the landing radar. These safety constraints restricted the first landing to one of the eastern maria on the equator, which put the primary landing site in either Mare Tranquilitatis or Mare Fecunditatis (Sea of Fertility), the backup on the meridian, and the reserves in the western hemisphere. But Fecunditatis was too far east to provide a comfortable margin for the final navigational update, and so Apollo 11 was assigned to Tranquilitatis, where there were two sites. The trajectory of Apollo 8 had been timed to view the most easterly site, ALS-1 (called 2P-2 by the Lunar Orbiter selectors), under ideal illumination. Apollo 10 made a low pass over ALS-2 (initially called 'A3', then 2P-6) and reported it to be generally suitable, although the far end of the 'landing ellipse' was rough.

There were a number of considerations which determined the launch windows for a lunar landing mission. These considerations included illumination conditions at launch, launch pad azimuth, translunar injection geometry, Sun elevation angle at the lunar landing site, illumination conditions at Earth splashdown, and the number and location of the lunar landing sites. The time of a lunar landing was determined by the location of the site and the acceptable range of Sun elevation angles. The range of these angles was from 5 to 14 deg and in a direction from east to west. Under these conditions, visible shadows of craters would aid the crew in recognizing topographical features. When the Sun angle approached the descent angle, the mean value of which was 16 deg, visual resolution would be degraded by a 'washout' phenomenon in which backward reflectance was high enough to eliminate contrast. Sun angles above the flight path were not as desirable since shadows would not be readily visible unless the Sun was significantly outside the descent plane. And higher Sun angles (exceeding 18 deg) could be eliminated from consideration by planning the landing one day earlier, with illumination of at least 5 deg. Because lunar sunlight incidence changed about 0.5 deg/hr, the Sun elevation angle restriction defined a 16-hr period, recurring every 29.5 days, when landing at a given site could be attempted. The number of Earth-launch opportunities for a given lunar month was equal to the number of candidate landing sites. The time of launch was primarily determined by the allowable variation in launch pad azimuth and the location of the Moon at spacecraft arrival. The spacecraft had to be launched into an orbital plane that contained the position of the Moon and its antipode at spacecraft arrival. A launch pad azimuth variation of 34 deg afforded a launch period of 4 hr 30 min. This period was called the 'daily launch window', the time when the direction of launch was within the required range to intercept the Moon. Two launch windows occurred each day. One was available for a translunar injection out of Earth orbit in the vicinity of the Pacific Ocean, and the other was in the vicinity of the Atlantic Ocean. The injection opportunity over the Pacific Ocean was preferred because it usually permitted a daytime launch.

LAUNCH PREPARATIONS

The terminal countdown started at T–28 hr, 21:00:00 GMT on 14 July. The scheduled holds of 11 hr at T–9 hr and 1 hour 32 min at T–3 hr 30 min were the only holds initiated.

The start of the S-II stage LH_2 loading was delayed 25 min due to a communications problem in the Pad Terminal Connection Room. However, the lost time was recovered during the planned countdown hold at T–3 hr 30 min. A high-pressure cell in the Atlantic Ocean off the North Carolina coast, along with a weak trough of low pressure located in the northeastern Gulf of Mexico, caused light southerly surface winds and brought moisture into the Cape Kennedy area. These circumstances contributed to the cloudy conditions and distant thunderstorms observed at launch time. Cumulus clouds covered 10% of the sky (base 2,400 ft), altocumulus covered 20% (base 15,000 ft), and cirrostratus covered 90% (base not recorded), the temperature was 84.9°F, the relative humidity was 73%, and the barometric pressure was 14.798 psi. The winds, as measured by the anemometer on the light pole 60.0 ft above ground at the launch site, were 6.4 kt at 175° from true north.

ASCENT PHASE

Apollo 11 was launched from Pad A of Launch Complex 39 at the Kennedy Space Center at a Range Zero time of 13:32:00 GMT (09:32:00 EDT) on 16 July 1969. The launch window extended to 17:54:00 GMT in order to take advantage of a Sun elevation angle on the lunar surface of 10.8 deg at the scheduled time of landing. Between T + 13.2 and T + 31.1 sec, the vehicle rolled from a pad azimuth of 90°E of N to a flight azimuth of 72.058°E of N. The maximum wind conditions encountered during ascent were 18.7 kt at 297° from true north at 37,400 ft, with a maximum wind shear of 0.0077/sec at 48,490 ft. The S-IC shut down at T + 161.63, followed by S-IC/S-II separation and S-II ignition. The S-II shut down at T + 548.22, followed by separation from the S-IVB, which ignited at T + 552.2. The first S-IVB cutoff occurred at T + 699.33, with deviations from the planned trajectory of only –0.6 ft/sec in velocity and only –0.1 nmi in altitude. The S-IC impacted the Atlantic Ocean at T + 543.70 at 30.212°N and 74.038°W, 357.1 nmi from the launch site. The S-II impacted the Atlantic Ocean at T + 1,213.7 at 31.535°N and 34.844°W, 2,371.8 nmi from the launch site. At insertion, at T + 709.33 (i.e. S-IVB cutoff plus 10 sec to account for engine tailoff and other transient effects), the parking orbit showed an apogee and perigee of 100.4 × 98.9 nmi, an inclination of 32.521 deg, a period of 88.18 min, and a velocity of 25,567.8 ft/sec. The apogee and perigee were based upon a spherical Earth with a radius of 3,443.934 nmi.

The COSPAR designation for the CSM upon achieving orbit was 1969-059A, and the S-IVB was designated 1969-059B. After undocking at the Moon, the LM ascent stage would be designated 1969-059C and the descent stage 1969-059D.

Apollo 11 preparation event	Date
Individual and combined CM and SM systems test completed at factory.	12 Oct 1968
LM-5 integrated test at factory.	21 Oct 1968
Integrated CM and SM systems test completed at factory.	06 Dec 1968
LM-5 final engineering evaluation acceptance test at factory.	13 Dec 1968
Ascent stage of LM-5 ready to ship from factory to KSC.	07 Jan 1969
Ascent stage of LM-5 delivered to KSC.	08 Jan 1969
Spacecraft/LM adapter SLA-14 delivered to KSC.	10 Jan 1969
Descent stage of LM-5 ready to ship from factory to KSC.	11 Jan 1969
Descent stage of LM-5 delivered to KSC.	12 Jan 1969
CSM-107 quads delivered to KSC.	15 Jan 1969
S-IVB-506 stage delivered to KSC.	19 Jan 1969
CM-107 and SM-107 ready to ship from factory to KSC.	22 Jan 1969
CM-107 and SM-107 delivered to KSC.	23 Jan 1969
CM-107 and SM-107 mated.	29 Jan 1969
S-II-6 stage delivered to KSC.	06 Feb 1969
LM-6 stages mated.	14 Feb 1969
CSM-107 combined systems test completed.	17 Feb 1969
LM-5 combined systems test completed.	17 Feb 1969
S-IC-6 stage delivered to KSC.	20 Feb 1969
S-IC-6 stage erected.	21 Feb 1969
Saturn V instrument unit IU-506 delivered to KSC.	27 Feb 1969
S-II-6 stage erected.	04 Mar 1969
S-IVB-506 stage erected.	05 Mar 1969
IU-506 erected.	05 Mar 1969
CSM-107 altitude test with prime crew completed.	18 Mar 1969
LM-5 altitude test with prime crew completed.	21 Mar 1969
CSM-107 altitude tests completed.	24 Mar 1969
LM-5 altitude tests completed.	25 Mar 1969
Launch vehicle propellant dispersion/malfunction overall test completed.	27 Mar 1969
CSM-107 moved to VAB	14 Apr 1969
Spacecraft erected.	14 Apr 1969
LM-5 combined systems test completed.	18 Apr 1969
CSM-107 integrated systems test completed.	22 Apr 1969
CSM-107 electrically mated to launch vehicle.	05 May 1969
Space vehicle overall test completed.	06 May 1969
Space vehicle overall test #1 (plugs in) completed.	14 May 1969
Space vehicle and MLP-1 transferred to Launch Complex 39A.	20 May 1969
Mobile service structure transferred to Launch Complex 39A.	22 May 1969
LM-5 flight readiness test completed.	02 Jun 1969
Space vehicle flight readiness test completed.	06 Jun 1969
S-IC-6 stage RP-1 fuel loading completed.	25 Jun 1969
Space vehicle countdown demonstration test (wet) completed.	02 Jul 1969
Space vehicle countdown demonstration test (dry) completed.	03 Jul 1969

Apollo 11 ascent phase event	GET (h:m:s)	Altitude (nmi)	Range (nmi)	Earth-fixed velocity (ft/sec)	Space-fixed velocity (ft/sec)	Event duration (sec)	Geocentric latitude (°N)	Longitude (°E)	Space-fixed flight path angle (deg)	Space-fixed heading angle (°E of N)
Liftoff	000:00:00.63	0.032	0.000	1.5	1,340.7	–	28.4470	–80.6041	0.06	90.00
Mach 1 achieved	000:01:06.3	4.236	1.044	1,054.1	2,023.9	–	28.4523	–80.5853	27.88	85.32
Maximum dynamic pressure	000:01:23.0	7.326	3.012	1,653.4	2,671.9	–	28.4624	–80.5499	29.23	82.41
S-IC center engine cutoff*	000:02:15.2	23.761	25.067	5,320.8	6,492.8	141.6	28.5739	–81.1517	22.957	76.315
S-IC outboard engine cutoff	000:02:41.63	35.701	50.529	7,851.9	9,068.6	168.03	28.7007	–79.6908	19.114	75.439
S-IC/S-II separation*	000:02:42.30	36.029	51.323	7,882.9	9,100.6	–	28.7046	–79.6764	19.020	75.436
S-II center engine cutoff	000:07:40.62	97.280	601.678	17,404.8	18,725.5	296.62	30.9513	–69.4309	0.897	79.646
S-II outboard engine cutoff	000:09:08.22	101.142	873.886	21,368.2	22,690.8	384.22	31.7089	–64.1983	0.619	82.396
S-II/S-IVB separation*	000:09:09.00	101.175	876.550	21,377.0	22,699.6	–	31.7152	–64.1467	0.611	82.426
S-IVB 1st burn cutoff	000:11:39.33	103.202	1,421.959	24,237.6	25,561.6	147.13	32.4865	–53.4588	0.011	88.414
Earth orbit insertion	000:11:49.33	103.176	1,460.697	24,243.9	25,567.9	–	32.5027	–52.6491	0.012	88.848

* Only the commanded time is available for this event.

EARTH ORBIT PHASE

Following in-flight systems checks, the 346.83-sec translunar injection maneuver (second S-IVB firing) was initiated at 002:44:16.20; the engine shut down at 002:50:03.03, and translunar injection occurred 10 sec later, after 1.5 Earth orbits lasting 2 hr 38 min 23.73 sec, at a velocity of 35,545.6 ft/sec.

Apollo 11 earth orbit phase event	GET (h:m:s)	Space-fixed velocity (ft/sec)	Event duration (sec)	Velocity change (ft/sec)	Apogee (nmi)	Perigee (nmi)	Period (min)	Inclination (deg)
Earth orbit insertion	000:11:49.33	25,567.8	–	–	100.4	98.9	88.18	32.521
S-IVB 2nd burn ignition	002:44:16.20	25,560.2	–	–	–	–	–	–
S-IVB 2nd burn cutoff	002:50:03.03	35,568.3	346.83	10,008.1	–	–	–	31.386

TRANSLUNAR PHASE

At 003:15:23.0, the CSM was separated from the S-IVB stage and transposed and docked with the LM at 003:24:03.7. The docked spacecraft were ejected from the S-IVB at 004:17:03.0, and a 2.93-sec separation maneuver was performed at 004:40:01.72. A ground command for propulsive venting of residual propellants targeted the S-IVB to go past the Moon and into solar orbit. The point of closest approach of the stage to the lunar surface at 078:42 was 1,825 nmi, at which point the lunar radius was 2,763 nmi. The orbital parameters after passing from the lunar sphere of influence resulted in a solar orbit with an aphelion and perihelion of 82,000,000 × 72,520,000 nmi, a semi-major axis of 77,260,000 nmi, an inclination to the ecliptic of 0.3836 deg, and a period of 342 days. The velocity increase relative to Earth from the lunar encounter was 0.367 nmi/sec.

Starting at 010:32 an unscheduled 16-min television transmission was recorded at the Goldstone Tracking Station in California, and starting at 011:26 the tape was replayed for transmission to Houston. Trajectory parameters after the translunar injection were nearly perfect. A 3.13-sec midcourse correction of 20.9 ft/sec was made at 026:44:58.64 during the translunar phase. During the remaining periods of free-attitude flight, passive thermal control, a rotating barbecue-like maneuver, was used to maintain spacecraft temperatures within desired limits. An unscheduled 50-min television transmission was accomplished at 030:28, and a 36-min scheduled transmission began at 033:59. The crew initiated a 96-min color television transmission at 055:08. The picture resolution and general quality were exceptional. The coverage included the interior of the CM and LM and views of the exterior of the vehicle and Earth. Excellent views of the crew accomplishing probe and drogue removal, spacecraft tunnel hatch opening, LM housekeeping, and equipment testing were broadcast. During this latter transmission, the commander and lunar module pilot transferred to the LM at 055:30 to make the initial inspection and preparations for the systems checks that would be made shortly after lunar orbit insertion. They returned to the CM at 057:55.

At 075:49:50.37, at an altitude of 86.7 nmi above the Moon, the service propulsion system engine was fired for 357.53 sec to insert the spacecraft into a lunar orbit of 169.7 × 60.0 nmi. The translunar coast had lasted 73 hr 5 min 34.83 sec.

Apollo 11 translunar phase event	GET (h:m:s)	Altitude (nmi)	Space-fixed velocity (ft/sec)	Event duration (sec)	Velocity change (ft/sec)	Space-fixed flight path angle (deg)	Space-fixed heading angle (°E of N)
Translunar injection	002:50:13.03	180.581	35,545.6	–	–	7.367	60.073
CSM separated from S-IVB	003:15:23.0	3,815.190	24,962.5	–	–	45.148	93.758
CSM docked with LM/S-IVB	003:24:03.7	5,317.6	22,662.5	–	–	44.94	99.57
CSM/LM evasive maneuver ignition	004:40:01.72	16,620.8	14,680.0	–	–	64.30	113.73
CSM/LM evasive maneuver cutoff	004:40:04.65	16,627.3	14,663.0	2.93	19.7	64.25	113.74
Midcourse correction ignition	026:44:58.64	109,475.3	5,025.0	–	–	77.05	120.88
Midcourse correction cutoff	026:45:01.77	109,477.2	5,010.0	3.13	20.9	76.88	120.87

LUNAR ORBIT AND LUNAR SURFACE PHASE

During the second lunar orbit, at 078:20, a scheduled live color television transmission was accomplished, providing spectacular views of the lunar surface and the approach path to Apollo Landing Site 2. After two revolutions and a navigation update, a second service propulsion system retrograde burn was performed. The 16.88-sec maneuver occurred at 080:11:36.75 and circularized the orbit at 66.1 × 54.5 nmi. The commander and lunar module pilot then transferred to the LM and, for about 2 hr, performed miscellaneous housekeeping functions, a voice and telemetry test, and an oxygen purge system check. LM functions and consumables checked out well. Additionally, both cameras were checked and verified operational. The pair then returned to the CSM. At 095:20, they returned to the LM to perform a thorough check of all LM systems in preparation for descent. Undocking occurred at 100:12:00 at an altitude of 62.9 nmi. This was followed by a CSM reaction control system 9.0-sec separation maneuver at 100:39:52.9 directed radially downward toward the center of the Moon as planned. The descent orbit insertion maneuver was performed with a 30.0-sec firing of the descent engine at 101:36:14.0, and placed the LM into an orbit of 58.5 × 7.8 nmi.

The 756.39-sec powered descent engine burn was initiated at 102:33:05.01. The time was as planned, but the position at which powered descent initiation occurred was about 4 nmi farther downrange than expected, which in turn resulted in the landing point being shifted downrange about 4 nmi. The first of five alarms occurred at 102:38:22 due to a computer overload, but Mission Control determined that it was safe to continue the descent. The crew checked the handling qualities of the LM at 102:41:53, and switched to automatic guidance 10 sec later. The landing radar switched to 'low-scale' at 102:42:19 as the LM passed below 2,500 ft altitude. The final alarm occurred at 102:42:58, followed by the red-line low-level fuel quantity light at 102:44:28, just 72 sec before landing. On the final approach, the commander noted that the landing point toward which the spacecraft was headed was in the center of a large crater that appeared extremely rugged, and was littered with boulders of 5 to 10 ft in diameter and larger. Consequently, he switched to manual attitude control and maneuvered manually 1,100 ft downrange, beyond the rough terrain area. The LM landed on the Moon at 102:45:39.9 (20:17:39 GMT, 16:17:39 EDT, on 20 July 1969). Engine shutdown occurred 1.5 sec later. The site was in Mare Tranquilitatis at selenographic coordinates 0.67408°N, 23.47297°E, some 22,500 ft west of the center of the landing ellipse. At engine cutoff, approximately 45 sec of firing time remained.[4]

For the first 2 hr on the lunar surface, the crew performed a checkout of all systems, configured the controls for lunar stay, and ate their first post-landing meal. A rest period had been planned to precede the extravehicular activity of exploring the lunar surface but was not needed. After donning the back-mounted portable life support and oxygen purge systems, the commander prepared to exit the LM. The forward hatch was opened at 109:07:33 and the commander exited at 109:19:16. While descending the LM ladder, he deployed the Modular Equipment Stowage Assembly from the descent stage. A camera in the module provided live television coverage of the leg with the ladder as he descended. His left boot made first contact with the lunar surface at 109:24:15 (02:56:15 GMT on 21 July; 22:56:15 EDT on 20 July), which he marked with the words, "That's one small step for a man, one giant leap for mankind." The commander made a brief check of the LM exterior, indicating that penetration of the footpads was only about 3 or 4 in, and collapse of the LM footpad strut was minimal. He reported his boot sinking into the fine, powdery surface material to a depth of only about 1/8 in, and that it adhered readily to his suit in a thin layer. There was no crater from the effects of the descent engine, and about 12 inches of clearance was observed between the engine bell and the lunar surface. He also reported that it was quite dark in the shadows of the LM, which made it difficult for him to see his footing. He then collected a contingency sample of lunar soil from the vicinity of the LM ladder. He reported that although loose material created a soft surface, as he dug down 6 or 8 in he encountered very hard, cohesive material.

[4] According to the Apollo 11 Mission Report (MSC-00171), post-flight analysis revealed that there was 45 sec of fuel remaining at lunar touchdown, not as little as 7 sec as indicated by other sources.

The commander then photographed the lunar module pilot as he exited at 109:37:57 and descended to the surface at 109:43:16. With both men on the surface, they unveiled a plaque mounted on the strut behind the ladder, and read its inscription to their worldwide television audience. The plaque read:

HERE MEN FROM THE PLANET EARTH
FIRST SET FOOT UPON THE MOON
JULY 1969, A.D.
WE CAME IN PEACE FOR ALL MANKIND.

The plaque featured the signatures of the three Apollo crew members and President Richard M. Nixon. Next, the commander removed the television camera from the descent stage, obtained a panorama, and placed the camera on its tripod in position to view the subsequent surface extravehicular operations. The lunar module pilot deployed the solar wind composition experiment on the lunar surface in direct sunlight and to the north of the LM as planned. At 110:09:43 they erected a 3-by-5-ft US flag on an 8-ft aluminum staff, with a strut to hold it out, and at 110:16:30 received a call from the White House in which Nixon congratulated them on their achievement and offered his good wishes.

During the environmental evaluation, the lunar module pilot indicated that he had to be careful of his center of mass in maintaining balance. He noted that the LM shadow had no significant effect on his backpack temperature. He also pointed out that his agility was better than expected, and that he was able to move with great ease. Both crew members indicated that their mobility during this period exceeded all expectations. In addition, the indications were that metabolic rates were much lower than pre-mission estimates.

The commander collected a 'bulk' sample, consisting of assorted surface material and rock chunks, and placed them in a sample return container. The crew then inspected the LM, finding the quads, struts, skirts, and antennas in satisfactory condition. The passive seismic experiment package and laser-ranging retroreflector were deployed south of the LM. Excellent PSEP data were obtained, including detection of the crew walking on the surface and later their movements inside the LM. The crew then collected more lunar samples, two core samples and about 20 lb of discretely selected material. The LMP had to exert considerable force to drive the core tubes 6 to 8 in into the lunar surface. Following 1 hr 17 min exposure, the solar wind experiment was retrieved and placed in a lunar sample container. The transfer to the LM of the lunar sample containers began at 111:23. The crew entered the LM and closed the hatch at 111:39:13, thus ending the first human exploration of the Moon. The total time spent outside the LM was 2 hr 31 min 40 sec; the total distance traveled was about 3,300 ft; and the collected samples totaled 47.51 lb (21.55 kg; the official metric total as determined by the Lunar Receiving Laboratory in Houston). The farthest point traveled from the LM was 200 ft, when the commander visited a crater 108 ft in diameter near the end of the extravehicular period.

Ignition of the ascent stage engine for lift off was at 124:22:00.79 (17:54:00 GMT on 21 July). The LM had been on the lunar surface for 21 hr 36 min 20.9 sec. An orbit of 48.0 × 9.4 nmi was achieved at 124:29:15.67, some 434.88 sec after liftoff. Several rendezvous maneuvers were required in order to effect docking 3.5 hr later. The 47.0-sec coelliptic sequence initiation at 125:19:35 raised the orbit to 49.3 × 45.7 nmi. The 17.8-sec constant differential height maneuver at 126:17:49.6 lowered the orbit to 47.4 × 42.1 nmi. The

On visiting the crater over which he had flown, Armstrong took this panoramic view looking back to the LM.

22.7-sec terminal phase initiate maneuver at 127:03:51.8 brought the ascent stage to an orbit of 61.7 × 43.7 nmi. The 28.4-sec terminal phase maneuver at 127:46:09.8 finalized the orbit at 63.0 × 56.5 nmi for docking of the ascent stage and the CSM at 128:03:00.0. The two craft had been undocked for exactly 27 hr 51 min. In the process of maneuvering the LM to docking attitude, while avoiding direct sunlight in the forward window, the platform inadvertently reached gimbal lock, causing a brief and unexpected tumbling motion of the LM. A quick recovery was made and the docking was completed using the abort guidance system for attitude control. Following transfer of the crew and samples to the CSM, the ascent stage was jettisoned at 130:09:31.2 at an altitude of 61.6 nmi, and the CSM was prepared for transearth injection. A 7.2-sec maneuver was made at 130:30:01.0 to separate the CM from the ascent stage; this resulted in an orbit of 62.7 × 54.0 nmi. The gravitational perturbations of the 'mascons' would cause the ascent stage to crash into the Moon.

The 151.41-sec transearth injection maneuver was performed at 135:23:42.28 at an altitude of 52.4 nmi. A nominal injection was achieved at 135:26:13.69 after 30 lunar orbits lasting 59 hr 30 min 25.79 sec, at a velocity of 8,589.0 ft/sec.

Apollo 11 lunar orbit phase event	GET (h:m:s)	Altitude (nmi)	Space-fixed velocity (ft/sec)	Event duration (sec)	Velocity change (ft/sec)	Apolune (nmi)	Perilune (nmi)
Lunar orbit insertion ignition	075:49:50.37	86.7	8,250.0	–	–	–	–
Lunar orbit insertion cutoff	075:55:47.90	60.1	5,479.0	357.53	2917.5	169.7	60.0
Lunar orbit circularization ignition	080:11:36.75	61.8	5,477.3	–	–	–	–
Lunar orbit circularization cutoff	080:11:53.63	61.6	5,338.3	16.88	158.8	66.1	54.5
CSM/LM undocked	100:12:00	62.9	5,333.8	–	–	–	–
CSM/LM separation ignition	100:39:52.9	62.7	5,332.7	–	–	–	–
CSM/LM separation cutoff	100:40:01.9	62.5	5,332.2	9.0	2.7	63.7	56.0
LM descent orbit insertion ignition	101:36:14.0	56.4	5,364.9	–	–	–	–
LM descent orbit insertion cutoff	101:36:44.0	57.8	5,284.9	30.0	76.4	64.3	55.6

Apollo 11 lunar orbit phase event – *continued*	GET (h:m:s)	Altitude (nmi)	Space-fixed velocity (ft/sec)	Event duration (sec)	Velocity change (ft/sec)	Apolune (nmi)	Perilune (nmi)
LM powered descent initiation	102:33:05.01	6.4	5,564.9	–	–	58.5	7.8
LM powered descent cutoff	102:45:41.40	–	–	756.39	–	–	–
LM lunar liftoff ignition	124:22:00.79	–	–	–	–	–	–
LM orbit insertion cutoff	124:29:15.67	10.0	5,537.9	434.88	6,070.1	48.0	9.4
LM coelliptic sequence initiation ignition	125:19:35.0	47.4	5,328.1	–	–	–	–
LM coelliptic sequence initiation cutoff	125:20:22.0	48.4	5,376.6	47.0	51.5	49.3	45.7
LM constant differential height ignition	126:17:49.6	–	–	–	–	–	–
LM constant differential height cutoff	126:18:07.4	–	–	17.8	19.9	47.4	42.1
LM terminal phase initiation ignition	127:03:51.8	44.1	5,391.5	–	–	–	–
LM terminal phase initiation cutoff	127:04:14.5	44.0	5,413.2	22.7	25.3	61.7	43.7
LM 1st midcourse correction	127:18:30.8	–	–	–	1.0	–	–
LM 2nd midcourse correction	127:33:30.8	–	–	–	1.5	–	–
LM terminal phase finalize ignition	127:46:09.8	7.6	5,339.7	–	–	–	–
LM terminal phase finalize cutoff	127:46:38.2	–	–	28.4	31.4	63.0	56.5
LM begin braking	127:36:57.3	–	–	–	–	–	–
LM begin stationkeeping	127:52:05.3	–	–	–	–	–	–
CSM/LM docked	128:03:00.0	60.6	5,341.5	–	–	–	–
LM ascent stage jettisoned	130:09:31.2	61.6	5,335.9	–	–	–	–
CSM/LM final separation ignition	130:30:01.0	62.7	5,330.1	–	–	–	–
CSM/LM final separation cutoff	130:30:08.1	62.7	5,326.9	7.2	2.2	62.7	54.0

TRANSEARTH PHASE

As in translunar flight, only one midcourse correction was required, a 10.0-sec, 4.8-ft/sec maneuver at 150:29:57.4. Passive thermal control was exercised for most of the transearth coast. An 18-min television transmission was initiated at 155:36; it featured a demonstration of the effect of weightlessness on food and water, as well as brief scenes of the Moon and Earth. The final color television broadcast was made at 177:32. The 12.5-min transmission featured a message of appreciation by each crew member to all the people who had helped to make the mission possible.

Apollo 11 transearth phase event	GET (h:m:s)	Altitude (nmi)	Space-fixed velocity (ft/sec)	Event duration (sec)	Velocity change (ft/sec)	Space-fixed flight path angle (deg)	Space-fixed heading angle (°E of N)
Transearth injection ignition	135:23:42.28	52.4	5,376.0	–	–	–0.03	–62.77
Transearth injection cutoff	135:26:13.69	58.1	8,589.0	151.41	3,279.0	5.13	–62.60
Midcourse correction ignition	150:29:57.4	169,087.2	4,075.0	–	–	–80.34	129.30
Midcourse correction cutoff	150:30:07.4	169,080.6	4,074.0	10.0	4.8	–80.41	129.30
CM/SM separation	194:49:12.7	1,778.3	29,615.5	–	–	–35.26	69.27

RECOVERY

Because of inclement weather in the planned recovery area, the splashdown point was relocated 215 nmi downrange. The weather in the new area was excellent, with visibility of 12 miles, waves to 3 ft, and wind speed of 16 kt. The service module was jettisoned at 194:49:12.7. The command module followed an automatic profile, entering the Earth's atmosphere (at the 400,000-ft altitude of the 'entry interface') at 195:03:05.7 at a velocity of 36,194.4 ft/sec, following a transearth coast of 59 hr 36 min 52.0 sec. The parachute system effected splashdown of the CM in the Pacific Ocean at 16:50:35 GMT on 24 July. The mission duration was 195:18:35. The impact point, estimated to be 13.30°N and 169.15°W, was 1.7 nmi from the target point and 13 nmi from the recovery ship USS *Hornet*. The CM assumed an apex-down flotation attitude but was successfully returned to the normal flotation position in 7 min 40 sec by the inflatable-bag-uprighting system. After splashdown, the crew donned biological isolation garments and exited the CM into a rubber boat, where they were scrubbed down with an iodine solution to protect against 'lunar germs'. They were then retrieved by helicopter and taken to the primary recovery ship, where they arrived 63 min after splashdown. The CM was recovered 125 min later. The estimated CM weight at splashdown was 10,873.0 lb, and the estimated distance traveled for the mission was 828,743 nmi. The two lunar sample containers were immediately retrieved from the CM and flown to Johnson Island, en route to the Lunar Receiving Laboratory in Houston, where they arrived on 25 July. The crew, the recovery physician, and one of the recovery technicians entered the Mobile Quarantine Facility on the *Hornet*. The CM and Mobile Quarantine Facility were offloaded in Hawaii at 00:15 GMT on 27 July. The Mobile Quarantine Facility was loaded onto a C-141 aircraft and flown to Houston, where it arrived at 06:00 GMT on 28 July. The crew arrived at the Lunar Receiving Laboratory 4 hr later. The safing of the CM pyrotechnics was completed at 02:05 GMT on 27 July. The CM was taken to Ford Island for deactivation, then it was transferred to Hickam Air Force Base, Hawaii, and flown on a C-133 aircraft to Houston, where it arrived at 23:17 GMT on 30 July. The spacecraft and crew were released from quarantine on 10 August. On 14 August the CM arrived at the North American Rockwell Space Division facility in Downey, California, for post-flight analysis; all spacecraft systems performed satisfactorily.

With the completion of the Apollo 11 mission, the national objective of landing humans on the Moon and returning them safely to Earth before the end of the decade was accomplished.

CONCLUSIONS

The Apollo 11 mission, including a manned lunar landing and surface exploration, was conducted with skill, precision, and relative ease. The excellent performance of the spacecraft in the preceding four missions and the thorough planning in all aspects of the program permitted the safe and efficient execution of this mission. The following conclusions were made from an analysis of post-mission data:

1. The effectiveness of pre-mission training was reflected in the skill and precision with which the crew executed the lunar landing. Manual control while maneuvering to the desired landing point was satisfactorily exercised.
2. The planned techniques involved in the guidance, navigation, and control of the descent trajectory were good. Performance of the landing radar met all expectations in providing the information required for descent.
3. The extravehicular mobility units were adequately designed to enable the crew to conduct the planned activities. Adaptation to 1/6 *g* was relatively quick, and mobility on the lunar surface was easy.
4. The two-person pre-launch checkout and countdown for ascent from the lunar surface were well planned and executed.
5. The timeline activities for all phases of the lunar landing mission were well within the crew's capability to perform the required tasks.
6. The quarantine operation from spacecraft landing until release of the crew, spacecraft, and lunar samples from the Lunar Receiving Laboratory was accomplished successfully and without any significant violation of the quarantine.
7. No microorganisms of extraterrestrial origin were recovered from either the crew or the spacecraft.
8. Hardware problems, as experienced on previous manned missions, did not unduly hamper the crew or compromise crew safety or mission objectives.
9. The Mission Control Center and the Manned Space Flight Network proved to be adequate for controlling and monitoring all phases of flight, including the descent, surface activities, and ascent phases of the mission.

On 25 July, the sample return containers arrived at the Lunar Receiving Laboratory. The Preliminary Examination Team processed the samples in vacuum, not only to protect the samples from the atmosphere but also to protect Earth from any 'bugs' robust enough to be able to survive on the radiation-soaked lunar surface. The risk of contamination was exceedingly low, but no chances were being taken for this 'first contact' situation. Once the lunar material had been catalogued, 150 principal investigators around the world were sent samples for study. The large rocks were found to be magnesium- and iron-rich (i.e. mafic) basalt, which isotopic dating showed to have crystallized 3.84 to 3.57 billion years ago. Compared to terrestrial basalt it was rich in titanium, but it was otherwise strikingly familiar in texture. This proved that the Moon had undergone thermal differentiation, with the lightweight aluminous minerals migrating to the surface, the heavier minerals sinking into the interior and later being erupted onto the surface. The titanium-bearing mineral, which was totally new, was dubbed 'armalcolite', in honor of the astronauts (it is the first letters of their surnames). The titanium is why the maria are dark. The absence of oxidized iron indicated that the lava formed in a reducing environment (that is, one devoid of oxygen). The most striking fact was the total absence of hydrous minerals. The lunar basalt was deficient in volatile metals such as sodium. The low-alkali (that is, sodium-depleted) lava would have had an extremely low viscosity, which is why it flowed so far and left so few 'positive-relief' flow fronts. The Moon accreted from the solar nebula 4.6 billion years ago, so the basins were excavated during the first billion years. In the case of Tranquilitatis, the dark plain we see today was built up by episodic volcanism over several hundred million years. The fact that the basalt was extruded long after the Moon had accreted implied the maria were not 'impact melt' from the formation of the Imbrium basin, as some had suggested. Another inconvenient fact for the idea that all of the maria were formed simultaneously was that *two* types of basalt were identified, implying either that there had been several magma reservoirs or that a single source had undergone chemical evolution over an extended time. Rock exposed at the surface was subjected to a process of mechanical erosion. The majority of the fragments in the 'soil' were pulverized basalt. There was little meteoritic material. This was consistent with the 'gardening' process in which a large impact dug up bedrock that was progressively worn down by smaller impacts to yield the seriate regolith. Many of the discrete samples turned out to be consolidated regolith. Such 'regolith breccias' had been created by the shock of an impact compressing the loose surface material. The 'glassy material' found in what was evidently a fresh crater proved to be regolith fused by the shock of a high-speed impact.

There was a small residue of the regolith that was startlingly different in character. On the basis of his analysis of data from Surveyor 7, which had landed in the southern highlands, Shoemaker had posited that 4% of the Tranquilitatis regolith would comprise minuscule fragments of light-colored rock. This was indeed the case. The light rock *was* plagioclase feldspar. Although this is one of the commonest minerals in the Earth's crust, terrestrial plagioclase is rich in sodium, whereas the lunar variety was depleted in sodium and enriched with calcium, making it calcic-plagioclase. Some of the fragments proved to be sufficiently pure to be called anorthosite (this being the name for a rock that is at least 90% plagioclase) but most were diluted by mafic minerals and more properly referred to as anorthositic gabbro (like the sample studied by Surveyor 7). Shoemaker's rationale for finding highland material in the mare regolith was based on the way in which the most recently formed highland craters sprayed out rays of material. Regarding the highlands, it was now clear that the primitive crust was composed of anorthositic rock. Nevertheless, the geologists were eager to find fragments of the 'light plains' in the highlands, but there was no evidence of it in the Tranquilitatis regolith. When the preliminary results were issued at the First Lunar Science Conference in Houston in January 1970, it was pointed out that if the non-mare regolith fragments were indeed highland rock, then their density of 2.9 gm/cm^3 compared with the 3.4 average for the Moon implied that the heat from the impacts during the accretion of the Moon had produced a 'magma ocean' of considerable depth, in which the heavier mafic minerals had sunk to create a mantle and the lighter plagioclase-rich minerals had floated to the surface and cooled to form a solid crust. This was a major insight into lunar history.

MISSION OBJECTIVES

Launch Vehicle Objectives

1. To launch on a variable 72–108°E of N flight azimuth and insert the S-IVB/instrument unit/spacecraft into a circular Earth parking orbit. *Achieved.*
2. To restart the S-IVB during either the second or third revolution, and injection of the S-IVB/instrument unit/spacecraft into the planned translunar trajectory. *Achieved.*
3. To provide the required attitude control for the S-IVB/instrument unit/spacecraft during the transposition, docking, and ejection maneuver. *Achieved.*
4. To use residual S-IVB propellants and auxiliary propulsion system after final launch vehicle/spacecraft separation, to safe the S-IVB, and to minimize the possibility of the following, in order of priority:
   (a) S-IVB/instrument unit recontact with the spacecraft. *Achieved.*
   (b) S-IVB/instrument unit Earth impact. *Achieved.*
   (c) S-IVB/instrument unit lunar impact. *Achieved.*

Spacecraft Primary Objective

1. To perform a manned lunar landing and return. *Achieved.*

Spacecraft Secondary Objectives

1. To perform selenological inspection and sampling.
   (a) Contingency sample collection. *Achieved.*
   (b) Lunar surface characteristics. *Achieved.*
   (c) Bulk sample collection. *Achieved.*
   (d) Lunar environment visibility. *Achieved.*
2. To obtain data to assess the capability and limitations of the astronaut and his equipment in the lunar surface environment.
   (a) Lunar surface extravehicular operations. *Achieved.*
   (b) Lunar surface operations with extravehicular mobility unit. *Achieved.*
   (c) Landing effects on lunar module. *Achieved.*
   (d) Location of landed lunar module. *Partially achieved. The LM crew was unable to make observations of lunar features during descent. The command module pilot was therefore unable to locate the lunar module through the sextant. Toward the end of the lunar surface stay, the location of the lunar module was determined from the lunar module rendezvous radar tracking data, which was confirmed post-mission using descent photographic data.*

The official portrait of the Apollo 11 crew: Neil Armstrong (left), Mike Collins and Buzz Aldrin.

Neil Armstrong and Buzz Aldrin in a LM simulator.

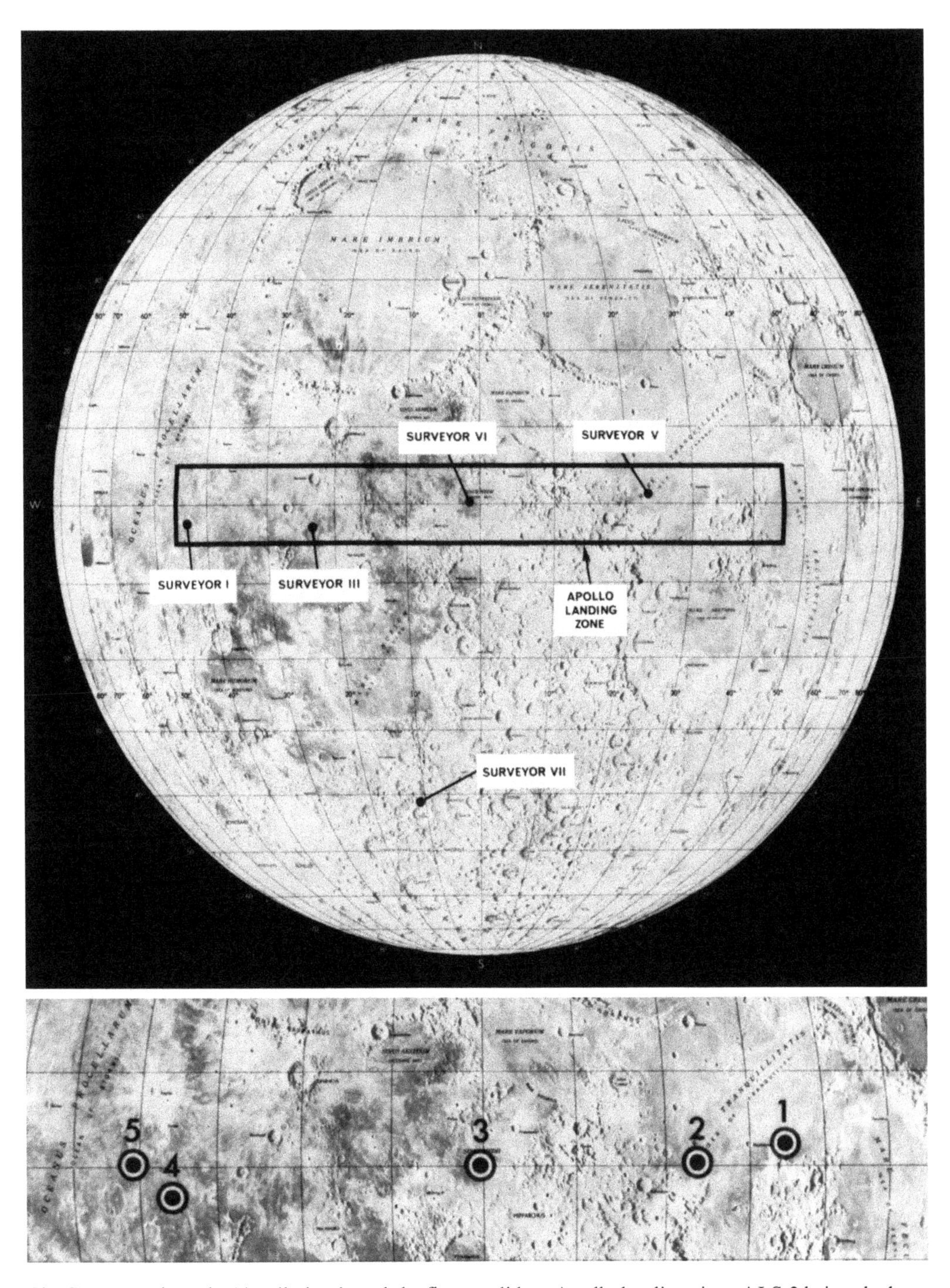

The Surveyor sites, the 'Apollo box', and the five candidate Apollo landing sites, ALS-2 being the best.

AS-506 emerges from the VAB.

The launch of Apollo 11 as viewed from the top of the tower.

As Apollo 11 ascended, the 360-ft-long vehicle was dwarfed by its tremendous exhaust plume.

Charles W. Mathews (left), Wernher von Braun, George E. Mueller and Samuel C. Phillips as Apollo 11 lifts off.

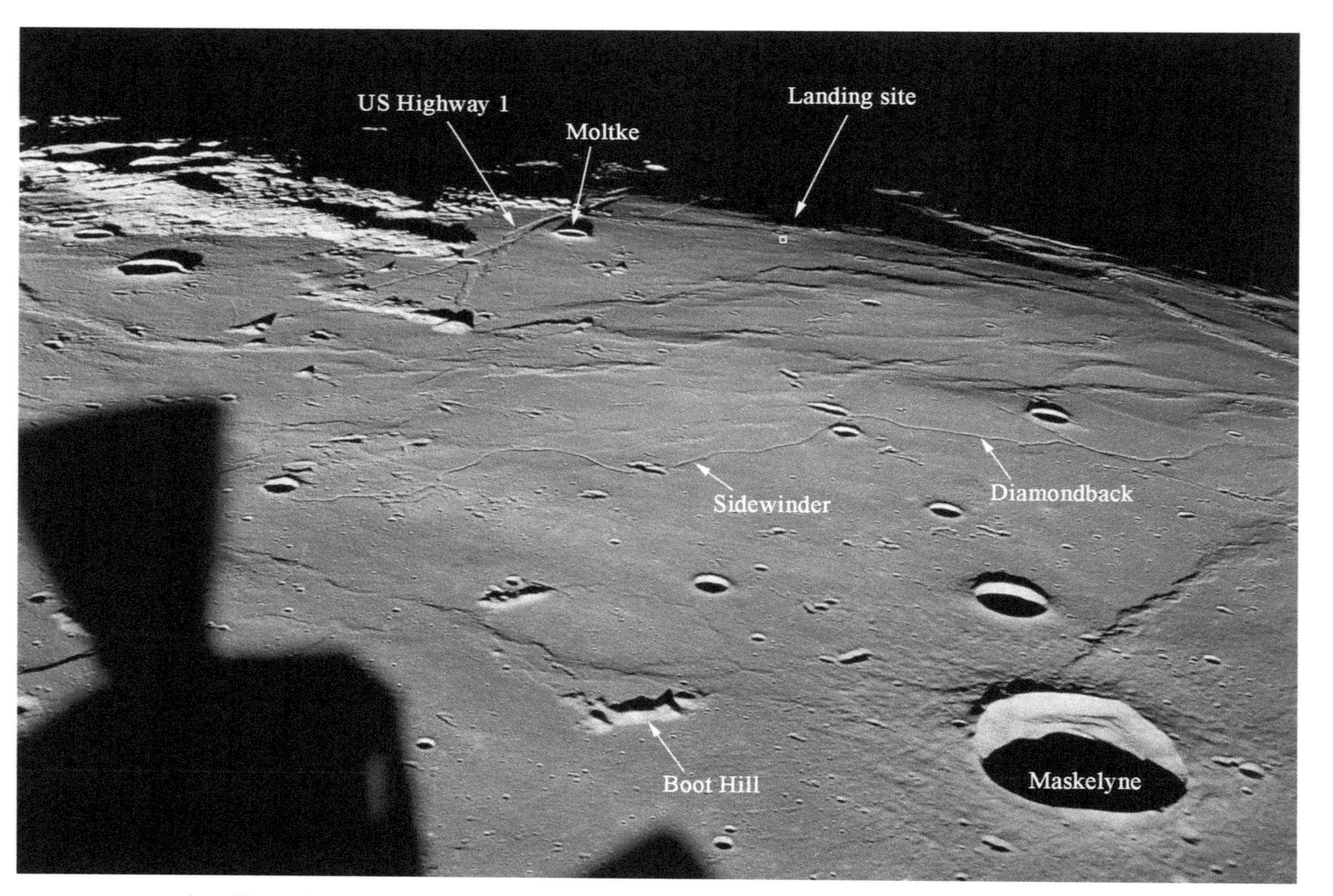

An oblique view of ALS-2 at local sunrise, taken by Apollo 11 soon after it had entered lunar orbit.

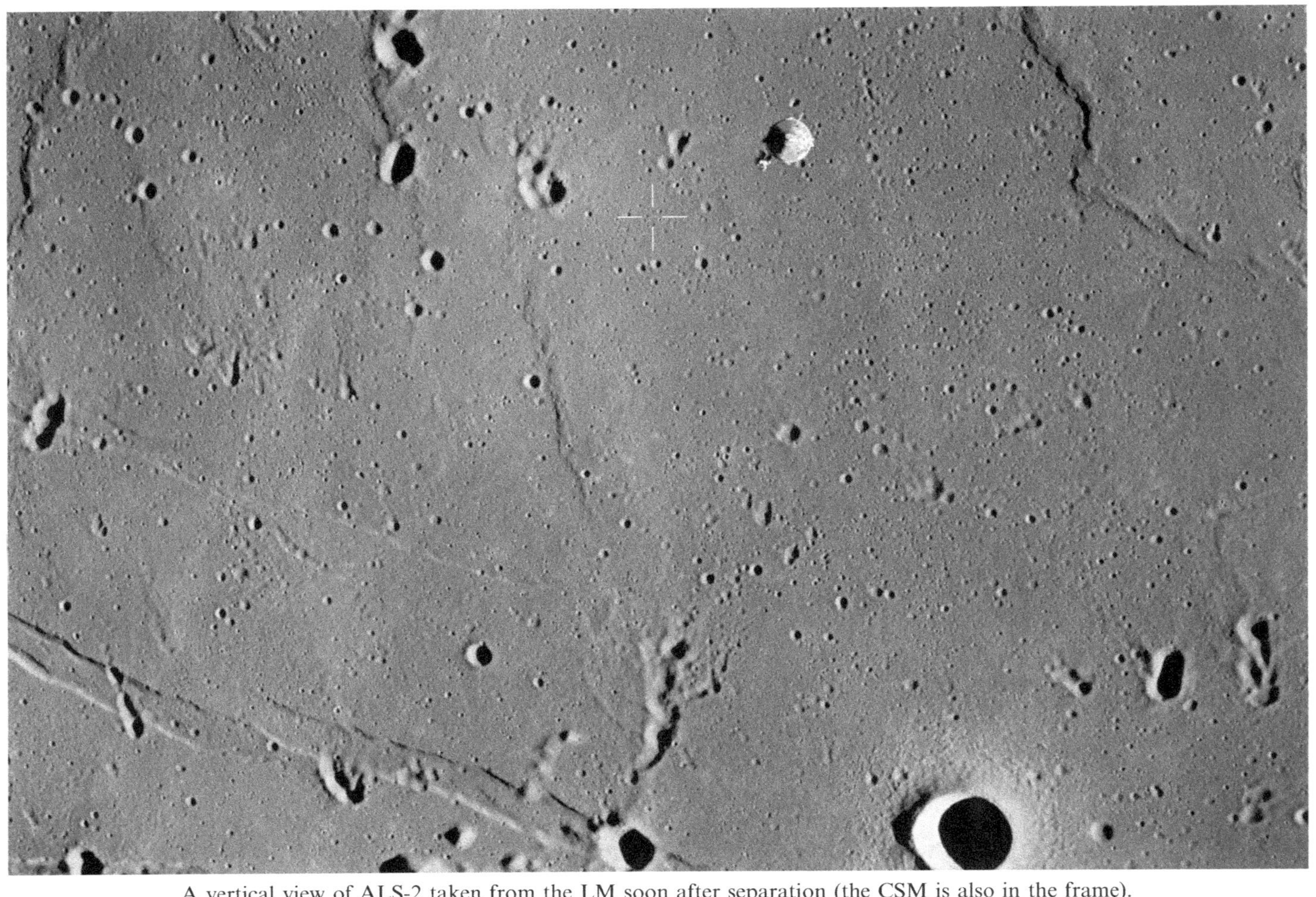

A vertical view of ALS-2 taken from the LM soon after separation (the CSM is also in the frame).

Buzz Aldrin descending the ladder to join Neil Armstrong on the lunar surface.

Buzz Aldrin documented his boot print.

Although done 'for all Mankind', landing a man on the Moon was an achievement by the United States of America.

A rare picture of Neil Armstrong on the lunar surface, here working at the MESA.

Buzz Aldrin unloads the Early Apollo Surface Experiment Package from the LM.

Buzz Aldrin with the deployed EASEP instruments.

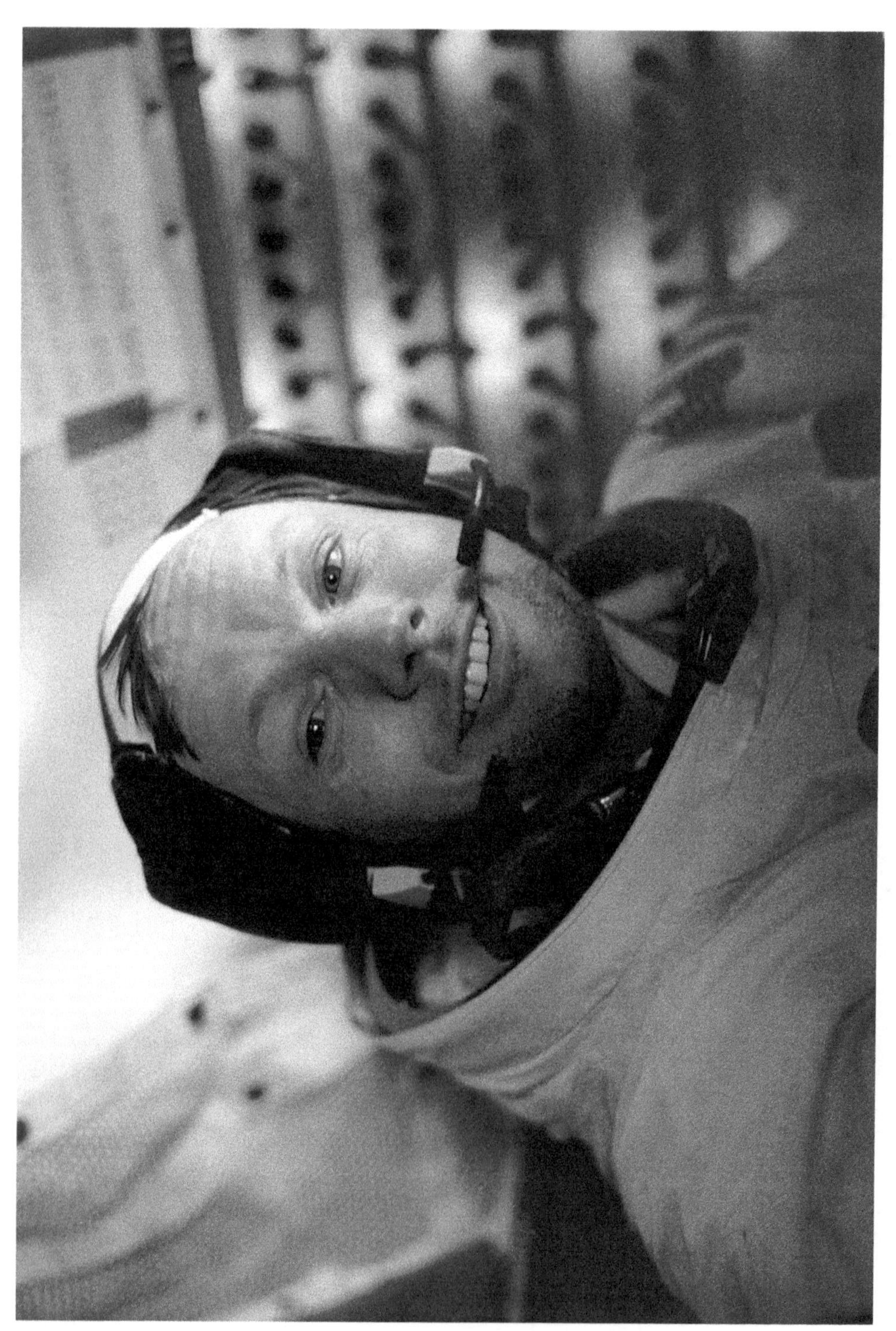

Neil Armstrong in the LM after the moonwalk.

Wearing Biological Isolation Garments, the Apollo 11 crew disembark from the recovery helicopter.

In the Mobile Quarantine Facility on the USS *Hornet*, the Apollo 11 crew chat with Richard Nixon.

A 'ticker tape' parade in New York for the Apollo 11 crew.

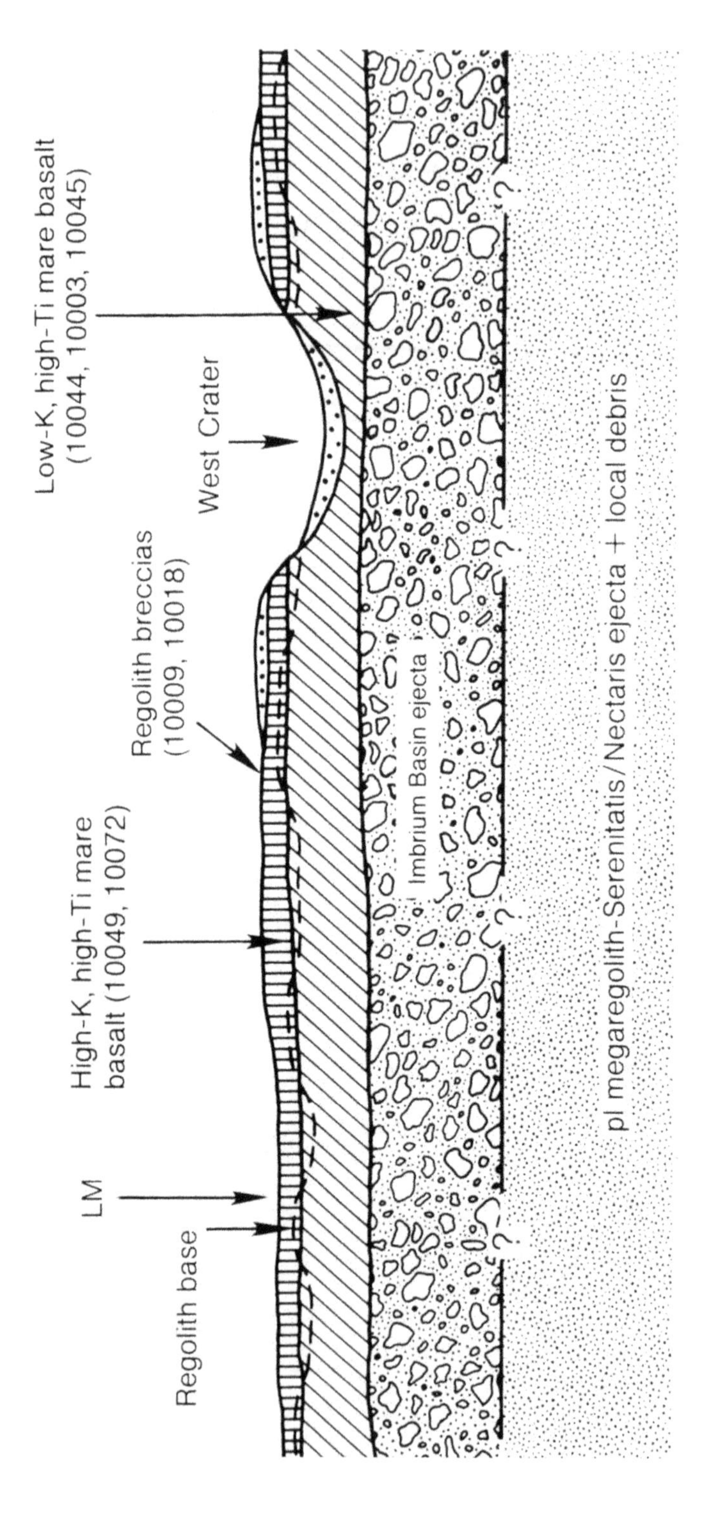

A schematic east–west cross-section through the Apollo 11 landing site showing at least two basaltic lava flows overlying older ejecta from major basins (modified after D.W. Beaty and A.L. Albee, 'The geology and petrology of the Apollo 11 landing site', *Proceedings of the 11th Lunar and Planetary Science Conference*, 1980). The presence of the Imbrium ejecta is inferred. Numbers refer to specific collected samples representative of the various units inferred to be present. The base of the regolith (dashed line) locally penetrates into the low-K (potassium) mare basalts. The abbreviation 'pI' stands for pre-Imbrian in the megaregolith. (Courtesy the Lunar and Planetary Institute and Cambridge University Press.)

(e) Assessment of contamination by lunar material. *Achieved.*
(f) Television coverage. *Achieved.*
(g) Photographic coverage. *Achieved.*
- Long-distance coverage from the command module.
- Lunar mapping photography from orbit.
- Landed lunar module location.
- Sequence photography during descent, lunar stay, and ascent.
- Still photographs through the lunar module window.
- Still photographs on the lunar surface.
- Close-up stereo photography.

Experiments

1. S-031: Passive seismic experiment. *Achieved.*
2. S-059: Lunar field geology. *Partially achieved. Although two core tube samples and 15 lb of additional lunar samples were obtained, time constraints precluded collection of these samples with the degree of documentation originally planned. In addition, time did not permit the collection of a lunar environmental sample or a gas analysis sample in the two special containers provided. It was, however, possible to obtain the desired results using other samples contained in the regular sample return containers.*
3. S-078: Laser-ranging retroreflector experiment. *Achieved.*
4. S-080: Solar wind composition. *Achieved.*
5. S-151: Cosmic ray detection. *Achieved.*
6. M-151: Pilot describing function. *Achieved.*

MISSION TIMELINE

Apollo 11 mission event	GET (h:m:s)	Date (GMT)	Time (h:m:s)
Terminal countdown started.	–028:00:00	14 Jul 1969	21:00:00
Scheduled 11-hr hold at T–9 hr.	–009:00:00	15 Jul 1969	16:00:00
Countdown resumed at T–9 hr.	–009:00:00	16 Jul 1969	03:00:00
Scheduled 1-hr 32-min hold at T–3 hr 30 min.	–003:30:00	16 Jul 1969	08:30:00
Countdown resumed at T–3 hr 30 min.	–003:30:00	16 Jul 1969	10:02:00
Guidance reference release.	–000:00:16.968	16 Jul 1969	13:31:43
S-IC engine start command.	–000:00:08.9	16 Jul 1969	13:31:51
S-IC engine ignition (#5).	–000:00:06.4	16 Jul 1969	13:31:53
All S-IC engines thrust OK.	–000:00:01.6	16 Jul 1969	13:31:58
Range zero.	000:00:00.00	16 Jul 1969	13:32:00
All holddown arms released (1st motion).	000:00:00.3	16 Jul 1969	13:32:00
Liftoff (umbilical disconnected) (1.07 g).	000:00:00.63	16 Jul 1969	13:32:00
Tower clearance yaw maneuver started.	000:00:01.7	16 Jul 1969	13:32:01
Yaw maneuver ended.	000:00:09.7	16 Jul 1969	13:32:09

Apollo 11 mission event – *continued*	GET (h:m:s)	Date (GMT)	Time (h:m:s)
Pitch and roll maneuver started.	000:00:13.2	16 Jul 1969	13:32:13
Roll maneuver ended.	000:00:31.1	16 Jul 1969	13:32:31
Mach 1 achieved.	000:01:06.3	16 Jul 1969	13:33:06
Maximum dynamic pressure (735.17 lb/ft^2).	000:01:23.0	16 Jul 1969	13:33:23
Maximum bending moment (33,200,000 lbf-in).	000:01:31.5	16 Jul 1969	13:33:31
S-IC center engine cutoff command.	000:02:15.2	16 Jul 1969	13:34:15
Pitch maneuver ended.	000:02:40.0	16 Jul 1969	13:34:40
S-IC outboard engine cutoff.	000:02:41.63	16 Jul 1969	13:34:41
S-IC maximum total inertial acceleration (3.94 g).	000:02:41.71	16 Jul 1969	13:34:41
S-IC maximum Earth-fixed velocity. S-IC/S-II separation command.	000:02:42.30	16 Jul 1969	13:34:42
S-II engine start command.	000:02:43.04	16 Jul 1969	13:34:43
S-II ignition.	000:02:44.0	16 Jul 1969	13:34:44
S-II aft interstage jettisoned.	000:03:12.3	16 Jul 1969	13:35:12
Launch escape tower jettisoned.	000:03:17.9	16 Jul 1969	13:35:17
Iterative guidance mode initiated.	000:03:24.1	16 Jul 1969	13:35:24
S-IC apex.	000:04:29.1	16 Jul 1969	13:36:29
S-II center engine cutoff.	000:07:40.62	16 Jul 1969	13:39:40
S-II maximum total inertial acceleration (1.82 g).	000:07:40.70	16 Jul 1969	13:39:40
S-IC impact (theoretical).	000:09:03.7	16 Jul 1969	13:41:03
S-II outboard engine cutoff.	000:09:08.22	16 Jul 1969	13:41:08
S-II maximum Earth-fixed velocity. S-II/S-IVB separation command.	000:09:09.00	16 Jul 1969	13:41:09
S-IVB 1st burn start command.	000:09:09.20	16 Jul 1969	13:41:09
S-IVB 1st burn ignition.	000:09:12.20	16 Jul 1969	13:41:12
S-IVB ullage case jettisoned.	000:09:21.0	16 Jul 1969	13:41:21
S-II apex.	000:09:47.0	16 Jul 1969	13:41:47
S-IVB 1st burn cutoff.	000:11:39.33	16 Jul 1969	13:43:39
S-IVB 1st burn maximum total inertial acceleration (0.69 g).	000:11:39.41	16 Jul 1969	13:43:39
Earth orbit insertion. S-IVB 1st burn maximum Earth-fixed velocity.	000:11:49.33	16 Jul 1969	13:43:49
Maneuver to local horizontal attitude started.	000:11:59.3	16 Jul 1969	13:43:59
Orbital navigation started.	000:13:21.1	16 Jul 1969	13:45:21
S-II impact (theoretical).	000:20:13.7	16 Jul 1969	13:52:13
S-IVB 2nd burn restart preparation.	002:34:38.2	16 Jul 1969	16:06:38
S-IVB 2nd burn restart command.	002:44:08.2	16 Jul 1969	16:16:08
S-IVB 2nd burn ignition (STDV open).	002:44:16.2	16 Jul 1969	16:16:16
S-IVB 2nd burn cutoff.	002:50:03.03	16 Jul 1969	16:22:03
S-IVB 2nd burn maximum total inertial acceleration (1.45 g).	002:50:03.11	16 Jul 1969	16:22:03
S-IVB 2nd burn maximum Earth-fixed velocity.	002:50:03.5	16 Jul 1969	16:22:03
S-IVB safing procedures started.	002:50:03.8	16 Jul 1969	16:22:03
Translunar injection.	002:50:13.03	16 Jul 1969	16:22:13
Maneuver to local horizontal attitude started.	002:50:23.0	16 Jul 1969	16:22:23
Orbital navigation started.	002:50:23.9	16 Jul 1969	16:22:23
Maneuver to transposition and docking attitude started.	003:05:03.9	16 Jul 1969	16:37:03

Apollo 11 mission event – *continued*	GET (h:m:s)	Date (GMT)	Time (h:m:s)
CSM separated from S-IVB.	003:15:23.0	16 Jul 1969	16:47:23
CSM separation maneuver ignition.	003:17:04.6	16 Jul 1969	16:49:04
CSM separation maneuver cutoff.	003:17:11.7	16 Jul 1969	16:49:11
CSM docked with LM/S-IVB.	003:24:03.7	16 Jul 1969	16:56:03
CSM/LM ejected from S-IVB.	004:17:03.0	16 Jul 1969	17:49:03
CSM/LM evasive maneuver from S-IVB ignition.	004:40:01.72	16 Jul 1969	18:12:01
CSM/LM evasive maneuver from S-IVB cutoff.	004:40:04.65	16 Jul 1969	18:12:04
S-IVB maneuver to lunar slingshot attitude initiated.	004:41:07.6	16 Jul 1969	18:13:07
S-IVB lunar slingshot maneuver – LH_2 tank CVS opened.	004:51:07.7	16 Jul 1969	18:23:07
S-IVB lunar slingshot maneuver – LOX dump started.	005:03:07.6	16 Jul 1969	18:35:07
S-IVB lunar slingshot maneuver – LOX dump ended.	005:04:55.8	16 Jul 1969	18:36:55
S-IVB lunar slingshot maneuver – APS ignition.	005:37:47.6	16 Jul 1969	19:09:47
S-IVB lunar slingshot maneuver – APS cutoff.	005:42:27.8	16 Jul 1969	19:14:27
S-IVB maneuver to communications attitude initiated.	005:42:48.8	16 Jul 1969	19:14:48
TV transmission started (recorded at Goldstone and transmitted to Houston at 011:26).	010:32	17 Jul 1969	00:04
TV transmission ended.	010:48	17 Jul 1969	00:20
Midcourse correction ignition.	026:44:58.64	17 Jul 1969	16:16:58
Midcourse correction cutoff.	026:45:01.77	17 Jul 1969	16:17:01
TV transmission started.	030:28	17 Jul 1969	20:00
TV transmission ended.	031:18	17 Jul 1969	20:50
TV transmission started.	033:59	17 Jul 1969	23:31
TV transmission ended.	034:35	18 Jul 1969	00:07
TV transmission started.	055:08	18 Jul 1969	20:40
CDR and LMP entered LM for initial inspection.	055:30	18 Jul 1969	21:02
TV transmission ended.	056:44	18 Jul 1969	22:16
CDR and LMP entered CM.	057:55	18 Jul 1969	23:27
Equigravisphere.	061:39:55	19 Jul 1969	03:11:55
Lunar orbit insertion ignition.	075:49:50.37	19 Jul 1969	17:21:50
Lunar orbit insertion cutoff.	075:55:47.90	19 Jul 1969	17:27:47
Sighting of an illumination in the Aristarchus region. 1st time, a lunar transient event sighted by an observer in space.	077:13	19 Jul 1969	18:45
TV transmission started.	078:20	19 Jul 1969	19:52
S-IVB closest approach to lunar surface.	078:42	19 Jul 1969	20:14
TV transmission ended.	079:00	19 Jul 1969	20:32
Lunar orbit circularization ignition.	080:11:36.75	19 Jul 1969	21:43:36
Lunar orbit circularization cutoff.	080:11:53.63	19 Jul 1969	21:43:53
LMP entered CM for initial power-up and system checks.	081:10	19 Jul 1969	22:42
LMP entered CM.	083:35	20 Jul 1969	01:07
CDR and LMP entered LM for final preparations for descent.	095:20	20 Jul 1969	12:52
LMP entered CM.	097:00	20 Jul 1969	14:32
LMP entered LM.	097:30	20 Jul 1969	15:02
LM system checks started.	097:45	20 Jul 1969	15:17
LM system checks ended.	100:00	20 Jul 1969	17:32
CSM/LM undocked.	100:12:00.0	20 Jul 1969	17:44:00

Apollo 11 mission event – *continued*	GET (h:m:s)	Date (GMT)	Time (h:m:s)
CSM/LM separation maneuver ignition.	100:39:52.9	20 Jul 1969	18:11:52
CSM/LM separation maneuver cutoff.	100:40:01.9	20 Jul 1969	18:12:01
LM descent orbit insertion ignition (LM SPS).	101:36:14	20 Jul 1969	19:08:14
LM descent orbit insertion cutoff.	101:36:44	20 Jul 1969	19:08:44
LM acquisition of data.	102:17:17	20 Jul 1969	19:49:17
LM landing radar on.	102:20:53	20 Jul 1969	19:52:53
LM abort guidance aligned to primary guidance.	102:24:40	20 Jul 1969	19:56:40
LM yaw maneuver to obtain improved communications.	102:27:32	20 Jul 1969	19:59:32
LM altitude 50,000 ft.	102:32:55	20 Jul 1969	20:04:55
LM propellant settling firing started.	102:32:58	20 Jul 1969	20:04:58
LM powered descent engine ignition.	102:33:05.01	20 Jul 1969	20:05:05
LM fixed throttle position.	102:33:31	20 Jul 1969	20:05:31
LM face-up maneuver completed.	102:37:59	20 Jul 1969	20:09:59
LM 1202 alarm.	102:38:22	20 Jul 1969	20:10:22
LM radar updates enabled.	102:38:45	20 Jul 1969	20:10:45
LM altitude less than 30,000 ft and velocity less than 2,000 ft/sec (landing radar velocity update started).	102:38:50	20 Jul 1969	20:10:50
LM 1202 alarm.	102:39:02	20 Jul 1969	20:11:02
LM throttle recovery.	102:39:31	20 Jul 1969	20:11:31
LM approach phase entered.	102:41:32	20 Jul 1969	20:13:32
LM landing radar antenna to position 2.	102:41:37	20 Jul 1969	20:13:37
LM attitude hold mode selected (check of LM handling qualities).	102:41:53	20 Jul 1969	20:13:53
LM automatic guidance enabled.	102:42:03	20 Jul 1969	20:14:03
LM 1201 alarm.	102:42:18	20 Jul 1969	20:14:18
LM landing radar switched to low scale.	102:42:19	20 Jul 1969	20:14:19
LM 1202 alarm.	102:42:43	20 Jul 1969	20:14:43
LM 1202 alarm.	102:42:58	20 Jul 1969	20:14:58
LM landing point redesignation.	102:43:09	20 Jul 1969	20:15:09
LM altitude hold.	102:43:13	20 Jul 1969	20:15:13
LM abort guidance attitude updated.	102:43:20	20 Jul 1969	20:15:20
LM rate of descent landing phase entered.	102:43:22	20 Jul 1969	20:15:22
LM landing radar data not good.	102:44:11	20 Jul 1969	20:16:11
LM landing data good.	102:44:21	20 Jul 1969	20:16:21
LM fuel low-level quantity light.	102:44:28	20 Jul 1969	20:16:28
LM landing radar data not good.	102:44:59	20 Jul 1969	20:16:59
LM landing radar data good.	102:45:03	20 Jul 1969	20:17:03
1st evidence of surface dust disturbed by descent engine.	102:44:35	20 Jul 1969	20:16:35
LM lunar landing.	102:45:39.9	20 Jul 1969	20:17:39
LM powered descent engine cutoff.	102:45:41.40	20 Jul 1969	20:17:41
Decision made to proceed with EVA prior to first rest period.	104:40:00	20 Jul 1969	22:12:00
Preparation for EVA started.	106:11:00	20 Jul 1969	23:43:00
EVA started (hatch open).	109:07:33	21 Jul 1969	02:39:33
CDR completely outside LM on porch.	109:19:16	21 Jul 1969	02:51:16
Modular equipment stowage assembly deployed (CDR).	109:21:18	21 Jul 1969	02:53:18
First clear TV picture received.	109:22:00	21 Jul 1969	02:54:00

Apollo 11 mission event – *continued*	GET (h:m:s)	Date (GMT)	Time (h:m:s)
CDR at foot of ladder (starts to report, then pauses to listen).	109:23:28	21 Jul 1969	02:55:28
CDR at foot of ladder and described surface as "almost like a powder".	109:23:38	21 Jul 1969	02:55:38
1st step taken lunar surface (CDR). "That's one small step for a man...one giant leap for mankind".	109:24:15	21 Jul 1969	02:56:15
CDR started surface examination and description, assessed mobility and described effects of LM descent engine.	109:24:48	21 Jul 1969	02:56:48
CDR ended surface examination. LMP started to send down camera.	109:26:54	21 Jul 1969	02:58:54
Camera installed on RCU bracket, LEC stored on secondary strut of LM landing gear.	109:30:23	21 Jul 1969	03:02:23
Surface photography (CDR).	109:30:53	21 Jul 1969	03:02:53
Contingency sample collection started (CDR).	109:33:58	21 Jul 1969	03:05:58
Contingency sample collection ended (CDR).	109:37:08	21 Jul 1969	03:09:08
LMP started egress from LM.	109:39:57	21 Jul 1969	03:11:57
LMP at top of ladder. Descent photographed by CDR.	109:41:56	21 Jul 1969	03:13:56
LMP on lunar surface.	109:43:16	21 Jul 1969	03:15:16
Surface examination and examination of landing effects on surface and on LM started (CDR, LMP).	109:43:47	21 Jul 1969	03:15:47
Insulation removed from modular equipment stowage assembly (CDR).	109:49:06	21 Jul 1969	03:21:06
TV camera focal distance adjusted (CDR).	109:51:35	21 Jul 1969	03:23:35
Plaque unveiled (CDR).	109:52:19	21 Jul 1969	03:24:19
Plaque read (CDR).	109:52:40	21 Jul 1969	03:24:40
TV camera redeployed. Panoramic TV view started (CDR).	109:59:28	21 Jul 1969	03:31:28
TV camera placed in final deployment position (CDR).	110:02:53	21 Jul 1969	03:34:53
Solar wind composition experiment deployed (LMP).	110:03:20	21 Jul 1969	03:35:20
US flag deployed (CDR, LMP).	110:09:43	21 Jul 1969	03:41:43
Evaluation of surface mobility started (LMP).	110:13:15	21 Jul 1969	03:45:15
Evaluation of surface mobility end (LMP).	110:16:02	21 Jul 1969	03:48:02
Presidential message from White House and response from CDR.	110:16:30	21 Jul 1969	03:48:30
Presidential message and CDR response ended.	110:18:21	21 Jul 1969	03:50:21
Evaluation of trajectory of lunar soil when kicked (LMP) and bulk sample collection started (CDR).	110:20:06	21 Jul 1969	03:52:06
Evaluation of visibility in lunar sunlight (LMP).	110:10:24	21 Jul 1969	03:42:24
Evaluation of thermal effects of Sun and shadow inside the suit (LMP).	110:25:09	21 Jul 1969	03:57:09
Evaluation of surface shadows and colors (LMP).	110:28:22	21 Jul 1969	04:00:22
LM landing gear inspection and photography (LMP).	110:34:13	21 Jul 1969	04:06:13
Bulk sample completed (CDR).	110:35:36	21 Jul 1969	04:07:36
LM landing gear inspection and photography (CDR, LMP).	110:46:36	21 Jul 1969	04:18:36
Scientific equipment bay doors opened.	110:53:38	21 Jul 1969	04:25:38
Passive seismometer deployed.	110:55:42	21 Jul 1969	04:27:42
Lunar ranging retroreflector deployed (CDR).	111:03:57	21 Jul 1969	04:35:57

Apollo 11 mission event – *continued*	GET (h:m:s)	Date (GMT)	Time (h:m:s)
1st passive seismic experiment data received on Earth.	111:08:39	21 Jul 1969	04:40:39
Collection of documented samples started (CDR/LMP).	111:11	21 Jul 1969	04:43
Solar wind composition experiment retrieved (LMP).	111:20	21 Jul 1969	04:52
LMP inside LM.	111:29:39	21 Jul 1969	05:01:39
Transfer of sample containers reported complete.	111:35:51	21 Jul 1969	05:07:51
CDR inside LM, assisted and monitored by LMP.	111:37:32	21 Jul 1969	05:09:32
EVA ended (hatch closed).	111:39:13	21 Jul 1969	05:11:13
LM equipment jettisoned.	114:05	21 Jul 1969	07:37
LM lunar liftoff ignition (LM APS).	124:22:00.79	21 Jul 1969	17:54:00
LM orbit insertion cutoff.	124:29:15.67	21 Jul 1969	18:01:15
Coelliptic sequence initiation ignition.	125:19:35	21 Jul 1969	18:51:35
Coelliptic sequence initiation cutoff.	125:20:22	21 Jul 1969	18:52:22
Constant differential height maneuver ignition.	126:17:49.6	21 Jul 1969	19:49:49
Constant differential height maneuver cutoff.	126:18:29.2	21 Jul 1969	19:50:29
Terminal phase initiation ignition.	127:03:51.8	21 Jul 1969	20:35:51
Terminal phase initiation cutoff.	127:04:14.5	21 Jul 1969	20:36:14
LM 1st midcourse correction.	127:18:30.8	21 Jul 1969	20:50:30
LM 2nd midcourse correction.	127:33:30.8	21 Jul 1969	21:05:30
Braking started.	127:36:57.3	21 Jul 1969	21:08:57
Terminal phase finalize ignition.	127:46:09.8	21 Jul 1969	21:18:09
Terminal phase finalize cutoff.	127:46:38.2	21 Jul 1969	21:18:38
Stationkeeping started.	127:52:05.3	21 Jul 1969	21:24:05
CSM/LM docked.	128:03:00	21 Jul 1969	21:35:00
CDR entered CM.	129:20	21 Jul 1969	22:52
LMP entered CM.	129:45	21 Jul 1969	23:17
LM ascent stage jettisoned.	130:09:31.2	21 Jul 1969	23:41:31
CSM/LM final separation ignition.	130:30:01.0	22 Jul 1969	00:02:01
CSM/LM final separation cutoff.	130:30:08.2	22 Jul 1969	00:02:08
Transearth injection ignition (SPS).	135:23:42.28	22 Jul 1969	04:55:42
Transearth injection cutoff.	135:26:13.69	22 Jul 1969	04:58:13
Midcourse correction ignition.	150:29:57.4	22 Jul 1969	20:01:57
Midcourse correction cutoff.	150:30:07.4	22 Jul 1969	20:02:07
TV transmission started.	155:36	23 Jul 1969	01:08
TV transmission ended.	155:54	23 Jul 1969	01:26
TV transmission started.	177:10	23 Jul 1969	22:42
TV transmission ended.	177:13	23 Jul 1969	22:45
TV transmission started.	177:32	23 Jul 1969	23:04
TV transmission ended.	177:44	23 Jul 1969	23:16
CM/SM separation.	194:49:12.7	24 Jul 1969	16:21:12
Entry.	195:03:05.7	24 Jul 1969	16:35:05
Drogue parachute deployed	195:12:06.9	24 Jul 1969	16:44:06
Visual contact with CM established by aircraft.	195:07	24 Jul 1969	16:39
Radar contact with CM established by recovery ship.	195:08	24 Jul 1969	16:40
VHF voice contact and recovery beacon contact established.	195:14	24 Jul 1969	16:46
Splashdown (went to apex-down).	195:18:35	24 Jul 1969	16:50:35

Apollo 11 mission event – *continued*	GET (h:m:s)	Date (GMT)	Time (h:m:s)
CM returned to apex-up position.	195:26:15	24 Jul 1969	16:58:15
Flotation collar inflated.	195:32	24 Jul 1969	17:04
Hatch opened for crew egress.	195:49	24 Jul 1969	17:21
Crew egress.	195:57	24 Jul 1969	17:29
Crew on board recovery ship.	196:21	24 Jul 1969	17:53
Crew entered mobile quarantine facility.	196:26	24 Jul 1969	17:58
CM lifted from water.	198:18	24 Jul 1969	19:50
CM secured to quarantine facility.	198:26	24 Jul 1969	19:58
CM hatch reopened.	198:33	24 Jul 1969	20:05
Sample return containers 1 and 2 removed from CM.	200:28	24 Jul 1969	22:00
Container 1 removed from mobile quarantine facility.	202:00	24 Jul 1969	23:32
Container 2 removed from mobile quarantine facility.	202:33	25 Jul 1969	00:05
Container 2 and film flown to Johnston Island.	207:43	25 Jul 1969	05:15
Container 1 flown to Hickam Air Force Base, Hawaii.	214:13	25 Jul 1969	11:45
Container 2 and film arrived in Houston, TX.	218:43	25 Jul 1969	16:15
Container 1, film, and biological samples arrived in Houston.	225:41	25 Jul 1969	23:13
CM decontaminated and hatch secured.	229:28	26 Jul 1969	03:00
Mobile quarantine facility secured.	231:03	26 Jul 1969	04:35
Mobile quarantine facility and CM offloaded.	250:43	27 Jul 1969	00:15
Safing of CM pyrotechnics completed.	252:33	27 Jul 1969	02:05
Mobile quarantine facility arrived at Ellington AFB, Houston.	280:28	28 Jul 1969	06:00
Crew in Lunar Receiving Laboratory, Houston.	284:28	28 Jul 1969	10:00
CM delivered to Lunar Receiving Laboratory.	345:45	30 Jul 1969	23:17
Passive seismic experiment turned off.	430:26:46	03 Aug 1969	11:58:46
Crew released from quarantine.	–	10 Aug 1969	–
CM delivered to contractor's facility in Downey, CA.	–	14 Aug 1969	–
EASEP turned off by ground command.	–	27 Aug 1969	–

Apollo 12

The sixth manned mission: the second lunar landing

14–24 November 1969

BACKGROUND

As the first of a series of Type 'H' missions, Apollo 12 was to demonstrate a precision manned lunar landing and systematic lunar exploration. It was the second landing on the Moon.

The primary objectives were:

- to perform selenological inspection, survey, and sampling in a mare area;
- to deploy the Apollo Lunar Surface Experiments Package (ALSEP);
- to develop techniques for a pin-point landing capability;
- to further develop human capability to work in the lunar environment; and
- to obtain photographs of candidate exploration sites.

The all-Navy crew included Commander Charles 'Pete' Conrad Jr (USN), commander; Commander Richard Francis 'Dick' Gordon Jr (USN), command module pilot; and Commander Alan LaVern Bean (USN), lunar module pilot. Selected as an astronaut in 1962, Conrad was making his third spaceflight. He had been pilot of Gemini 5 and command pilot of Gemini 11. Born on 2 June 1930 in Philadelphia, Pennsylvania, Conrad was 39 years old at the time of the Apollo 12 mission. He received a BS in aeronautical engineering from Princeton University in 1953.[5] His backup was Colonel David Randolph Scott (USAF). Gordon had been pilot of Gemini 11. Born on 5 October 1929 in Seattle, Washington, he was 40 years old at the time of the Apollo 12 mission. Gordon received a BS in chemistry from the University of Washington in 1951, and was selected as an astronaut in 1963. His backup was Major Alfred Merrill Worden (USAF). Bean was making his first spaceflight. Born on 15 March 1932 in Wheeler, Texas, he was 37 years old at the time of the Apollo 12 mission. Bean received a BS in aeronautical engineering from the University of Texas in 1955, and was selected as an astronaut in 1963. His backup was Lieutenant Colonel James Benson Irwin (USAF). The capsule communicators

[5] Conrad died 8 July 1999 in Ojai, CA, as a result of injuries sustained in a motorcycle accident.

(CAPCOMs) for the mission were Lieutenant Colonel Gerald Paul Carr (USMC), Edward George Gibson PhD, Commander Paul Joseph Weitz (USN), Don Leslie Lind PhD, Scott, Worden, and Irwin. For this mission, there were also four civilian backup CAPCOMs: Dickie K. Warren, James O. Rippey, James L. Lewis, and Michael R. Wash. The support crew members were Carr, Weitz, and Gibson. The flight directors were Gerald D. Griffin (first shift), M.P. 'Pete' Frank (second shift), Clifford E. Charlesworth (third shift), and Milton L. Windler (fourth shift).

The Apollo 12 launch vehicle was a Saturn V, designated AS-507. The mission also carried the designation Eastern Test Range #2793. The CSM was designated CSM-108, and had the call-sign 'Yankee Clipper'. The lunar module was designated LM-6, and had the call-sign 'Intrepid'.

LANDING SITE

Apollo 11 had proven the LM's ability to land on the Moon, but the fact that it had strayed off target was embarrassing. The ability to land within a few hundred yards of a given location was a prerequisite to being able to follow a specific geological traverse. The flight dynamics team had devised a simple method to correct for the perturbations of the mascons: the primary guidance, navigation and control system would be told to head for a 'different' landing site, which redirection would really take it to the intended target. The engineers were so confident this would work that they reduced the size of the target ellipse. In addition, it was decided to reduce the requirement from two backup sites to one. There had been five prime sites on the short list for the first landing. The easterly ALS-1 and ALS-2 sites in Tranquilitatis had been backed up by ALS-3 in Sinus Medii, with ALS-4 and ALS-5 in Oceanus Procellarum in reserve in case of a prolonged launch delay. It would have been natural to send Apollo 12 to one of these sites, but the conservative constraints of the initial landing had resulted in the choice of open sites, and the geologists were eager to sample the ejecta of a sizeable crater. In fact, even before Apollo 11 flew, the site selectors had drawn up a list of craters for this eventuality. In principle, it should be a simple matter of re-examining the sites that had been rejected for the first landing because of the *inconvenient* proximity of a crater. Although a number of pin-point sites were nominated in the vicinity of ALS-5, they were pretty bland. A site in the large crater Hipparchus was deemed to be insufficiently documented. The hummocky terrain north of the crater Fra Mauro was rejected as too demanding. The relaxation of the operational constraints enabled ALS-6 to be reinstated. This was in the Flamsteed Ring, the large crater that had been almost completely flooded by a Procellarum lava flow. However, the fact that ALS-6 was not far from where Surveyor 1 had landed prompted the flight dynamics team to propose making this the target because to set down within walking distance of another spacecraft would be a powerful demonstration of precision. Unfortunately, because this site was so far west, it had no back up. But there was another spacecraft in Procellarum to the east that could readily accommodate a westerly back up. The Lunar Orbiter site selectors had designated the Surveyor 3 site 3P-9 and, using their conservative criteria, had dismissed it as too rough for Apollo, but it was now redesignated as ALS-7.

In addition to scouting out the blandest sites for the first Apollo landing, the Lunar Orbiters had photographed many 'feature' sites. Nevertheless, this program had been so dedicated to locating wide open sites for the first landing that the coverage of more demanding sites was too cursory to certify them. It was evident, therefore, that if Apollo was to exploit the relaxation of the flight dynamics constraints, the early missions would have to reconnoiter 'out of area' sites for their successors. Once pin-point accuracy had been demonstrated, it would be practicable to assign 'feature' rather than 'area' targets. Apollo 12 was to open this door. The selection of ALS-7 for Apollo 12 would reconnoiter Fra Mauro, the Descartes area, and the Davy Rille, which was a chain of craters believed, by some, to be volcanic vents along a geological fault.

LAUNCH PREPARATIONS

The terminal countdown started at T–28 hr at 02:00:00 GMT on 13 November 1969. Two holds were planned at T–9 hr for 9 hr 22 min and at T–3 hr 30 min for one hour. However, while loading cryogenic fuel cell reactants on 12 November, a leak developed in the CSM LH_2 tank No. 2. The tank was drained and replaced using a tank from CSM-109 (which was assigned to Apollo 13). An unscheduled hold was initiated on 13 November at T–17 hr (12:00:00 GMT) for retanking cryogenics in the CSM. Loading was completed in 6 hours and the count resumed at 19:00:00 GMT. The scheduled hold at T–9 hr was reduced by 6 hours, thereby averting a launch delay. A cold front was moving slowly southward through the central section of Florida. This front produced the rain showers and overcast conditions that existed over the pad at launch time. Stratocumulus clouds covered 100% of the sky (base 2,100 ft), the temperature was 68.0°F, the relative humidity was 92%, and the barometric pressure was 14.621 psi. Winds, as measured by the anemometer on the light pole 60.0 ft above ground at the launch site, measured 13.2 kt at 280° from true north.

ASCENT PHASE

Apollo 12 was launched from Pad A of Launch Complex 39 at the Kennedy Space Center at a Range Zero time of 16:22:00 GMT (11:22:00 EST) on 14 November 1969. The launch window extended to 19:26:00 GMT in order to take advantage of a Sun elevation angle on the lunar surface of 5.1 deg at the scheduled time of landing.

Apollo 12 was the first Saturn V to be launched during a rainstorm, following the decision to waive Manned Space Flight Center Launch Mission Rule 1-404, which stated:

> "The vehicle will not be launched when its flight path will carry it through a cumulonimbus (thunderstorm) cloud formation."

The reason for the rule was that the Saturn V had not been designed to withstand thunderstorm conditions during launch.

Between T + 12.8 and T + 32.3 sec, the vehicle rolled from a pad azimuth of 90°E of N to a flight azimuth of 72.029°E of N. At T + 36.5 there were numerous space vehicle

Apollo 12 preparation event	Date
LM-6 integrated test at factory.	31 Dec 1968
Individual and combined CM and SM systems test completed at factory.	20 Jan 1969
Integrated CM and SM systems test completed at factory.	03 Feb 1969
LM-6 final engineering evaluation acceptance test at factory.	18 Feb 1969
S-IVB-507 stage delivered to KSC.	10 Mar 1969
Descent stage of LM-6 ready to ship from factory to KSC.	22 Mar 1969
Ascent stage of LM-6 ready to ship from factory to KSC.	23 Mar 1969
LM-6 stages delivered to KSC.	24 Mar 1969
CM-108 and SM-108 ready to ship from factory to KSC.	27 Mar 1969
CM-108 and SM-108 delivered to KSC.	28 Mar 1969
CM-108 and SM-108 mated.	02 Apr 1969
CSM-108 combined systems test completed.	21 Apr 1969
S-II-7 stage delivered to KSC.	21 Apr 1969
LM-6 stages mated.	28 Apr 1969
LM-6 combined systems test completed.	01 May 1969
S-IC-7 stage delivered to KSC.	03 May 1969
Spacecraft/LM adapter SLA-15 delivered to KSC.	06 May 1969
S-IC-7 stage erected on MLP-2.	07 May 1969
Saturn V instrument unit IU-507 delivered to KSC.	08 May 1969
S-II-7 stage erected.	21 May 1969
S-IVB-507 stage and IU-107 erected.	22 May 1969
CSM-108 altitude test with prime crew completed.	07 Jun 1969
CSM-108 altitude tests completed.	09 Jun 1969
CSM-108 altitude test with backup crew completed.	10 Jun 1969
Launch vehicle propellant dispersion/malfunction overall test completed.	12 Jun 1969
LM-6 altitude test with backup crew completed.	13 Jun 1969
LM-6 altitude test with prime crew completed.	16 Jun 1969
Spacecraft moved to VAB.	20 Jun 1969
LM-6 landing gear installed.	22 Jun 1969
LM-6 mated to SLA-15.	23 Jun 1969
CSM-108 mated to SLA-15.	27 Jun 1969
CSM-108 moved to VAB.	30 Jun 1969
Spacecraft erected.	01 Jul 1969
LM-6 combined systems test completed.	05 Jul 1969
CSM-108 integrated systems test completed.	07 Jul 1969
CSM-108 electrically mated to launch vehicle.	16 Jul 1969
Space vehicle overall test completed.	17 Jul 1969
Space vehicle electrically mated.	17 Aug 1969
Space vehicle overall test #1 (plugs in) completed.	21 Aug 1969
Space vehicle and MLP-2 transferred to Launch Complex 39A.	08 Sep 1969
Mobile service structure transferred to Launch Complex 39A.	10 Sep 1969
LM-6 flight readiness test completed.	18 Sep 1969
Space vehicle flight readiness test completed.	30 Sep 1969
S-IC-7 stage RP-1 fuel loading completed.	20 Oct 1969
Space vehicle countdown demonstration test (wet) completed.	28 Oct 1969
Space vehicle countdown demonstration test (dry) completed.	29 Oct 1969

indications of a massive electrical disturbance, followed by a second disturbance at T + 52. The crew reported that, in their opinion, the vehicle had been struck by lightning, and that the fuel cells in the service module had been disconnected and that all A/C power in the spacecraft was lost. Numerous indicator lamps were illuminated at this time. Ground camera data, telemetered data, and launch computers later showed that the vehicle had indeed been hit by lightning. Virtually no discernible effects were noted on the launch vehicle during the second disturbance. Atmospheric electrical factors, and the fact that the vehicle did not have the capacitance to store sufficient energy to produce the effects noted, indicated that the first discharge was triggered by the vehicle. The second disturbance may have been due to a lesser lightning discharge. The launch vehicle hardware and software suffered no significant effects, and the mission proceeded as scheduled. Because the lightning was self-induced, and because the vehicle did not fly through cumulonimbus clouds, it was determined that Rule 1-404 had not been violated. The maximum wind conditions encountered during ascent were 92.5 kt at 245° from true north at 46,670 ft, with a maximum wind shear of 0.0183/sec at 46,750 ft. The S-IC shut down at T + 161.74, followed by S-IC/S-II separation, and S-II ignition. The S-II shut down at T + 552.34 followed by separation from the S-IVB, which ignited at T + 556.60. The first S-IVB cutoff occurred at T + 693.91, with deviations from the planned trajectory of only –1.9 ft/sec in velocity and only 0.2 nmi in altitude. The S-IC impacted the Atlantic Ocean at T + 554.5 at 30.273°N and 73.895°W, 365.2 nmi from the launch site. The S-II impacted the Atlantic Ocean at T + 1,221.6 at 31.465°N and 34.214°W, 2,404.4 nmi from the launch site. At insertion, at T + 703.91 (i.e. S-IVB cutoff plus 10 sec to account for engine tailoff and other transient effects), the parking orbit showed an apogee and perigee of 100.1 × 97.8 nmi, an inclination of 32.540 deg, a period of 88.16 min, and a velocity of 25,565.9 ft/sec. The apogee and perigee were based upon a spherical Earth with a radius of 3,443.934 nmi.

The COSPAR designation for the CSM upon achieving orbit was 1969-099A and the S-IVB was designated 1969-099B. After undocking at the Moon, the LM ascent stage would be designated 1969-099C and the descent stage 1969-099D.

Apollo 12 ascent phase event	GET (h:m:s)	Altitude (nmi)	Range (nmi)	Earth-fixed velocity (ft/sec)	Space-fixed velocity (ft/sec)	Event duration (sec)	Geocentric latitude (°N)	Longitude (°E)	Space-fixed flight path angle (deg)	Space-fixed heading angle (°E of N)
Liftoff	000:00:00.68	0.032	0.000	0.0	1,340.7	–	28.4470	–80.6041	0.07	90.00
1st lightning strike	000:00:36.5	1.053	0.062	387.9	1,445.7	–	28.4469	–80.6030	15.40	89.29
2nd lightning strike	000:00:52	2.374	0.399	692.1	1,690.4	–	28.4487	–80.5968	22.74	87.32
Mach 1 achieved	000:01:06.1	4.215	1.228	1,067.6	2,057.7	–	28.4532	–80.5820	27.13	84.84
Maximum dynamic pressure	000:01:21.1	6.934	3.019	1,601.4	2,617.3	–	28.4627	–80.5498	29.02	82.10
S-IC center engine cutoff	000:02:15.24	24.158	25.441	5,334.5	6,494.4	141.7	28.5794	–80.1463	23.944	76.115
S-IC outboard engine cutoff	000:02:41.74	36.773	50.616	7,821.4	9,024.5	168.2	28.7069	–79.6913	20.513	75.231
S-IC/S-II separation	000:02:42.4	37.118	51.338	7,850.3	9,054.2	–	28.7107	–79.6773	20.430	75.228
S-II center engine cutoff	000:07:40.75	100.463	599.172	17,453.5	18,775.3	297.55	30.9599	–69.4827	0.502	79.632
S-II outboard engine cutoff	000:09:12.34	102.801	884.711	21,508.8	22,831.7	389.14	31.7508	–63.9914	0.442	82.501
S-II/S-IVB separation	000:09:13.20	102.827	887.667	21,517.8	22,840.7	–	31.7576	–63.9341	0.432	82.533
S-IVB 1st burn cutoff	000:11:33.91	103.093	1,399.874	24,236.6	25,560.2	137.31	32.4933	–53.8956	–0.015	88.146
Earth orbit insertion	000:11:43.91	103.086	1,438.608	24,242.3	25,565.9	–	32.5128	–53.1311	–0.014	88.580

EARTH ORBIT PHASE

After in-flight systems checks, made with extra care because of the two lightning strikes, the 341.14-sec translunar injection maneuver (second S-IVB firing) was performed at 002:47:22.80. The S-IVB engine shut down at 002:53:03.94 and translunar injection occurred 10 sec later, at a velocity of 35,389.9 ft/sec after 1.5 Earth orbits lasting 2 hr 41 min 30.03 sec.

Apollo 12 earth orbit phase event	GET (h:m:s)	Space-fixed velocity (ft/sec)	Event duration (sec)	Velocity change (ft/sec)	Apogee (nmi)	Perigee (nmi)	Period (min)	Inclination (deg)
Earth orbit insertion	000:11:43.91	25,565.9	–	–	100.1	97.8	88.16	32.540
S-IVB 2nd burn ignition	002:47:22.80	25,555.4	–	–	–	–	–	–
S-IVB 2nd burn cutoff	002:53:03.94	35,419.3	341.14	10,515	–	–	–	30.360

TRANSLUNAR PHASE

Whereas previous Apollo lunar missions flew a free-return translunar coast, a trajectory that would achieve satisfactory Earth entry within the velocity correction capability of the spacecraft's reaction control system, Apollo 12 flew a 'hybrid' free-return that increased mission planning flexibility. It permitted a daylight launch to the planned landing site on a western mare, and a greater performance margin for the service propulsion system. The hybrid profile was constrained so that a safe return using the descent propulsion system could be made after a failure to enter lunar orbit. At 003:18:04.9, the CSM was separated from the S-IVB stage, transposed, and docked with the LM at 003:26:53.3. Television transmitted from 003:25 to 004:28 clearly showed the docking. The docked combination was ejected from the S-IVB at 004:13:00.9. An S-IVB auxiliary propulsion system evasive maneuver was performed at 004:26.40, and was also observed on television. A ground command for propulsive venting of residual propellants targeted the S-IVB to pass the Moon and into solar orbit, but an excessively long ullage engine burn opened the distance of closest approach to the Moon and prevented sufficient energy being derived from the 'slingshot' to enable the S-IVB to escape the Earth–Moon system, and it entered an elliptical orbit around Earth and the Moon instead. However, the objectives of having it not strike the spacecraft, Earth, or the Moon were achieved. The closest approach of the S-IVB to the Moon was 3,082 nmi at 085:48.

To ensure that the electrical transients experienced during launch had not affected the LM systems, the commander and lunar module pilot entered the LM earlier than planned, at 007:20, to perform some of the housekeeping and systems checks. The checks indicated that the LM systems were satisfactory. One midcourse correction was required during translunar coast, a 9.19-sec, 61.8-ft/sec maneuver at 030:52:44.36. It placed the spacecraft on the desired hybrid, non-free-return circumlunar trajectory. Good-quality television coverage of the preparations for this burn was received for 47 min, starting at 030:18. A 56-min television transmission began at 062:52. It provided excellent color pictures of the CM, intravehicular transfer, the LM interior, and brief shots of the Earth and the Moon. At 083:25:23.36, at an altitude of 83.91 nmi above the Moon, the service propulsion system was fired for 352.25 sec to insert the spacecraft into a lunar orbit of 170.20 × 61.66 nmi. The translunar coast had lasted 80 hr 38 min 1.67 sec.

Apollo 12 translunar phase event	GET (h:m:s)	Altitude (nmi)	Space-fixed velocity (ft/sec)	Event duration (sec)	Velocity change (ft/sec)	Space-fixed flight path angle (deg)	Space-fixed heading angle (°E of N)
Translunar injection	002:53:13.94	199.023	35,389.8	–	–	8.584	63.902
CSM separated from S-IVB	003:18:04.9	3,819.258	24,865.5	–	–	45.092	100.194
CSM docked with LM/S-IVB	003:26:53.3	5,337.7	22,534	–	–	49.896	105.29
CSM/LM ejected from S-IVB	004:13:00.9	12,506.3	16,451.1	–	–	60.941	114.52
S-IVB APS evasive maneuver	004:29:21.4	–	–	80.0	9.5	–	–
Midcourse correction ignition	030:52:44.36	116,929.1	4,317.4	–	–	75.833	120.80
Midcourse correction cutoff	030:52:53.55	116,935.4	4,297.5	9.19	61.8	76.597	120.05

LUNAR ORBIT AND LUNAR SURFACE PHASE

During the first lunar orbit, good-quality television coverage of the surface was received for about 33 min, beginning at 084:00. The crew provided excellent descriptions of the lunar features while transmitting sharp pictures back to Earth. Two revolutions later, at 087:48:48.08, a 16.91-sec maneuver was performed to circularize the orbit at 66.10 × 54.59 nmi. On the next revolution, the LM crew transferred to the LM to perform various housekeeping chores and communication checks. At 104:20, the commander entered the LM, followed by the lunar module pilot at 105:00 to prepare for descent to the lunar surface. The two spacecraft were undocked at 107:54:02.3 at an altitude of 63.02 nmi, followed by a 14.4-sec separation maneuver at 108:24:36.8. At 109:23:39.9, a 29.0-sec descent orbit insertion maneuver placed the LM into an orbit of 61.53 × 8.70 nmi. The 717.0-sec powered descent was initiated at 7.96 nmi at 110:20:38.1. The landing was at 110:32:36.2 (06:54:36 GMT on 19 November) at selenographic coordinates 3.01239°S,

23.42157°W, just 535 ft from Surveyor 3.[6] On noting the illumination of the contact light, the commander shut down the engine 1.1 sec before landing, and at cutoff approximately 103 sec of firing time remained. On the next CSM revolution, the commander reported a visual sighting of the spacecraft passing overhead. On the next revolution, the command module pilot reported sighting Surveyor 3 in its crater and the LM to the northwest.

Three hours after landing, the LM crew members began preparations for egress. The commander began to exit the hatch at 115:10:35. While on the porch, he pulled the cable to deploy the modularized equipment stowage assembly, which automatically activated a color television camera to permit his actions to be viewed on Earth. Before reaching the surface, he reported seeing Surveyor 3 and said that the LM had landed about 25 ft from the lip of a crater. He was on the lunar surface at 115:22:22. His description indicated that the surface was quite soft and loosely packed, causing his boots to dig in somewhat as he walked. The lunar module pilot descended to the lunar surface at 115:51:50. Shortly after the television camera was removed from its bracket on the LM, it was pointed at the Sun and transmission was lost. Lithium hydroxide canisters and the contingency sample were transferred to the LM cabin as planned. The S-band erectable antenna and solar wind composition experiment were deployed, and the US flag was erected at 116:19:31. Apart from a minor difficulty removing the radioisotope thermoelectric generator fuel element from its storage cask, the removal of the Apollo Lunar Surface Experiments Package (ALSEP), transport, and deployment were nominal. The deployment site was estimated to be 600 to 700 ft from the LM. The passive seismometer detected and transmitted to Earth the crew's footsteps as they returned to the LM. On the return traverse, the crew collected a core tube sample and additional surface samples. They entered the LM and the hatch was closed at 119:06:36. The first extravehicular activity period lasted 3 hr 56 min 3 sec. The crew walked a total distance of about 3,300 ft and collected 36.82 lb of lunar samples.

At 119:47:13.23, the CSM performed a plane change maneuver of 18.23 sec which changed the orbit to 62.50 × 57.60 nmi.

The second extravehicular activity period began at 131:32:45, after a 7-hr rest period. The crew first cut the cable and stored the inoperative LM TV camera in the equipment transfer bag for return to Earth and failure analysis. The commander then went to the ALSEP site to check the leveling of the lunar atmosphere detector. As he approached the instrument, it recorded a higher atmosphere, which was attributed to the outgassing of his suit. Astronaut movement on the lunar surface was recorded by the passive seismometer and the lunar surface magnetometer. Although the commander rolled a grapefruit-sized rock down the wall of Head Crater, about 300 to 400 ft from the passive seismometer, no significant response was detected on any of the four axes. During the geological traverse toward Surveyor 3, the crew got the desired photographic panoramas, stereo photographs, core samples (two single-length tubes and one double-length tube), an 8-inch-deep trench sample, lunar environment samples, and assorted rock, dirt and 'glassy' samples. They reported seeing fine dust buildup on all sides of larger rocks, and that fine-grained surface material

[6] The COSPAR designation for Surveyor 3 was 1967-035A. The NORAD designation was 02756. It landed in Oceanus Procellarum on 20 April 1967.

seemed to become lighter the deeper they excavated. On reaching Surveyor 3, they photographed it and removed several parts, including its soil scoop and camera. They reported that the Surveyor footpad marks were still visible and that the entire spacecraft had a brown appearance. The glass parts did not break off, they just warped slightly on their mountings, and therefore were not retrieved. After the return traverse, the crew retrieved the solar wind composition experiment after 18 hr 42 min exposure. The Apollo lunar surface close-up camera was used to take stereo pictures in the vicinity of the LM during the final few minutes of surface activity. Prior to re-entering the LM, the crew members dusted one another off. The lunar module pilot entered the LM at 135:08, hauled up the lunar samples, Surveyor parts and other apparatus. The commander re-entered at 135:20. Crew ingress was complete at 135:22:00, thus ending the second human exploration of the Moon. After the expendable apparatus was jettisoned at 136:55, the cabin was repressurized. The second extravehicular activity period lasted 3 hr 49 min 15 sec. The distance traveled was about 4,300 ft and 38.80 lb of samples were collected. Mobility and portable life support system operation were excellent throughout both extravehicular periods. The total time spent outside the LM was 7 hr 45 min 18 sec, the total distance traveled was about 7,600 ft, and the collected samples totaled 75.73 lb (34.35 kg; the official metric total as determined by the Lunar Receiving Laboratory in Houston). The farthest point traveled from the LM was 1,350 ft. During the lunar surface stay, the S-158 lunar multispectral photography experiment was completed by the command module pilot. In addition, photography was obtained of two possible future landing sites at Fra Mauro and Descartes.

Ignition of the ascent stage engine for lunar liftoff occurred at 142:03:47.78. The LM had been on the lunar surface for 31 hr 31 min 11.6 sec. The 434-sec burn was 1.2 sec longer than planned and at 142:10:59.9 inserted the spacecraft into an orbit of 51.93 × 9.21 nmi. Several rendezvous maneuvers were required in order to effect docking 3.5 hr later. The 41.1-sec coelliptic sequence burn at 143:01:51.0 raised the orbit to 51.49 × 41.76 nmi. The 13.0-sec constant differential height maneuver at 144:00:02.6 lowered the orbit to 44.4 × 40.4 nmi. The 26.0-sec terminal phase initiate maneuver occurred at 144:36:26 and brought the ascent stage to an orbit of 60.2 × 43.8 nmi. Finally, the ascent stage made a 38.0-sec maneuver at 145:19:29.3 to finalize the orbit at 62.3 × 58.3 nmi for docking with the CM at 145:36:20.2 at an altitude of 58.14 nmi. The two craft had been undocked for 37 hr 42 min 17.9 sec. Good quality television was transmitted from the CSM for 24 min during the final portions of the rendezvous sequence. After the transfer of the crew and samples to the CSM, the ascent stage was jettisoned at 147:59:31.6, and the CSM was prepared for transearth injection. A 5.4-sec maneuver was made at 148:04:30.9 to separate the CSM from the ascent stage, and resulted in an orbit of 62.0 × 57.5 nmi. The ascent stage was then maneuvered by remote control to impact the lunar surface. An 82.1-sec ascent stage deorbit firing was made at 149:28:14.8 at 57.62 nmi altitude. The firing depleted the ascent stage propellants, and impact occurred at 149:55:17.7, at a point estimated at selenographic coordinates 3.42°S, 19.67°W, some 39 nmi east southeast of the Apollo 12 landing site and 5 nmi from the aim point. During the final lunar orbits, extensive landmark tracking and photography from lunar orbit were conducted. In particular, a 500-mm long focal length lens was used to obtain mapping and training data for future missions.

Prior to transearth injection, a 19.25-sec plane change maneuver at 159:04:45.47 altered the CSM orbit to 64.66 × 56.81 nmi. Following a 130.32-sec maneuver at 63.60 nmi altitude at 172:27:16.81, transearth injection was achieved at 172:29:27.13 at a velocity of 8,350.4 ft/sec after 45 lunar orbits lasting 88 hr 58 min 11.52 sec. Good-quality television of the receding Moon and the spacecraft interior was received for about 38 min, beginning about 20 min after transearth injection.

Apollo 12 lunar orbit phase event	GET (h:m:s)	Altitude (nmi)	Space-fixed velocity (ft/sec)	Event duration (sec)	Velocity change (ft/sec)	Apolune (nmi)	Perilune (nmi)
Lunar orbit insertion ignition	083:25:23.36	83.91	8,173.6	–	–	–	64.94
Lunar orbit insertion cutoff	083:31:15.61	62.91	5,470.1	352.25	2,889.5	170.20	61.66
Lunar orbit circularization ignition	087:48:48.08	62.79	5,470.6	–	–	170.37	61.42
Lunar orbit circularization cutoff	087:49:04.99	62.74	5,331.4	16.91	165.2	66.10	54.59
CSM/LM undocked	107:54:02.3	63.02	5,329.0	–	–	63.08	56.91
CSM/LM separation ignition	108:24:36.8	59.22	5,350.0	–	–	63.91	56.99
CSM/LM separation cutoff	108:24:51.2	59.15	5,350.5	14.4	2.4	64.06	56.58
LM descent orbit insertion ignition	109:23:39.9	60.52	5,343.0	–	–	63.27	57.25
LM descent orbit insertion cutoff	109:24:08.9	61.52	5,268.0	29.0	72.4	61.53	8.70
LM powered descent initiation	110:20:38.1	7.96	5,566.4	–	–	62.30	7.96
LM powered descent cutoff	110:32:35.1	–	–	717.0	–	–	–
CSM plane change ignition	119:47:13.23	62.20	5,333.5	–	–	62.50	57.61
CSM plane change cutoff	119:47:31.46	62.20	5,683.4	18.23	349.9	62.50	57.60
LM lunar liftoff ignition	142:03:47.78	–	–	–	–	–	–
LM orbit insertion	142:10:59.9	9.97	5,542.5		6,057	51.93	9.21
LM ascent stage cutoff	142:11:01.78	–	–	434	–	–	–
LM coelliptic sequence initiation ignition	143:01:51.0	51.46	5,310.3	–	–	52.51	9.94
LM coelliptic sequence initiation cutoff	143:02:32.1	51.48	5,354.9	41.1	45	51.49	41.76
LM constant differential height ignition	144:00:02.6	–	–	–	–	–	–
LM constant differential height cutoff	144:00:15.6	–	–	13.0	13.8	44.40	40.40
LM terminal phase initiation ignition	144:36:26	44.50	5,382.5	–	–	44.73	40.91
LM terminal phase initiation cutoff	144:36:52	–	–	26.0	29	60.20	43.80
LM 1st midcourse correction	144:51:29	–	–	–	–	–	–
LM 2nd midcourse correction	145:06:29	–	–	–	–	–	–
LM terminal phase finalize ignition	145:19:29.3	–	–	–	–	–	–
LM terminal phase finalize cutoff	145:20:07.3	–	–	38.0	40	62.30	58.30
CSM/LM docked	145:36:20.2	58.14	5,357.1	–	–	63.43	58.04
LM ascent stage jettison	147:59:31.6	–	–	–	–	–	–
CSM/LM separation ignition	148:04:30.9	59.94	5,347.4	–	–	64.66	59.08
CSM/LM separation cutoff	148:04:36.3	–	–	5.4	1.0	62.00	57.50
LM ascent stage deorbit ignition	149:28:14.8	57.62	5,361.8	–	–	63.52	57.94
LM ascent stage deorbit cutoff*	149:29:36.9	57.42	5,176.8	82.1	196.2	57.59	–63.15
CSM orbit plane change ignition	159:04:45.47	58.70	5,353.2	–	–	64.23	56.58
CSM orbit plane change cutoff	159:05:04.72	58.90	5,353.0	19.25	381.8	64.66	56.81

* The negative perilune indicates that the ascent stage was placed on a surface-intersecting trajectory.

TRANSEARTH PHASE

A small midcourse correction was made at 188:27:15.8. It was a 4.4-sec, 2.0-ft/sec burn, delayed one hour to allow additional crew rest. The final television transmission included the spacecraft interior and a question and answer period with scientists and members of the press. It began at 224:07 and lasted for approximately 37 min. The final midcourse correction, a 5.7-sec, 2.4-ft/sec maneuver, was made at 241:21:59.7.

Apollo 12 transearth phase event	GET (h:m:s)	Altitude (nmi)	Space-fixed velocity (ft/sec)	Event duration (sec)	Velocity change (ft/sec)	Space-fixed flight path angle (deg)	Space-fixed heading angle (°E of N)
Transearth injection ignition	172:27:16.81	63.6	5,322.9	–	–	–0.202	–115.73
Transearth injection cutoff	172:29:27.13	66.0	8,350.4	130.32	3,042.0	2.718	–116.45
Midcourse correction ignition	188:27:15.8	180,031.2	3,035.6	–	–	–78.444	91.35
Midcourse correction cutoff	188:27:20.2	180,029.0	3,036.0	4.4	2.0	–78.404	91.36
Midcourse correction ignition	241:21:59.7	25,059.0	12,082.9	–	–	–68.547	96.00
Midcourse correction cutoff	241:22:05.4	25,048.3	12,084.7	5.7	2.4	–68.547	96.01
CM/SM separation	244:07:20.1	1,949.5	29,029.1	–	–	–36.454	98.17

RECOVERY

The service module was jettisoned at 244:07:20.1, and command module entry (at the 400,000-ft altitude of the 'entry interface') occurred at 244:22:19.09 at 36,116.6 ft/sec, after a transearth coast of 71 hr, 52 min and 52.0 sec. Following separation from the CM, the SM reaction control system was fired to depletion. However, no radar acquisition nor visual sightings by the crew or recovery personnel were made, and it is believed that the SM became unstable during the depletion firing and did not execute the velocity change required to skip out of Earth's atmosphere into the planned high-apogee orbit. Instead, it probably impacted before detection. Sea-state conditions in the recovery zone were fairly rough, and the parachute system effected an extremely hard splashdown of the CM in the Pacific Ocean at 20:58:25 GMT on 24 November 1969. The force of the impact, about 15 *g*, not only knocked loose portions of the heat shield, but caused the 16-mm sequence camera to separate from its bracket and strike the LMP above his right eye. The mission duration was 244:36:25. The impact point, estimated to be 15.78°S and 165.15°W, was about 2.0 nmi from the target point and 3.91 nmi from the recovery ship USS *Hornet*. The CM assumed an apex-down attitude, but was successfully returned to the normal flotation position in 4 min 26 sec by the inflatable-bag-uprighting system. Biological isolation precautions similar to Apollo 11 were taken. The crew was retrieved by helicopter and was on the recovery ship 60 min after splashdown. The crew immediately entered the mobile quarantine facility. The CM was recovered 48 min later. The estimated CM weight at splashdown was 11,050.2 lb, and the estimated distance traveled for the mission was 828,134 nmi. The mobile quarantine facility was offloaded from the *Hornet* in Hawaii at 02:18 GMT on 29 November, followed shortly by the CM. The mobile quarantine facility was loaded onto a C-141 aircraft and flown to Ellington Air Force Base, Houston, Texas, where it arrived at 11:50 GMT. The crew entered the Lunar Receiving Laboratory 2 hr later. The CM was taken to Hickam Air Force Base, Hawaii, for deactivation. On completion of deactivation, at 14:15 GMT on 1 December, the CM was flown to Ellington Air Force Base on a C-133 aircraft, and delivered to the Lunar Receiving Laboratory at 19:30 GMT on 2 December. The crew was released from quarantine on 10 December. The CM was released soon after, and on 11 January was delivered to the North American Rockwell Space Division facility in Downey, California, for post-flight analysis.

CONCLUSIONS

The Apollo 12 mission demonstrated the capability for performing a precision lunar landing, which was a requirement for the success of future lunar surface explorations. The excellent performance of the spacecraft, the crew, and the supporting ground elements resulted in a wealth of scientific information. The following conclusions were made from an analysis of post-mission data:

1. The effectiveness of crew training, mission planning, and real-time navigation from the ground resulted in a precision landing near a previously landed Surveyor spacecraft and well within the desired landing footprint.
2. A hybrid non-free-return translunar profile was flown to demonstrate a capability for additional maneuvering which would be required for future landings at greater selenographic latitudes.
3. The timeline activities and metabolic loads associated with the extended lunar surface scientific exploration were within the capability of the crew and the portable life support systems.
4. An ALSEP was deployed for the first time and, despite some operating anomalies, returned valuable scientific data in a variety of study areas.

A pre-mission investigation of the morphology of the craters at the landing site had indicated a relatively thin 6-ft regolith; so, therefore, as expected, most of the craters had excavated bedrock. In contrast to the Apollo 11 samples, in which breccias and basalts were represented equally by number, only two of the 34 rocks returned were breccias, the rest were crystalline. The crystalline rock was coarser and more texturally diverse than that from Tranquilitatis. As before, it was low in volatiles, but this time it was also low in titanium. Upon reflection, therefore, it seemed that the Tranquilitatis basalt might be unusually *enriched* in this element. This chemical difference confirmed that the maria were not all from the same source. Indeed, the fact that four kinds of basalt were identified at the Procellarum site (which could bc characterized in terms terrestrial equivalents as olivine basalt, pyroxene basalt, ilmenite basalt and feldspathic basalt) indicated that there had been several distinct flows in this local area. However, the crystallization dates clustered within a fairly narrow window. The combination of the visually distinct patchwork of flows and the chemical variation within a short interval suggested the extrusions had resulted from partial melting of distinct pockets of rock, rather than a succession of flows from a single reservoir deep in the mantle. The initial results were confusing, but it was immediately evident that something profound had been discovered about early lunar history. The first measurement produced an age of 2.7 ($\pm$0.2) billion years, which meant that fully *a billion years* elapsed between the Tranquilitatis and Procellarum extrusions. Even the most ardent critics of the proposal that the maria formed simultaneously were taken aback by such an extended interval of volcanic activity. The next result pushed this up to 3.4 billion years, but as the data came in it converged on 3.2 billion years. This 500 million year span in ages indicated the driving process had been both endogenic and well established. The Moon was definitely not an unevolved body that had been splashed by melted rock from a massive impact soon after it had accreted; it was a planetary body possessing an differentiated crust.

Geochemist Paul W. Gast of the Lamont–Doherty Geological Observatory, Palisades, NY, made a surprising discovery in the Apollo 12 basalts in the form of an abundance of potassium, phosphorus and a variety of the 'rare earth' elements. Linking their chemical symbols, he coined the label 'KREEP'. On attempting to isolate this material, he realized that it was not present as a distinct mineral. Consequently, the term did not indicate a new type of rock; it was an *adjective*, and so it is more correct to describe the Apollo 12 basalts as being KREEPy. By way of an 'instant science' explanation, Gast suggested that the additive might have been picked up from the ancient crust that some scientists believed formed the 'basement' of the maria. He even predicted that it might prove to be associated with the putative light-toned basalt believed to be prevalent in the highlands. However, when the KREEPy basalts proved to be rich in radioactive elements, in particular thorium and uranium, it was realized that this material could not be typical of the crust, since the radioactive heating would have prevented the crust cooling sufficiently to inhibit volcanism. The mystery was not resolved until other such rocks were returned by subsequent missions.

After being discarded, the ascent stage of the LM was deliberately crashed into the Moon; it hit at a point 42 nmi east-southeast of the landing site. The ALSEP recorded the lunar crust "ringing" for almost an hour with a seismic signature that was quite different to a terrestrial signal. At the First Lunar Science Conference, in January 1970, Gary Latham, who was principal investigator of the Passive Seismic Experiment, reported the Apollo 11 results and the early data from Apollo 12. It had initially been difficult to distinguish between a moonquake and an impact, but crashing the ascent stage had served to calibrate the system, and it was found that surprisingly few of the 150 events on record were true quakes. It seemed that the lunar crust was brecciated to a depth of some 18 nmi. As further data accumulated, patterns became evident. Although the *average* rate of events was about one per day, there was a spate when the Moon was at perigee due to events being triggered by the tidal forces of the Earth's gravity disturbing the interior of the Moon. Also, there were fewer impacts than expected. The instrument was sufficiently sensitive to detect the impact of a grapefruit-sized meteoroid within a radius of 200 nmi. It reported only one such event per month, on average. It was also evident that there had not been an impact anywhere on the Moon comparable to crashing the ascent stage of the LM. In order to probe the structure of the lunar crust, it was decided that on future missions the spent S-IVB stage should be made to impact the Moon.

MISSION OBJECTIVES

Launch Vehicle Objectives

1. To launch on a 72–96°E of N flight azimuth, and insert the S-IVB/instrument unit/ spacecraft into a circular Earth parking orbit. *Achieved.*
2. To restart the S-IVB during either the second or third revolution and injection of the S-IVB/instrument unit/spacecraft into the planned translunar trajectory. *Achieved.*
3. To provide the required attitude control for the S-IVB/instrument unit/spacecraft during the transposition, docking, and ejection maneuver. *Achieved.*
4. To use the S-IVB auxiliary propulsion system burn to execute a launch vehicle evasive maneuver after ejection of the command and service module/lunar module from the S-IVB/instrument unit. *Achieved.*
5. To use residual S-IVB propellants and auxiliary propulsion system to maneuver to a trajectory that utilizes lunar gravity to insert the expended S-IVB/instrument unit into a solar orbit ('slingshot'). *Not achieved. The S-IVB/instrument unit failed to achieved solar orbit.*
6. To vent and dump all remaining gases and liquids to safe the S-IVB/instrument unit. *Achieved.*

Spacecraft Primary Objectives

1. To perform selenological inspection, survey, and sampling in a mare area. *Achieved.*
2. To deploy the ALSEP consistent with a seismic event. *Achieved.*
   (a) S-031: Passive seismic experiment.
   (b) S-034: Lunar surface magnetometer experiment.
   (c) S-035: Solar wind spectrometer experiment.
   (d) S-036: Suprathermal ion detector experiment.
   (e) S-058: Cold cathode ion gauge experiment.
3. To develop techniques for a point landing capability. *Achieved.*
4. To further develop man's capability to work in the lunar environment. *Achieved.*
5. To obtain photographs of candidate exploration sites. *Achieved.*

Detailed Objectives

1. Contingency sample collection. *Achieved.*
2. Lunar surface extravehicular operations. *Achieved.*
3. Portable life-support system recharge. *Achieved.*
4. Selected sample collection. *Achieved.*
5. Photographs of candidate exploration sites.
   (a) 70-mm stereoscopic photography of the ground track from terminator to terminator during two passes over the three sites, with concurrent 16-mm sextant

The Apollo 12 crew: Pete Conrad (left), Dick Gordon and Al Bean.

Apollo 12 lifts off in heavy rain.

Pete Conrad sets off down the ladder.

Al Bean descends the ladder.

Al Bean attaches the ALSEP instruments to the carrying bar.

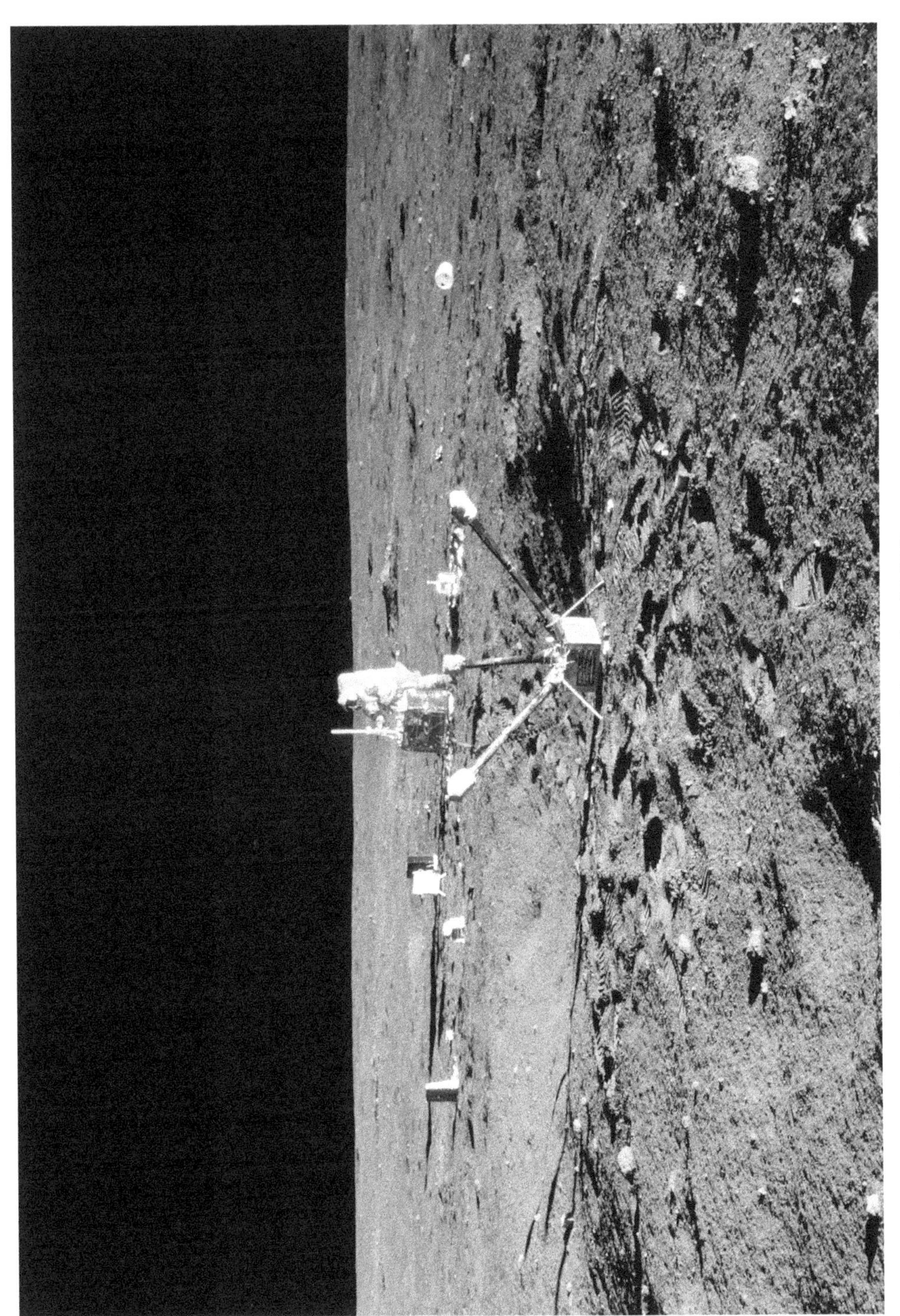

Al Bean deploys the ALSEP.

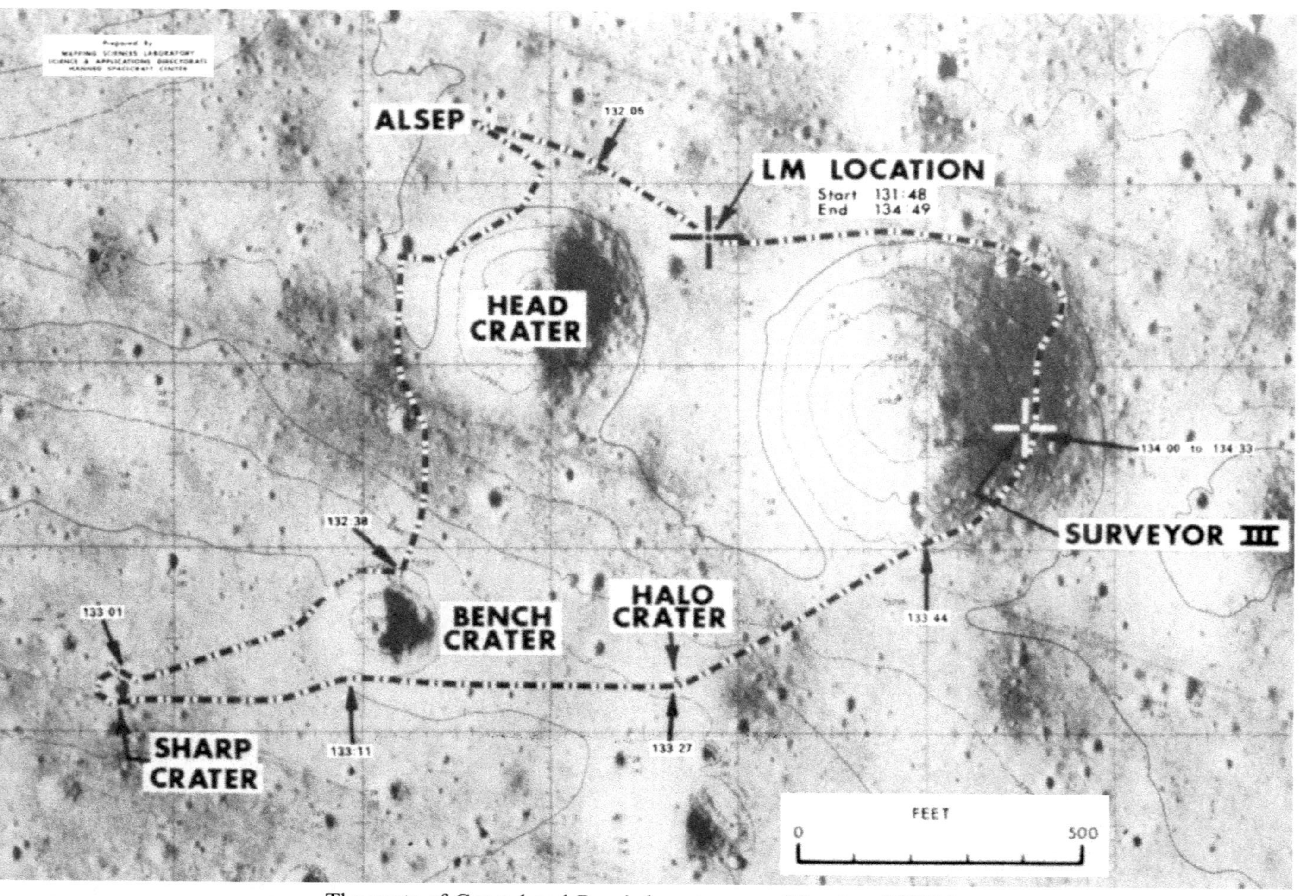

The route of Conrad and Bean's long traverse. (Courtesy USGS.)

Al Bean with the hand-tool carrier collecting samples on the traverse.

Pete Conrad alongside Surveyor 3, with the LM standing outside the crater's rim.

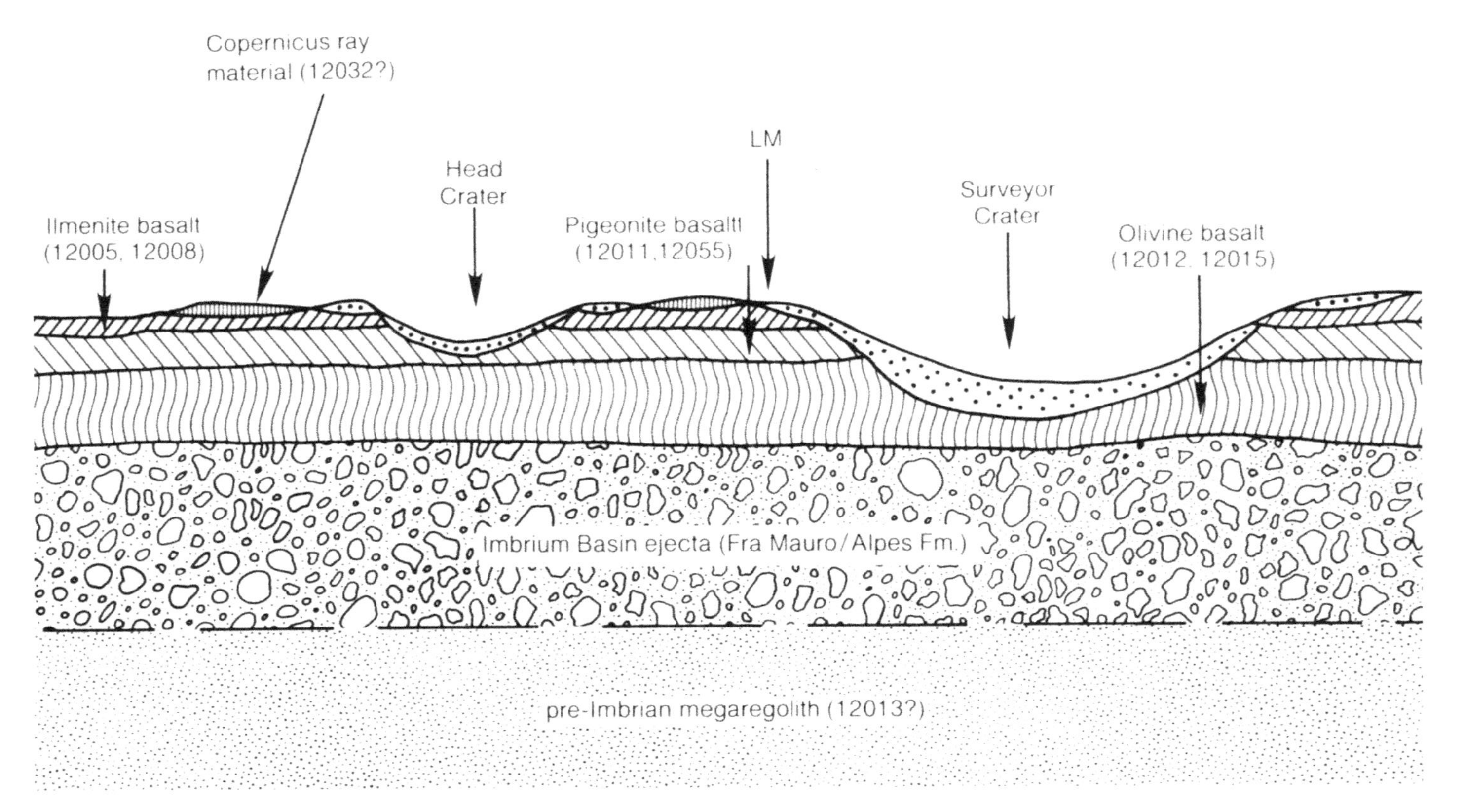

A schematic east–west cross-section through the Apollo 12 landing site showing several basaltic lava flows overlying older ejecta from major basins (modified after J.M. Rhodes *et al.*, 'Chemistry of the Apollo 12 mare basalts: magma types and fractionation processes', *Proceedings of the 8th Lunar and Planetary Science Conference*, 1977; D.E. Wilhelms, 'The Moon', in M.H, Carr (Ed.), *The Geology of the Terrestrial Planets*, SP-469, NASA, 1984). The presence of the Imbrium ejecta is inferred, as is the pre-Imbrian megaregolith beneath it. Numbers refer to specific collected samples representative of the various units inferred to be present. 'LM' shows where the Lunar Module landed. (Courtesy the Lunar and Planetary Institute and Cambridge University Press.)

sequence photography during the first pass. *Partially achieved. The first 70-mm pass and the concurrent 16-mm sextant sequence were accomplished. However, the necessity to repeat high-resolution photography did not provide sufficient time to complete the second stereoscopic pass.*

(b) Landmark tracking of a series of four landmarks bracketing the three sites included in the stereoscopic photography, and performed during two subsequent, successive orbits. *Partially achieved. First series accomplished. However, necessity to repeat high-resolution photography did not provide sufficient time to complete second series. Real-time decision assigning higher priority to landmark tracking allowed tracking of two landmarks associated with Fra Mauro and Descartes and completion of about one-fourth of the second stereoscopic pass.*

(c) High-resolution photographs using a 500-mm lens, and additional high-resolution oblique photography. *Partially achieved. The first photographs were of Herschel instead of Lalande due to crew error. A first attempt to obtain high-resolution photographs of Fra Mauro and Descartes was unsuccessful because of a camera malfunction. On a second attempt, photographs were obtained of Fra Mauro and an area slightly east of Descartes.*

6. Lunar surface characteristics. *Achieved.*
7. Lunar environment visibility. *Achieved.*
8. Landed lunar module location. *Achieved.*
9. Photographic coverage. *Achieved.*
10. Television coverage.
    (a) A crewman descending to the lunar surface. *Achieved.*
    (b) An external view of the landed lunar module. *Not achieved. The camera was damaged immediately after it was removed from its stowage compartment.*
    (c) The lunar surface in the general vicinity of the lunar module. *Not achieved. The camera was damaged immediately after it was removed from its stowage compartment.*
    (d) Panoramic coverage of distant terrain features. *Not achieved. The camera was damaged immediately after it was removed from its stowage compartment.*
    (e) A crewman during extravehicular activity. *Not achieved. The camera was damaged immediately after it was removed from its stowage compartment.*
11. Surveyor 3 investigation. *Achieved.*
12. Selenodetic reference point update. *Achieved.*

Experiments

1. S-059: Lunar field geology. *Achieved.*
2. S-080: Solar wind composition. *Achieved.*
3. S-158: Lunar multispectral photography. *Achieved.*
4. T-029: Pilot describing function. *Achieved.*
5. M-515: Lunar dust detector. *Achieved.*

MISSION TIMELINE

Apollo 12 mission event	GET (h:m:s)	Date (GMT)	Time (h:m:s)
Countdown started.	–098:00:00	09 Nov 1969	00:00:00
Scheduled 12-hr hold at T–66 hr.	–066:00:00	10 Nov 1969	07:00:00
Countdown resumed at T–66 hr.	–066:00:00	10 Nov 1969	20:00:00
Scheduled 16-hr hold at T–48 hr.	–048:00:00	11 Nov 1969	08:00:00
Countdown resumed at T–48 hr.	–048:00:00	12 Nov 1969	00:00:00
Terminal countdown started.	–028:00:00	13 Nov 1969	02:00:00
Unscheduled 6-hr hold at T–17 hr to replace CSM LH2 tank #2 due to leak.	–017:00:00	13 Nov 1969	13:00:00
Countdown resumed at T–17 hr.	–017:00:00	13 Nov 1969	19:00:00
Scheduled 9-hr 22-min hold at T–9 hr (shortened by 6 hr to avert launch delay).	–009:00:00	14 Nov 1969	03:00:00
Countdown resumed at T–9 hr.	–009:00:00	14 Nov 1969	06:22:00
Scheduled 1-hr hold at T–3 hr 30 min.	–003:30:00	14 Nov 1969	11:52:00
Countdown resumed at T–3 hr 30 min.	–003:30:00	14 Nov 1969	12:52:00
Guidance reference release.	–000:00:16.968	14 Nov 1969	16:21:43
S-IC engine start command.	–000:00:08.9	14 Nov 1969	16:21:51
S-IC engine ignition (#5).	–000:00:06.5	14 Nov 1969	16:21:53
All S-IC engines thrust OK.	–000:00:01.4	14 Nov 1969	16:21:58
Range zero.	000:00:00.00	14 Nov 1969	16:22:00
All holddown arms released (1st motion) (1.09 *g*).	000:00:00.25	14 Nov 1969	16:22:00
Liftoff (umbilical disconnected).	000:00:00.68	14 Nov 1969	16:22:00
Tower clearance yaw maneuver started.	000:00:02.4	14 Nov 1969	16:22:02
Yaw maneuver ended.	000:00:10.2	14 Nov 1969	16:22:10
Pitch and roll maneuver started.	000:00:12.8	14 Nov 1969	16:22:12
Roll maneuver ended.	000:00:32.3	14 Nov 1969	16:22:32
1st electrical disturbance (lightning).	000:00:36.5	14 Nov 1969	16:22:36
2nd electrical disturbance (lightning).	000:00:52	14 Nov 1969	16:22:52
Mach 1 achieved.	000:01:06.1	14 Nov 1969	16:23:06
Maximum bending moment achieved (37,000,000 lbf-in).	000:01:17.5	14 Nov 1969	16:23:17
Maximum dynamic pressure (682.95 lb/ft^2).	000:01:21.1	14 Nov 1969	16:23:21
S-IC center engine cutoff command.	000:02:15.24	14 Nov 1969	16:24:15
Fuel cell power restored to buses.	000:02:22	14 Nov 1969	16:24:22
Pitch maneuver ended.	000:02:38.1	14 Nov 1969	16:24:38
S-IC outboard engine cutoff.	000:02:41.74	14 Nov 1969	16:24:41
S-IC maximum total inertial acceleration (3.91 *g*).	000:02:41.82	14 Nov 1969	16:24:41
S-IC maximum Earth-fixed velocity.	000:02:42.18	14 Nov 1969	16:24:42
S-IC/S-II separation command.	000:02:42.4	14 Nov 1969	16:24:42
S-II engine start command.	000:02:43.17	14 Nov 1969	16:24:43
S-II ignition.	000:02:43.2	14 Nov 1969	16:24:43
S-II aft interstage jettisoned.	000:03:12.4	14 Nov 1969	16:25:12
Launch escape tower jettisoned.	000:03:17.9	14 Nov 1969	16:25:17
Iterative guidance mode initiated.	000:03:22.5	14 Nov 1969	16:25:22
S-IC apex.	000:04:35.6	14 Nov 1969	16:26:35

Apollo 12 mission event – *continued*	GET (h:m:s)	Date (GMT)	Time (h:m:s)
S-II center engine cutoff.	000:07:40.75	14 Nov 1969	16:29:40
S-II maximum total inertial acceleration (1.83 g).	000:07:40.83	14 Nov 1969	16:29:40
S-II outboard engine cutoff.	000:09:12.34	14 Nov 1969	16:31:12
S-II maximum Earth-fixed velocity. S-II/S-IVB separation command.	000:09:13.20	14 Nov 1969	16:31:13
S-IVB 1st burn start command.	000:09:13.30	14 Nov 1969	16:31:13
S-IC impact (theoretical).	000:09:14.5	14 Nov 1969	16:31:14
S-IVB 1st burn ignition.	000:09:16.60	14 Nov 1969	16:31:16
S-IVB ullage case jettisoned.	000:09:25.1	14 Nov 1969	16:31:25
S-II apex.	000:09:41.7	14 Nov 1969	16:31:41
S-IVB 1st burn cutoff.	000:11:33.91	14 Nov 1969	16:33:33
S-IVB 1st burn maximum total inertial acceleration (0.69 g).	000:11:33.99	14 Nov 1969	16:33:34
Earth orbit insertion; S-IVB 1st burn maximum Earth-fixed velocity.	000:11:43.91	14 Nov 1969	16:33:43
Maneuver to local horizontal attitude started.	000:11:54.2	14 Nov 1969	16:33:54
Orbital navigation started.	000:13:15.1	14 Nov 1969	16:35:15
S-II impact (theoretical).	000:20:21.6	14 Nov 1969	16:42:21
S-IVB 2nd burn restart preparation.	002:37:44.50	14 Nov 1969	18:59:44
S-IVB 2nd burn restart command.	002:47:15.10	14 Nov 1969	19:09:15
S-IVB 2nd burn ignition.	002:47:22.80	14 Nov 1969	19:09:22
S-IVB 2nd burn cutoff.	002:53:03.94	14 Nov 1969	19:15:03
S-IVB 2nd burn maximum total inertial acceleration (1.48 g).	002:53:04.02	14 Nov 1969	19:15:04
S-IVB 2nd burn maximum Earth-fixed velocity.	002:53:04.32	14 Nov 1969	19:15:04
S-IVB safing procedures started.	002:53:04.6	14 Nov 1969	19:15:04
Translunar injection.	002:53:13.94	14 Nov 1969	19:15:13
Maneuver to local horizontal attitude and orbital navigation started.	002:53:24.4	14 Nov 1969	19:15:24
1st LH_2 tank CVS open.	002:54:20	14 Nov 1969	19:16:20
Cold helium dump start.	002:54:55	14 Nov 1969	19:16:55
1st LOX tank NPV valve closed.	002:57:05	14 Nov 1969	19:19:05
2nd LH_2 tank latching valve open.	003:04:00	14 Nov 1969	19:26:00
1st LH_2 tank latching valve closed.	003:08:03.9	14 Nov 1969	19:30:03
Maneuver to transposition and docking attitude started.	003:08:05.0	14 Nov 1969	19:30:05
Cold helium dump stop.	003:08:30	14 Nov 1969	19:30:30
1st LH_2 tank CVS closed.	003:09:05	14 Nov 1969	19:31:05
CSM separated from S-IVB.	003:18:04.9	14 Nov 1969	19:40:04
TV transmission started.	003:25	14 Nov 1969	19:47
CSM docked with LM/S-IVB.	003:26:53.3	14 Nov 1969	19:48:53
Engine start-tank dump start.	003:53:04.9	14 Nov 1969	20:15:04
Ambient repressurization helium dump start.	003:53:05	14 Nov 1969	20:15:05
Cold helium dump start.	003:54:00	14 Nov 1969	20:16:00
Ambient repressurization helium dump stop.	003:54:07	14 Nov 1969	20:16:07
Engine start-tank dump stop.	003:55:34.9	14 Nov 1969	20:17:34
2nd LH_2 tank latching valve closed.	003:56:35	14 Nov 1969	20:18:35
Cold helium dump stop.	004:08:35	14 Nov 1969	20:30:35

Apollo 12 mission event – *continued*	GET (h:m:s)	Date (GMT)	Time (h:m:s)
CSM/LM ejected from S-IVB.	004:13:00.9	14 Nov 1969	20:35:00
Observation and photography of two ventings from the S-IVB burner area started.	004:19:20	14 Nov 1969	20:41:20
Maneuver to evasive attitude start.	004:20:00	14 Nov 1969	20:42:00
Maneuver to evasive attitude stop.	004:23:20	14 Nov 1969	20:45:20
1st APS evasive maneuver ignition/S-IVB stage control helium dump start.	004:26:40	14 Nov 1969	20:48:40
Observation and photography of S-IVB ventings ended.	004:26:40	14 Nov 1969	20:48:40
Cold helium dump start.	004:26:41.2	14 Nov 1969	20:48:41
TV transmission ended.	004:28	14 Nov 1969	20:50
1st APS evasive maneuver cutoff.	004:28:00	14 Nov 1969	20:50:00
S-IVB APS ullage evasive maneuver started.	004:28:01.4	14 Nov 1969	20:50:01
S-IVB APS ullage evasive maneuver ended.	004:29:21.4	14 Nov 1969	20:51:21
S-IVB slingshot maneuver – Propulsive LH_2 vent (CVS).	004:36:20.4	14 Nov 1969	20:58:20
S-IVB maneuver to lunar slingshot attitude for solar orbit initiated.	004:36:21.0	14 Nov 1969	20:58:21
S-IVB slingshot maneuver – LOX dump started.	004:48:00.2	14 Nov 1969	21:10:00
S-IVB slingshot maneuver – LOX dump ended.	004:48:58.2	14 Nov 1969	21:10:58
2nd LOX tank NPV valve open.	004:49:00	14 Nov 1969	21:11:00
3rd LH_2 tank latching valve open.	004:50:07.2	14 Nov 1969	21:12:07
Cold helium dump stop.	004:50:50	14 Nov 1969	21:12:50
Engine control helium dump start.	004:58:00	14 Nov 1969	21:20:00
Engine control helium dump stop.	005:05:30	14 Nov 1969	21:27:30
S-IVB slingshot maneuver – APS ullage ignition (planned).	005:23:20.4	14 Nov 1969	21:45:20
Stage control helium dump stop.	005:26:40	14 Nov 1969	21:48:40
S-IVB slingshot maneuver – APS ullage cutoff.	005:28:20.4	14 Nov 1969	21:50:20
Ground commanded APS ignition.	005:29:10	14 Nov 1969	21:51:00
S-IVB slingshot maneuver – APS ullage ignition (unplanned).	005:29:13.2	14 Nov 1969	21:51:13
Ground commanded APS cutoff.	005:33:40	14 Nov 1969	21:55:40
S-IVB slingshot maneuver – APS ullage cutoff.	005:33:43.2	14 Nov 1969	21:55:43
S-IVB maneuver to communications attitude initiated.	005:36:37.0	14 Nov 1969	21:58:37
LMP entered LM.	007:20	14 Nov 1969	23:42
LM inspection.	007:30	14 Nov 1969	23:52
LMP entered CM.	008:10	15 Nov 1969	00:32
LMP entered LM.	010:40	15 Nov 1969	03:02
LMP entered CM.	010:50	15 Nov 1969	03:12
TV transmission started.	030:18	15 Nov 1969	22:40
Midcourse correction ignition.	030:52:44.36	15 Nov 1969	23:14:44
Midcourse correction cutoff.	030:52:53.55	15 Nov 1969	23:14:53
TV transmission ended.	031:05	15 Nov 1969	23:27
TV transmission started.	062:52	17 Nov 1969	07:14
LMP entered LM.	063:10	17 Nov 1969	07:32
LMP entered CM.	063:45	17 Nov 1969	08:07
TV transmission ended.	063:48	17 Nov 1969	08:10
Equigravisphere.	068:30:00	17 Nov 1969	12:52:00

Apollo 12 mission event – *continued*	GET (h:m:s)	Date (GMT)	Time (h:m:s)
Rendezvous transponder activation and self-test.	079:35	17 Nov 1969	23:57
System checks for lunar orbit insertion maneuver.	082:00	18 Nov 1969	02:22
Lunar orbit insertion ignition.	083:25:23.36	18 Nov 1969	03:47:23
Lunar orbit insertion cutoff.	083:31:15.61	18 Nov 1969	03:53:15
TV transmission started.	084:00	18 Nov 1969	04:22
TV transmission ended.	084:33	18 Nov 1969	04:55
S-IVB closest approach to lunar surface.	085:48	18 Nov 1969	06:10
System checks for lunar orbit circularization maneuver.	086:30	18 Nov 1969	06:52
Lunar orbit circularization ignition.	087:48:48.08	18 Nov 1969	08:10:48
Lunar orbit circularization cutoff.	087:49:04.99	18 Nov 1969	08:11:05
LMP entered LM.	089:20	18 Nov 1969	09:42
LM activation and checkout.	089:45	18 Nov 1969	10:07
LM deactivation and LMP transferred back to CM.	090:30	18 Nov 1969	10:52
LMP entered LM.	103:45	19 Nov 1969	00:07
LM system checks.	104:04	19 Nov 1969	00:26
CDR entered LM.	104:20	19 Nov 1969	00:42
LM system checks.	104:30	19 Nov 1969	00:52
LMP entered CM.	104:40	19 Nov 1969	01:02
LMP entered LM. System checks.	105:00	19 Nov 1969	01:22
TV transmission started.	107:50	19 Nov 1969	04:12
CSM/LM undocked.	107:54:02.3	19 Nov 1969	04:16:02
CSM/LM separation maneuver ignition.	108:24:36.8	19 Nov 1969	04:46:36
CSM/LM separation maneuver cutoff.	108:24:51.2	19 Nov 1969	04:46:51
TV transmission ended.	108:30	19 Nov 1969	04:52
LM descent orbit insertion ignition (SPS).	109:23:39.9	19 Nov 1969	05:45:39
LM descent orbit insertion cutoff.	109:24:08.9	19 Nov 1969	05:46:08
LM powered descent engine ignition.	110:20:38.1	19 Nov 1969	06:42:38
LM throttle up.	110:21:05	19 Nov 1969	06:43:05
LM landing site correction initiated.	110:22:03	19 Nov 1969	06:44:03
LM landing site correction entered.	110:22:27	19 Nov 1969	06:44:27
LM landing radar altitude lock.	110:24:00	19 Nov 1969	06:46:00
LM landing radar velocity lock.	110:24:04	19 Nov 1969	06:46:04
LM "permit landing radar updates" entered.	110:24:09	19 Nov 1969	06:46:09
LM state-vector update allowed.	110:24:25	19 Nov 1969	06:46:25
LM "permit landing radar updates" exited.	110:24:31	19 Nov 1969	06:46:31
LM abort guidance system altitude updated.	110:26:08	19 Nov 1969	06:48:08
LM velocity update initiated.	110:26:24	19 Nov 1969	06:48:24
LM X-axis override inhibited.	110:26:39	19 Nov 1969	06:48:39
LM throttle recovery.	110:27:01	19 Nov 1969	06:49:01
LM abort guidance system altitude updated.	110:27:26	19 Nov 1969	06:49:26
LM approach phase entered.	110:29:11	19 Nov 1969	06:51:11
LM landing point designator enabled.	110:29:14	19 Nov 1969	06:51:14
LM landing radar antenna to position 2.	110:29:18	19 Nov 1969	06:51:18
LM abort guidance system altitude updated.	110:29:20	19 Nov 1969	06:51:20
LM redesignation right.	110:29:44	19 Nov 1969	06:51:44

Apollo 12 mission event – *continued*	GET (h:m:s)	Date (GMT)	Time (h:m:s)
LM landing radar switched to low scale.	110:29:47	19 Nov 1969	06:51:47
LM redesignation long.	110:30:02	19 Nov 1969	06:52:02
LM redesignation long.	110:30:06	19 Nov 1969	06:52:06
LM redesignation right.	110:30:12	19 Nov 1969	06:52:12
LM redesignation short.	110:30:30	18 Nov 1969	06:52:30
LM redesignation right.	110:30:42	19 Nov 1969	06:52:42
LM attitude hold mode selected.	110:30:46	19 Nov 1969	06:52:46
LM rate of descent landing phase entered.	110:30:50	19 Nov 1969	06:52:50
LM landing radar data dropout.	110:31:18	19 Nov 1969	06:53:18
LM landing radar data recovery.	110:31:24	19 Nov 1969	06:53:24
LM landing radar data dropout.	110:31:27	19 Nov 1969	06:53:27
LM landing radar data recovery.	110:31:37	19 Nov 1969	06:53:37
1st photographic evidence of surface dust disturbed by descent engine.	110:31:44	19 Nov 1969	06:53:44
LM premature low level fuel light on tank #2.	110:31:59.6	19 Nov 1969	06:53:59
LM landing radar data dropout.	110:32:00	19 Nov 1969	06:54:00
LM landing radar data recovery.	110:32:04	19 Nov 1969	06:54:04
Lunar dust completely obscured landing site.	110:32:11	19 Nov 1969	06:54:11
LM powered descent engine cutoff.	110:32:35.1	19 Nov 1969	06:54:35
LM lunar landing.	110:32:36.2	19 Nov 1969	06:54:36
1st EVA started (egress).	115:10:35	19 Nov 1969	11:32:35
CDR on lunar surface. Environmental familiarization.	115:22:22	19 Nov 1969	11:44:22
Contingency sample collected (CDR). CDR activities photographed (LMP).	115:25:41	19 Nov 1969	11:47:41
Equipment bag transferred (LMP to CDR).	115:38:53	19 Nov 1969	12:00:53
Contingency sample site photographed (CDR).	115:46:57	19 Nov 1969	12:08:57
LMP egress.	115:49:41	19 Nov 1969	12:11:41
LMP on lunar surface.	115:51:50	19 Nov 1969	12:13:50
S-band antenna deployed (CDR).	116:09:38	19 Nov 1969	12:31:38
Solar wind composition experiment deployed (LMP).	116:13:17	19 Nov 1969	12:35:17
US flag deployed (CDR).	116:19:31	19 Nov 1969	12:41:31
Panoramic photography complete (CDR).	116:25:51	19 Nov 1969	12:47:51
LM inspection complete (LMP).	116:31:46	19 Nov 1969	12:53:46
Experiment package unloaded (CDR, LMP).	116:32	19 Nov 1969	12:54
Experiment package transferred (CDR, LMP).	116:52	19 Nov 1969	13:14
Experiment package deployed (CDR) and activated (LMP).	117:01	19 Nov 1969	13:23
Return traverse started (CDR, LMP).	118:00	19 Nov 1969	14:22
Sample container packing started (CDR).	118:27	19 Nov 1969	14:49
Core tube sample gathered (LMP).	118:35	19 Nov 1969	14:57
LMP on ladder for ingress.	118:50:46	19 Nov 1969	15:12:46
LMP inside LM.	118:52:18	19 Nov 1969	15:14:18
Equipment transfer bag in LM (CDR to LMP).	118:56:19	19 Nov 1969	15:18:19
Sample return container in LM (CDR to LMP).	118:58:30	19 Nov 1969	15:20:30
CDR on LM footpad.	119:02:11	19 Nov 1969	15:24:11
CDR inside LM.	119:05:17	19 Nov 1969	15:27:17

Apollo 12 mission event – *continued*	GET (h:m:s)	Date (GMT)	Time (h:m:s)
1st EVA ended (hatch closed).	119:06:38	19 Nov 1969	15:28:38
CSM plane change ignition (SPS).	119:47:13.23	19 Nov 1969	16:09:13
CSM plane change cutoff.	119:47:31.46	19 Nov 1969	16:09:31
Debriefing for 1st EVA.	120:45	19 Nov 1969	17:07
2nd EVA started (egress).	131:32:45	20 Nov 1969	03:54:45
Safety monitoring of CDR descent to surface by LMP.	131:35	20 Nov 1969	03:57
CDR set foot on lunar surface.	131:37	20 Nov 1969	03:59
CDR transferred equipment bag.	131:39	20 Nov 1969	04:01
CDR prepared for traverse. LMP began egress.	131:44	20 Nov 1969	04:06
Contrast chart photographs taken by LMP.	131:49	20 Nov 1969	04:11
Initial geological traverse started (CDR).	132:00	20 Nov 1969	04:22
Initial geological traverse started (LMP).	132:11	20 Nov 1969	04:33
Core tube sample gathered (CDR).	133:23	20 Nov 1969	05:45
Final geological traverse started (CDR).	133:36	20 Nov 1969	05:58
Surveyor spacecraft inspected (CDR, LMP).	133:53	20 Nov 1969	06:15
Sample container packing and close-up photographs (LMP).	134:46	20 Nov 1969	07:08
Solar wind composition experiment retrieved.	134:55	20 Nov 1969	07:17
Ingress (LMP).	135:08	20 Nov 1969	07:30
Equipment transferred (CDR to LMP).	135:11	20 Nov 1969	07:33
Ingress (CDR) started.	135:20	20 Nov 1969	07:42
2nd EVA ended (ingress completed).	135:22:00	20 Nov 1969	07:44:00
LM equipment jettisoned.	136:55	20 Nov 1969	90:17
Debriefing for 2nd EVA.	138:20	20 Nov 1969	10:42
LM lunar liftoff ignition (LM APS).	142:03:47.78	20 Nov 1969	14:25:47
LM ascent stage orbit insertion.	142:10:59.9	20 Nov 1969	14:32:59
LM ascent stage cutoff.	142:11:01.78	20 Nov 1969	14:33:01
LM RCS trim burn (due to overburn on ascent) cutoff.	142:11:51.78	20 Nov 1969	14:33:51
LM coelliptic sequence initiation ignition.	143:01:51.0	20 Nov 1969	15:23:51
LM coelliptic sequence initiation cutoff.	143:02:32.1	20 Nov 1969	15:24:32
LM constant differential height maneuver ignition.	144:00:02.6	20 Nov 1969	16:22:02
LM constant differential height maneuver cutoff.	144:00:15.6	20 Nov 1969	16:22:15
LM terminal phase initiation ignition.	144:36:26	20 Nov 1969	16:58:26
LM terminal phase initiation cutoff.	144:36:52	20 Nov 1969	16:58:52
LM 1st midcourse correction.	144:51:29	20 Nov 1969	17:13:29
LM 2nd midcourse correction.	145:06:29	20 Nov 1969	17:28:29
LM terminal phase finalize ignition.	145:19:29.3	20 Nov 1969	17:41:29
LM terminal phase finalize cutoff.	145:20:07.3	20 Nov 1969	17:42:07
CSM/LM docked.	145:36:20.2	20 Nov 1969	17:58:20
CDR entered CM.	147:05	20 Nov 1969	19:27
LMP entered CM.	147:20	20 Nov 1969	19:42
LM ascent stage jettisoned.	147:59:31.6	20 Nov 1969	20:21:31
LM ascent stage separation maneuver ignition.	148:04:30.9	20 Nov 1969	20:26:30
LM ascent stage separation maneuver cutoff.	148:04:36.3	20 Nov 1969	20:26:36
LM ascent stage deorbit ignition.	149:28:14.8	20 Nov 1969	21:50:14
LM ascent stage deorbit cutoff.	149:29:36.9	20 Nov 1969	21:51:36

Apollo 12 mission event – *continued*	GET (h:m:s)	Date (GMT)	Time (h:m:s)
LM ascent stage impact on lunar surface.	149:55:17.7	20 Nov 1969	22:17:17
CSM lunar orbit plane change ignition.	159:04:45.47	21 Nov 1969	07:26:45
CSM lunar orbit plane change cutoff.	159:05:04.72	21 Nov 1969	07:27:04
CSM landmark tracking and photography.	160:15	21 Nov 1969	08:37
CSM landmark tracking and photography.	165:05	21 Nov 1969	13:27
CSM landmark tracking and photography.	166:50	21 Nov 1969	15:12
CSM landmark tracking and photography.	171:20	21 Nov 1969	19:42
Transearth injection ignition (SPS).	172:27:16.81	21 Nov 1969	20:49:16
Transearth injection cutoff.	172:29:27.13	21 Nov 1969	20:51:27
TV transmission started.	172:45	21 Nov 1969	21:07
TV transmission ended.	173:23	21 Nov 1969	21:45
Midcourse correction ignition.	188:27:15.8	22 Nov 1969	12:49:15
Midcourse correction cutoff.	188:27:20.2	22 Nov 1969	12:49:20
High-gain antenna test started.	191:15	22 Nov 1969	15:37
High-gain antenna test ended.	194:00	22 Nov 1969	18:22
High-gain antenna test started.	214:00	23 Nov 1969	14:22
High-gain antenna test ended.	216:40	23 Nov 1969	17:02
TV transmission started.	224:07	24 Nov 1969	00:29
TV transmission ended.	224:44	24 Nov 1969	01:06
Midcourse correction ignition.	241:21:59.7	24 Nov 1969	17:43:59
Midcourse correction cutoff.	241:22:05.4	24 Nov 1969	17:44:05
CM/SM separation.	244:07:20.1	24 Nov 1969	20:29:20
Entry.	244:22:19.09	24 Nov 1969	20:44:19
Radar contact with CM established by recovery ship.	244:24	24 Nov 1969	20:46
S-band contact with CM established by rescue aircraft.	244:28	24 Nov 1969	20:50
Drogue parachute deployed	244:30:39.7	24 Nov 1969	20:52:39
VHF recovery beacon contact established with CM by recovery forces.	244:31	24 Nov 1969	20:53
Main parachute deployed.	244:31:30.2	24 Nov 1969	20:53:30
VHF voice contact with CM established by aircraft and recovery ship.	244:32	24 Nov 1969	20:54
Splashdown (went to apex-down).	244:36:25	24 Nov 1969	20:58:25
CM returned to apex-up position.	244:40:51	24 Nov 1969	21:02:51
Swimmers deployed to CM.	244:46	24 Nov 1969	21:08
Flotation collar inflated.	244:53	24 Nov 1969	21:15
Hatch opened for respirator transfer.	245:14	24 Nov 1969	21:36
Hatch opened for crew egress.	245:18	24 Nov 1969	21:40
Crew on board recovery ship.	245:36	24 Nov 1969	21:58
Crew entered mobile quarantine facility.	245:44	24 Nov 1969	22:06
CM lifted from water.	246:24	24 Nov 1969	22:46
CM secured to quarantine facility.	247:53	25 Nov 1969	00:15
CM hatch opened.	248:18	25 Nov 1969	00:40
Sample containers 1 and 2 removed from CM.	249:30	25 Nov 1969	01:52
Container 1 removed from mobile quarantine facility.	250:52	25 Nov 1969	03:14

Apollo 12 mission event – *continued*	GET (h:m:s)	Date (GMT)	Time (h:m:s)
Container 1, controlled temperature shipping container 1, and film flown to Samoa.	254:18	25 Nov 1969	06:40
Container 2 removed from mobile quarantine facility.	255:49	25 Nov 1969	08:11
Container 2, remainder of biological samples and film flown to Samoa.	259:08	25 Nov 1969	11:30
Container 1, controlled temperature shipping container 1, and film arrived in Houston, TX.	268:23	25 Nov 1969	20:45
CM hatch secured and decontaminated.	270:01	25 Nov 1969	22:23
Mobile quarantine facility secured after removal of transfer tunnel.	271:08	25 Nov 1969	23:30
Container 2, remainder of biological samples, and film arrived in Houston.	276:26	26 Nov 1969	04:48
Mobile quarantine facility and CM offloaded in Hawaii.	345:56	29 Nov 1969	02:18
Safing of CM pyrotechnics completed.	352:18	29 Nov 1969	08:40
Mobile quarantine facility arrived at Ellington AFB, Houston.	355:28	29 Nov 1969	11:50
Crew in Lunar Receiving Laboratory, Houston.	357:28	29 Nov 1969	13:50
Deactivation of CM fuel and oxidizer completed at Hickam AFB, Hawaii.	405:53	01 Dec 1969	14:15
CM delivered to Lunar Receiving Laboratory.	435:08	02 Dec 1969	19:30
Crew released from quarantine.	–	12 Dec 1969	–
CM delivered to contractor's facility in Downey, CA.	–	11 Jan 1969	–
ALSEP central station turned off by ground command.	–	30 Sep 1977	–

Apollo 13

The seventh manned mission: in-flight abort

11–17 April 1970

BACKGROUND

Apollo 13 was to be a Type 'H' mission with a precision manned lunar landing and systematic lunar exploration. It was, however, aborted during translunar flight owing to the loss of all the oxygen stored in two tanks in the service module.

The primary objectives were:

- to perform selenological inspection, survey, and sampling of materials at a preselected site;
- to deploy and activate an Apollo lunar surface experiments package;
- to further develop human capability to work in the lunar environment; and
- to obtain photographs of candidate exploration sites.

The crew members were Captain James Arthur Lovell Jr (USN), commander; John Leonard 'Jack' Swigert Jr [SWY-girt], command module pilot; and Fred Wallace Haise Jr, lunar module pilot. Swigert was backup command module pilot, but Lieutenant Commander Thomas Kenneth 'Ken' Mattingly II (USN), the prime command module pilot, had been exposed to rubella (German measles) by a member of the backup crew,[7] 8 days before the scheduled launch date, and results of his pre-mission physical examination revealed he had no immunity to the disease. Consequently, on April 10, the day prior to launch, after several days of intense training with the prime crew, Swigert was named to replace Mattingly. Selected as an astronaut in 1962, Lovell was making his fourth spaceflight and second trip to the Moon, the first person ever to achieve those milestones. He had been pilot of Gemini 7, command pilot of Gemini 12, and command module pilot of Apollo 8, the first manned mission to the Moon. Lovell was born on 25 March 1928 in Cleveland, Ohio, and was 42 years old at the time of the Apollo 13 mission. He received a BS from the US Naval Academy in 1952. His backup for the mission was Commander John Watts Young (USN). The original command module pilot, Mattingly would have

[7] Major Charles Moss Duke Jr (USAF).

been making his first spaceflight. Born on 17 March 1936 in Chicago, Illinois, he was 34 years old at the time of the Apollo 13 mission. He received a BS in aeronautical engineering from Auburn University in 1958, and was selected as an astronaut in 1966. Swigert was making his first spaceflight. Born on 30 August 1931 in Denver, Colorado, he was 38 years old at the time of Apollo 13. Swigert received a BS in mechanical engineering from the University of Colorado in 1953, an MS in aerospace science from Rensselaer Polytechnic Institute in 1965, and an MBA from the University of Hartford in 1967. He was selected as an astronaut in 1966.[8] Haise was also making his first spaceflight. Born in Biloxi, Mississippi, on 14 November 1933, he was 36 years old at the time of the Apollo 13 mission. Haise received a BS in aeronautical engineering from the University of Oklahoma in 1959, and was selected as an astronaut in 1966. His backup was Major Charles Moss Duke Jr (USAF). The capsule communicators (CAPCOMs) for the mission were Commander Joseph Peter Kerwin (USN/MD/MC), Vance DeVoe Brand, Major Jack Robert Lousma (USMC), Young, and Mattingly. The support crew were Lousma, Brand, and Major William Reid Pogue (USAF). The flight directors were Milton L. Windler (first shift), Gerald D. Griffin (second shift), Eugene F. Kranz (third shift), and Glynn S. Lunney (fourth shift).

The Apollo 13 launch vehicle was a Saturn V, designated AS-508. The mission also carried the designation Eastern Test Range #3381. The CSM was designated CSM-109, and had the call-sign 'Odyssey'. The lunar module was designated LM-7, and had the call-sign 'Aquarius'.

LANDING SITE

Having accomplished a pin-point landing with Apollo 12, the flight dynamics team felt sufficiently confident to reduce the target ellipse to enable the next mission to aim for a more confined site in rougher terrain. Since an inclined orbit would be more expensive than an equatorial orbit, the propellant margins were relaxed in order to escape the confinement of the equatorial zone. Furthermore, the requirement for a backup site was deleted. From now on, therefore, a launch delay of up to 3 days would be accommodated by making the landing at a less-favorable Sun angle. This relaxation of the constraints did not 'open up' the Moon, since high-latitude sites were still out of bounds, but it offered a welcome degree of flexibility. It was the shrinking of the target ellipse and discarding the requirement that the landing site be free of terrain relief which permitted more interesting sites to be considered. Some geologists advocated landing inside a large crater such as Hipparchus or Censorinus, but the consensus was for the terrain to the north of the crater Fra Mauro. In 1962, E.M. Shoemaker and R.J. Hackman had published a stratigraphic

[8] Swigert died 27 December 1982 in Washington, DC, of complications from bone marrow cancer treatments, one week before being sworn in as a member of the US House of Representatives. On 2 November, he had been elected to Colorado's new Sixth Congressional District, receiving 64% of the vote.

study of the Imbrium basin. In extending this map across the near side, R.E. Eggleton decided to list the peripheral hummocky terrain as Imbrium ejecta, and named it the Fra Mauro Formation. Although a single unit, it was distributed in isolated patches around the periphery of the basin. In fact, if added up, it was the largest distinct stratigraphic unit on the near side. Contemporary understanding of early lunar history was based on the manner in which the Imbrium ejecta had splattered across thousands of miles, sculpting ruts and groves. Dating Imbrium was the single most important item on the agenda, since it would 'lock in' many other structures. It was not just a matter of learning about the Moon. The lunar basins indicated that the early Solar System had been a very *dynamic* environment. If the Moon had undergone such bombardment, so too must the Earth. Studying the Moon would yield insight into the early history of our own planet. The Earth's record of this period is missing, because the crust is recycled by plate tectonics. The Moon, in contrast, is so endogenically inert that its face has remained essentially unchanged for billions of years. The task was to find a crater in the hummocky terrain that had a clear line of approach from the east, offered a landing site within a mile or so, and had a *very* blocky rim. A 1,200-ft-diameter pit located 22 nmi north of Fra Mauro was selected. As this was precisely the kind of terrain that had been avoided while seeking 'safe' sites, the selectors faced the task of certifying a site for which they possessed only *four* high-resolution photographs – these having been taken by Lunar Orbiter 3 for 'scientific' interest. However, because it was on the eastern shore of Procellarum, the site was well illuminated when Apollo 12 flew, and it had been possible to obtain additional pictures. On 10 December 1969, therefore, Fra Mauro was confirmed as the landing site for Apollo 13. In view of its shape, the 'drill hole' crater to be sampled was named Cone. The best terrain for a landing was the relatively flat plain 1,000 yards west of the crater, but since this was too close to the fringe of Cone's debris field for comfort the target for the computerized descent was specified twice as far from the crater. So vital was Cone to the geological objectives that there would be little merit to a landing beyond walking distance of the crater. This site perfectly matched the single-feature 'H'-mission criteria. It could be stratigraphically related to the larger-scale character of the near side, but all that would be needed to sample it would be to make a radial approach through the ejecta blanket. Following the 'H'-mission format, the ALSEP would be deployed on first 4-hr excursion, in order that the second day could be devoted to the lengthy traverse.

LAUNCH PREPARATIONS

The terminal countdown was picked up at T–28 hr at 05:00:00 GMT on 10 April, with planned holds of 9-hr 13-min duration at T–9 hr and 1-hr duration at T–3 hr 30 min. At launch time, a cold front extended from a low-pressure cell in the North Atlantic, becoming stationary through northern Florida and along the Gulf Coast to a low-pressure area located in southern Louisiana. The frontal intensity was weak in northern Florida but became stronger in the northwestern Gulf of Mexico/Louisiana area. Surface winds in the Kennedy Space Center area were light and variable. Generally, winds in the lower part of the troposphere were light, permitting the sea breeze to switch the surface wind to the east–southeast by early afternoon. Altocumulus clouds covered 40% of the sky (base 19,000 ft) and cirrostratus 100% (base 26,000 ft), the temperature was 75.9°F, the relative humidity was 57%, and the barometric pressure was 14.676 psi. The winds, as measured by the anemometer on the light pole 60.0 ft above ground at the launch site were 12.2 kt at 105° from true north.

Apollo 13 preparation event	Date
Individual and combined CM and SM systems test completed at factory.	16 Mar 1969
Integrated CM and SM systems test completed at factory.	08 Apr 1969
LM-7 final engineering evaluation acceptance test at factory.	18 May 1969
LM-7 integrated test at factory.	18 May 1969
S-IVB-508 stage delivered to KSC.	13 Jun 1969
S-IC-8 stage delivered to KSC.	16 Jun 1969
S-IC-8 stage erected on MLP #3.	18 Jun 1969
Ascent stage of LM-7 ready to ship from factory to KSC.	24 Jun 1969
CM-109 and SM-109 ready to ship from factory to KSC.	25 Jun 1969
Descent stage of LM-7 ready to ship from factory to KSC.	25 Jun 1969
CM-109 and SM-109 delivered to KSC.	26 Jun 1969
Ascent stage of LM-7 delivered to KSC.	27 Jun 1969
Descent stage of LM-7 delivered to KSC.	28 Jun 1969
S-II-8 stage delivered to KSC.	29 Jun 1969
CM-109 and SM-109 mated.	30 Jun 1969
CSM-109 combined systems test completed.	07 Jul 1969
Saturn V instrument unit IU-508 delivered to KSC.	07 Jul 1969
LM-7 stages mated.	15 Jul 1969
S-II-8 stage erected.	17 Jul 1969
Spacecraft/LM adapter SLA-16 delivered to KSC.	18 Jul 1969
LM-7 combined systems test completed.	22 Jul 1969
S-IVB-508 stage erected.	31 Jul 1969
IU-508 erected.	01 Aug 1969
Launch vehicle electrical systems test completed.	29 Aug 1969
CSM-109 altitude tests completed.	12 Sep 1969
LM-7 altitude tests completed.	20 Sep 1969
Launch vehicle propellant dispersion/malfunction overall test completed.	21 Oct 1969
Launch vehicle service arm overall test completed.	04 Dec 1969
CSM-109 moved to VAB.	09 Dec 1969
Spacecraft erected.	10 Dec 1969
Space vehicle and MLP-3 transferred to Launch Complex 39A.	15 Dec 1969
CSM-109 integrated systems test completed.	05 Jan 1970
LM-7 combined systems test completed.	05 Jan 1970
CSM-109 electrically mated to launch vehicle.	18 Jan 1970
Space vehicle overall test #1 (plugs in) completed.	20 Jan 1970
LM-7 flight readiness test completed.	24 Feb 1970
Space vehicle flight readiness test completed.	26 Feb 1970
S-IC-8 stage RP-1 fuel loading completed.	16 Mar 1970
Space vehicle countdown demonstration test (wet) completed.	25 Mar 1970
Space vehicle countdown demonstration test (dry) completed.	26 Mar 1970

ASCENT PHASE

Apollo 13 was launched from Pad A of Launch Complex 39 at the Kennedy Space Center at a Range Zero time of 19:13:00 GMT (14:13:00 EST) on 11 April 1970. The launch window extended to 22:36:00 GMT in order to take advantage of a Sun elevation angle on the lunar surface of 10.0 deg at the scheduled time of landing. Between T + 12.6 and T + 32.1 sec, the vehicle rolled from a pad azimuth of 90°E of N to a flight azimuth of 72.043°E of N. The maximum wind conditions encountered during ascent were 108.13 kt at 252° from true north at 44,540 ft, with a maximum wind shear of 0.0166/sec at 50,610 ft. The S-IC shut down at T + 163.6, followed by S-IC/S-II separation, and S-II ignition. As a result of high amplitude oscillations in the propulsion/structural system, an effect known as 'pogo', the S-II center engine cut off at T + 330.64, 132 sec earlier than planned. This caused considerable deviations from the intended trajectory. The altitude at shutdown was 10.7 nmi lower and the velocity was 5,685.3 ft/sec slower than expected. The remaining S-II engines burned 34 sec longer than planned and shut down at T + 592.64. After separation, the S-IVB ignited at T + 596.90 and burned 9 sec longer than planned, shutting down at T + 749.83, only −1.9 ft/sec in velocity and + 0.2 nmi in altitude from the planned trajectory. The S-IC impacted the Atlantic Ocean at T + 546.9 at 30.177°N and 74.065°W, 355.3 nmi from the launch site. The S-II impacted the Atlantic Ocean at T + 1,258.1 at 32.320°N and 33.289°W, 2,452.6 nmi from the launch site. Despite the early shutdown of the S-II center engine, at insertion, at T + 759.83 (i.e. S-IVB cutoff plus 10 sec to account for engine tailoff and other transient effects), the parking orbit showed a nearly nominal apogee and perigee of 100.3 × 99.3 nmi, a period of 88.19 min, an inclination of 32.547 deg, and a velocity of 25,566.1 ft/sec. The apogee and perigee were based upon a spherical Earth with a radius of 3,443.934 nmi.

The COSPAR designation for the CSM upon achieving orbit was 1970-029A and the S-IVB was designated 1970-029B. After undocking prior to Earth entry, the LM would be designated 1970-029C.

Apollo 13 ascent phase event	GET (h:m:s)	Altitude (nmi)	Range (nmi)	Earth-fixed velocity (ft/sec)	Space-fixed velocity (ft/sec)	Event duration (sec)	Geocentric latitude (°N)	Longitude (°E)	Space-fixed flight path angle (deg)	Space-fixed heading angle (°E of N)
Liftoff	000:00:00.61	0.032	0.000	0.9	1,340.7	–	28.4470	–80.6041	0.04	90.00
Mach 1 achieved	000:01:08.4	4.394	1.310	1,095.2	2,087.5	–	28.4533	–80.5804	27.34	85.14
Maximum dynamic pressure	000:01:21.3	6.727	2.829	1,550.6	2,566.2	–	28.4608	–80.5529	28.98	82.96
S-IC center engine cutoff*	000:02:15.18	23.464	24.266	5,162.8	6,328.2	141.9	28.5677	–80.1654	23.612	76.609
S-IC outboard engine cutoff	000:02:43.6	36.392	50.991	7,787.3	9,002.5	170.3	28.6989	–79.6810	19.480	75.696
S-IC/S-II separation*	000:02:44.3	36.739	51.815	7,820.8	9,036.3	–	28.7029	–79.6660	19.383	75.693
S-II center engine #5 cutoff	000:05:30.64	86.183	298.100	11,566.6	12,859.6	164.64	29.8167	–75.1433	4.158	76.956
S-II to complete CECO*	000:07:42.6	97.450	580.109	15,583.8	16,904.3	132.00	30.8785	–69.8409	0.77	79.40
S-II outboard engine cutoff	000:09:52.64	102.112	964.578	21,288.0	22,610.8	426.64	31.9133	–62.4374	0.657	83.348
S-II/S-IVB separation*	000:09:53.50	102.150	967.505	21,301.6	22,624.5	–	31.9193	–62.3805	0.650	83.380
S-IVB 1st burn cutoff	000:12:29.83	103.469	1,533.571	24,236.4	25,560.4	152.93	32.5241	–51.2552	0.004	89.713
Earth orbit insertion	000:12:39.83	103.472	1,572.300	24,242.1	25,566.1	–	32.5249	–50.4902	0.005	90.148

* Only the commanded time is available for this event.

EARTH ORBIT PHASE

After orbital insertion, all launch vehicle and spacecraft systems were verified and preparations were made for translunar injection. Onboard television was initiated at 001:35 for about 5.5 min. The 350.85-sec translunar injection maneuver (second S-IVB firing) was performed at 002:35:46.30. The S-IVB engine shut down at 002:41:37.15 and translunar injection occurred 10 sec later, after 1.5 Earth orbits lasting 2 hr 29 min 7.3 sec, at a velocity of 35,538.4 ft/sec.

Apollo 13 earth orbit phase event	GET (h:m:s)	Space-fixed velocity (ft/sec)	Event duration (sec)	Velocity change (ft/sec)	Apogee (nmi)	Perigee (nmi)	Period (min)	Inclination (deg)
Earth orbit insertion	000:12:39.83	25,566.1	–	–	100.3	99.3	88.19	32.547
S-IVB 2nd burn ignition	002:35:46.30	25,573.2	–	–	–	–	–	–
S-IVB 2nd burn cutoff	002:41:37.15	35,562.6	350.85	10,039.0	–	–	–	31.818

TRANSLUNAR PHASE

At 003:06:38.9, the CSM was separated from the S-IVB stage and onboard television was initiated at 003:09 for about 72 min to show the docking, ejection, and interior and exterior views of the CM. Transposition and docking with the LM occurred at 003:19:08.8. The docked spacecraft were ejected from the S-IVB at 004:01:00.8, and an 80.2-sec separation maneuver was initiated by the S-IVB auxiliary propulsion system at 004:18:00.6. On previous lunar missions, the S-IVB had been maneuvered by ground command into a trajectory that would take it by the Moon and on into a solar orbit. For Apollo 13, the S-IVB was targeted to hit the Moon so that the vibrations resulting from the impact could be sensed by the Apollo 12 seismic station and telemetered to Earth for study. A 217.2-sec lunar impact maneuver was made at 005:59:59.5. The stage impacted the lunar surface at 077:56:40.0. The seismic signals lasted for 3 hr 20 min, and were so strong that the Apollo 12 seismometer gain had to be reduced to keep the recording on the scale. The suprathermal ion detector recorded a jump in the number of ions from zero at impact to 2,500 and then back to zero. It was theorized that the impact drove particles from the lunar surface up to 200,000 ft above the Moon, where they were ionized by sunlight. The impact was at selenographic coordinates 2.75°S, 27.86°W, some 35.4 nmi from the aim point and 73 nmi from the Apollo 12 seismometer. At impact, the S-IVB weighed 29,599 lb and was traveling 8,465 ft/sec.

Photographs of Earth were taken during the early part of translunar coast in support of an analysis of atmospheric winds. Good-quality television coverage of the preparations and performance of the second midcourse correction maneuver was received for 49 min, starting at 030:13. At 030:40:49.65, a 3.49-sec midcourse correction lowered the closest point of spacecraft approach to the Moon to an altitude of 60 nmi. Before this maneuver, the spacecraft had been on a free-return trajectory in which it would have looped around the Moon and returned to Earth without requiring a major maneuver.

ABORT

During the first 46 hr of the flight, telemetered data and crew observations indicated that the performance of oxygen tank 2 was normal. At 046:40:02, the crew routinely activated the fans in the oxygen tanks. Within 3 sec, the oxygen tank 2 quantity indication changed from a normal reading of about 82% full to an obviously false 'off-scale-high' reading in excess of 100%. Analysis of the electrical wiring of the quantity gauge revealed that this erroneous reading could have been caused by either a short circuit or an open circuit in the gauge wiring or a short circuit between the gauge plates. Subsequent events indicated that a short was the more likely failure mode. At 047:54:50 and at 051:07:44, the fans were operated again, with no apparent adverse effects. The quantity gauge continued to read off-scale-high. At 053:27, after a rest period, the commander and lunar module pilot were cleared to enter the LM to start in-flight inspection. A television transmission of the spacecraft interior started at 055:14 and ended at 055:46. The crew returned to the CM and at 055:50 the LM hatch was closed. At 055:52:31, a master alarm on the CM caution and warning system alerted the crew to a low-pressure indication in cryogenic hydrogen tank 1. This tank had reached the low end of its normal operating pressure range several times previously in the flight. At 055:52:58, flight controllers requested the crew to turn on the cryogenic system fans and heaters. The command module pilot acknowledged the fan cycle request at 55:53:06, and data indicated that current was applied to the oxygen tank 2 fan motors at 055:53:20, followed by a power transient in the stabilization control system.

About 90 sec later, at 055:54:53.555, telemetry from the spacecraft was lost almost totally for 1.8 sec. During the period of data loss, the caution and warning system alerted the crew to a low-voltage condition on DC main bus B. A loud bang and a shudder at about the same time indicated that a problem existed in the spacecraft. When the crew heard the bang and got the master alarm for low DC main bus B voltage, the commander was in the lower equipment bay of the command module, stowing the television camera that had just been in use. The lunar module pilot was in the tunnel between the CSM and the LM, returning to the CSM. The command module pilot was in the left-hand couch, monitoring spacecraft performance. Because of the master alarm indicating low voltage, the command module pilot moved across to the right-hand couch where CSM voltages could be observed. He reported that voltages were "looking good" at 055:56:10 and also reported hearing "a pretty good bang" a few seconds before. At this time, DC main bus B had recovered and fuel cell 3 did not fail for another 90 sec. He also reported fluctuations in the oxygen tank 2 quantity, followed by a return to the off-scale-high position.

The commander reported, "We're venting something ... into space..." at 056:09:07, followed at 056:09:58 by the lunar module pilot's report that fuel cell 1 was off-line. Less than half an hour later, he reported fuel cell 3 to be off-line also. When electrical output readings for fuel cells 1 and 3 went to zero, the ground controllers could not be certain that these cells had not somehow been disconnected from their respective buses, and were otherwise functioning normally. Attention continued to be focused on electrical problems. Five minutes after the accident, controllers asked the crew to connect fuel cell 3 to DC main bus B in order to be sure that the configuration was known. When it was realized that fuel cells 1 and 3 were not functioning, the crew was directed to perform an emergency powerdown to lower the load on the remaining fuel cell. Fuel cell 2 was shut down at 058:00, followed 10 min later by powerdown of the CM computer and platform. Noting the rapid decay in oxygen tank 1 pressure, controllers asked the crew to switch power to the oxygen tank 2 instrumentation. When this was done, and it was realized that oxygen tank 2 had failed, the extreme seriousness of the situation became clear. Several attempts were then made to save the remaining oxygen in oxygen tank 1, but the pressure continued to decrease. It was obvious by about 90 min after the accident that the oxygen tank 1 leak could not be stopped, and that shortly it would be necessary to use the LM as a 'lifeboat' for the remainder of the mission. The resultant loss of oxygen made the three fuel cells inoperative. This left the CM batteries, normally used only during re-entry, as the sole power source. The only oxygen left was contained in a surge tank and repressurization packages used to replenish the CM after cabin venting. The LM became the only source of sufficient electrical power and oxygen to permit a safe return to Earth, and led to the decision to abort the Apollo 13 mission. By 058:40, the LM had been activated, the inertial guidance reference transferred from the CSM guidance system to the LM guidance system, and the CSM systems were turned off.

The remainder of the mission was characterized by two main activities: planning and conducting the necessary propulsion maneuvers to return the spacecraft to Earth, and managing the use of consumables in such a way that the LM, having been designed for a basic mission with two crewmen for a relatively short duration, could support three crew members and serve as the actual control vehicle for the time required. A number of propulsion options were developed and considered. It was necessary to return the spacecraft to a free-return trajectory and make any required midcourse corrections. Normally, the SM service propulsion system would be used for such maneuvers. However, because of the high electrical power requirements for that engine, and in view of its uncertain condition and the uncertain nature of the structure of the SM after the accident, it was decided to use the LM descent engine. The vehicle was maneuvered back onto a free-return trajectory at 061:29:43.49 by firing the LM descent engine for 34.23 sec. It then looped behind the Moon and was out of contact with the terrestrial tracking stations between 077:08:35 and 077:33:10, a total of 24 min 35 sec.[9] Flight controllers calculated that the minimum practical return time was 133 hours total mission time to the Atlantic Ocean, and the maximum was 152 hours to the Indian Ocean. Since recovery forces were deployed in the Pacific, a return path was selected for splashdown there at 142:40.

[9] The source of these lunar occultation times is unknown, but they appear to be more accurate expressions of times in Apollo 13 Mission Operations Report, p. III-26. The 1992 *Guinness Book of World Records*, page 118, states that Apollo 13 holds the record for the farthest distance traveled from Earth: 248,655 statute miles at 01:21 British Daylight Time 15 April 1970 at 158 miles above the Moon, the equivalent of 216,075 nmi at 00:21 GMT 15 April 1970 (19:21 EST, 14 April) at an altitude of 137 nmi.

Apollo 13 translunar phase event	GET (h:m:s)	Altitude (nmi)	Space-fixed velocity (ft/sec)	Event duration (sec)	Velocity change (ft/sec)	Space-fixed flight path angle (deg)	Space-fixed heading angle (°E of N)
Translunar injection	002:41:47.15	182.445	35,538.4	–	–	7.635	59.318
CSM separated from S-IVB	003:06:38.9	3,778.582	25,029.2	–	–	45.030	72.315
CSM docked with LM/S-IVB	003:19:08.8	5,934.90	21,881.4	–	–	51.507	79.351
CSM/LM ejected from S-IVB	004:01:00.8	12,455.83	16,619.0	–	–	61.092	91.491
Midcourse correction ignition (CM SPS)	030:40:49.65	121,381.93	4,682.5	–	–	77.464	112.843
Midcourse correction cutoff	030:40:53.14	121,385.43	4,685.6	3.49	23.2	77.743	112.751
Midcourse correction ignition (LM DPS)	061:29:43.49	188,371.38	3,065.8	–	–	79.364	115.464
Midcourse correction cutoff	061:30:17.72	188,393.19	3,093.2	34.23	37.8	79.934	116.54

A 263.82-sec transearth injection maneuver using the LM descent propulsion system was made at 079:27:38.95 to speed up the return to Earth by 860.5 ft/sec after the docked spacecraft had looped around the far side of the Moon. Guidance errors during this maneuver necessitated a 14.0-sec midcourse correction of 7.8 ft/sec using the descent propulsion system at 105:18:42.0 to bring the projected entry flight path angle within the specified limits. During the transearth coast, the docked spacecraft were maneuvered into a passive thermal control mode.

The most critical consumables were water, used to cool the CSM and LM systems during use; CSM and LM battery power, the CSM batteries being for use during re-entry and the LM batteries being needed for the rest of the mission; LM oxygen for breathing; and lithium hydroxide (LiOH) filter canisters used to remove carbon dioxide from the spacecraft cabin atmosphere. These consumables, and in particular the water and LiOH canisters, appeared to be extremely marginal in quantity shortly after the accident, but once the LM was powered down to conserve electric power and to generate less heat and thus use less water, the situation improved greatly. Engineers in Houston also developed a method that allowed the crew to use materials on board to fashion a device allowing use of the CM LiOH canisters in the LM cabin atmosphere cleaning system. At splashdown, many hours of each consumable remained available.

The unprecedented powered-down state of the CM required several new procedures for entry. The CM was briefly powered up to assess the operational capability of critical systems. Also, the CM entry batteries were charged through the umbilical connectors that had supplied power from the LM while the CM was powered down. Approximately 6 hr before entry, the passive thermal control mode was discontinued, and a final midcourse correction was made using the LM reaction control system to refine the flight path angle slightly. The 21.5-sec maneuver of 3.2 ft/sec was made at 137:40:13.00. Less than half an hour later, at 138:01:48.0, the service module was jettisoned, which afforded the crew an opportunity to observe and photograph the damage caused by the failed oxygen tank. The crew viewed the SM and reported that an entire panel was missing near the S-band high-gain antenna, the fuel cells on the shelf above the oxygen tanks were tilted, the high-gain antenna was damaged, and a great deal of debris was exposed. The LM was retained until 141:30:00.2, about 70 min prior to entry, to minimize usage of CM electrical power. At undocking, normal tunnel pressure provided the necessary force to separate the two spacecraft. All other events were the same as on a normal mission.

Apollo 13 transearth phase event	GET (h:m:s)	Altitude (nmi)	Space-fixed velocity (ft/sec)	Event duration (sec)	Velocity change (ft/sec)	Space-fixed flight path angle (deg)	Space-fixed heading angle (°E of N)
Transearth injection ignition (LM DPS)	079:27:38.95	5,465.26	4,547.7	–	–	72.645	–116.308
Transearth injection cutoff	079:32:02.77	5,658.68	5,020.2	263.82	860.5	64.784	–117.886
Midcourse correction ignition (LM DPS)	105:18:28.0	152,224.32	4,457.8	–	–	–79.673	114.134
Midcourse correction cutoff	105:18:42.0	152,215.52	4,456.6	14.0	7.8	–79.765	114.242
Midcourse correction ignition (LM RCS)	137:39:51.5	37,808.58	10,109.1	–	–	–72.369	118.663
Midcourse correction cutoff	137:40:13.0	37,776.05	10,114.6	21.5	3.2	–72.373	118.660
SM separation	138:01:48.0	35,694.93	10,405.9	–	–	–71.941	118.824
LM jettisoned	141:30:00.2	11,257.48	17,465.9	–	–	–60.548	120.621

RECOVERY

The command module re-entered the Earth's atmosphere (at the 400,000-ft altitude of the 'entry interface') at 142:40:45.7 at 36,210.6 ft/sec, after a transearth coast of 63 hr 8 min 42.9 sec. Some pieces of the LM survived entry and projected trajectory data indicated that they struck the open sea between Samoa and New Zealand. The parachute system effected splashdown of the CM in the Pacific Ocean at 18:07:41 GMT on 17 April. The mission duration was 142:54:41. The impact point at 21.63°S and 165.37°W was about 1.0 nmi from the target point and 3.5 nmi from the recovery ship USS *Iwo Jima*. The CM assumed an apex-up flotation attitude. The crew was retrieved by helicopter and was on board the recovery vessel 45 min after splashdown. The CM was recovered 43 min later. The estimated CM weight at splashdown was 11,132.9 lb, and the estimated distance traveled for the mission was 541,103 nmi. The crew departed the *Iwo Jima* by aircraft at 18:20 GMT on 18 April and arrived in Houston 03:30 GMT on 20 April. The *Iwo Jima* arrived with the CM at Hawaii at 19:30 GMT on 24 April. On completion of the deactivation procedure on 26 April, the CM was delivered to the North American Rockwell Space

Division facility in Downey, California, for post-flight analysis, arriving at 14:00 GMT on 27 April.

CONCLUSIONS

The Apollo 13 accident was nearly catastrophic. Only outstanding performances by the crew and ground support personnel and the excellent performance of the LM systems made a safe return possible.

The following conclusions were made from an analysis of post-mission data:

1. The mission was aborted because of the total loss of primary oxygen in the service module. This loss resulted from an incompatibility between the design of a switch and pre-mission procedures, a condition that, when combined with an abnormal pre-mission detanking procedure, caused an in-flight shorting and a rapid oxidation within one of two redundant storage tanks. The oxidation then resulted in a loss of pressure integrity in the related tank and eventually in the remaining tank.
2. The concept of a backup crew was proven for the first time when, 3 days prior to launch, the backup command module pilot was substituted for his prime crew counterpart, who was exposed and found susceptible to rubella (German measles).
3. The performance of lunar module systems demonstrated an emergency operational capability. Lunar module systems supported the crew for a period that was twice the intended design lifetime.
4. The effectiveness of pre-mission crew training, especially in conjunction with ground personnel, was reflected in the skill and precision with which the crew responded to the emergency.
5. Although the mission was not a complete success, a lunar flyby mission, including three planned experiments (lightning phenomena, Earth photography, and S-IVB lunar impact), were completed and data were derived with respect to the capabilities of the lunar module.

INVESTIGATION

On 17 April 1970, NASA Administrator Thomas O. Paine established the Apollo 13 Review Board, naming Edgar M. Cortright, director of the NASA Langley Research Center, as chairman. Cortright's 8-member panel met for nearly 2 months, and submitted their final report on 15 June. Neil Armstrong, commander of the recent Apollo 11 mission, was the only astronaut on the Board. William Anders, lunar module pilot of Apollo 8, and executive secretary of the National Aeronautics and Space Council, was one of three observers.

The evidence pointed strongly to an electrical short circuit with arcing as the initiating event. About 2.7 sec after the fans were turned on in the SM oxygen tanks, an 11.1-ampere current spike and a simultaneous voltage drop were recorded in the spacecraft electrical system. Immediately thereafter, current drawn from the fuel cells decreased by an amount consistent with the loss of power to one fan. No other changes in spacecraft power were being made at the time. The power to the heaters in the tanks was not on (the quantity gauge and temperature sensor were very low power devices). The next anomalous event recorded was the beginning, 13 sec later, of a pressure rise in oxygen tank 2. Such a time lag was possible with the low-level of the combustion at that time. All this pointed to the likelihood that an electrical short circuit with arcing occurred in the fan motor or its wires to initiate the accident sequence. The energy available from the short circuit was probably 10 to 20 joules. Tests conducted for the investigation showed this energy to be more than adequate to ignite Teflon of the type contained within the tank. The prospect of electrical initiation was enhanced by the high probability that the electrical wires within the tank were damaged during abnormal tanking operations at KSC prior to launch.

Data were not adequate to determine precisely the way in which the oxygen tank 2 system lost its integrity. However, available information, analyses, and tests performed during the investigation, indicated that most probably combustion within the pressure vessel ultimately led to localized heating and failure at the pressure vessel closure. It is at this point, the upper end of the quantity probe, that the Inconel conduit was located, through which the Teflon-insulated wires entered the pressure vessel. It is likely that the combustion progressed along the wire insulation and reached this location where all of the wires came together. This, possibly augmented by ignition of the metal in the upper end of the probe, led to weakening and failure of the closure or the conduit, or both. Failure at this point would lead immediately to pressurization of the tank dome, which was fitted with a rupture disc rated at about 75 psi. Rupture of this disc or of the entire dome would then release the oxygen, accompanied by combustion products, into bay 4. Vehicle accelerations recorded at this time were probably caused by this release. Release of the oxygen then began to pressurize the oxygen shelf space of bay 4. If the holes formed in the pressure vessel were large enough and formed rapidly enough, the escaping oxygen alone would be adequate to blow off the bay 4 panel. However, it is also quite possible that the escape of oxygen was accompanied by combustion of Mylar and Kapton (used extensively as thermal insulation in the oxygen shelf compartment and in the tank dome), which would augment the pressure caused by the oxygen itself. The slight temperature increases recorded at various SM locations indicated that combustion external to the tank probably took place. The ejected panel then struck the high-gain antenna, disrupting communications from the spacecraft for the 1.8 sec.

How the Problem Occurred

Following is a list of factors that led to the accident:

- After assembly and acceptance testing, oxygen tank 2, assigned to Apollo 13, was shipped from Beech Aircraft Corporation to North American Rockwell (NR) in apparently satisfactory condition.
- It is now known, however, that the tank contained two protective thermostatic switches on the heater assembly, which were inadequate and would subsequently fail during ground test operations at KSC.
- In addition, it is probable that the tank contained a loosely fitting fill tube assembly. This assembly was probably displaced during subsequent handling, which included an incident at the prime contractor's arc plant in which the tank was jarred.

- In itself, the displaced fill tube assembly was not particularly serious, but it led to the use of improvised detanking procedures at KSC which almost certainly set the stage for the accident.
- Although Beech did not encounter any problem in detanking during acceptance tests, it was not possible to detank oxygen tank 2 using normal procedures at KSC. Tests and analyses indicated that this was due to gas leakage through the displaced fill tube assembly.
- The special detanking procedures at KSC subjected the tank to an extended period of heater operation and pressure cycling. These procedures had not been used before, and the tank had not been qualified by test for the conditions experienced. However, the procedures did not violate the specifications that governed the operation of the heaters at KSC.
- In reviewing these procedures before the flight, officials of NASA, NR, and Beech did not recognize the possibility of damage due to overheating. Many of these officials were not aware of the extended heater operation. In any event, adequate thermostatic switches might have been expected to protect the tank.
- A number of factors contributed to the presence of inadequate thermostatic switches in the heater assembly. The original 1962 specifications from NR to Beech for the tank and heater assembly specified the use of 28-V DC power, which was used in the spacecraft. In 1965, NR issued a revised specification which stated that the heaters should use a 65-volt DC power supply for tank pressurization; this was the power supply used at KSC to reduce pressurization time. Beech ordered switches for the Block II tanks but did not change the switch specifications to be compatible with 65-volt DC.
- The thermostatic switch discrepancy was not detected by NASA, NR, or Beech in their review of documentation, nor did tests identify the incompatibility of the switches with the ground support equipment at KSC, since neither qualification nor acceptance testing required switch cycling under load as should have been done. It was a serious oversight in which all parties shared.
- The thermostatic switches could accommodate 65 volts DC during tank pressurization because they normally remained cool and closed, but they could not open without damage with 65-volt DC power applied. They were never required to do so until the special detanking operation. During this procedure, as the switches started to open when they reached their upper temperature limit, they were welded permanently closed by the resulting arc and were rendered inoperative as protective thermostats.
- Failure of the thermostatic switches to open could have been detected at KSC if switch operation had been checked by observing heater current readings on the oxygen tank heater control panel. Although it was not recognized at that time, the tank temperature readings indicated that the heaters had reached their temperature limit and switch opening should have been expected.
- As shown by subsequent tests, failure of the thermostatic switches probably permitted the temperature of the heater tube assembly to reach about 1,000°F in spots during the continuous 8-hr period of heater operation. Such heating was shown in tests to severely damage the Teflon insulation on the fan motor wires in the

vicinity of the heater assembly. From that time on, including pad occupancy, oxygen tank 2 was in a hazardous condition when filled with oxygen and electrically powered.

- It was not until nearly 56 hours into the mission, however, that the fan motor wiring, possibly moved by the fan stirring, short circuited and ignited its insulation by means of an electric arc. The resulting combustion in the oxygen tank probably overheated and caused a failure in the wiring conduit where it entered the tank and possibly in a portion of the tank itself.
- The rapid expulsion of high-pressure oxygen that followed, possibly augmented by combustion of insulation in the space surrounding the tank, blew off the outer panel into bay 4 of the SM, caused a leak in the high-pressure system of oxygen tank 1, damaged the high-gain antenna, caused other miscellaneous damage, and aborted the mission.

MISSION OBJECTIVES

Launch Vehicle Objectives

1. To launch on a 72–96°E of N flight azimuth, and insert the S-IVB/instrument unit/spacecraft into the planned circular Earth parking orbit. *Achieved, despite premature shutdown of the S-II inboard engine.*
2. To restart the S-IVB during either the second or third revolution and inject the S-IVB/instrument unit/spacecraft into the planned translunar trajectory. *Achieved.*
3. To provide the required attitude control for the S-IVB/instrument unit/spacecraft during transposition, docking, and ejection. *Achieved.*
4. To perform an evasive maneuver after ejection of the command and service module/lunar module from the S-IVB/instrument unit. *Achieved.*
5. To attempt to impact the S-IVB/instrument unit on the lunar surface within 189 nmi (350 km) of selenographic coordinates 3°S, 30°W. *Achieved.*
6. To determine actual impact point within 2.7 nmi (5.0 km) and time of impact within 1 sec. *Achieved.*
7. To vent and dump the remaining gases and propellants to safe the S-IVB/instrument unit. *Achieved.*

Spacecraft Primary Objectives

1. To perform selenological inspection, survey, and sampling of materials in a preselected region of the Fra Mauro Formation. *Not attempted.*
2. To deploy and activate an Apollo lunar surface experiments package. *Not attempted.*
3. To further develop human capability to work in the lunar environment. *Not attempted.*
4. To obtain photographs of candidate exploration sites. *Not attempted.*

Detailed Objectives

1. Television coverage. *Not attempted following accident.*

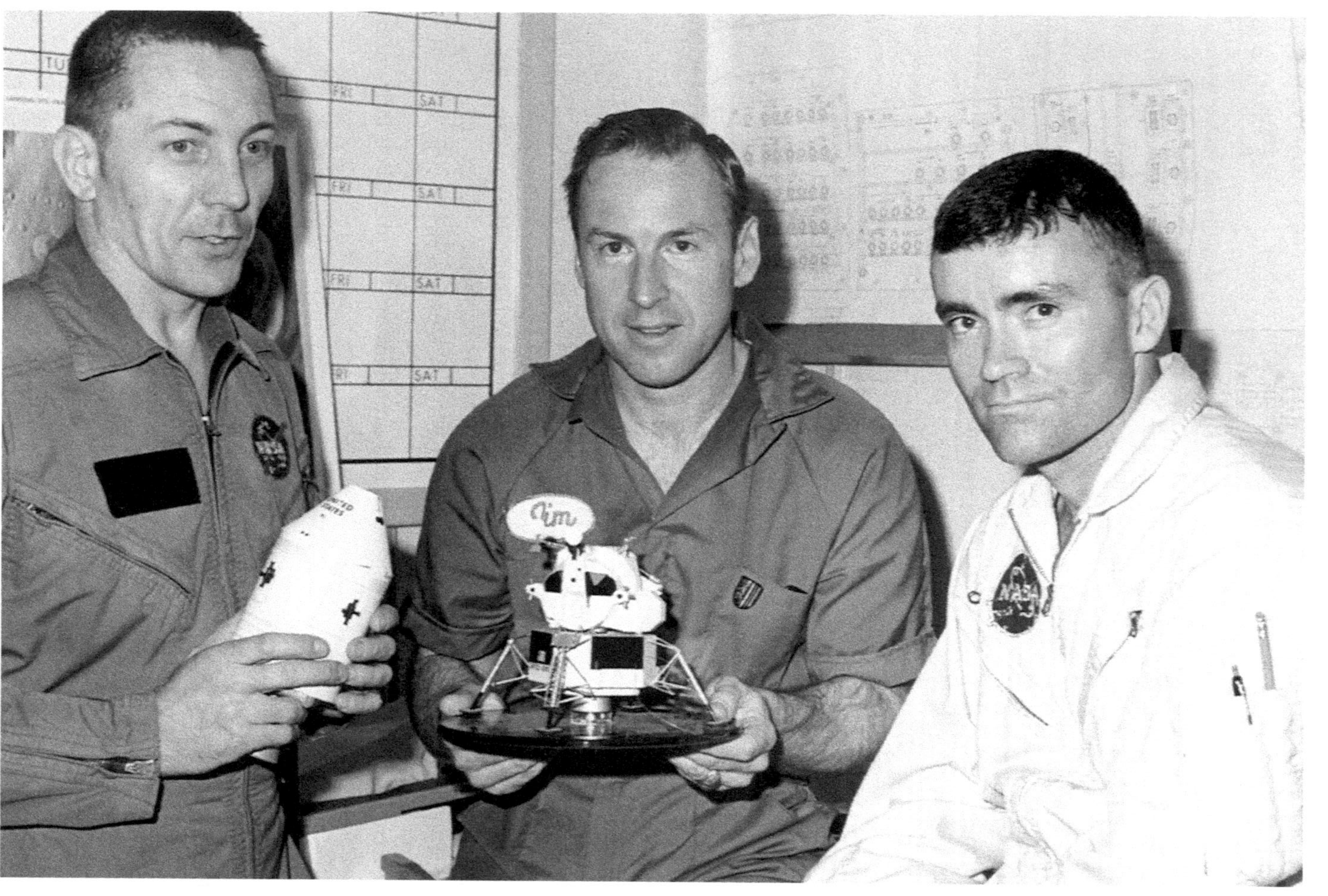

The Apollo 13 crew, Jack Swigert (left), Jim Lovell and Fred Haise, on the day before launch.

AS-508 roll out.

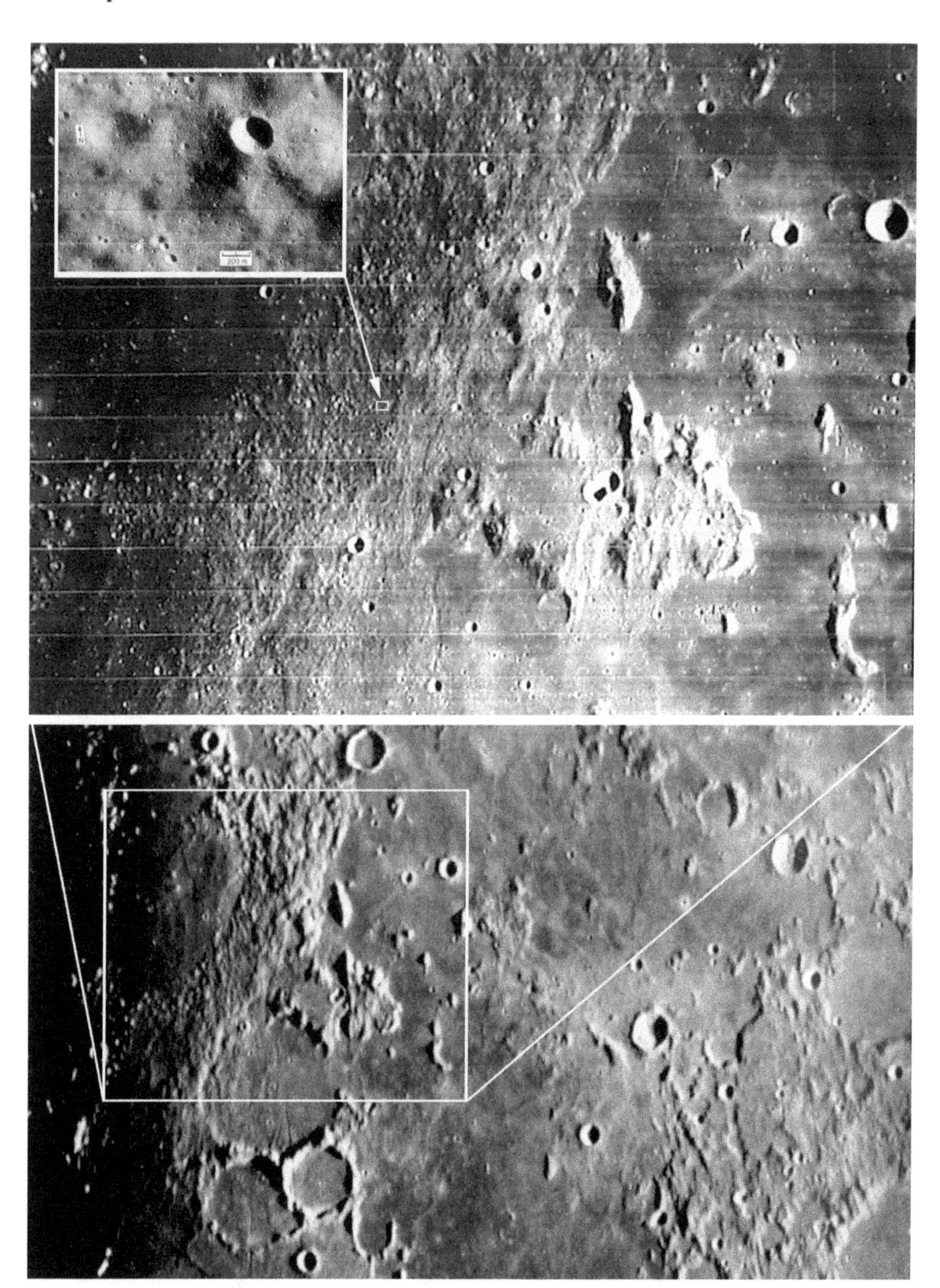

The objective was west of a small crater in the hummocky Fra Mauro Formation.

Eugene F. Kranz (foreground) watches Apollo 13's final TV transmission.

As the crisis develops, astronauts cluster around the CapCom's console.

When the SM was jettisoned, the full extent of the damage became evident.

The Apollo 13 crew disembark from the recovery helicopter.

Mission Control celebrates Apollo 13's return.

Odyssey command module is recovered.

2. Contingency sample collection. *Not attempted.*
3. Selected sample collection. *Not attempted.*
4. Landing accuracy improvement techniques. *Not attempted.*
5. Photographs of candidate exploration sites. *Not attempted.*
6. Extravehicular communication system performance. *Not attempted.*
7. S-200: Lunar soil mechanics. *Not attempted.*
8. Dim light photography. *Not attempted.*
9. Selenodetic reference point update. *Not attempted.*
10. CSM orbital science photography. *Not attempted as planned following accident.*
11. Transearth lunar photography. *Not attempted as planned following accident.*
12. EMU water consumption measurement. *Not attempted.*
13. Thermal coating degradation. *Not attempted*

Experiments

1. ALSEP III: Apollo Lunar Surface Experiments Package. *Not Attempted.*
   (a) S-031: Passive seismic experiment.
   (b) S-037:Heat flow experiment.
   (c) S-038: Charged particle lunar environment experiment.
   (d) S-058: Cold cathode ion gauge experiment.
   (e) M-515: Lunar dust detection.
2. S-059: Lunar field geology. *Not attempted.*
3. S-080: Solar wind composition. *Not attempted.*
4. S-164: S-band transponder exercise. *Not attempted.*
5. S-170: Downlink bistatic radar observations of the Moon. *Not attempted.*
6. S-178: Gegenschein from lunar orbit. *Not attempted.*
7. S-184: Lunar surface close-up photography. *Not attempted.*
8. T-029: Pilot describing function. *Achieved.*

MISSION TIMELINE

Apollo 13 mission event	GET (h:m:s)	Date (GMT)	Time (h:m:s)
Terminal countdown started at T–28 hr.	–028:00:00	10 Apr 1970	05:00:00
Scheduled 9-hr 13-min hold at T–9 hr.	–009:00:00	11 Apr 1970	00:00:00
Countdown resumed at T–9 hr.	–009:00:00	11 Apr 1970	09:13:00
Scheduled 1-hr hold at T–3 hr 30 min.	–003:30:00	11 Apr 1970	14:43:00
Countdown resumed at T–3 hr 30 min.	–003:30:00	11 Apr 1970	15:43:00
Guidance reference release.	–000:00:16.961	11 Apr 1970	19:12:43
S-IC engine start command.	–000:00:08.9	11 Apr 1970	19:12:51
S-IC engine ignition (#5).	–000:00:06.7	11 Apr 1970	19:12:53
All S-IC engines thrust OK.	–000:00:01.4	11 Apr 1970	19:12:58
Range zero.	000:00:00.00	11 Apr 1970	19:13:00
All holddown arms released (1st motion) (1.06 *g*).	000:00:00.3	11 Apr 1970	19:13:00

Apollo 13 mission event – *continued*	GET (h:m:s)	Date (GMT)	Time (h:m:s)
Liftoff (umbilical disconnected).	000:00:00.61	11 Apr 1970	19:13:00
Tower clearance yaw maneuver started.	000:00:02.3	11 Apr 1970	19:13:02
Yaw maneuver ended.	000:00:10.0	11 Apr 1970	19:13:10
Pitch and roll maneuver started.	000:00:12.6	11 Apr 1970	19:13:12
Roll maneuver ended.	000:00:32.1	11 Apr 1970	19:13:32
Mach 1 achieved.	000:01:08.4	11 Apr 1970	19:14:08
Maximum bending moment achieved (69,000,000 lbf-in).	000:01:16	11 Apr 1970	19:14:16
Maximum dynamic pressure (651.63 lb/ft^2).	000:01:21.3	11 Apr 1970	19:14:21
S-IC center engine cutoff command.	000:02:15.18	11 Apr 1970	19:15:15
Pitch maneuver ended.	000:02:43.3	11 Apr 1970	19:15:43
S-IC outboard engine cutoff.	000:02:43.60	11 Apr 1970	19:15:43
S-IC maximum total inertial acceleration (3.83 *g*).	000:02:43.70	11 Apr 1970	19:15:43
S-IC maximum Earth-fixed velocity.	000:02:44.10	11 Apr 1970	19:15:44
S-IC/S-II separation command.	000:02:44.3	11 Apr 1970	19:15:44
S-II engine start command.	000:02:45.0	11 Apr 1970	19:15:45
S-II ignition.	000:02:46.0	11 Apr 1970	19:15:46
S-II aft interstage jettisoned.	000:03:14.3	11 Apr 1970	19:16:14
Launch escape tower jettisoned.	000:03:21.0	11 Apr 1970	19:16:21
Iterative guidance mode initiated.	000:03:24.5	11 Apr 1970	19:16:24
S-IC apex.	000:04:31.7	11 Apr 1970	19:17:31
S-II center (or 'inboard') engine cutoff (S-II engine #5 cutoff 132.36 sec early).	000:05:30.64	11 Apr 1970	19:18:30
CMP (Swigert): "Inboard."	000:05:32	11 Apr 1970	19:20
CAPCOM (Joe Kerwin): "We confirm inboard out."	000:05:36	11 Apr 1970	19:24
CDR (Lovell): "That shouldn't have happened."	000:05:38	11 Apr 1970	19:26
CMP: "No, that's 7:42. That's two minutes early." Swigert meant that engine cutoff should have occurred at 7 min 42 sec into the flight, not 5 min 30 sec.	000:05:40	11 Apr 1970	19:28
S-II command to complete CECO.	000:07:42.6	11 Apr 1970	19:20:42
S-II maximum total inertial acceleration (1.66 *g*).	000:08:57.00	11 Apr 1970	19:21:57
S-IC impact (theoretical).	000:09:06.9	11 Apr 1970	19:22:06
S-II outboard engine cutoff (34.53 sec later than planned).	000:09:52.64	11 Apr 1970	19:22:52
S-II maximum Earth-fixed velocity; S-II/S-IVB separation command.	000:09:53.50	11 Apr 1970	19:22:53
S-IVB 1st burn start command.	000:09:53.60	11 Apr 1970	19:22:53
S-IVB 1st burn ignition.	000:09:56.90	11 Apr 1970	19:22:56
S-IVB ullage case jettisoned.	000:10:05.4	11 Apr 1970	19:23:05
S-II apex.	000:10:32.2	11 Apr 1970	19:23:32
S-IVB 1st burn cutoff (9 sec later than planned).	000:12:29.83	11 Apr 1970	19:25:29
S-IVB 1st burn maximum total inertial acceleration (0.58 *g*).	000:12:30.00	11 Apr 1970	19:25:30
S-IVB 1st burn maximum Earth-fixed velocity.	000:12:30.50	11 Apr 1970	19:25:30
Earth orbit insertion.	000:12:39.83	11 Apr 1970	19:25:39
Maneuver to local horizontal attitude started.	000:12:50.1	11 Apr 1970	19:25:50
Orbital navigation started.	000:14:10.4	11 Apr 1970	19:27:10
S-II impact (theoretical).	000:20:58.1	11 Apr 1970	19:33:58

Apollo 13 mission event – *continued*	GET (h:m:s)	Date (GMT)	Time (h:m:s)
TV transmission started.	001:37	11 Apr 1970	20:50
TV transmission ended.	001:43	11 Apr 1970	20:56
S-IVB 2nd burn restart preparation.	002:26:08.10	11 Apr 1970	21:39:08
S-IVB 2nd burn restart command.	002:35:38.10	11 Apr 1970	21:48:38
S-IVB 2nd burn ignition.	002:35:46.30	11 Apr 1970	21:48:46
S-IVB 2nd burn cutoff.	002:41:37.15	11 Apr 1970	21:54:37
S-IVB 2nd burn maximum total inertial acceleration (1.44 g).	002:41:37.23	11 Apr 1970	21:54:37
S-IVB 2nd burn maximum Earth-fixed velocity.	002:41:37.80	11 Apr 1970	21:54:37
S-IVB safing procedures started.	002:41:37.9	11 Apr 1970	21:54:37
Translunar injection.	002:41:47.15	11 Apr 1970	21:54:47
Maneuver to local horizontal attitude and orbital navigation started.	002:44:08	11 Apr 1970	21:57:08
Maneuver to transposition and docking attitude started.	002:56:38.3	11 Apr 1970	22:09:38
CSM separated from S-IVB.	003:06:38.9	11 Apr 1970	22:19:38
TV transmission started.	003:09	11 Apr 1970	22:22
CSM docked with LM/S-IVB.	003:19:08.8	11 Apr 1970	22:32:08
CSM/LM ejected from S-IVB.	004:01:00.8	11 Apr 1970	23:14:00
S-IVB maneuver to evasive APS burn attitude.	004:09:00	11 Apr 1970	23:22:00
S-IVB APS evasive maneuver ignition.	004:18:00.6	11 Apr 1970	23:31:00
S-IVB APS evasive maneuver cutoff.	004:19:20.8	11 Apr 1970	23:32:20
TV transmission ended.	004:20	11 Apr 1970	23:33
S-IVB maneuver to LOX dump attitude initiated.	004:27:40.0	11 Apr 1970	23:40:40
S-IVB lunar impact maneuver – CVS vent opened.	004:34:39.4	11 Apr 1970	23:47:39
S-IVB lunar impact maneuver – LOX dump started.	004:39:19.4	11 Apr 1970	23:52:19
S-IVB lunar impact maneuver – CVS vent opened.	004:39:39.4	11 Apr 1970	23:52:39
S-IVB lunar impact maneuver – LOX dump ended.	004:40:07.4	11 Apr 1970	23:53:07
Maneuver to attitude for final S-IVB APS burn initiated.	005:48:07.8	12 Apr 1970	01:01:07
S-IVB lunar impact maneuver – APS ignition.	005:59:59.5	12 Apr 1970	01:12:59
S-IVB lunar impact maneuver – APS cutoff.	006:03:36.7	12 Apr 1970	01:16:36
Earth weather photography started.	007:17:14	12 Apr 1970	02:30:14
Unsuccessful passive thermal control attempt.	007:43:02	12 Apr 1970	02:56:02
Earth weather photography ended.	011:17:19	12 Apr 1970	06:30:19
2nd S-IVB transposition maneuver (unplanned) initiated by launch vehicle digital computer.	013:42:33	12 Apr 1970	08:55:33
Unplanned S-IVB velocity increase of 5 ft/sec which altered lunar impact trajectory closer to target point.	019:29:10	12 Apr 1970	14:42:10
TV transmission started.	030:13	13 Apr 1970	01:26
Midcourse correction ignition (SPS) – transfer to hybrid non-free return trajectory.	030:40:49.65	13 Apr 1970	01:53:49
Midcourse correction cutoff.	030:40:53.14	13 Apr 1970	01:53:53
TV transmission ended.	031:02	13 Apr 1970	02:15
Photography of Comet Bennett.	031:50	13 Apr 1970	03:03
Unsuccessful passive thermal control attempt.	032:21:49	13 Apr 1970	03:34:49
Crew turned on fans in oxygen tank #2 (routine procedure).	046:40:02	13 Apr 1970	17:15

Apollo 13 mission event – *continued*	GET (h:m:s)	Date (GMT)	Time (h:m:s)
Cryogenic oxygen tank #2 quantity gauge indicated 'off-scale-high', of over 100% (probably due to short circuit). First indication of a problem.	046:40:05	13 Apr 1970	17:18:05
Cryogenic oxygen tank #2 quantity probe short circuited.	046:40:08	13 Apr 1970	17:21:08
Oxygen tank #2 fans turned on again with no apparent adverse affects. Quantity gauge continued to read 'off-scale-high'.	047:54:50	13 Apr 1970	19:03:50
Oxygen tank #2 fans turned on again with no apparent adverse affects. Quantity gauge continued to read 'off-scale-high'.	051:07:44	13 Apr 1970	22:57:44
CDR and LMP cleared to enter the LM to commence in-flight inspection.	053:27	14 Apr 1970	00:40
LMP entered LM.	054:20	14 Apr 1970	01:33
CDR entered LM.	054:25	14 Apr 1970	01:38
LM system checks.	054:40	14 Apr 1970	01:53
TV transmission started.	055:14	14 Apr 1970	02:27
CDR and LMP entered LM.	055:30	14 Apr 1970	02:43
TV transmission ended.	055:46:30	14 Apr 1970	02:59:30
Tunnel hatch closed.	055:50	14 Apr 1970	03:03
Master caution and warning triggered by low hydrogen pressure in tank #1. Alarm turned off after 4 sec.	055:52:31	14 Apr 1970	03:05:31
CAPCOM (Jack Lousma): "13, we've got one more item for you, when you get a chance. We'd like you to stir up your cryo tanks. In addition, I have shaft and trunnion..."	055:52:58	14 Apr 1970	03:05:58
CMP (Swigert): "Okay."	055:53:06	14 Apr 1970	03:06:06
CAPCOM: "...for looking at the Comet Bennett, if you need it."	055:53:07	14 Apr 1970	03:06:07
CMP: "Okay. Stand by."	055:53:12	14 Apr 1970	03:06:12
Oxygen tank #1 fans on.	055:53:18	14 Apr 1970	03:06:18
Oxygen tank #1 pressure decreased 8 psi due to normal destratification. Space-craft current increased by 1 ampere.	055:53:19	14 Apr 1970	03:06:19
Oxygen tank #2 fans on. Stabilization control system electrical disturbance indicated a power transient.	055:53:20	14 Apr 1970	03:06:20
Oxygen tank #2 pressure decreased 4 psi.	055:53:21	14 Apr 1970	03:06:21
Electrical short in tank #2 (stabilization control system electrical disturbance indicated a power transient).	055:53:22.718	14 Apr 1970	03:06:22
1.2-volt decrease in AC bus #2 voltage.	055:53:22.757	14 Apr 1970	03:06:22
11.1 ampere 'spike' recorded in fuel cell #3 current followed by drop in current and rise in voltage typical of removal of power from one fan motor, indicating opening of motor circuit.	055:53:22.772	14 Apr 1970	03:06:22
Oxygen tank #2 pressure started to rise for 24 sec.	055:53:36	14 Apr 1970	03:06:36
11-volt decrease in AC bus #2 voltage for one sample.	055:53:38.057	14 Apr 1970	03:06:38
Stabilization control system electrical disturbance indicated a power transient.	055:53:38.085	14 Apr 1970	03:06:38
22.9-ampere 'spike' recorded in fuel cell #3 current, followed by drop in current and rising voltage typical of one fan motor, indicating opening of another motor circuit.	055:53:41.172	14 Apr 1970	03:06:41

Apollo 13 mission event – *continued*	GET (h:m:s)	Date (GMT)	Time (h:m:s)
Stabilization control system electrical disturbance indicated a power transient.	055:53:41.192	14 Apr 1970	03:06:41
Oxygen tank #2 pressure rise ended at a pressure of 953.8 psia.	055:54:00	14 Apr 1970	03:07:00
Oxygen tank #2 pressure started to rise again.	055:54:15	14 Apr 1970	03:07:15
Oxygen tank #2 quantity dropped from full scale (to which it had failed at 046:40) for 2 sec and then read 75.3% full. This indicated the gauge short circuit may have corrected itself.	055:54:30	14 Apr 1970	03:07:30
Oxygen tank #2 temperature started to rise rapidly.	055:54:31	14 Apr 1970	03:07:31
Flow rate of oxygen to all three fuel cells started to decrease.	055:54:43	14 Apr 1970	03:07:43
Oxygen tank #2 pressure reached maximum value of 1,008.3 psia.	055:54:45	14 Apr 1970	03:07:45
Oxygen tank #2 temperature rises 40°F for one sample (invalid reading).	055:54:48	14 Apr 1970	03:07:48
Oxygen tank #2 quantity jumped to off-scale-high and then started to drop until the time of telemetry loss, indicating a failed sensor.	055:54:51	14 Apr 1970	03:07:51
Oxygen tank #2 temperature read –151.3°F. Last valid indication.	055:54:52	14 Apr 1970	03:07:52
Oxygen tank #2 temperature suddenly went off-scale-low, indicating a failed sensor.	055:54:52.703	14 Apr 1970	03:07:52
Last telemetered pressure from oxygen tank #2 before telemetry loss was 995.7 psia.	055:54:52.763	14 Apr 1970	03:07:52
Sudden accelerometer activity on X, Y, and Z axes.	055:54:53.182	14 Apr 1970	03:07:53
Body-mounted roll, pitch, and yaw rate gyros showed low-level activity for 1/4 sec.	055:54:53.220	14 Apr 1970	03:07:53
Oxygen tank #1 pressure dropped 4.2 psi.	055:54:53.323	14 Apr 1970	03:07:53
2.8-amp rise in total fuel cell current.	055:54:53.5	14 Apr 1970	03:07:53
X, Y, and Z accelerations in CM indicate 1.17 g, 0.65 g, and 0.65 g.	055:54:53.542	14 Apr 1970	03:07:53
Telemetry loss for 1.8 sec. Master caution and warning triggered by DC main bus B undervoltage. Alarm turned off in 6 sec. Indications were that the cryogenic oxygen tank #2 lost pressure during this time period and the panel separated. It was at this time that the crew heard a loud bang.	055:54:53.555	14 Apr 1970	03:07:53
Nitrogen pressure in fuel cell #1 went off-scale-low indicating a failed sensor.	055:54:54.741	14 Apr 1970	03:07:54
Telemetry recovered.	055:54:55.35	14 Apr 1970	03:07:55
Service propulsion system engine valve body temperature started a rise of 1.65°F in 7 sec. DC main bus A decreased 0.9 volt to 28.5 volts and DC main bus B decreased 0.9 volt to 29.0 volts. Total fuel cell current was 15 amps higher than he final value before telemetry loss. High current continued			

Apollo 13 mission event – *continued*	GET (h:m:s)	Date (GMT)	Time (h:m:s)
for 19 sec. Oxygen tank #2 temperature read off-scale-high after telemetry recovery, probably indicating failed sensors. Oxygen tank #2 pressure read off-scale-low following telemetry recovery, indicating a broken supply line, a tank pressure below 19 psia, or a failed sensor. Oxygen tank #1 pressure read 781.9 psia and started to drop steadily. Pressure dropped over a period of 130 min to the point at which it was insufficient to sustain operation of fuel cell #2.	055:54:56	14 Apr 1970	03:07:56
Oxygen tank #2 quantity read off-scale-high following telemetry recovery indicating a failed sensor.	055:54:57	14 Apr 1970	03:07:57
The reaction control system helium tank C temperature began a 1.66°F increase in 36 sec.	055:54:59	14 Apr 1970	03:07:59
Oxygen flow rates to fuel cells #1 and #3 approached zero after decreasing for 7 sec.	055:55:01	14 Apr 1970	03:08:01
Surface temperature of SM oxidizer tank in bay #3 started a 3.8°F increase in a 15-sec period. Service propulsion system helium tank temperature started a 3.8°F increase in a 32-sec period.	055:55:02	14 Apr 1970	03:08:02
DC main bus A voltage recovered to 29.0 volts. DC main bus B recovered to 28.8.	055:55:09	14 Apr 1970	03:08:09
CMP: "Okay, Houston, we've had a problem here."	055:55:20	14 Apr 1970	03:08:20
CAPCOM: "This is Houston. Say again, please."	055:55:28	14 Apr 1970	03:08:28
CDR: "Houston, we've had a problem. We've had a main B bus undervolt."	055:55:35	14 Apr 1970	03:08:35
CAPCOM: "Roger. Main B bus undervolt."	055:55:42	14 Apr 1970	03:08:42
Oxygen tank #2 temperature started steady drop lasting 59 sec, indicating a failed sensor.	055:55:49	14 Apr 1970	03:08:49
LMP (Haise): "Okay. Right now, Houston, the voltage is – is looking good. And we had a pretty large bang associated with the caution and warning there. And as I recall, main B was the one that had an amp spike on it once before."	055:56:10	14 Apr 1970	03:09:10
CAPCOM: "Roger, Fred."	055:56:30	14 Apr 1970	03:09:30
Oxygen tank #2 quantity became erratic for 69 sec before assuming an off-scale-low state, indicating a failed sensor.	055:56:38	14 Apr 1970	03:09:38
LMP: "In the interim here, we're starting to go ahead and button up the tunnel again."	055:56:54	14 Apr 1970	03:09:54
LMP: "That jolt must have rocked the sensor on – see now – oh-two oxygen quantity 2. It was oscillating down around 20 to 60%. Now it's full-scale-high."	055:57:04	14 Apr 1970	03:10:04
Master caution and warning triggered by DC main bus B undervoltage. Alarm was turned off in 6 sec.	055:57:39	14 Apr 1970	03:10:39
DC main bus B dropped below 26.25 volts and continued to fall rapidly.	055:57:40	14 Apr 1970	03:10:40
CDR: "Okay. And we're looking at our service module RCS helium 1. We have – B is barber poled and D is barber			

Apollo 13 mission event – *continued*	GET (h:m:s)	Date (GMT)	Time (h:m:s)
poled, helium 2, D is barber pole, and secondary propellants, I have A and C barber pole." AC bus fails within 2 sec.	055:57:44	14 Apr 1970	03:10:44
Fuel cell #3 failed.	055:57:45	14 Apr 1970	03:10:45
Fuel cell current started to decrease.	055:57:59	14 Apr 1970	03:10:59
Master caution and warning caused by AC bus #2 being reset.	055:58:02	14 Apr 1970	03:11:02
Master caution and warning triggered by DC main bus A undervoltage.	055:58:06	14 Apr 1970	03:11:06
DC main bus A dropped below 26.25 volts and in the next few seconds leveled off at 25.5 volts.	055:58:07	14 Apr 1970	03:11:07
LMP: "AC #2 is showing zip."	055:58:07	14 Apr 1970	03:11:07
LMP: "Yes, we got a main bus A undervolt now, too, showing. It's reading about 25 and a half. Main B is reading zip right now."	055:58:25	14 Apr 1970	03:11:25
Master caution and warning triggered by high hydrogen flow rate to fuel cell #2.	056:00:06	14 Apr 1970	03:13:06
CDR: "...It looks to me, looking out the hatch, that we are venting something. We are venting something out into the – into space."	056:09:07	14 Apr 1970	03:22:07
LMP reported fuel cell #1 off line.	056:09:58	14 Apr 1970	03:22:58
Emergency power-down.	056:33:49	14 Apr 1970	03:46:49
LMP reported fuel cell #3 off line.	056:34:46	14 Apr 1970	03:47:46
CDR and LMP entered LM.	057:43	14 Apr 1970	04:56
Shutdown of fuel cell #2.	058:00	14 Apr 1970	05:13
CM computer and platform powered down.	058:10	14 Apr 1970	05:23
CSM systems powered down. LM systems powered up.	058:40	14 Apr 1970	05:53:00
Midcourse correction ignition to free-return trajectory (LM DPS).	061:29:43.49	14 Apr 1970	08:42:43
Midcourse correction cutoff.	061:30:17.72	14 Apr 1970	08:43:17
LM systems powered down.	062:50	14 Apr 1970	10:03
Lunar occultation entered.	077:08:35	15 Apr 1970	00:21:35
Lunar occultation exited.	077:33:10	15 Apr 1970	00:46:10
S-IVB impact on lunar surface.	077:56:40.0	15 Apr 1970	01:09:40
LM systems powered up.	078:00	15 Apr 1970	01:13
Abort guidance system to primary guidance system aligned.	078:10	15 Apr 1970	01:23
Transearth injection ignition (LM DPS).	079:27:38.95	15 Apr 1970	02:40:39
Transearth injection cutoff.	079:32:02.77	15 Apr 1970	02:45:02
LM systems powered down.	082:10	15 Apr 1970	05:23
Apparent short-circuit in LM electrical system, accompanied by a "thump" in vicinity of descent stage and observation of venting for several min in area of LM descent batteries #1 and #2.	097:13:53	15 Apr 1970	20:26:53
LM configured for midcourse correction.	100:00	15 Apr 1970	23:13
CSM power configuration for telemetry established.	101:20	16 Apr 1970	00:33
CM powered up.	101:53	16 Apr 1970	01:06
LM systems powered up.	104:50	16 Apr 1970	04:03
Midcourse correction ignition (LM DPS).	105:18:28.0	16 Apr 1970	04:31:28

Apollo 13 mission event – *continued*	GET (h:m:s)	Date (GMT)	Time (h:m:s)
Midcourse correction cutoff.	105:18:42.0	16 Apr 1970	04:31:42
Passive thermal control started.	105:20	16 Apr 1970	04:33
LM systems powered down.	105:50	16 Apr 1970	05:03
LM power transferred to CSM.	112:05	16 Apr 1970	11:18
Battery A charge initiated.	112:20	16 Apr 1970	11:33
Battery A charge terminated. Battery B charge initiated.	126:10	17 Apr 1970	01:23
Battery B charge terminated.	128:10	17 Apr 1970	03:23
LM systems powered up.	133:35	17 Apr 1970	08:48
Platform aligned.	134:40	17 Apr 1970	09:53
Preparation for midcourse correction.	136:30	17 Apr 1970	11:43
Midcourse correction ignition (LM RCS).	137:39:51.5	17 Apr 1970	12:52:51
Midcourse correction cutoff.	137:40:13.0	17 Apr 1970	12:53:13
SM separation.	138:01:48.0	17 Apr 1970	13:14:48
SM photographed.	138:15	17 Apr 1970	13:28
CM powered up.	140:10	17 Apr 1970	15:23
Platform aligned.	140:40	17 Apr 1970	15:53
LM maneuvered to undocking attitude.	140:50	17 Apr 1970	16:03
LM jettisoned.	141:30:00.2	17 Apr 1970	16:43:00
Entry.	142:40:45.7	17 Apr 1970	17:53:45
S-band contact with CM established by recovery aircraft.	142:48	17 Apr 1970	18:01
Visual contact with CM established by recovery helicopters.	142:49	17 Apr 1970	18:02
Visual contact with CM established by recovery ship. Voice contact with CM established by recovery helicopters.	142:50	17 Apr 1970	18:03
Splashdown (went to apex-up).	142:54:41	17 Apr 1970	18:07:41
Swimmers deployed to retrieve main parachutes.	142:56	17 Apr 1970	18:09
1st swimmer deployed to CM.	143:03	17 Apr 1970	18:16
Flotation collar inflated.	143:11	17 Apr 1970	18:24
Life preserver unit delivered to lead swimmer.	143:18	17 Apr 1970	18:31
CM hatch opened for crew egress.	143:19	17 Apr 1970	18:32
Crew egress.	143:22	17 Apr 1970	18:35
Crew on board recovery helicopter.	143:29	17 Apr 1970	18:42
Crew on board recovery ship.	143:40	17 Apr 1970	18:53
CM on board recovery ship.	144:23	17 Apr 1970	19:36
Crew departed recovery ship for Samoa and then Hawaii.	167:07	18 Apr 1970	18:20
Crew arrived in Hawaii.	199:22	19 Apr 1970	02:35
Crew arrived at Ellington AFB, Houston, TX.	224:17	20 Apr 1970	03:30
Recovery ship arrived in Hawaii.	312:17	24 Apr 1970	19:30
Safing of CM pyrotechnics completed.	343:22	26 Apr 1970	02:35
Deactivation of fuel and oxidizer completed.	360:15	26 Apr 1970	19:28
CM arrived at contractor's facility in Downey, CA.	378:47	27 Apr 1970	14:00

Apollo 14

The eighth manned mission: the third lunar landing

31 January–9 February 1971

BACKGROUND

Apollo 14 was to be a Type 'H' mission with a precision manned lunar landing and systematic lunar exploration. In view of the importance of the Fra Mauro Formation, the sampling of which had been assigned to Apollo 13, which had unfortunately not made it to the Moon, this site was assigned to Apollo 14, which became the third successful lunar landing.

The primary objectives were:

- to perform selenological inspection, survey, and sampling of materials at a preselected site;
- to deploy and activate the Apollo lunar surface experiments package;
- to develop human capability of working in the lunar environment;
- to obtain photographs of candidate exploration sites.

Although the primary mission objectives for Apollo 14 were the same as those for Apollo 13, provisions were made for returning a significantly greater quantity of lunar material and scientific data than had been possible previously. An innovation that allowed an increase in the range of lunar surface exploration and in the amount of material collected was the provision of a collapsible two-wheeled cart, the modular equipment transporter (MET), for carrying tools, cameras, a portable magnetometer, and lunar samples. An investigation into the cause of the Apollo 13 cryogenic oxygen tank failure prompted three significant changes in the CSM cryogenic oxygen storage and electrical power systems for Apollo 14. The internal construction of the oxygen tanks was modified, a third oxygen tank was added, and an auxiliary battery was installed. These changes were also incorporated into all subsequent spacecraft.

The crew members were Captain Alan Bartlett Shepard Jr (USN), commander; Major Stuart Allen Roosa (USAF), command module pilot; and Commander Edgar Dean Mitchell (USN), lunar module pilot. Selected as one of the original astronauts in 1959, Shepard became the first American in space when he manned the initial Mercury suborbital mission (MR-3). Shepard subsequently developed an ear disorder, Meniere's syndrome, which caused the Navy to forbid him to fly solo in jet planes and forced NASA to ground him. He then became chief of the astronaut office. In 1969, however, Shepard underwent experimental surgery that corrected the problem. He was restored to full status in May and assigned to command Apollo 14 in August. Shepard was born on 18 November 1923 in East Derry, New Hampshire, and at 47 years old, he would become the oldest person to walk on the Moon. He received a BS from the US Naval Academy in 1944.[1] His backup for the mission was Captain Eugene Andrew 'Gene' Cernan (USN). Roosa and Mitchell were making their first space flights. Roosa, born on 16 August 1933 in Durango, Colorado, was 37 years old at the time of the Apollo 14 mission. He received a BS in aeronautical engineering from the University of Colorado in 1960 and was selected as an astronaut in 1966.[2] His backup was Commander Ronald Ellwin Evans (USN). Mitchell, born on 17 September 1930 in Hereford, Texas, was 40 years old. He received a BS in industrial management from the Carnegie Institute of Technology in 1952, a BS in aeronautical engineering from the US Naval Postgraduate School in 1961, and an ScD in aeronautics and astronautics from the Massachusetts Institute of Technology in 1964. He was selected as an astronaut in 1966. His backup was Lieutenant Colonel Joe Henry Engle (USAF). The capsule communicators (CAPCOMs) for the mission were Major Charles Gordon Fullerton (USAF), Lieutenant Commander Bruce McCandless II (USN), Fred Wallace Haise Jr, and Evans. The support crew were McCandless, Lieutenant Colonel William Reid Pogue (USAF), Fullerton, and Phillip Kenyon Chapman ScD. The flight directors were M.P. 'Pete' Frank and Glynn S. Lunney (first shift), Milton L. Windler (second shift), Gerald D. Griffin (third shift), and Glynn S. Lunney (fourth shift).

The Apollo 14 launch vehicle was a Saturn V, designated AS-509. The mission also carried the designation Eastern Test Range #7194. The CSM was designated CSM-110 and had the call-sign 'Kitty Hawk'. The lunar module was designated LM-8 and had the call-sign 'Antares'.

LAUNCH PREPARATIONS

The terminal countdown was picked up at T–28 hr at 06:00:00 GMT on 30 January 1971. Scheduled holds were initiated at T–9 hr for 9 hr 23 min and at T–3 hr 30 min for one hour. At launch time, a cold front extended through northern Florida. Scattered rain shower activity existed to the south of this front throughout the morning of launch, but the showers did not reach the launch area until just before the scheduled launch time. A band of cumulus congestus clouds with showers developed about 30 min before scheduled launch time along a line extending from Orlando toward the northern Merritt Island Launch Area (MILA). This, and the threat of lightning, necessitated a 40-min 2-sec hold at T–8 min until the showers had moved a sufficient distance from the launch complex.

[1] Shepard died of leukemia 21 July 1998 in Community Hospital, Monterey, CA.

[2] Roosa died of complications from pancreatitis 12 December 1994 in Washington, DC. (NASA Headquarters Release No. 94-210).

Apollo 14 preparation event	Date
Individual and combined CM and SM systems test completed at factory.	02 Apr 1969
Integrated CM and SM systems test completed at factory.	07 May 1969
LM-8 final engineering evaluation acceptance test at factory.	25 Aug 1969
LM-8 integrated test at factory.	25 Aug 1969
Ascent stage of LM-8 ready to ship from factory to KSC.	08 Nov 1969
Descent stage of LM-8 ready to ship from factory to KSC.	13 Nov 1969
CM-110 and SM-110 ready to ship from factory to KSC.	17 Nov 1969
CM-110 and SM-110 delivered to KSC.	19 Nov 1969
CM-110 and SM-110 mated.	24 Nov 1969
Ascent stage of LM-8 delivered to KSC.	24 Nov 1969
Descent stage of LM-8 delivered to KSC.	24 Nov 1969
S-IC-9 stage delivered to KSC.	11 Jan 1970
S-IC-9 stage erected on MLP-2.	14 Jan 1970
LM-8 stages mated.	20 Jan 1970
S-IVB-509 stage delivered to KSC.	20 Jan 1970
S-II-9 stage delivered to KSC.	21 Jan 1970
LM-8 combined systems test completed.	22 Jan 1970
CSM-110 combined systems test completed.	02 Feb 1970
Spacecraft/LM adapter SLA-17 delivered to KSC.	31 Mar 1970
Saturn V instrument unit IU-509 delivered to KSC.	06 May 1970
S-II-9 stage erected.	12 May 1970
S-IVB-509 stage erected.	13 May 1970
IU-509 erected.	14 May 1970
Launch vehicle electrical systems test completed.	04 Jun 1970
LM-8 altitude tests completed.	22 Jun 1970
Launch vehicle propellant dispersion/malfunction overall test completed.	07 Jul 1970
CSM-110 altitude tests completed.	01 Sep 1970
Launch vehicle service arm overall test completed.	21 Oct 1970
CSM-110 moved to VAB.	04 Nov 1970
Spacecraft erected.	04 Nov 1970
Space vehicle and MLP-2 transferred to Launch Complex 39A.	09 Nov 1970
LM-8 combined systems test completed.	16 Nov 1970
CSM-110 integrated systems test completed.	18 Nov 1970
CSM-110 electrically mated to launch vehicle.	13 Dec 1970
LM-8 flight readiness test completed.	14 Dec 1970
Space vehicle overall test #1 (plugs in) completed.	14 Dec 1970
Space vehicle flight readiness test completed.	19 Dec 1970
S-IC-9 stage RP-1 fuel loading completed.	08 Jan 1971
Space vehicle countdown demonstration test (wet) completed.	18 Jan 1971
Space vehicle countdown demonstration test (dry) completed.	19 Jan 1971

Although it was raining prior to launch, there was no rain at the pad at the time of launch, but the vehicle did travel through the cloud decks. Surface winds in the Cape Canaveral area were fairly light and westerly. Cumulus clouds covered 70% of the sky (base 4,000 ft) and altocumulus covered 20% (base 8,000 ft), the temperature was 71.1°F, the relative humidity was 86%, and the barometric pressure was 14.652 psi. The winds, as measured by the anemometer on the light pole 60.0 ft above ground at the launch site, were 9.7 kt at 255° from true north; the winds at 530 ft were 16.5 kt at 275° from true north. The weather delay required the flight azimuth to be changed from 72.067°E of N to 75.558°E of N.

ASCENT PHASE

Apollo 14 was launched from Pad A of Launch Complex 39 at the Kennedy Space Center at a Range Zero time of 21:03:02 GMT (16:03:02 EST) on 31 January 1971. The launch window extended from 20:23:00 GMT on 31 January to 00:12:00 GMT on 1 February in order to take advantage of a Sun elevation angle on the lunar surface of 10.3 deg at the scheduled time of landing. Between T+12.81 and T+28.00 sec, the vehicle rolled from a pad azimuth of 90°E or N to a flight azimuth of 75.558°E of N. The maximum wind conditions encountered during ascent were 102.6 kt at 255° from true north at 43,270 ft, with a maximum wind shear of 0.0201/sec at 43,720 ft. The S-IC shut down at T+164.10, followed by S-IC/S-II separation and S-II ignition. The S-II shut down at T+559.05, followed by separation from the S-IVB, which ignited at T+563.4. The first S-IVB cutoff occurred at T+700.56, with deviations from the planned trajectory of only –2.6 ft/sec in velocity; altitude was exactly as planned. At insertion, at T+710.56 (i.e. S-IVB cutoff plus 10 sec to account for engine tailoff and other transient effects), the parking orbit showed an apogee and perigee of 100.1 × 98.9 nmi, an inclination of 31.120 deg, a period of 88.18 min, and a velocity of 25,565.8 ft/sec. The apogee and perigee were based upon a spherical Earth with a radius of 3,443.934 nmi.

The COSPAR designation for the CSM upon achieving orbit was 1971-008A, and the S-IVB was designated 1971-008B. After undocking at the Moon, the LM ascent stage would be designated 1971-008C and the descent stage 1971-008D.

Apollo 14 ascent phase event	GET (h:m:s)	Altitude (nmi)	Range (nmi)	Earth-fixed velocity (ft/sec)	Space-fixed velocity (ft/sec)	Event duration (sec)	Geocentric latitude (°N)	Longitude (°E)	Space-fixed flight path angle (deg)	Space-fixed heading angle (°E of N)
Liftoff	000:00:00.57	0.060	0.000	1.1	1,340.7	–	28.4470	–80.6041	0.05	90.00
Mach 1 achieved	000:01:08.0	4.337	1.379	1,077.0	2,082.4	–	28.4521	–80.5787	26.80	86.06
Maximum dynamic pressure	000:01:21.0	6.649	2.886	1,524.6	2,540.5	–	28.4580	–80.5509	28.77	84.61
S-IC center engine cutoff	000:02:15.14	23.202	24.169	5,103.0	6,283.6	141.6	28.5441	–80.1598	23.554	79.228
S-IC outboard engine cutoff	000:02:44.10	36.317	51.132	7,741.7	8,972.5	170.6	28.6516	–79.6634	19.584	78.468
S-IC/S-II separation	000:02:44.8	36.663	51.947	7,773.0	9,004.8	–	28.6548	–79.6484	19.489	78.468
S-II center engine cutoff	000:07:43.09	98.091	594.709	17,212.7	18,554.4	296.59	30.3347	–69.4425	0.829	82.809
S-II outboard engine cutoff	000:09:19.05	101.556	890.920	21,562.5	22,905.8	392.55	30.8611	–63.7444	0.621	85.784
S-II/S-IVB separation	000:09:20.00	101.596	894.194	21,573.8	22,917.2	–	30.8654	–63.6810	0.612	85.818
S-IVB 1st burn cutoff	000:11:40.56	103.091	1,406.287	24,215.6	25,559.9	137.16	31.0978	–53.7349	–0.004	91.245
Earth orbit insertion	000:11:50.56	103.086	1,444.989	24,221.6	25,565.8	–	31.0806	–52.9826	–0.003	91.656

EARTH ORBIT PHASE

After in-flight systems checks, the 350.84-sec translunar injection maneuver (second S-IVB firing) was performed at 002:28:32.40. The S-IVB engine shut down at 002:34:23.24 and translunar injection occurred 10 sec later, at a velocity of 35,511.6 ft/sec after 1.5 Earth orbits lasting 2 hr 22 min 42.68 sec.

Apollo 14 earth orbit phase event	GET (h:m:s)	Space-fixed velocity (ft/sec)	Event duration (sec)	Velocity change (ft/sec)	Apogee (nmi)	Perigee (nmi)	Period (min)	Inclination (deg)
Earth orbit insertion	000:11:50.56	25,565.8	–	–	100.1	98.9	88.18	31.120
S-IVB 2nd burn ignition	002:28:32.40	25,579.0	–	–	–	–	–	–
S-IVB 2nd burn cutoff	002:34:23.24	35,535.5	350.84	10,366.5	–	–	–	30.835

TRANSLUNAR PHASE

At 003:02:29.4, the CSM was separated from the S-IVB stage. Transposition occurred normally; however, six docking attempts were required before the CSM was successfully docked with the LM at 004:56:56.7. The docked spacecraft were ejected from the S-IVB by a 6.9-sec maneuver at 005:47:14.4, and an 80.2-sec separation maneuver was performed at 006:04:01.7. In-flight examination of the docking probe revealed no problems. It was therefore assumed that the capture-latch assembly must not have been in the locked configuration during the first five attempts. As on Apollo 13, the S-IVB stage was targeted to impact the Moon within a prescribed area to supply seismic data. A 252.2-sec auxiliary propulsion system lunar impact maneuver was performed at 008:59:59.0 to accomplish that objective. The S-IVB impacted the lunar surface at 082:37:53.4. The impact point was 8.09°S and 26.02°W, 159 nmi from the target point, and 94 nmi southwest of the Apollo 12 seismometer. The seismometer recorded the impact 37 sec later and responded to vibrations for more than 3 hr. At impact, the S-IVB weighed 30,836 lb and was traveling 8,333 ft/sec.

Translunar activities included star and Earth horizon calibration sightings in preparation for a cislunar navigation exercise to be performed during transearth coast, and dim-light photography of the Earth. A 10.19-sec midcourse correction was made at 030:36:07.91. At 060:30, the commander and lunar module pilot transferred to the LM for about 2 hr of housekeeping and systems checks. While there, the crew photographed a wastewater dump from the CM to obtain data for a particle contamination study being conducted for the Skylab program. A second midcourse correction, a 0.65-sec maneuver, was made at 076:58:11.98 to achieve the final trajectory desired for lunar orbit insertion. At 054:53:36, a clock update was performed to compensate for the weather hold during the launch countdown. This procedure, which added 40 min 2.9 sec to the mission time clock, was an aid to the command module pilot while in lunar orbit as it eliminated the need for numerous updates to his flight log. At 081:56:40.70, at an altitude of 87.4 nmi above the Moon, the service propulsion system was fired for 370.84 sec to insert the spacecraft into a lunar orbit of 169.0 × 58.1 nmi. The translunar coast had lasted 79 hr 28 min 18.30 sec.

Apollo 14 translunar phase event	GET (h:m:s)	Altitude (nmi)	Space-fixed velocity (ft/sec)	Event duration (sec)	Velocity change (ft/sec)	Space-fixed flight path angle (deg)	Space-fixed heading angle (°E of N)
Translunar injection	002:34:33.24	179.544	35,511.6	–	–	7.480	65.583
CSM separated from S-IVB	003:02:29.4	4,289.341	24,102.3	–	–	46.810	65.369
CSM docked with LM/S-IVB	004:56:56.7	20,603.4	13,204.1	–	–	66.31	84.77
CSM/LM ejection ignition	005:47:14.4	26,299.6	11,723.5	–	–	68.54	87.76
CSM/LM ejection cutoff	005:47:21.3	–	–	6.9	0.8	–	–
Midcourse correction ignition	030:36:07.91	118,515	4,437.9	–	–	76.47	101.98
Midcourse correction cutoff	030:36:18.10	118,522.1	4,367.2	10.19	71.1	76.95	102.23
Midcourse correction ignition	076:58:11.98	11,900.3	3,711.4	–	–	–80.1	295.57
Midcourse correction cutoff	076:58:12.63	11,899.7	3,713.1	0.65	3.5	–80.1	295.65

LUNAR ORBIT AND LUNAR SURFACE PHASE

At 086:10:52.97, a 20.81-sec service propulsion system maneuver was performed and established the descent orbit of 58.8 × 9.1 nmi in preparation for undocking of the LM. On previous missions, the descent orbit insertion maneuver had been performed with the LM descent propulsion system. Because the landing site was more rugged than previous ones, a greater margin of LM propellant was provided.[3] The commander and lunar module pilot entered their spacecraft at 101:20 to perform system checks and prepare for undocking. A 2.7-sec firing of the service module reaction control system separated the CM from the LM at 103:47:41.6 and resulted in an orbit of 60.2 × 7.8 nmi. A 4.02-sec maneuver at 105:11:46.11 circularized the CSM to 63.9 × 56.0 nmi. Following vehicle separation and before powered descent, ground personnel detected the presence of an abort command at a computer input channel although the crew had not depressed the abort switch. The failure was isolated to the abort switch, and to prevent an unwanted abort, a workaround procedure was developed. The procedure was followed and the

[3] If Apollo 13 had not aborted prior to reaching lunar orbit, it would have been the first mission to perform this maneuver.

764.61-sec powered descent was initiated successfully at 108:02:26.52 at an altitude of 7.8 nmi. Some 6 min after initial actuation of the landing radar, the system switched to the low-range scale, forcing the trackers into the narrow-band mode of operation. This ranging scale problem would have prevented acquisition of radar data until late in the descent – and prevented a lunar landing – but was corrected by cycling the circuit breaker on and off manually. The left pad surface contact occurred at 108:15:09.3, the engine shut down 1.83 sec later, and the remaining pads settled at 108:15:11.4 (09:18:13 GMT on 5 February). The landing was at selenographic coordinates 3.64530°S, 17.47136°W, which, being 0.6 nmi west of Cone Crater, was by design slightly closer to this objective than the target selected for its predecessor.[4] At engine cutoff, approximately 68 sec of firing time remained.

Preparations for the initial period of lunar surface exploration began 2 hr following landing, and cabin depressurization began at 113:39:11. The first EVA began 49 min late due to intermittent PLSS communications during the EVA preparations. The cause was believed to have been a LM configuration problem. A recycling of the audio circuit breaker cleared the problem. The commander exited at 113:47. He was followed 8 min later by the lunar module pilot, whose first task was to collect the contingency sample. In the first extravehicular period the crew deployed the television, S-band antenna, and solar wind experiment; deployed and loaded the modularized equipment; collected samples; and photographed activities, panoramas, and equipment. At 115:46, they began their trip to the Apollo lunar surface experiments package deployment site, about 500 ft west of the LM. They also deployed the laser-ranging retroreflector 100 ft west of the ALSEP. The first ALSEP data were received on Earth at 116:47:58. Several problems were encountered during the deployment of the ALSEP package. They were as follows: difficulty in releasing the Boyd bolt on the suprathermal ion detector; stiffness in the cable between the suprathermal ion detector and the cold cathode ion gauge, which caused the cold cathode ion gauge to fall over; low transmitter strength on the central station; noisy data from the suprathermal ion detector experiment; and failure of five of the active seismic experiment 'thumper' initiators to fire. Although communications were nominal during this period, gradual degradation of the television picture resolution was noted during the latter part of the EVA. Once the crew had re-entered the LM, the cabin was repressurized at 118:27:01. The first extravehicular activity period lasted 4 hr 47 min 50 sec. The distance traveled was about 3,300 ft; 45.19 lb of samples were collected. During the lunar surface operations, the CSM made an 18.50-sec plane change maneuver at 117:29:33.17, which adjusted the orbit to 62.1 × 57.7 nmi.

The second extravehicular activity began with cabin depressurization at 131:08:13, 27 min earlier than planned, commander egress at 131:13 and lunar module pilot egress 7 min later. In preparation for an excursion to the area of Cone Crater, on a low ridge some 1,400 ft east–northeast of the lander, they prepared and loaded the two-wheeled modular equipment transporter. They experienced difficulties in navigating the slopes and fell 30

[4] In fact, having set down within 150 ft of the designated target point, this was the most accurate landing of the entire program.

min behind schedule. As a result, they only reached a point within 50 ft from the rim of the crater. Nevertheless, the objectives associated with reaching the vicinity of this crater were achieved. Photographs, various samples, and terrain descriptions were obtained en route. Rock and soil samples were collected in a blocky field near the rim of the crater. On the return trip, the crew also obtained magnetometer measurements at two sites. An estimated 1.5-ft trench was excavated and some samples were taken. An attempt to obtain a triple-length core sample was unsuccessful, but other containerized samples were able to be collected. Back at the LM, an alignment adjustment was made to the ALSEP central station's antenna just prior to crew ingress in order to improve the signal strength being received at the Manned Space Flight Network ground stations. This improved the signal strength approximately 0.5 dB, which enabled data to be received using a 30-ft antenna. Before re-entering the LM, the CDR dropped a golf ball onto the surface. Using the head of a '6-iron' attached to the handle of the contingency sample collector, he attempted to strike the ball but struck mostly soil and barely moved the ball. The second swing sent the ball a few feet to the right. He dropped a second ball and hit this into a crater about 50 ft away. The LMP later threw the staff from the solar wind experiment into the same crater. The second extravehicular activity period lasted 4 hr 34 min 41 sec. The distance traveled was about 9,800 ft, and 49.16 lb of samples were collected. The crew re-entered the LM and the cabin was repressurized at 135:42:54, thereby ending the third human exploration of the Moon. The total time spent outside the LM was 9 hr 22 min 31 sec, the total distance traveled was more than 13,100 ft, and the collected samples totaled 93.21 lb (42.28 kg; the official metric total as determined by the Lunar Receiving Laboratory in Houston). The farthest point traveled from the LM was 4,770 ft.

While the landing crew was on the surface, the CMP performed tasks to obtain data for scientific analyses and future mission planning. These tasks included photography of a proposed landing site in the Moon's central highlands, the lunar surface illuminated at a high Sun angle, the Gegenschein and Moulton Point regions of the sky and miscellaneous faint astronomical sources.

Ignition of the ascent stage engine for lunar liftoff occurred at 141:45:40 (18:48:42 GMT on 6 February). The LM had been on the lunar surface for 33 hr 30 min 29 sec. The 432.1-sec firing of the ascent engine placed the vehicle directly into an orbit of 51.7 × 8.5 nmi, the first use of a direct lunar orbit rendezvous in the program. However, a 12.1-sec vernier adjustment was required at 141:56:49.4 and altered the orbit to 51.2 × 8.4 nmi. A 3.6-sec terminal phase initiate burn at 142:30:51.1 and two small midcourse corrections brought the ascent stage to an orbit of 60.1 × 46.0 nmi. The ascent stage made a 26.7-sec maneuver at 143:13:29.1 to finalize the orbit at 61.5 × 58.2 nmi for docking with the CSM at 143:32:50.5 at an altitude of 58.6 nmi. The two craft had been undocked for 39 hr 45 min 8.9 sec. During the braking phase for docking, telemetry indicated that the abort guidance system had failed, but no caution and warning signals were on. A cycling of all circuit breakers and switches did not remedy this condition. Although no probe/drogue problems were experienced during docking, the probe was returned to Earth for post-flight analysis. Television during rendezvous and docking was excellent and clearly showed the docking maneuver. After transfer of the crew and samples to the CM, the ascent stage was jettisoned at 145:44:58.0, and the CSM was prepared for transearth injection. A 15.8-sec maneuver was made at 145:49:42.5 to separate the CM from the ascent stage, and resulted in an orbit of 63.4 × 56.8 nmi. The ascent stage was then maneuvered by remote control to impact the lunar surface. A 76.2-sec deorbit firing at 57.2 nmi altitude depleted the ascent stage propellants, and impact occurred at 147:42:23.7, at selenographic coordinates 3.42°S, 19.67°W, some 36 nmi west of the Apollo 14 landing site, 62 nmi from the Apollo 12 landing site, and 7 nmi from the aim point.

On previous missions, dust adhering to equipment being returned to Earth had created a problem. On Apollo 14, special dust control procedures were used to reduce the amount of dust in the cabins. After a 149.23-sec maneuver at 148:36:02.30, transearth injection was achieved at 148:38:31.53 at a velocity of 8,505 ft/sec after 34 lunar orbits lasting 66 hr 35 min 39.99 sec.

Apollo 14 lunar orbit phase event	GET (h:m:s)	Altitude (nmi)	Space-fixed velocity (ft/sec)	Event duration (sec)	Velocity change (ft/sec)	Apolune (nmi)	Perilune (nmi)
Lunar orbit insertion ignition	081:56:40.70	87.4	8,061.4	–	–	–	–
Lunar orbit insertion cutoff	082:02:51.54	64.2	5,458.5	370.84	3,022.4	169.0	58.1
Descent orbit insertion ignition	086:10:52.97	59.2	5,484.8	–	–	–	–
Descent orbit insertion cutoff	086:11:13.78	59	5,279.5	20.81	205.7	58.8	9.1
CSM/LM undocking/separation ignition	103:47:41.6	30.5	5,435.8	–	–	–	–
CSM/LM undocking/separation cutoff	103:47:44.3	–	–	2.7	0.8	60.2	7.8
CSM orbit circularization ignition	105:11:46.11	60.5	5,271.3	–	–	–	–
CSM orbit circularization cutoff	105:11:50.13	60.3	5,342.1	4.02	77.2	63.9	56.0
LM powered descent initiation	108:02:26.52	7.8	5,565.6	–	–	–	–
LM powered descent cutoff	108:15:11.13	–	–	764.61	–	–	–
CSM plane change ignition	117:29:33.17	62.1	5,333.1	–	–	–	–
CSM plane change cutoff	117:29:51.67	62.1	5,333.3	18.50	370.5	62.1	57.7
LM lunar liftoff ignition	141:45:40	–	–	–	–	–	–
Lunar ascent orbit cutoff	141:52:52.1	–	–	432.1	6,066.1	51.7	8.5
LM vernier adjustment ignition	141:56:49.4	11.1	5,548.5	–	–	–	–
LM vernier adjustment cutoff	141:57:01.5			12.1	10.3	51.2	8.4
LM terminal phase initiation ignition	142:30:51.1	44.8	5,396.6	–	–	–	–
LM terminal phase initiation cutoff	142:30:54.7			3.6	88.5	60.1	46.0
LM terminal phase finalize ignition	143:13:29.1	58.8	5,365.5	–	–	–	–
LM terminal phase finalize cutoff	143:13:55.8	–	–	26.7	32	61.5	58.2
CSM/LM docked	143:32:50.5	58.6	5,353.5	–	–	–	–
LM ascent stage jettisoned	145:44:58.0	59.9	5,344.6	–	–	–	–
CSM/LM final separation ignition	145:49:42.5	60.6	5,341.7	–	–	–	–
CSM/LM final separation cutoff	145:49:58.3	–	–	15.8	3.4	63.4	56.8
LM ascent stage deorbit ignition	147:14:16.9	57.2	5,358.7	–	–	–	–
LM ascent stage fuel depletion*	147:15:33.1	57.2	5,177	76.2	186.1	56.7	–59.8

* The negative perilune indicates that the ascent stage was placed on a surface-intersecting trajectory.

TRANSEARTH PHASE

During the transearth coast, a 3.00-sec midcourse correction of 0.5 ft/sec was made at 165:34:56.69 using the service module reaction control system. In addition, a special flow-rate test was performed to evaluate the oxygen system for planned extravehicular activities on subsequent missions, and a navigation exercise simulating a return to Earth without ground control was conducted using only the guidance and navigation system. Scientific investigations included televised demonstrations of electrophoretic separation, liquid transfer, heat flow and convection, and composite casting under zero-gravity conditions.

Apollo 14 transearth phase event	GET (h:m:s)	Altitude (nmi)	Space-fixed velocity (ft/sec)	Event duration (sec)	Velocity change (ft/sec)	Space-fixed flight path angle (deg)	Space-fixed heading angle (°E of N)
Transearth injection ignition	148:36:02.30	60.9	5,340.6	–	–	–0.17	260.81
Transearth injection cutoff	148:38:31.53	66.5	8,505	149.23	3,460.6	5.29	266.89
Midcourse correction ignition	165:34:56.69	176,713.8	3,593.2	–	–	–79.61	124.88
Midcourse correction cutoff	165:34:59.69	–	–	3.00	0.5	–	–
CM/SM separation	215:32:42.2	1,965	29,050.8	–	–	–36.62	117.11

RECOVERY

The service module was jettisoned at 215:32:42.2, and the CM followed a normal entry profile. The command module re-entered the Earth's atmosphere (at the 400,000-ft altitude of the 'entry interface') at 215:47:45.3 at 36,170.2 ft/sec, after a transearth coast of 67 hr 9 min 13.8 sec. The parachute system effected splashdown of the CM in the Pacific Ocean at 21:05:00 GMT on 9 February. The mission duration was 216:01:58.1. The impact point, estimated to be 27.02°S and 172.67°W, was about 0.6 nmi from the target point and 3.8 nmi from the recovery ship USS *New Orleans*. The service module should have entered the atmosphere and impacted some 650 nmi southwest of the CM, but no radar data or sightings confirmed its entry or impact. The CM assumed an apex-up flotation attitude. The crew was retrieved by helicopter and was on board the recovery ship 48 min after splashdown. The CM was recovered 76 min later. The estimated CM weight at splashdown was 11,481.2 lb, and the estimated distance traveled for the mission was 1,000,279 nmi. The crew remained aboard the *New Orleans* in the mobile quarantine facility until they departed by aircraft for Pago Pago, Samoa, at 17:46 GMT on 11 February. They were then transferred to a second mobile quarantine facility on a C-141 aircraft and flown to Ellington Air Force Base, Houston, where they arrived at 09:34 GMT on 12 February, following a refueling stop at Norton Air Force Base, California, and the crew entered the lunar receiving laboratory at 11:35 GMT the same day. At 21:30 GMT on 17 February the CM and first mobile quarantine facility were offloaded by the *New Orleans* at Hawaii. The first mobile quarantine facility was sent by aircraft to Houston, where it arrived at 07:40 GMT on 18 February. The CM was taken to Hickam Air Force Base, Hawaii, for deactivation. Upon completion of deactivation, at 23:00 GMT on 19 February, it was transferred to Ellington Air Force Base via C-133 aircraft, where it arrived at 21:45 GMT on 22 February. The crew and medical support personnel were released from quarantine on 27 February, and the CM and lunar samples were released on 4 April. The tests showed no evidence of lunar microorganisms at the three sites explored, and this was considered to be sufficient justification for discontinuing the quarantine procedures on future missions.

CONCLUSIONS

The Apollo 14 mission was the third successful lunar landing and demonstrated excellent performance of all contributing elements, thereby resulting in the collection of a wealth of scientific information. All of the objectives and experiment operations were accomplished satisfactorily except for some desired photography that could not be obtained.

The following conclusions were made from an analysis of post-mission data:

1. Cryogenic oxygen system hardware modifications and changes made as a result of the Apollo 13 failure satisfied, within safe limits, all system requirements for future missions, including extravehicular activity.
2. The advantages of manned space flight were again clearly demonstrated on this mission by the crew's ability to diagnose and work around hardware problems and malfunctions which otherwise might have resulted in mission termination.
3. Navigation was the most difficult lunar surface task because of problems in finding and recognizing small features, reduced visibility in the up-Sun and down-Sun directions, and the inability to judge distances.
4. Rendezvous within one orbit of lunar ascent was demonstrated for the first time in the Apollo program. This type of rendezvous reduced the time between lunar liftoff and docking by approximately 2 hr from that required on previous missions. The timeline activities, however, were greatly compressed.
5. On previous lunar missions, lunar surface dust adhering to equipment being returned to Earth had created a problem in both spacecraft. The special dust control procedures and equipment used on this mission were effective in lowering the overall level of dust.
6. Onboard navigation without air-to-ground communications was successfully demonstrated during the transearth phase of the mission to be sufficiently accurate for use as a contingency mode of operation during future missions.
7. Launching through cumulus clouds with tops up to 10,000 ft was demonstrated to be a safe launch restriction for the prevention of triggered lightning. The cloud conditions at liftoff were at the limit of this restriction and no triggered lightning was recorded during the launch phase.

The 42 kg or so of lunar material returned by Apollo 14 included 33 individual rocks. Because the Fra Mauro rocks were consolidations of shattered precursors, their analysis was rather more complicated than for earlier missions. Although one objective was to date when the fragments had been bound together, in order to date the impact that applied the shock, this was complicated by the fact that the isotopic 'clocks' used to measure the formation date are 'reset' when a rock is melted. This was not an issue in the mare basalts, but the analysis of breccias involved dating the individual clasts. It was soon noted that the samples tended to cluster into two age ranges, one spanning the interval 3.96 to 3.87 billion years and the other 3.85 to 3.82 billion years. It was therefore inferred that the breccias formed 3.84 billion years ago in a splash of ejecta from the Imbrium impact. This date was the 'ground truth' that Apollo 14 had been sent to determine. The older dates for some clasts provided the formation ages of the rocks that had been shattered by that impact. While the primary objective had been achieved, many subsidiary issues remained unresolved. It had been hoped that samples from Cone's rim would characterize the basement on which the Fra Mauro Formation resided, which was expected to be volcanic. At first, several intriguing samples did indeed appear as if they might represent ancient volcanism, but they turned out to be the first instances of another type of breccia. In fact, it was a while before it was discovered that there were several kinds of breccia. The term 'fragmental breccia' was coined for clasts of shattered rock bound up in a matrix of pulverized rock. As more samples were studied it was found that in some cases fragments of *minerals* were caught up in breccias, indicating that not all clasts are lithic. Also, because breccias can be caught up in impacts, it is possible to get 'breccias of breccias', in which the clasts in one breccia are pieces of previous breccias, and the term 'one-rock' and 'two-rock' were introduced to reflect this complex history.

The samples that were initially taken to be ancient basalt were found to be yet another type of breccia in which clasts were bound up in impact-melt. (It resembled basalt to the extent that it was a solidified rock melt, but basalt is homogeneous.) Despite the violence of the shock-melting, the resulting breccias contained very fragile crystals that could only have built-up by diffusion processes as mineral-rich vapor escaped from the ejecta. This crystallization process was similar to sulphur encrustation of a volcanic vent on Earth, but the gas was produced by the ejecta itself rather than the material on which the ejecta sat. The discovery of these crystals, some of which were metallic iron, showed that the rubble had been hot when it was deposited, and had fused as it congealed. Several of the clasts were igneous. At 3.96 billion years old, one basaltic clast was not only the oldest sample yet found, it indicated both that there had been volcanism prior to the basin in-fill period and that this earlier magma was considerably more aluminous than the dark mare, which prompted the term 'non-mare' basalt. One intriguing aspect of the Fra Mauro impact-melt breccias was that they were KREEPy, meaning that they were rich in the 'rare earth' elements. When such material was first noted in samples from Procellarum, it was briefly suspected of representing the ancient crust, but the accompanying radioactives ruled this out. Analysis of the Fra Mauro samples showed it to have originally been a gabbro that solidified from the magma ocean. During crystallization, an element is accepted or rejected according to whether it fits the regular crystalline structure; the elements that do not fit are 'incompatibles'. As the trace elements tend not to participate in mineralization, they remain in the melt as the compatible elements are extracted, with the result that they become progressively more concentrated. The radioactives at depth helped to keep this concentrated reservoir molten, and then were locked in when it finally solidified. The impact that created the Imbrium basin had excavated deep enough to scatter some of this material across the surface.

When Fra Mauro was assigned to Apollo 13, it was deemed to be ideal for an 'H'-mission, but it had turned out to be just barely manageable, not for the task of finding the landing site, which was not difficult, but for the surprisingly undulatory nature of the 'plain'. Although the landing site had been relocated closer to Cone for Apollo 14, the trek up the ridge and back involved a round trip of 10,000 ft. If Lovell and Haise had landed at the more distant site, it is unlikely they would have reached crest of the ridge. Although Shepard and Mitchell did reach the summit, the crater eluded them. Navigation was difficult, but the fundamental limitation was *time*. Another hour on Cone's rim would have fully justified the overhead of reaching it. The portable life-support system was rated for a total of 7 hr, but this included the pre-egress time and a 2-hr post-ingress margin for troubleshooting, and the mission rules obliged the crew to re-enter the LM rather than spend an hour or so sampling the landing site. In effect, therefore, the time at Cone's rim was constrained by the need to preserve the *margin*. Apollo 14 showed the 'H'-mission format to be severely limited. Certainly, it could not do justice to the multiple-objective sites that were being proposed for the remaining missions. The 'exploration imperative' had outpaced the transportation system rather sooner than expected. It was impractical to devote so much of the limited time to traveling to and from a sample site; what was really needed was a *vehicle* that would not only minimize the transit time, and hence maximize sampling time, but also enable more tools to be carried and more rocks to be collected, all of which would increase the overall productivity of a traverse. The visit to Fra Mauro had achieved its primary objective, but it was clear that foot-traverses were no way to explore the Moon.

MISSION OBJECTIVES

Launch Vehicle Objectives

1. To launch on a 72–96°E of N flight azimth and insert the S-IVB/instrument unit/spacecraft into the planned circular Earth parking orbit. *Achieved.*
2. To restart the S-IVB during either the second or third revolution and inject the S-IVB/instrument unit/spacecraft into the planned translunar trajectory. *Achieved.*
3. To provide the required attitude control for the S-IVB/instrument unit/spacecraft during transposition, docking, and ejection. *Achieved.*
4. To perform an evasive maneuver after ejection of the command and service module/lunar module from the S-IVB/instrument unit. *Achieved.*
5. To attempt to impact the S-IVB/instrument unit on the lunar surface within 189 nmi (350 km) of latitude 01°35′06″S, longitude 33°15′W. *Achieved.*
6. To determine actual impact point within 2.7 nmi (5.0 km) and time of impact within 1 sec. *Achieved.*
7. To vent and dump the remaining gases and propellants to safe the S-IVB/instrument unit after final launch vehicle/spacecraft separation. *Achieved.*

8. To verify the operation of the liquid oxygen feedline accumulator systems installed on the S-II stage center engine. *Achieved.*

Spacecraft Primary Objectives

1. To perform selenological inspection, survey, and sampling of materials in a preselected region of the Fra Mauro Formation. *Achieved.*
2. To deploy and activate the Apollo lunar surface experiments package. *Achieved.*
3. To develop human capability to work in the lunar environment. *Achieved.*
4. To obtain photographs of candidate exploration sites. *Achieved.*

Detailed Objectives

1. Contingency sample collection. *Achieved.*
2. Photography of a candidate exploration site. *Partially achieved. On the low-altitude pass (fourth revolution), the Hiflex lunar terrain camera malfunctioned and no usable photography was obtained of Descartes–Cayley. During the stereo strip photographic pass, the S-band high-gain antenna malfunctioned and consequently no camera shutter-open data were obtained.*
3. Visibility at high Sun angles. *Partially achieved. The last of four sets of observations was deleted to provide another opportunity to photograph the Descartes area; however sufficient data were collected to verify that the visibility analytical model could be used for Apollo planning purposes.*
4. Modular equipment transporter evaluation. *Achieved.*
5. Selenodetic reference point update. *Achieved.*
6. Command and service module orbital science photography. *Partially achieved. The lunar topographic camera malfunctioned, and the Hasselblad 70-mm camera with the 500-mm lens was substituted. The photography was excellent, but the resolution was considerably lower than possible with the lunar topographic camera.*
7. Assessment of extravehicular activity operation limits. *Achieved.*
8. Command and service module oxygen flow rate. *Achieved.*
9. Transearth lunar photography. *Partially achieved. Excellent photography of the lunar surface was obtained, but no lunar topographic photography was obtained because of a camera malfunction.*
10. Thermal coating degradation. *Achieved.*
11. Dim-light photography. *Achieved.*

Detailed Objectives Added In-Flight

1. S-IVB photography. *Not Achieved. The S-IVB could not be identified on the film during post-mission analysis.*
2. Command and service module water-dump photography. *Partially achieved. Although some water particles were seen on photographs of the water dump, there was no indication of the "snow storm" described by the crew.*

Experiments

1. ALSEP IV: Apollo Lunar Surface Experiments Package.
   (a) S-031: Lunar passive seismology. *Achieved.*

   (b) S-033:Lunar active seismology. *Achieved.*
   (c) S-036: Suprathermal ion detector. *Achieved.*
   (d) S-038: Charged particle lunar environment. *Achieved.*
   (e) S-058: Cold cathode ion gauge. *Achieved.*
   (f) M-515: Lunar dust detector. *Achieved.*
2. S-059: Lunar geology investigation. *Achieved.*
3. S-078: Laser-ranging retroreflector. *Achieved.*
4. S-080: Solar wind composition. *Achieved.*
5. S-164: S-band transponder. *Achieved.*
6. S-170: Downlink bistatic radar observation of the Moon. *Achieved.*
7. S-176: Apollo window meteoroid experiment. *Achieved.*
8. S-178: Gegenschein from lunar orbit. *Achieved.*
9. S-198: Portable magnetometer. *Achieved.*
10. S-200: Soil mechanics. *Achieved.*
11. M-078: Bone mineral measurement. *Achieved.*

In-flight Demonstrations
1. Electrophoretic separation (Marshall Space Flight Center). *Achieved.*
2. Heat flow and convection (Marshall Space Flight Center). *Achieved.*
3. Liquid transfer (Lewis Research Center). *Achieved.*
4. Composite casting (Marshall Space Flight Center). *Achieved.*

Operational Tests
1. For Manned Spacecraft Center:
   (a) Lunar gravity measurement (using the lunar module primary guidance system). *Achieved.*
   (b) Hydrogen maser test (a Network and unified S-band investigation sponsored by the Goddard Space Flight Center). *Achieved.*
2. For Department of Defense:
   (a) Chapel Bell (classified Department of Defense test). *Results classified.*
   (b) Radar skin tracking. *Results classified*
   (c) Ionospheric disturbance from missiles. *Results classified*
   (d) Acoustic measurement of missile exhaust noise. *Results classified*
   (e) Army acoustic test. *Results classified*
   (f) Long-focal-length optical system. *Results classified.*

The Apollo 14 crew: Stu Roosa (left), Al Shepard and Ed Mitchell.

The launch of Apollo 14.

During the traverse with the MET, Al Shepard works on a core-sample tube.

Although the astronauts did not realize it, the rocks on the near-horizon marked the rim of Cone crater.

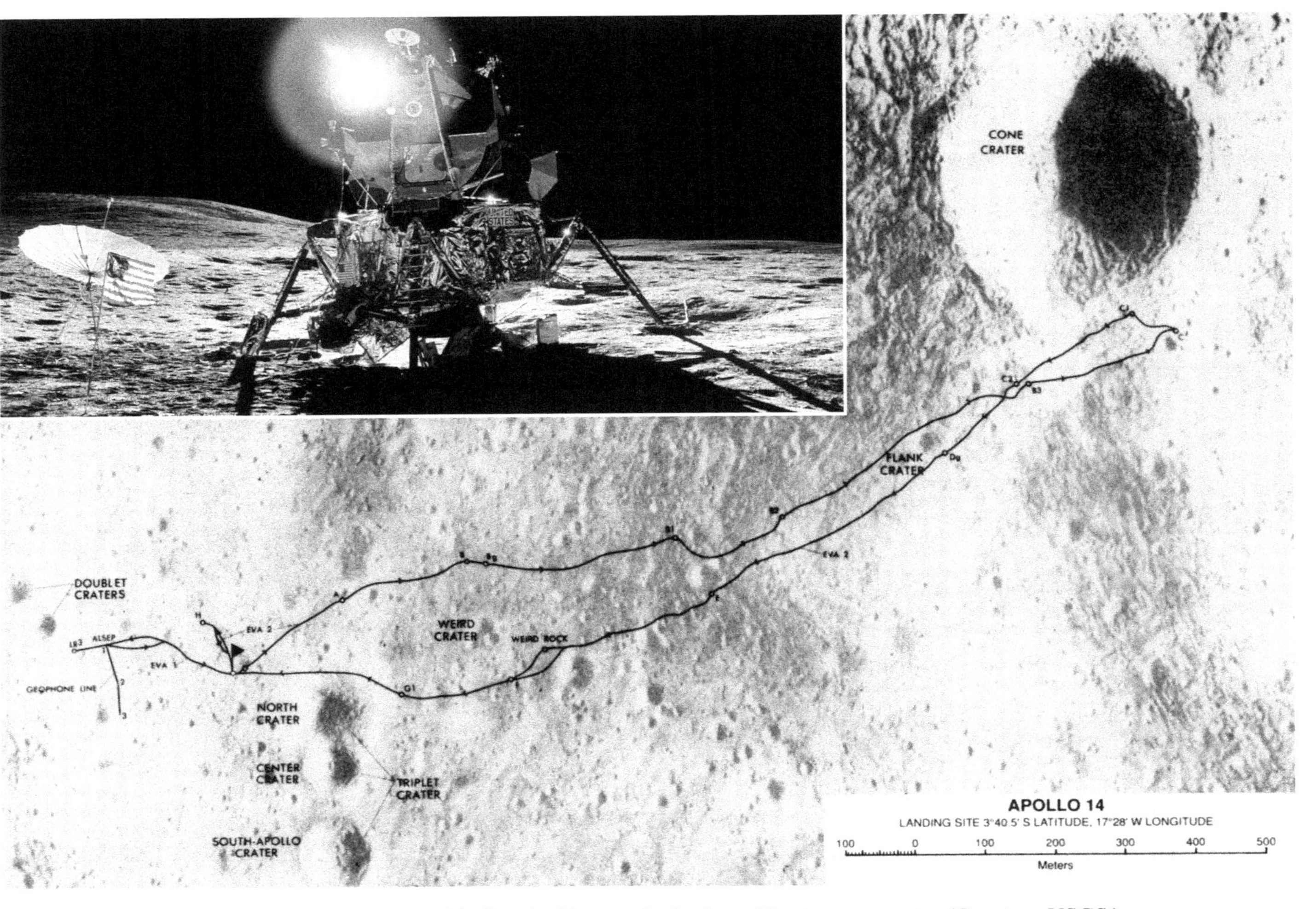

A view east from the LM with Cone's ridge on the horizon. The traverse route. (Courtesy USGS.)

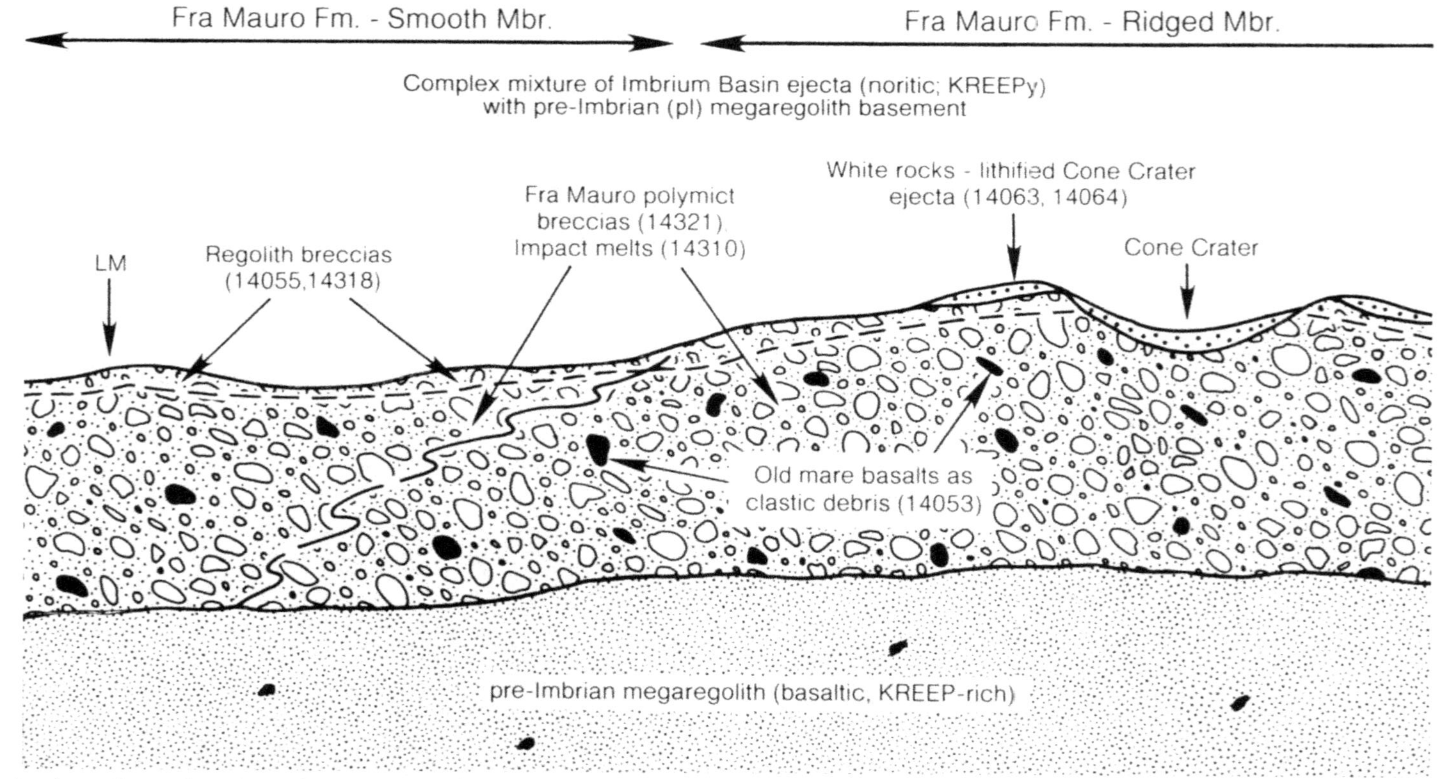

A schematic southwest–northeast cross-section through the Apollo 14 landing site showing the contact between the 'smooth' and 'ridged' members (Mbr) of the Fra Mauro Formation, which is an assemblage of complex impact-produced breccias (after G.A. Swann *et al.*, 'Geology of the Apollo 14 landing site in the Fra Mauro highlands', professional paper no. 800, USGS, 1977). Numbers refer to specific collected samples representative of the various units inferred to be present. 'LM' shows where the Lunar Module landed. (Courtesy the Lunar and Planetary Institute and Cambridge University Press.)

MISSION TIMELINE

Apollo 14 mission event	GET (h:m:s)	Date (GMT)	Time (h:m:s)
Terminal countdown started.	–028:00:00	30 Jan 1971	06:00:00
Scheduled 9-hr 23-min hold at T–9 hr.	–009:00:00	31 Jan 1971	01:00:00
Countdown resumed at T–9 hr.	–009:00:00	31 Jan 1971	10:23:00
Scheduled 1-hr hold at T–3 hr 30 min.	–003:30:00	31 Jan 1971	15:53:00
Countdown resumed at T–3 hr 30 min.	–003:30:00	31 Jan 1971	16:53:00
Unscheduled 40-min 2-sec weather hold at T–8 min.	–000:08:00	31 Jan 1971	20:15:00
Countdown resumed at T–8 min.	–000:08:00	31 Jan 1971	20:55:02
Guidance reference release.	–000:00:16.960	31 Jan 1971	21:02:45
S-IC engine start command.	–000:00:08.9	31 Jan 1971	21:02:53
S-IC engine ignition (#5).	–000:00:06.5	31 Jan 1971	21:02:55
All S-IC engines thrust OK.	–000:00:01.6	31 Jan 1971	21:03:00
Range zero.	000:00:00.00	31 Jan 1971	21:03:02
All holddown arms released (1st motion) (1.05 *g*).	000:00:00.2	31 Jan 1971	21:03:02
Liftoff (umbilical disconnected).	000:00:00.57	31 Jan 1971	21:03:02
Tower clearance yaw maneuver started.	000:00:01.958	31 Jan 1971	21:03:04
Yaw maneuver ended.	000:00:09.896	31 Jan 1971	21:03:11
Pitch and roll maneuver started.	000:00:12.814	31 Jan 1971	21:03:14
Roll maneuver ended.	000:00:28.000	31 Jan 1971	21:03:30
Mach 1 achieved.	000:01:08.0	31 Jan 1971	21:04:10
Maximum bending moment (116,000,000 lbf-in).	000:01:16	31 Jan 1971	21:04:18
Maximum dynamic pressure (655.80 lb/ft^2).	000:01:21.0	31 Jan 1971	21:04:23
S-IC center engine cutoff command.	000:02:15.14	31 Jan 1971	21:05:17
Pitch maneuver ended.	000:02:44.088	31 Jan 1971	21:05:46
S-IC outboard engine cutoff.	000:02:44.10	31 Jan 1971	21:05:46
S-IC maximum total inertial acceleration (3.82 *g*).	000:02:44.18	31 Jan 1971	21:05:46
S-IC maximum Earth-fixed velocity.	000:02:44.59	31 Jan 1971	21:05:46
S-IC/S-II separation command.	000:02:44.8	31 Jan 1971	21:05:46
S-II engine start command.	000:02:45.5	31 Jan 1971	21:05:47
S-II ignition.	000:02:46.5	31 Jan 1971	21:05:48
S-II aft interstage jettisoned.	000:03:14.8	31 Jan 1971	21:06:16
Launch escape tower jettisoned.	000:03:20.7	31 Jan 1971	21:06:22
Iterative guidance mode initiated.	000:03:25.912	31 Jan 1971	21:06:27
S-IC apex.	000:04:31.8	31 Jan 1971	21:07:33
S-II center engine cutoff.	000:07:43.09	31 Jan 1971	21:10:45
S-II maximum total inertial acceleration (1.81 *g*).	000:07:43.17	31 Jan 1971	21:10:45
S-IC impact (theoretical).	000:09:06.2	31 Jan 1971	21:12:08
S-II outboard engine cutoff.	000:09:19.05	31 Jan 1971	21:12:21
S-II/S-IVB separation command.	000:09:20.00	31 Jan 1971	21:12:22
S-II maximum Earth-fixed velocity.	000:09:20.07	31 Jan 1971	21:12:22
S-IVB 1st burn start command.	000:09:20.1	31 Jan 1971	21:12:22
S-IVB 1st burn ignition.	000:09:23.4	31 Jan 1971	21:12:25
S-IVB ullage case jettisoned.	000:09:31.8	31 Jan 1971	21:12:33
S-II apex.	000:10:00.2	31 Jan 1971	21:13:02

Apollo 14 mission event – *continued*	GET (h:m:s)	Date (GMT)	Time (h:m:s)
S-IVB 1st burn cutoff command.	000:11:40.56	31 Jan 1971	21:14:42
S-IVB 1st burn maximum total inertial acceleration (0.67 *g*).	000:11:40.66	31 Jan 1971	21:14:42
Earth orbit insertion. S-IVB 1st burn maximum Earth-fixed velocity.	000:11:50.56	31 Jan 1971	21:14:52
Maneuver to local horizontal attitude started.	000:12:02.092	31 Jan 1971	21:15:04
Orbital navigation started.	000:13:22.323	31 Jan 1971	21:16:24
S-II impact (theoretical).	000:20:46.3	31 Jan 1971	21:23:48
S-IVB 2nd burn restart preparation.	002:18:54.20	31 Jan 1971	23:21:56
S-IVB 2nd burn restart command.	002:28:24.10	31 Jan 1971	23:31:26
S-IVB 2nd burn ignition.	002:28:32.40	31 Jan 1971	23:31:34
S-IVB 2nd burn cutoff.	002:34:23.24	31 Jan 1971	23:37:25
S-IVB 2nd burn maximum total inertial acceleration (1.45 *g*).	002:34:23.34	31 Jan 1971	23:37:25
S-IVB 2nd burn maximum Earth-fixed velocity.	002:34:23.67	31 Jan 1971	23:37:25
S-IVB safing procedures started.	002:34:23.9	31 Jan 1971	23:37:25
Translunar injection.	002:34:33.24	31 Jan 1971	23:37:35
Orbital navigation started.	002:36:54.841	31 Jan 1971	23:39:56
Maneuver to local horizontal attitude started.	002:36:55.064	31 Jan 1971	23:39:57
Maneuver to transposition and docking attitude started.	002:51:04.339	31 Jan 1971	23:54:06
Maneuver to transposition and docking attitude ended.	002:55:23.37	31 Jan 1971	23:58:25
CSM separated from S-IVB.	003:02:29.4	01 Feb 1971	00:05:31
TV transmission started.	003:05	01 Feb 1971	00:08
1st docking attempt – 1st contact.	003:13:53.7	01 Feb 1971	00:16:55
1st docking attempt – 2nd contact.	003:14:01.5	01 Feb 1971	00:17:03
1st docking attempt – 3rd contact.	003:14:04.45	01 Feb 1971	00:17:06
1st docking attempt – 4th contact.	003:14:09.0	01 Feb 1971	00:17:11
2nd docking attempt.	003:14:43.7	01 Feb 1971	00:17:45
3rd docking attempt.	003:16:43.4	01 Feb 1971	00:19:45
4th docking attempt.	003:23:41.7	01 Feb 1971	00:26:43
5th docking attempt.	004:32:29.3	01 Feb 1971	01:35:31
6th docking attempt.	004:56:44.9	01 Feb 1971	01:59:46
CSM docked with LM/S-IVB (initial docking latch triggered).	004:56:56.7	01 Feb 1971	01:59:58
TV transmission ended.	005:00	01 Feb 1971	02:03
CSM/LM ejected from S-IVB (RCS ignition).	005:47:14.4	01 Feb 1971	02:50:16
CSM/LM ejected from S-IVB (RCS cutoff).	005:47:21.3	01 Feb 1971	02:50:23
Maneuver to attitude for S-IVB APS evasive burn initiated.	005:55:30	01 Feb 1971	02:58:32
S-IVB APS evasive maneuver ignition.	006:04:01.7	01 Feb 1971	03:07:03
S-IVB APS evasive maneuver cutoff.	006:05:21.9	01 Feb 1971	03:08:23
Maneuver to S-IVB LOX dump attitude initiated.	006:13:43.0	01 Feb 1971	03:16:45
S-IVB lunar impact maneuver – CVS vent for lunar targeting velocity change started.	006:20:40.5	01 Feb 1971	03:23:42
S-IVB lunar impact maneuver – LOX dump started.	006:25:20.5	01 Feb 1971	03:28:22
S-IVB lunar impact maneuver – CVS vent for lunar targeting velocity change started.	006:25:40.5	01 Feb 1971	03:28:42
S-IVB lunar impact maneuver – LOX dump ended.	006:26:08.5	01 Feb 1971	03:29:10
Maneuver to attitude required for final S-IVB APS burn initiated.	008:43:41.0	01 Feb 1971	05:46:43

Apollo 14 mission event – *continued*	GET (h:m:s)	Date (GMT)	Time (h:m:s)
S-IVB lunar impact maneuver – APS ignition.	008:59:59.0	01 Feb 1971	06:03:01
S-IVB lunar impact maneuver – APS cutoff.	009:04:11.2	01 Feb 1971	06:07:13
TV transmission started.	011:00	01 Feb 1971	08:03
Hatch, probe, and drogue removed for inspection.	011:30	01 Feb 1971	08:33
TV transmission ended.	012:12	01 Feb 1971	09:15:02
Midcourse correction ignition (SPS).	030:36:07.91	02 Feb 1971	03:39:09
Midcourse correction cutoff.	030:36:18.10	02 Feb 1971	03:39:20
Earth darkside dim-light photography.	031:00	02 Feb 1971	04:03
S-IVB photography.	034:00	02 Feb 1971	07:03
Lunar topographic camera unstowed and checked out.	034:15	02 Feb 1971	07:18
Bistatic radar frequency check.	052:00	03 Feb 1971	01:03
Mission clock updated (000:040:02.9 added).	054:53:36	03 Feb 1971	03:56:38
LM pressurization started.	059:50	03 Feb 1971	08:53
TV transmission started.	060:05	03 Feb 1971	09:08
Preparation for LM ingress.	060:10	03 Feb 1971	09:13
CDR and LMP entered LM.	060:30	03 Feb 1971	09:33
TV transmission ended.	060:42	03 Feb 1971	09:45
LM system checks.	061:40	03 Feb 1971	10:43
Water dump photography.	061:50	03 Feb 1971	10:53
CDR and LMP entered CM.	062:20	03 Feb 1971	11:23
Equigravisphere.	066:09:01	03 Feb 1971	15:12:03
LM cabin pressurized.	075:20	04 Feb 1971	00:23
Midcourse correction ignition (SPS).	076:58:11.98	04 Feb 1971	02:01:14
Midcourse correction cutoff.	076:58:12.63	04 Feb 1971	02:01:14
LM ascent battery test started.	078:20	04 Feb 1971	03:23
LM ascent battery test ended.	080:20	04 Feb 1971	05:23
Lunar orbit insertion ignition (SPS).	081:56:40.70	04 Feb 1971	06:59:42
Lunar orbit insertion cutoff.	082:02:51.54	04 Feb 1971	07:05:53
S-IVB impact on lunar surface.	082:37:53.4	04 Feb 1971	07:40:55
CSM landmark tracking.	085:10	04 Feb 1971	10:13
Descent orbit insertion ignition (SPS).	086:10:52.97	04 Feb 1971	11:13:55
Descent orbit insertion cutoff.	086:11:13.78	04 Feb 1971	11:14:15
CSM landmark tracking.	087:10	04 Feb 1971	12:13
Descartes photographed.	088:50	04 Feb 1971	13:53
LM pressurized.	101:05	05 Feb 1971	02:08
Docking tunnel opened. CDR and LMP entered LM.	101:20	05 Feb 1971	02:23
LM activation and system checks.	101:30	05 Feb 1971	02:33
CSM/LM undocking and separation ignition (SM RCS).	103:47:41.6	05 Feb 1971	04:50:43
CSM/LM undocking and separation cutoff.	103:47:44.3	05 Feb 1971	04:50:46
CSM landmark tracking.	104:20	05 Feb 1971	05:23
LM landing site observation.	104:30	05 Feb 1971	05:33
CSM orbit circularization ignition (SPS).	105:11:46.11	05 Feb 1971	06:14:48
CSM orbit circularization cutoff.	105:11:50.13	05 Feb 1971	06:14:52
Checkout of LM descent propulsion system and landing radar.	105:40	05 Feb 1971	06:43
CSM landmark tracking.	106:20	05 Feb 1971	07:23

Apollo 14 mission event – *continued*	GET (h:m:s)	Date (GMT)	Time (h:m:s)
CSM orbital science photography.	107:50	05 Feb 1971	08:53
LM landing radar on.	107:51:18.66	05 Feb 1971	08:54:20
LM false 'data good' indications from landing radar.	107:52:46.66	05 Feb 1971	08:55:48
LM landing radar switched to low scale.	107:57:34.66	05 Feb 1971	09:00:36
LM loading abort bit work-around routine started.	107:58:13.80	05 Feb 1971	09:01:15
LM ullage on.	108:02:19.12	05 Feb 1971	09:05:21
LM powered descent engine ignition (DPS).	108:02:26.52	05 Feb 1971	09:05:28
LM manual throttle-up to full throttle position.	108:02:53.80	05 Feb 1971	09:05:55
LM manual target (landing site) update.	108:04:49.80	05 Feb 1971	09:07:51
LM throttle down.	108:08:47.68	05 Feb 1971	09:11:49
LM landing radar switched to high scale.	108:08:50.66	05 Feb 1971	09:11:52
LM landing radar velocity data good.	108:09:10.66	05 Feb 1971	09:12:12
LM landing radar range data good.	108:09:12.66	05 Feb 1971	09:12:14
LM altitude updates enabled.	108:09:35.80	05 Feb 1971	09:12:37
LM approach phase program selected.	108:11:09.80	05 Feb 1971	09:14:11
LM pitchover started.	108:11:10.42	05 Feb 1971	09:14:12
LM landing radar redesignation enabled.	108:11:51.56	05 Feb 1971	09:14:53
LM radar antenna to position 2.	108:11:52.66	05 Feb 1971	09:14:54
LM attitude hold mode selected.	108:13:07.86	05 Feb 1971	09:16:09
LM landing phase program selected.	108:13:09.80	05 Feb 1971	09:16:11
LM lunar landing (left pad touchdown).	108:15:09.30	05 Feb 1971	09:18:11
LM powered descent engine cutoff.	108:15:11.13	05 Feb 1971	09:18:13
LM right, forward, and aft pad touchdown.	108:15:11.40	05 Feb 1971	09:18:13
CSM landmark tracking.	109:30	05 Feb 1971	10:33
LM lunar surface navigation.	110:00	05 Feb 1971	11:03
CSM Gegenschein photography.	110:40	05 Feb 1971	11:43
CSM backward-looking zero phase observations.	111:20	05 Feb 1971	12:23
CSM forward-looking zero phase observations.	112:20	05 Feb 1971	13:23
CSM zodiacal light photography.	112:50	05 Feb 1971	13:53
1st EVA started (LM cabin depressurization started).	113:39:11	05 Feb 1971	14:42:13
Egress started CDR). Pre-egress operations started (LMP).	113:47	05 Feb 1971	14:50
1st EVA television transmission started.	113:50	05 Feb 1971	14:53
CDR on lunar surface. Environmental familiarization, modular equipment transporter unloading, and television deployment (CDR).	113:51	05 Feb 1971	14:54
LMP egress.	113:55	05 Feb 1971	14:58
Environmental familiarization, contingency sample collection (LMP).	113:57	05 Feb 1971	15:00
CSM tracking of landed LM.	114:10	05 Feb 1971	15:13
S-band antenna deployment started (CDR).	114:12	05 Feb 1971	15:15
Solar wind composition experiment deployed (LMP).	114:13	05 Feb 1971	15:16
Laser-ranging retroreflector unloading started (LMP).	114:14	05 Feb 1971	15:17
Expendables transferred (CDR).	114:22	05 Feb 1971	15:25
LMP ingress.	114:23	05 Feb 1971	15:26
S-band antenna switching (LMP).	114:25	05 Feb 1971	15:28

Apollo 14 mission event – *continued*	GET (h:m:s)	Date (GMT)	Time (h:m:s)
LMP egress.	114:37	05 Feb 1971	15:40
Camera setup (LMP).	114:39	05 Feb 1971	15:42
US flag deployed and photographed.	114:41	05 Feb 1971	15:44
LM and site inspection (CDR). Traverse to television started (LMP).	114:47	05 Feb 1971	15:50
TV panorama (LMP).	114:50	05 Feb 1971	15:53
Modular equipment transporter deployment (LMP). CSM landmark tracking.	115:00	05 Feb 1971	16:03
TV transfer to scientific equipment bay (CDR).	115:05	05 Feb 1971	16:08
Experiment package offloading started (CDR and LMP).	115:08	05 Feb 1971	16:11
TV positioning (CDR).	115:22	05 Feb 1971	16:25
Modular equipment transporter loading (CDR).	115:25	05 Feb 1971	16:28
Traverse to experiment package deployment site (CDR, LMP).	115:46	05 Feb 1971	16:49
Experiment package system interconnect, thumper and geophone unloading started (LMP).	116:03	05 Feb 1971	17:06
Experiment package system interconnect, passive seismic experiment offloading, laser-ranging retroreflector deployment.	116:04	05 Feb 1971	17:07
Mortar offloaded (LMP).	116:26	05 Feb 1971	17:29
Charged particle lunar environment experiment deployment (CDR).	116:30	05 Feb 1971	17:33
Suprathermal ion detector experiment unloading and deployment (LMP).	116:34	05 Feb 1971	17:37
Deployment of experiment package antenna, passive seismic experiment, and laser-ranging retroreflector and sample collection (CDR).	116:35	05 Feb 1971	17:38
CSM galactic survey photography.	116:40	05 Feb 1971	17:43
Penetrometer activity (LMP).	116:45	05 Feb 1971	17:48
Geophone deployment started (LMP).	116:47	05 Feb 1971	17:50
1st ALSEP data received on Earth.	116:47:58	05 Feb 1971	17:51:00
Thumper activity (LMP).	117:02	05 Feb 1971	18:05
CSM plane change ignition (SPS).	117:29:33.17	05 Feb 1971	18:32:35
CSM plane change cutoff.	117:29:51.67	05 Feb 1971	18:32:53
Mortar pack arming started (LMP).	117:37	05 Feb 1971	18:40
Return traverse started (CDR).	117:38	05 Feb 1971	18:41
Return traverse started (LMP).	117:42	05 Feb 1971	18:45
EVA closeout (LMP).	117:54	05 Feb 1971	18:57
Sample collection (CDR).	118:00	05 Feb 1971	19:03
EVA closeout (CDR).	118:03	05 Feb 1971	19:06
CSM Earthshine photography.	118:10	05 Feb 1971	19:13
LMP ingress.	118:15	05 Feb 1971	19:18
EVA ended (LMP).	118:18	05 Feb 1971	19:21
CDR ingress.	118:19	05 Feb 1971	19:22
1st EVA television transmission ended.	118:20	05 Feb 1971	19:23
1st EVA ended (cabin pressurization started).	118:27:01	05 Feb 1971	19:30:03
VHF bistatic radar test started.	119:10	05 Feb 1971	20:13

Apollo 14 mission event – *continued*	GET (h:m:s)	Date (GMT)	Time (h:m:s)
CSM orbital science photography.	129:30	06 Feb 1971	06:33
S-band bistatic radar test started.	129:45	06 Feb 1971	06:48
VHF and S-band bistatic radar tests terminated.	130:20	06 Feb 1971	07:23
2nd EVA started (cabin depressurization started).	131:08:13	06 Feb 1971	08:11:15
CDR egress.	131:13	06 Feb 1971	08:16
Familiarization and transfer of equipment transfer bag (CDR). LMP egress. CSM vertical and orbital science photography.	131:20	06 Feb 1971	08:23
Modular equipment transporter preparation (LMP).	131:21	06 Feb 1971	08:24
Modular equipment transporter loading (CDR).	131:28	06 Feb 1971	08:31
Lunar portable magnetometer offloading (CDR).	131:38	06 Feb 1971	08:41
Lunar portable magnetometer offloading (LMP).	131:39	06 Feb 1971	08:42
2nd EVA television transmission started.	131:40	06 Feb 1971	08:43
Evaluation of modular equipment transporter track (CDR).	131:43	06 Feb 1971	08:46
Lunar portable magnetometer operation (LMP).	131:44	06 Feb 1971	08:47
Departed LM for station A (CDR).	131:46	06 Feb 1971	08:49
Departed LM for station A (LMP).	131:48	06 Feb 1971	08:51
Station A activity (CDR/LMP).	131:54	06 Feb 1971	08:57
CSM galactic survey photography.	132:25	06 Feb 1971	09:28
Departed station A for station B (CDR/LMP).	132:26	06 Feb 1971	09:29
Station B activity (CDR/LMP).	132:34	06 Feb 1971	09:37
CSM lunar libration photography.	132:35	06 Feb 1971	09:38
Departed station B for station Delta (CDR/LMP).	132:39	06 Feb 1971	09:42
Station Delta activity (CDR/LMP).	132:42	06 Feb 1971	09:45
Departed station Delta for station B1 (LMP).	132:44	06 Feb 1971	09:47
Departed station Delta for station B1 (CDR/LMP).	132:45	06 Feb 1971	09:48
Station B1 activity (CDR/LMP).	132:48	06 Feb 1971	09:51
Departed station B1 for station B2 (CDR/LMP).	132:52	06 Feb 1971	09:55
Station B2 activity (CDR/LMP).	132:57	06 Feb 1971	10:00
Departed station B2 for station B3 (CDR/LMP).	133:00	06 Feb 1971	10:03
Station B3 activity (CDR/LMP).	133:14	06 Feb 1971	10:17
Departed station B2 for station C prime (CDR/LMP).	133:16	06 Feb 1971	10:19
Station C prime activity (CDR/LMP).	133:22	06 Feb 1971	10:25
Departed station C prime for station C1 (CDR/LMP).	133:38	06 Feb 1971	10:41
Station C1 activity (CDR/LMP).	133:40	06 Feb 1971	10:43
Departed station C1 for station C2 (CDR/LMP).	133:46	06 Feb 1971	10:49
Station C2 activity (CDR/LMP).	133:52	06 Feb 1971	10:55
Departed station C2 for station E (CDR/LMP).	133:54	06 Feb 1971	10:57
Station E activity (CDR/LMP).	134:00	06 Feb 1971	11:03
Departed station E for station F (CDR/LMP).	134:02	06 Feb 1971	11:05
Station F activity (CDR/LMP).	134:06	06 Feb 1971	11:09
Departed station F for station G (CDR/LMP).	134:09	06 Feb 1971	11:12
Station G activity (CDR/LMP).	134:11	06 Feb 1971	11:14
Departed station G for station G1 (CDR/LMP).	134:47	06 Feb 1971	11:50
Station G1 activity (CDR/LMP).	134:49	06 Feb 1971	11:52
Departed station G1 for LM (CDR/LMP).	134:52	06 Feb 1971	11:55

Apollo 14 mission event – *continued*	GET (h:m:s)	Date (GMT)	Time (h:m:s)
EVA closeout (CDR).	134:55	06 Feb 1971	11:58
EVA closeout (LMP).	134:57	06 Feb 1971	12:00
Solar wind composition experiment retrieved.	135:13	06 Feb 1971	12:16
CSM contingency photography of Descartes.	135:20	06 Feb 1971	12:23
EVA ended (LMP).	135:25	06 Feb 1971	12:28
EVA ended (CDR). Post-EVA activity operations prior to LM cabin repressurization (LMP).	135:35	06 Feb 1971	12:38
Post-EVA activity operations prior to LM cabin repressurization (CDR).	135:41	06 Feb 1971	12:44
2nd EVA ended (cabin repressurization started).	135:42:54	06 Feb 1971	12:45:56
LM cabin depressurized, equipment jettisoned, cabin repressurized.	136:40	06 Feb 1971	13:43
CSM landmark tracking started.	137:10	06 Feb 1971	14:13
CSM landmark tracking ended.	137:55	06 Feb 1971	14:58
Rendezvous radar activation and self-test.	138:40	06 Feb 1971	15:43
CSM backward-looking zero phase observations and orbital science photography.	139:00	06 Feb 1971	16:03
CSM forward-looking zero phase observations.	139:55	06 Feb 1971	16:58
LM lunar liftoff ignition (LM APS).	141:45:40	06 Feb 1971	18:48:42
Lunar ascent orbit cutoff.	141:52:52.1	06 Feb 1971	18:55:54
Vernier adjustment ignition (LM RCS).	141:56:49.4	06 Feb 1971	18:59:51
Vernier adjustment cutoff.	141:57:01.5	06 Feb 1971	19:00:03
Terminal phase initiation ignition.	142:30:51.1	06 Feb 1971	19:33:53
Terminal phase initiation cutoff.	142:30:54.7	06 Feb 1971	19:33:56
LM 1st midcourse correction.	142:45	06 Feb 1971	19:48
LM 2nd midcourse correction.	143:00	06 Feb 1971	20:03
Terminal phase finalize ignition.	143:13:29.1	06 Feb 1971	20:16:31
Terminal phase finalize cutoff.	143:13:55.8	06 Feb 1971	20:16:57
TV transmission started.	143:15	06 Feb 1971	20:18
TV transmission ended.	143:20	06 Feb 1971	20:23
TV transmission started.	143:28	06 Feb 1971	20:31
CSM/LM docked.	143:32:50.5	06 Feb 1971	20:35:52
TV transmission ended.	143:35	06 Feb 1971	20:38
Equipment and samples transferred to CM.	144:00	06 Feb 1971	21:03
LM ascent stage jettisoned.	145:44:58.0	06 Feb 1971	22:48:00
CSM/LM final separation ignition (SM RCS).	145:49:42.5	06 Feb 1971	22:52:44
CSM/LM final separation cutoff.	145:49:58.3	06 Feb 1971	22:53:00
Contamination control.	146:20	06 Feb 1971	23:23
LM ascent stage deorbit ignition (LM RCS).	147:14:16.9	07 Feb 1971	00:17:18
LM ascent stage fuel depletion.	147:15:33.1	07 Feb 1971	00:18:35
LM ascent stage impact on lunar surface.	147:42:23.7	07 Feb 1971	00:45:25
Apollo 12 LM impact point and Apollo 13 and Apollo 14 S-IVB impact points photographed.	147:45	07 Feb 1971	00:48
Transearth injection ignition (SPS).	148:36:02.30	07 Feb 1971	01:39:04
Transearth injection cutoff.	148:38:31.53	07 Feb 1971	01:41:33

Apollo 14 mission event – *continued*	GET (h:m:s)	Date (GMT)	Time (h:m:s)
Lunar photography.	148:55	07 Feb 1971	01:58
Cislunar navigation started.	163:30	07 Feb 1971	16:33
Cislunar navigation ended.	164:20	07 Feb 1971	17:23
Midcourse correction ignition (SM RCS).	165:34:56.69	07 Feb 1971	18:37:58
Midcourse correction cutoff.	165:34:59.69	07 Feb 1971	18:38:01
Cislunar navigation started.	165:40	07 Feb 1971	18:43
Cislunar navigation ended.	166:50	07 Feb 1971	19:53
Oxygen flow rate test attitude started.	167:25	07 Feb 1971	20:28
Oxygen flow rate test started.	167:50	07 Feb 1971	20:53
Oxygen flow rate test ended.	169:00	07 Feb 1971	22:03
Oxygen flow rate test attitude ended.	170:40	07 Feb 1971	23:43
Contamination control.	171:20	08 Feb 1971	00:23
TV transmission started.	171:30	08 Feb 1971	00:33
In-flight demonstrations started.	171:50	08 Feb 1971	00:53
In-flight demonstrations ended.	172:09	08 Feb 1971	01:12
TV transmission ended.	172:20	08 Feb 1971	01:23
Light flash experiment started.	190:50	08 Feb 1971	19:53
Light flash experiment ended.	191:50	08 Feb 1971	20:53
TV transmission started.	194:29	08 Feb 1971	23:32
TV transmission ended.	194:52	08 Feb 1971	23:55
Earth darkside dim-light photography.	197:44	09 Feb 1971	02:47
CM/SM separation.	215:32:42.2	09 Feb 1971	20:35:44
Entry.	215:47:45.3	09 Feb 1971	20:50:47
Communication blackout started.	215:48:02	09 Feb 1971	20:51:04
Communication blackout ended.	215:51:19	09 Feb 1971	20:54:21
S-band contact with CM established by recovery forces.	215:52	09 Feb 1971	20:55
Radar contact with CM established by recovery ship.	215:53	09 Feb 1971	20:56
Drogue parachute deployed.	215:56:08	09 Feb 1971	20:59:10
Visual contact with CM established by recovery helicopter.	215:57	09 Feb 1971	21:00
Voice contact with CM established by recovery ship.	215:58	09 Feb 1971	21:01
Splashdown (went to apex-up).	216:01:58.1	09 Feb 1971	21:05:00
VHF beacon contact established with CM by recovery helicopter.	216:04	09 Feb 1971	21:07
Swimmers deployed to CM.	216:09	09 Feb 1971	21:12
Flotation collar inflated.	216:17	09 Feb 1971	21:20
Decontamination swimmer deployed.	216:24	09 Feb 1971	21:27
Hatch opened for crew egress.	216:37	09 Feb 1971	21:40
Crew in life raft.	216:38	09 Feb 1971	21:41
Crew on board recovery helicopter.	216:45	09 Feb 1971	21:48
Crew on board recovery ship.	216:50	09 Feb 1971	21:53
Crew entered mobile quarantine facility.	217:00	09 Feb 1971	22:03
CM on board recovery ship.	218:06	09 Feb 1971	23:09
1st sample flight departed recovery ship.	246:52	11 Feb 1971	03:55
Crew departed recovery ship.	260:43	11 Feb 1971	17:46
1st sample flight arrived in Houston, TX.	263:54	11 Feb 1971	20:57

Apollo 14 mission event – *continued*	GET (h:m:s)	Date (GMT)	Time (h:m:s)
Crew arrived at Ellington AFB, Houston.	276:31	12 Feb 1971	09:34
Crew in Lunar Receiving Laboratory, Houston.	278:32	12 Feb 1971	11:35
Mobile quarantine facility and CM offloaded in Hawaii.	408:27	17 Feb 1971	21:30
Mobile quarantine facility arrived in Houston.	418:37	18 Feb 1971	07:40
CM RCS deactivation completed.	457:57	19 Feb 1971	23:00
CM arrived in Houston.	528:42	22 Feb 1971	21:45
CM delivered to Lunar Receiving Laboratory.	530:27	22 Feb 1971	23:30
Crew released from quarantine.	–	27 Feb 1971	–
CM arrived at contractor's facility in Downey, CA.	–	08 Apr 1971	–
Lunar samples released.	–	04 Apr 1971	–
ALSEP central station turned off by ground command.	–	30 Sep 1977	–

Apollo 15

The ninth manned mission: the fourth lunar landing

26 July–7 August 1971

BACKGROUND

Apollo 15 was the first of a series of Type 'J' missions, consisting of extensive scientific investigations of the Moon on the lunar surface and from lunar orbit. It was designed to conduct exploration of the Moon over longer periods, over greater ranges, and with more instruments for scientific data acquisition than on previous Apollo missions. Major modifications and augmentations to the basic Apollo hardware were made. The most significant was the installation of an instrument module in one of the service module bays to conduct scientific investigations from lunar orbit, and a scientific subsatellite that was to be deployed into lunar orbit. Other hardware changes consisted of LM modifications to accommodate a greater payload and permit a longer stay on the lunar surface, the provision of a lunar roving vehicle (LRV). Planned to be used on this and the next two lunar missions, the LRV was a four-wheeled, lightweight vehicle designed to greatly extend the area that could be explored on the lunar surface. It had five major systems: mobility, crew station, navigation, power, and thermal control. Auxiliary equipment included the lunar communications relay unit with high- and low-gain antennas, ground control television assembly, a motion picture film camera, scientific equipment, astronaut tools, and sample stowage bags. It was 122 in long and 44.8 in high, with a 90-in wheelbase and 72-in tread width. Two 36-volt batteries provided power, although one alone was sufficient to power all LRV systems. Its Earth weight was 462 lb, and it had a payload capacity of 1,080 lb – including two astronauts and their life-support equipment (about 800 lb), communication equipment (100 lb), scientific equipment, photographic gear (120 lb), and lunar samples (60 lb). For the flight to the Moon, the LRV was folded flat and stowed in Quad 1 of the descent stage. Following landing, the astronauts were to manually deploy the vehicle and prepare it for cargo loading and operation. It was designed to operate for 78 hr during the lunar day, and could travel a cumulative distance of 35 nmi, within a 5-nmi radius of the LM.

The primary objectives for Apollo 15 were:

- to perform selenological inspection, survey, and sampling of materials and surface features in a preselected area of the Moon;

- to emplace and activate surface experiments;
- to evaluate the capability of the Apollo equipment to provide extended lunar surface stay time, increased extravehicular operations, and surface mobility; and
- to conduct in-flight experiments and photographic tasks from lunar orbit.

The all-Air Force crew included Colonel David Randolph Scott (USAF), commander; Major Alfred Merrill Worden [WARD-in] (USAF), command module pilot; and Lieutenant Colonel James Benson Irwin (USAF), lunar module pilot. Selected as an astronaut in 1963, Scott had been pilot of Gemini 8, the first docking of two vehicles in space, and command module pilot of Apollo 9, the first flight test of the LM. Born on 6 June 1932 in San Antonio, Texas, he was 39 years old at the time of the Apollo 15 mission. Scott received a BS from the US Military Academy in 1954 and a MS in aeronautics and astronautics from the Massachusetts Institute of Technology in 1962. His backup for the mission was Captain Richard Francis 'Dick' Gordon Jr (USN). Worden and Irwin were making their first spaceflights. Worden was born on 7 February 1932 in Jackson, Michigan, and was 39 years old at the time of the Apollo 15 mission. He received a BS in military science from the US Military Academy in 1955, a MS in astronautical and aeronautical engineering and a MS in instrumentation engineering from the University of Michigan in 1963, and was selected as an astronaut in 1966. His backup was Vance DeVoe Brand. Born on 17 March 1930 in Pittsburgh, Pennsylvania, Irwin was 41 years old at the time of the Apollo 15 mission. He received a BS in naval science from the US Naval Academy in 1951, and a MS in aeronautical engineering and an MS in instrumentation engineering from the University of Michigan in 1957, and was selected as an astronaut in 1966.[5] His backup was Harrison Hagan 'Jack' Schmitt PhD. The capsule communicators (CAPCOMs) for the mission were Joseph Percival Allen IV PhD, Major Charles Gordon Fullerton (USAF), Karl Gordon Henize PhD, Commander Edgar Dean Mitchell (USN/ScD), Robert Alan Ridley Parker PhD, Schmitt, Captain Alan Bartlett Shepard Jr (USN), Gordon, and Brand. The support crew were Henize, Allen, and Parker. The flight directors were Gerald D. Griffin (first shift), Milton L. Windler (second shift), and Glynn S. Lunney and Eugene F. Kranz (third shift).

The Apollo 15 launch vehicle was a Saturn V, designated AS-510. The mission also carried the designation Eastern Test Range #7744. The CSM was designated CSM-112, and had the call-sign 'Endeavour'. The lunar module was designated LM-10 and had the call-sign 'Falcon'.

LANDING SITE

With eastern and western maria and the Fra Mauro Formation sampled, the geologists were eager to attempt a highland site, but none of the 'feature' sites had been surveyed sufficiently to certify them for a landing. However, there was no shortage of attainable

[5] Irwin, who had a history of heart trouble, died of a heart attack on 8 August 1991 in Glenwood Springs, Colorado.

alternatives. In addition to the seemingly volcanic Marius Hills and Davy Rille, there was a system of 'dark mantled' ridges and rilles in Mare Serenitatis (Sea of Serenity) which Apollo 14 had been set to sample before it was redirected to Fra Mauro. Initially, Apollo 15 was to have been an 'H'-mission, but on 2 September 1970 it was decided to recast it as the first advanced 'J'-mission, and a 'multiple feature' site was really required to exploit the tremendous increase in exploratory capacity offered by the LRV. The favorite was a new contender: Hadley Rille, on the eastern rim of the Imbrium basin. Of two types of lunar rille – linear and sinuous – Hadley is one of the most impressive of the sinuous ones. It starts in an arcuate cleft in the basin-facing side of the Montes Apenninus (Apennine Mountains) and then exploits a system of radial and peripheral fractures to run north for 60 nmi, parallel with the mare shoreline. At the base of Mount Hadley Delta (a peak south of Mount Hadley) it transects the mouth of the embayed valley named Palus Putredinis (Marsh of Decay) before finally petering out.

Hadley–Apennine promised to be a rich venue for Apollo's study of the formation of the Imbrium basin, which was the basis for the stratigraphic study of the Moon, but landing there would be operationally demanding as it would involve passing low over the Apennines, whose peaks are amongst the highest on the near side, and then making a plunging descent, nearly twice as steep as previous missions, to set down in the valley nestling between a pair of massifs and a rille. There was no scope for a significant departure from the ideal approach track. It could not have been visited earlier, since it required the site selection rules to be relaxed. Heretofore, to be certified, both the site itself and the approach from the east must have been documented at high resolution. But Hadley–Apennine had been imaged by a Lunar Orbiter only as a 'scientific' site with the barely acceptable resolution of 66 ft. Nevertheless, it was the topography which confined Hadley–Apennine that made it so attractive to the scientists. The availability of a rille on a mare plain that had flooded the valley in a range of mountains on the periphery of the Imbrium basin made Apollo 15 the first 'multiple-objective' mission. It would have been a waste to send an earlier flight to a site as rich as this, because the LRV was essential to exploiting this potential.

LAUNCH PREPARATIONS

The terminal countdown was picked up at T–28 hr at 23:00:00 GMT on 24 July 1971. Scheduled holds were initiated at T–9 hr for 9 hr 34 min and at T–3 hr 30 min for 1 hour. At launch time, the Cape Kennedy launch area was experiencing fair weather resulting from a ridge of high pressure extending westward, from the Bermuda High, through central Florida. Cirrus clouds covered 70% of the sky (base 25,000 ft), the temperature was 85.6°F, the relative humidity was 68%, and the barometric pressure was 14.788 psi. The winds, as measured by the anemometer on the light pole 60.0 ft above ground at the launch site, were 9.9 kt at 156° from true north; the winds at 530 ft were 10.5 kt at 158° from true north.

Apollo 15 preparation event	Date
Individual and combined CM and SM systems test completed at factory.	05 Nov 1969
S-II-10 stage delivered to KSC.	18 May 1970
S-IVB-510 stage delivered to KSC.	13 Jun 1970
Saturn V instrument unit IU-510 delivered to KSC.	26 Jun 1970
S-IC-10 stage delivered to KSC.	06 Jul 1970
S-IC-10 stage erected on MLP-3.	08 Jul 1970
Spacecraft/LM adapter SLA-19 delivered to KSC.	08 Jul 1970
S-II-10 stage erected.	15 Sep 1970
S-IVB-510 stage erected.	16 Sep 1970
IU-510 erected.	17 Sep 1970
LM-10 final engineering evaluation acceptance test at factory.	21 Sep 1970
LM-10 integrated test at factory.	21 Sep 1970
LM-10 delivered to KSC and launch electrical systems tests completed.	17 Nov 1970
Launch vehicle electrical systems tests completed.	17 Nov 1970
Integrated CM and SM systems test completed at factory.	24 Nov 1970
CM-112 and SM-112 ready to ship from factory to KSC.	11 Jan 1971
CM-112 and SM-112 delivered to KSC.	14 Jan 1971
CM-112 and SM-112 mated.	18 Jan 1971
LM-10 stages mated.	09 Feb 1971
LM-10 combined systems test completed.	12 Feb 1971
CSM-112 combined systems test completed.	08 Mar 1971
LRV-1 delivered to KSC.	15 Mar 1971
LM-10 altitude tests completed.	06 Apr 1971
CSM-112 altitude tests completed.	09 Apr 1971
Launch vehicle propellant dispersion/malfunction overall test completed.	15 Apr 1971
Launch vehicle service arm overall test completed.	27 Apr 1971
LRV-1 installed.	28 Apr 1971
CSM-112 moved to VAB.	08 May 1971
Spacecraft erected.	08 May 1971
Space vehicle and MLP-3 transferred to Launch Complex 39A.	11 May 1971
LM-10 combined systems test completed.	17 May 1971
CSM-112 integrated systems test completed.	18 May 1971
CSM-112 electrically mated to launch vehicle.	07 Jun 1971
Space vehicle overall test #1 (plugs in) completed.	09 Jun 1971
LM-10 flight readiness test completed.	10 Jun 1971
Space vehicle flight readiness test completed.	22 Jun 1971
S-IC-10 stage RP-1 fuel loading completed.	06 Jul 1971
Space vehicle countdown demonstration test (wet) completed.	13 Jul 1971
Space vehicle countdown demonstration test (dry) completed.	14 Jul 1971

Apollo 15 ascent phase event	GET (h:m:s)	Altitude (nmi)	Range (nmi)	Earth-fixed velocity (ft/sec)	Space-fixed velocity (ft/sec)	Event duration (sec)	Geocentric latitude (°N)	Longitude (°E)	Space-fixed flight path angle (deg)	Space-fixed heading angle (°E of N)
Liftoff	000:00:00.58	0.060	0.000	1.5	1,340.7	–	28.4470	–80.6041	0.07	90.00
Mach 1 achieved	000:01:05.0	4.224	1.004	1,052.0	2,028.1	–	28.4497	–80.5854	27.86	87.36
Maximum dynamic pressure	000:01:22.0	7.401	2.970	1,661.1	2,681.3	–	28.4555	–80.5847	29.80	85.77
S-IC center engine cutoff*	000:02:15.96	25.271	25.987	5,518.4	6,708.5	142.46	28.5203	–80.1190	24.217	82.494
S-IC outboard engine cutoff	000:02:39.56	36.947	48.610	7,811.3	9,043.3	166.06	28.5824	–79.6961	21.266	82.129
S-IC/S-II separation*	000:02:41.2	37.830	596.012	7,827.6	9,062.2	–	28.5876	–79.6605	21.021	82.144
S-II outboard engine cutoff	000:09:09.06	95.184	874.532	21,588.4	22,949.6	386.06	29.6810	–63.9910	0.059	89.863
S-II/S-IVB separation*	000:09:10.1	95.187	878.126	21,601.2	22,962.5	–	29.6811	–63.9221	0.047	89.900
S-IVB 1st burn cutoff	000:11:34.67	93.215	1,406.808	24,236.4	25,596.7	141.47	29.2688	–53.8183	0.013	95.149
Earth orbit insertion	000:11:44.67	93.215	1,445.652	24,242.4	25,602.6	–	29.2052	–53.0807	0.015	95.531

* Only the commanded time is available for this event.

ASCENT PHASE

Apollo 15 was launched from Pad A of Launch Complex 39 at the Kennedy Space Center at a Range Zero time of 13:34:00 GMT (09:34:00 EDT) on 26 July 1971. The launch window extended to 16:11:00 GMT in order to take advantage of a Sun elevation angle on the lunar surface of 12.0 deg at the scheduled time of landing. Between T + 12.21 and T + 23.02 sec, the vehicle rolled from a pad azimuth of 90°E of N to a flight azimuth of 80.088°E of N. The maximum wind conditions encountered during ascent were 36.2 kt at 63° from true north at 45,110 ft with a maximum wind shear of 0.0110/sec at 36,830 ft. The S-IC shut down at T + 159.56, followed by S-IC/S-II separation, and S-II ignition. The S-II shut down at T + 549.06 followed by separation from the S-IVB, which ignited at T + 553.20. The first S-IVB cutoff occurred at T + 694.67, with deviations from the planned trajectory of only –2.0 ft/sec in velocity and only 0.4 nmi in altitude. At insertion, at T + 704.67 (i.e. S-IVB cutoff plus 10 sec to account for engine tailoff and other transient effects), the parking orbit showed an apogee and perigee of 91.5 × 89.6 nmi, an inclination of 29.679 deg, a period of 87.84 min, and a velocity of 25,602.6 ft/sec. The apogee and perigee were based upon a spherical Earth with a radius of 3,443.934 nmi.

The COSPAR designation for the CSM upon achieving orbit was 1971-063A and the S-IVB was designated 1971-063B. After undocking at the Moon, the LM ascent stage would be designated 1971-063C, the descent stage 1971-063E, and the subsatellite 1971-063D.

EARTH ORBIT PHASE

After in-flight systems checks, the 350.71-sec translunar injection maneuver (second S-IVB firing) was performed at 002:50:02.90. The S-IVB engine shut down at 002:55:53.61 and translunar injection occurred 10 sec later at a velocity of 35,579.1 ft/sec after 1.5 Earth orbits lasting 2 hr 44 min 18.94 sec.

Apollo 15 earth orbit phase event	GET (h:m:s)	Space-fixed velocity (ft/sec)	Event duration (sec)	Velocity change (ft/sec)	Apogee (nmi)	Perigee (nmi)	Period (min)	Inclination (deg)
Earth orbit insertion	000:11:44.67	25,602.6	–	–	91.5	89.6	87.84	29.679
S-IVB 2nd burn ignition	002:50:02.90	25,597.1	–	–	–	–	–	–
S-IVB 2nd burn cutoff	002:55:53.61	35,603.0	350.71	10,414.7	–	–	–	–

TRANSLUNAR PHASE

At 003:22:27.2, the CSM was separated from the S-IVB stage, transposed, and docked at 003:33:49.5. The onboard color television camera covered the docking. The docked spacecraft were ejected from the S-IVB at 004:18:01.2, and an 80.2-sec separation maneuver was performed at 004:40:01.8. At 005:46:00.7 the S-IVB tanks were vented and the auxiliary propulsion system was fired for 241.2 sec to target the S-IVB for a lunar impact. An additional 71-sec maneuver was made at 010:00:01, about 30 min later than planned. The late burn provided additional tracking time to compensate for any trajectory perturbations introduced by liquid oxygen and liquid hydrogen tank venting. The S-IVB impacted the lunar surface at 079:24:42.9. The impact point was 1.51°S and 11.81°W, 83 nmi from the target point, 192 nmi from the Apollo 12 seismometer, and 99 nmi from the Apollo 14 seismometer. At impact, the S-IVB weighed 30,880 lb and was traveling 8,465 ft/sec.

Two minor CSM midcourse corrections were required during translunar flight to assure proper lunar orbit insertion. The first was a 0.8-sec maneuver at 028:40:22.0 that produced a change in velocity of 5.3 ft/sec. The second midcourse correction was performed with the service propulsion system bank A in order to provide better analysis of an apparent intermittent short. Because power could still be applied to the valve with a downstream short, bank A could be operated satisfactorily in the manual mode for subsequent firings. The redundant bank B system was nominal and could be used for automatic starting and shutdown. The commander and lunar module pilot entered the LM at 033:56 for checkout, some 50 min earlier than scheduled. LM communications checks were performed between 034:21 and 034:45. Good-quality voice and data were received even though the Goldstone tracking station in California was not yet configured correctly during the initial portion of the down-voice backup checks. Approximately 15 min later the downlink carrier lock was lost for a 90 sec, but because other stations were tracking, data loss was reduced to just a few seconds. A television transmission of the CSM and LM interiors was broadcast between 034:55 and 035:46. Camera operation was nominal, but the picture quality varied with the lighting of the scene observed. During the checkout of the LM, the crew discovered the range/range-rate exterior cover glass was broken, thus removing the protective helium barrier. Subsequent ground testing qualified the unprotected meter for use during the remainder of the mission in the spacecraft ambient atmosphere. Intravehicular transfer and LM housekeeping began at 056:26, about 1.5 hours earlier than scheduled. The crew vacuumed the LM to remove broken glass from the damaged range/range-rate meter. LM checkout was completed as planned. Based on the first midcourse correction burn test data, it was decided to perform all service propulsion system maneuvers except lunar orbit insertion and transearth injection using bank B only. The insertion and injection maneuvers would be dual bank burns employing modified procedures to permit automatic start and shutdown on bank B. The second midcourse correction, using this propulsion system, was made at 073:31:14.81 for 0.91 sec and changed the velocity by 5.4 ft/sec. The scientific instrument bay door was jettisoned at 074:06:47.1. The lunar module pilot photographed the jettisoned door and visually observed it slowly tumbling through space away from the CSM and eventually into heliocentric orbit.

At 078:31:46.70, at an altitude of 86.7 nmi above the Moon, the service propulsion system was fired for 398.36 sec, inserting the spacecraft into a lunar orbit of 170.1 × 57.7 nmi. The translunar coast had lasted 75 hr 42 min 21.45 sec. During the burn, bank A was shut down 32 sec before planned cutoff to obtain performance data on bank B for future single bank burns.

Apollo 15 translunar phase event	GET (h:m:s)	Altitude (nmi)	Space-fixed velocity (ft/sec)	Event duration (sec)	Velocity change (ft/sec)	Space-fixed flight path angle (deg)	Space-fixed heading angle (°E of N)
Translunar injection	002:56:03.61	173.679	35,579.1	–	–	7.430	73.173
CSM separated from S-IVB	003:22:27.2	4,028.139	24,586.6	–	–	46.015	112.493
CSM docked with LM/S-IVB	003:33:49.5	5,985.4	21,811.0	–	–	51.66	115.86
CSM/LM ejected from S-IVB	004:18:01.2	12,826.9	16,402.2	–	–	61.45	119.20
Midcourse correction ignition	028:40:22.0	114,783.2	4,849.8	–	–	77.22	116.83
Midcourse correction cutoff	028:40:22.8	114,784.0	4,845.6	0.8	5.3	77.18	116.76
Midcourse correction ignition	073:31:14.81	12,618.4	3,963.1	–	–	–81.08	–139.68
Midcourse correction cutoff	073:31:15.72	12,617.7	3,966.8	0.91	5.4	–81.10	–140.00

LUNAR ORBIT AND LUNAR SURFACE PHASE

At 082:39:49:09, a 24.53-sec service propulsion system maneuver was performed to establish the descent orbit of 58.5 × 9.6 nmi in preparation for undocking of the LM. A 30.40-sec orbit trim maneuver was performed at 095:56:44.70 and adjusted the orbit to 60.3 × 8.8 nmi. During the 12th lunar revolution, at about 100:14, while on the far side of the Moon, the CSM/LM undocking and separation maneuver was initiated, but without effect. The crew and ground control decided that the probe instrumentation LM/CSM umbilical was either loose or disconnected. The command module pilot went into the tunnel to inspect the connection and confirmed the umbilical plug to be loose. After reconnecting the plug and adjusting the spacecraft attitude, undocking and separation were achieved approximately 25 min late at 100:39:16.2 at an altitude of 7.4 nmi. A 3.67-sec SPS maneuver at 101:38:58.98 circularized the CSM orbit to 65.2 × 54.8 nmi in preparation for the acquisition of scientific data. The powered descent was initiated at 104:30:09.4 at an altitude of 5.8 nmi, and ended 739.2 sec later, some 0.7 sec prior to the landing at 104:42:29.3 (22:16:29 GMT on 30 July). It was at selenographic coordinates 26.13222°N, 3.63386°E, near the eastern rim of Hadley Rille and 1,800 ft northwest of the planned landing point. At engine cutoff, approximately 103 sec of firing time remained.

At 106:42:49, 2 hours after landing, the cabin was depressurized and the commander opened the LM top hatch to describe the geological character of the landing site. During this 'stand-up EVA' (SEVA), which lasted 33 min 7 sec, he took a series of panoramic photos.

The first lunar surface extravehicular activity was initiated at 119:39:17, when the cabin of the LM was depressurized. On the way down the ladder, the commander deployed the modularized equipment stowage assembly (MESA). The television in the MESA was activated and the pictures of the commander's remaining descent to the lunar surface were excellent. The lunar module pilot then exited to the surface. While the commander removed the television camera from the MESA and deployed it on a tripod, the lunar module pilot collected the contingency sample. At 120:18:31, they offloaded the LRV. It was successfully deployed 13 min later. They unstowed the third Apollo lunar surface experiments package (ALSEP) and other equipment, and configured the LRV for lunar surface operations. Some problems were experienced in deploying and checking out the LRV but these problems were worked out. During checkout of the LRV, it was found that the front steering mechanism was inoperative. Additionally, there were no readouts on the LRV battery #2 ampere/voltmeter. After minor troubleshooting, a decision was made to perform the first extravehicular activity (EVA-1) with only the LRV rear wheel steering activated. At 121:44:55, the crew set-off driving for Elbow Crater, where they collected documented samples and gave an enthusiastic and informative commentary on lunar features. The Mission Control Center provided television control during various stops. After obtaining additional samples and photographs near St. George Crater, the crew returned to the LM using the LRV navigation system. The crew then proceeded to the selected Apollo lunar surface experiments package deployment site, 360 ft west-northwest of the LM. There, the experiments were deployed essentially as planned, except that the second heat flow experiment probe was not emplaced because drilling was more difficult than expected and the hole was not completed. The crew entered the LM and the cabin was repressurized at 126:11:59. The first EVA lasted 6 hr 32 min 42 sec, about 27 min less than planned because of higher than anticipated oxygen usage by the commander. The distance traveled in the LRV was 5.6 nmi, the vehicle drive time was 1 hr 2 min, parked time was 1 hr 14 min, and 31.97 lb of samples were collected.

Between the first and second extravehicular periods, the crew spent 16 hours in the LM. The second period began at 142:14:48 when the cabin was depressurized. After the crew left the LM, they rechecked the LRV and prepared it for the second traverse. During this checkout, they recycled the circuit breakers and the front steering became completely operational. The crew started their traverse at 143:10:43, heading south to the Apennine front, just east of the first traverse. Stops were made at Spur Crater and other points along the base of the front, as well as at Dune Crater on the return trip. Television transmission was very good. Their return closely followed the outbound route. Documented samples, a core sample, and a comprehensive sample were collected, and photographs were taken.

After reaching the LM at 148:32:17, they returned to the ALSEP site where, while the lunar module pilot performed various soil mechanics tasks, the commander completed drilling the second hole for the heat flow experiment and emplaced the probe. Although he was able to drill the deep-core sample, extracting the tube proved difficult and it had to be left overnight. They then returned to the LM and deployed the US flag. The sample container and film were stowed in the LM. The crew entered the LM and the cabin was repressurized at 149:27:02. The second extravehicular activity period lasted 7 hr 12 min 14 sec. The distance traveled in the LRV was 6.7 nmi, the vehicle drive time was 1 hr 23 min, the vehicle was parked for 2 hr 34 min, and 76.94 lb of samples were collected.

The crew spent almost 14 hours in the LM before the cabin was depressurized for the third extravehicular period at 163:18:14. The third extravehicular activity began 1 hour 45 min later than planned due to cumulative changes in the surface activities timeline. As a result of this delay, and the time spent attempting to extract and disassemble the core tube of the heat flow experiment, the planned trip to the North Complex was deleted. The first stop was the ALSEP site at 164:09:00 in order to retrieve drill core stem samples left during EVA-2. Two core sections were disengaged and placed in the LRV. The drill and the remaining four sections could not be separated and were left for later retrieval. The third geologic traverse took a westerly direction and included stops at Scarp Crater, Rim Crater, and 'The Terrace' near Rim Crater. Extensive samples and a double-length core tube sample were obtained. Photographs were taken of the west wall of Hadley Rille, where exposed layering was observed. The return trip was east toward the LM with a stop at the ALSEP site at 166:43:40 to retrieve the remaining sections of the deep-core sample. One more section was separated, and the remaining three sections were returned in one piece. During sample collecting, the commander had tripped over a rock and fallen, but experienced no difficulty in getting up. After returning to the LM, the LRV was unloaded and parked at 167:35:24 for ground-controlled television coverage of the LM ascent. The commander selected a site slightly closer to the LM than planned in order to take advantage of more elevated terrain for better television coverage of the ascent. They re-entered the LM and the cabin was repressurized at 168:08:04, thus ending the fourth human exploration of the Moon. The third extravehicular period had lasted 4 hr 49 min 50 sec. The distance traveled in the LRV was 2.7 nmi; the vehicle drive time was 35 min and it was parked for 1 hr 22 min and 60.19 lb of samples were collected.

For the mission, the total time spent outside the LM was 18 hr 34 min 46 sec, the total distance traveled in the LRV was about 15.1 nmi, the vehicle drive time was 3 hr 0 min, the vehicle was parked during extravehicular activities for 5 hr 10 min, and the collected samples totaled 170.44 lb (77.31 kg; the official metric total as determined by the Lunar Receiving Laboratory in Houston). The farthest point traveled from the LM was 16,470 ft.

While the LM was on the surface, the command module pilot completed 34 lunar orbits, conducting scientific instrument module experiments and operating cameras to obtain data concerning the lunar surface and the lunar environment. Some scientific tasks accomplished during this time were photographing the sunlit lunar surface, gathering x-ray and gamma-ray data needed for mapping the bulk chemical composition of the lunar surface and for determining the geometry of the Moon along the ground track, visually surveying regions of the Moon to assist in identifying processes that formed geologic features, and obtaining lunar atmospheric data. High-resolution photographs were obtained with the panoramic and mapping cameras. An 18.31-sec CSM plane change maneuver had been conducted at 165:11:32.74 and resulted in an orbit of 64.5 × 53.6 nmi.

Ignition of the ascent stage engine for lunar liftoff occurred at 171:37:23.2 (17:11:23 GMT on 2 August). The LM had been on the lunar surface for 66 hr 54 min 53.9 sec. The 431.0-sec firing achieved the initial lunar orbit of 42.5 × 9.0 nmi. Several rendezvous sequence maneuvers were required before docking could occur approximately 2 hours later. A 2.6-sec terminal phase initiate maneuver at 172:29:40.0 adjusted the ascent stage orbit to 64.4 × 38.7 nmi. The ascent stage and the CSM docked at 173:36:25.5 at an altitude of 57 nmi. The two craft had been undocked for 72 hr 57 min 9.3 sec. After transfer of the crew and samples to the CSM, the ascent stage was jettisoned at 179:30:01.4, and the CSM was prepared for transearth injection. Jettison had been delayed one revolution because of difficulty verifying the spacecraft tunnel sealing and astronaut pressure suit integrity. At 181:04:19.8 and 61.5 nmi altitude, the ascent stage was maneuvered to impact the lunar surface by firing the engine to depletion, which occurred 83.0 sec after ignition. Impact was at 181:29:37.0 at selenographic coordinates 26.36°N, 0.25°E, some 12.7 nmi from the planned point and 50 nmi west of the Apollo 15 landing site. The impact was recorded by the Apollo 12, 14, and 15 seismic stations.

In preparation for the launch of a subsatellite into lunar orbit, a 3.42-sec orbit-shaping maneuver at 221:20:48.02 altered the CSM orbit to 76.0 × 54.3 nmi. The subsatellite was then spring-ejected from the scientific instrument module bay at 222:39:29.1 during the 74th revolution into an orbit of 76.3 × 55.1 nmi at an inclination of −28.7 deg to the lunar equator. The subsatellite was instrumented to measure plasma and energetic-particle fluxes, vector magnetic fields, and subsatellite velocity from which lunar gravitational anomalies could be determined. All systems operated as expected.

Following a 140.90-sec maneuver at 67.6 nmi altitude at 223:48:45.84, transearth injection was achieved at 223:51:06.74 at a velocity of 8,272.4 ft/sec after 74 lunar orbits lasting 145 hr 12 min 41.68 sec.

Apollo 15 lunar orbit phase event	GET (h:m:s)	Altitude (nmi)	Space-fixed velocity (ft/sec)	Event duration (sec)	Velocity change (ft/sec)	Apolune (nmi)	Perilune (nmi)
Lunar orbit insertion ignition	078:31:46.70	86.7	8,188.6	–	–	–	–
Lunar orbit insertion cutoff	078:38:25.06	74.1	5,407.5	398.36	3,000.1	170.1	57.7
Descent orbit insertion ignition	082:39:49.09	55.3	5,491.7	–	–	–	–
Descent orbit insertion cutoff	082:40:13.62	54.9	5,285	24.53	213.9	58.5	9.6
Descent orbit trim ignition	095:56:44.70	56.4	5,276.9	–	–	–	–
Descent orbit trim cutoff	095:57:15.10	50.1	5,314.8	30.40	3.2	60.3	8.8
LM undocking and separation	100:39:16.2	7.4	5,553.6	–	–	–	–
CSM orbit circularization ignition	101:38:58.98	57.1	5,276.5	–	–	–	–
CSM orbit circularization cutoff	101:39:02.65	55.8	5,352.3	3.67	68.3	65.2	54.8
LM powered descent initiation	104:30:09.4	5.8	5,560.2	–	–	–	–
LM powered descent cutoff	104:42:28.6	–	–	739.2	6813	–	–
CSM plane change ignition	165:11:32.74	61.8	5,318.1	–	–	–	–
CSM plane change cutoff	165:11:51.05	62	5,318.8	18.31	330.6	64.5	53.6
LM lunar liftoff ignition	171:37:23.2	54.8	5,357.1	–	–	–	–
LM ascent orbit cutoff	171:44:34.2	–	–	431.0	6,059	42.5	9.0
LM terminal phase initiation ignition	172:29:40.0	34.2	5,368.8	–	–	–	–
LM terminal phase initiation cutoff	172:29:42.6	–	–	2.6	72.7	64.4	38.7
CSM/LM docked	173:36:25.5	57	5,345.8	–	–	–	–
LM ascent stage jettisoned	179:30:01.4	57.5	5,342.1	–	–	–	–
CSM separation from LM	179:50	–	–	–	2	–	–
LM ascent stage deorbit ignition	181:04:19.8	61.5	5,318.9	–	–	–	–
LM ascent stage deorbit cutoff	181:05:42.8	61.8	5,196.0	83.0	200.3	–	–
CSM orbit shaping maneuver ignition	221:20:48.02	53.6	5,362.9	–	–	–	–
CSM orbit shaping maneuver cutoff	221:20:51.44	53.7	5,379.2	3.42	66.4	76.0	54.3
Subsatellite deployed	222:39:29.1	62.6	5,331.6	–	–	76.3	55.1

TRANSEARTH PHASE

At 241:57:12, the command module pilot began a transearth coast extravehicular activity. Television coverage was provided for the 39-min 7-sec extravehicular period, during which he retrieved panoramic and mapping camera film cassettes from the scientific instrument module bay. Three excursions were made to the bay. The film cassettes were retrieved during the first two trips. The third trip was used to observe and report the general condition of the instruments, in particular the mapping camera. He reported no evidence of the cause for the mapping camera extend/retract mechanism failure in the extended position and no observable reason for the pan camera velocity/altitude sensor failure. He also reported that the mass spectrometer boom was not fully retracted. The EVA was completed at 242:36:19. This brought the total extravehicular activity for the mission to 19 hr 46 min 59 sec.

A 22.30-sec midcourse correction of 5.6 ft/sec was performed 291:56:49.91 to put the CSM on a proper track for Earth entry.

Apollo 15 transearth phase event	GET (h:m:s)	Altitude (nmi)	Space-fixed velocity (ft/sec)	Event duration (sec)	Velocity change (ft/sec)	Space-fixed flight path angle (deg)	Space-fixed heading angle (°E of N)
Transearth injection ignition	223:48:45.84	67.6	5,305.9	–	–	0.52	–128.90
Transearth injection cutoff	223:51:06.74	71.8	8,272.4	140.90	3,046.8	4.43	–129.08
Midcourse correction ignition	291:56:49.91	25,190.3	11,994.6	–	–	–68.47	103.11
Midcourse correction cutoff	291:57:12.21	25,149.3	12,002.4	22.30	5.6	–68.49	103.09
CM/SM separation	294:43:55.2	1,951.8	29,001.7	–	–	–36.44	56.65

RECOVERY

The service module was jettisoned at 294:43:55.2, and CM entry followed a normal profile. The command module re-entered the Earth's atmosphere (at the 400,000-ft altitude of the 'entry interface') at 294:58:54.7 at 36,096.4 ft/sec, after a transearth coast of 71 hr 7 min 48 sec. The parachute system, with two main parachutes properly inflated and one collapsed, effected splashdown of the CM in the Pacific Ocean at 20:45:53 GMT on 7 August. The mission duration was 295:11:53.0. The impact point, estimated to be 26.13°N and 158.13°W, was about 1.0 nmi from the target point and 5 nmi from the recovery ship USS *Okinawa*. The collapsed parachute contributed to the fastest entry time in the Apollo program, just 778.3 sec from entry to splashdown. The CM assumed an apex-up flotation attitude. The crew was retrieved by helicopter and was on the ship 39 min after splashdown. The CM was recovered 55 min later. The estimated CM weight at splashdown was 11,731 lb, and the estimated distance traveled for the mission was 1,107,945 nmi.

CONCLUSIONS

The mission accomplished all its primary objectives, and provided scientists with a large amount of new information concerning the Moon and its characteristics. It demonstrated that in addition to providing a means of transportation, the Apollo system excelled as an operational scientific facility. The following conclusions were made from an analysis of post-mission data:

1. The Apollo 15 mission demonstrated that, with the addition of consumables and the installation of scientific instruments, the CSM is an effective means of gathering scientific data. Real-time data allowed participation by scientists with the crew in planning and making decisions to maximize scientific results.
2. The mission demonstrated that the modified launch vehicle, spacecraft, and life-support system configurations could successfully transport larger payloads and safely extend the time spent on the Moon.
3. The modified pressure garment and portable life support systems provided better mobility and extended the lunar surface extravehicular time.
4. The ground-controlled mobile television camera allowed greater real-time participation by Earth-bound scientists and operational personnel during lunar surface extravehicular activity.
5. The practicality of the LRV was demonstrated by greatly increasing load-carrying capability and range of exploration of the lunar surface.
6. The lunar communications relay unit provided the capability for continuous communications en route to and at the extended ranges made possible by the LRV.
7. Landing site visibility was improved by the use of a steeper landing trajectory.
8. Apollo 15 demonstrated that the crew could operate to a greater degree as scientific observers and investigators and rely more on the ground support team for systems monitoring.
9. The value of human space flight was further demonstrated by the unique human capability to observe and think creatively, as shown in the supplementation and redirection of many tasks by the crew to enhance scientific data return.
10. The mission confirmed that, in order to maximize mission success, crews should train with actual flight equipment or equipment with equal reliability.

For this first 'multiple-objective' mission, the geological objectives were, in order of priority:

1) to sample the Apennines at the base of Mount Hadley Delta,
2) to sample the rim of Hadley Rille,
3) to sample the craters of the South Cluster, which had excavated the mare plain, and
4) to investigate the possibly volcanic nature of the North Complex.

The Apennines

On Earth, the process of orogeny (mountain building) is the result of the process of plate tectonics, and it takes millions of years to create a mountain range. The lunar surface, in contrast, is shaped by bombardment, and the mountains that ring the impact basins were thrust up instantaneously as the shock shattered the crust into a set of intersecting radial and peripheral faults, and the mountains are individual massifs. The Apennines form the south-eastern rim of the Imbrium basin, and the steep basin-facing slope of this range is known as the Apennine front. To appreciate the significance of this mountain range, it is necessary to consider its position between the Imbrium and Serenitatis basins. The Moon can be studied by stratigraphic analysis, which is based on the Principle of Superposition. This record shows that prior to the Imbrium event, the terrain where the Apennines stand was the inner part of the Serenitatis equivalent of the Fra Mauro Formation. As a result, there was a layer of Serenitatis ejecta deposited on the pre-Serenitatis crust, and when this terrain was faulted by the formation of the Imbrium basin the underlying crustal blocks were projected through this overburden. It is possible that much of this material remained on the summits, but much of it would have slipped downslope and accumulated in the valleys. In the formation of an impact crater, the most deeply excavated material is left on the rim. As a basin forms, there is also a 'base surge' of semi-molten material excavated from deep beneath the surface crust, and this would have 'mantled' the Apennines. The rocks from Fra Mauro revealed the Imbrium basin to have been formed 3.85 billion years ago, but the Fra Mauro Formation represented only the crustal ejecta. It was the prospect of a large crater on the flank of an Apennine massif yielding a sample of the ancient crust beneath this overburden, or a boulder that had rolled down from the summit proving to be Serenitatis ejecta, that made the Apennine front Apollo 15's primary geological objective.

The rim of St. George, the 6,000-ft-diameter crater located on the northern flank of Mount Hadley Delta, was found to be surprisingly clear of ejecta. This crater formed after the massif had been up-thrust, but before the Palus Putredinis embayment of the adjacent valley. It had been hoped to find pieces of the massif that it had excavated, but this debris had been broken down and mixed into the regolith by the ongoing process of 'gardening'. The solitary boulder sampled was clearly an interloper. Further around the mountain, to the east, the open flank was similarly lacking in large rocks. It was the absence of suitable samples that prompted curtailing the traverse before its nominal turning point, in order to focus on Spur crater. This was where the crew found what they were looking for, in the shape of what the media dubbed the 'Genesis Rock'. Although it was the eagerly sought anorthosite, it had been subjected to intense shock and the age of 4.15 ($\pm$0.1) billion years marked when it had recrystallized; it was very likely a fragment of the crust of the magma ocean. The fact that most of the rocks were breccias was in no way a disappointment, since most had clasts of a coarsely grained feldspathic rock. The fact that these were simpler than the 'two-rock' breccias found by Apollo 14 confirmed that whereas the Fra Mauro Formation was shallow crustal material that was already processed by impacts, the homogeneous 'base surge' that mantled the Apennine front derived from much deeper. A 'green boulder' found on the lower flank of Mount Hadley Delta had acquired a coating of microscopic droplets of shock-cooled magma that had issued from a 'fire fountain'. In retrospect, it was clear that fine glass would form near vents on the periphery of a basin, because the faults beneath the massifs would provide routes for deep lava reservoirs to reach the surface. A fire fountain could be expected to form as a precursor to an upwelling of magma, because, as the material in the reservoir made its way up through the crust it would liberate the volatiles in solution, and these would blast out under pressure from fissures, and the mist of magma would chill as it fell back to coat the material on the surface. This glass, which was green due to the silicates it contained, provided a sample of the olivine and pyroxene reservoirs of the mantle, since the material in a fire fountain would have ascended so rapidly that it would not have had time to evolve chemically. Although 3.3 billion years old, the magnesium-rich pyroclastic proved to be the most 'pristine' igneous sample ever recovered from the Moon.

The rille and plain

In studying the sinuous rilles, photo-geologists had used Lunar Orbiter imagery to chart both their topographic and stratigraphic structure. The fact that they cut through the pre-existing terrain, rather than simply 'dropping' it, together with the fact that the inner and outer rim arcs at the turns in the sinuous path did not have the same radius, meant that they were not caused by extensional stress. Their resemblance to terrestrial river channels suggested that they were cut by flowing lava. It was evident though, that if they *were* lava channels, the fact that they were much larger than their terrestrial counterparts indicated that the sources must have given rise to enormous outpourings of lava. Terrestrial lava flows tend to follow pre-existing valleys, run for short distances, and clog their channels. In contrast, Hadley Rille had *excavated* a 60-nmi-long trench that was 4,000 ft wide and 1,000 ft deep. A low-rate terrestrial lava flow, if laminar, yields its thermal energy to the surrounding rock and rapidly congeals, but a vastly more effusive lunar vent would be a *turbulent* flow that would not only not cool, it would melt the wall of its channel, scour it clean, and excavate a rille.

There was a shallow slope down from the plain to the rille, as the regolith diminished from its 16-ft depth on the plain to essentially zero at the very lip of the rille. This zone along the rim marked both an increase in the coverage of the rocks and an increase in the size of the rocks in comparison to the plain. The increase in rock coverage and size was a consequence of the regolith growing progressively thinner. The thinning regolith enabled smaller and smaller craters to punch through to excavate the bedrock, and made it easier for rocks liberated from the bedrock to reach the surface. However, there was insufficient energy in small impacts to do more than shatter the bedrock, and the blocks found on the rims of small craters in this transition zone were almost certainly indigenous. Along the rim itself, there were fractured outcrops that were still essentially *in situ*, where variations in chemical composition could be unambiguously related to the stratigraphic record. The bedding in the opposite wall was essentially horizontal, indicating that there had been no significant subsequent tectonic forces. The outcrops were heavily fractured, and large blocks of rock had broken off and slipped down the slope. Most of the blocks had become lodged in the finer talus that had accumulated on the lower slope. Some large blocks had reached the floor of the rille, but it was remarkably clear. The absence of outcropping on the wall below St. George showed this face of the rille to be scoured massif rather than mare in-fill. Indeed, there was evidence in the wall below the crater Elbow of the contact where the mare in-fill had overridden the Serenitatis ejecta that had slipped off the mountain soon after its creation. The ridge along the rille's eastern wall was pronounced near Elbow. The fact that this bank was on the 'outer' rim at the corner, suggested that the 'shoulders' had been built-up by surges of lava spilling over the lip while turning the corner. The evidence of the 'high lava' mark on Mount Hadley suggested that massive surges of lava flowing down the rille had flooded out across the plain, and then drained away again leaving only a thin scum on the mountains surrounding the bay. Remarkably, these flood marks were 250 ft above the current level of the plain. However, this likely includes the fact that that the plain has settled isostatically since it formed. A withdrawal is suggested by the trench that runs along the base of Hadley Delta and around the base of the Swann Range to the east. This trench is conspicuously absent where the rim has been built up in the vicinity of Elbow. If the trench is the result of the newly solidified plain shrinking and withdrawing, the fact that it still exists indicates that there has been little slippage off the mountains in the intervening eons. Nevertheless, the fact that the LRV was able to drive across it without difficulty meant that there had been just enough in-fill to smooth the profile of the trench.

Stratigraphic analysis revealed the relative ages of features (it was clear that the rille post-dated the Palus Putredinis plain, because the rille cut across the plain, for example), but did not yield absolute ages. Those who favored 'recent volcanism' found comfort in the possibility that the rille was a recent addition. What was required was a little 'ground truth' in the form of rocks that were clearly associated with the rille. It turned out that the vents that produced the lava that cut the rille were ancient, ranging from 3.3 to 3.4 billion years old. Most of it was pyroxene-enriched, but some darker material on the very lip was olivine-enriched, implying that just before the vent dried up it issued a chemically distinct magma. The crater Dune, several miles away on the plain, gave a check on the horizontal variation of the lava flow. This crater had penetrated the bedrock to a depth of about 300 ft (although the pit was not that deep, it would have excavated rock from that depth) and there were basaltic blocks on its rim. The mare yielded both extrusive and hypabyssal (i.e. magma crystallized at shallow depth) basalts. The pyroxene-rich basalt was found at each site, indicating that this covered a wide area, but the olivine-rich basalt was only on the rim of the rille, indicating that this late extrusion had barely spilled over the lip. Due to lack of time, the putative pyroclastic deposit on the North Complex was not able to be visited.

The Apollo 15 seismometer, coordinating with those left by Apollo 12 and Apollo 14, provided a three-station network capable of triangulating on the source of an event. Most seismicity was only 1 or 2 on the Richter scale, and on Earth would be ignored as part of the ongoing rumble in the extremely dynamic crust, but the Moon is so inert that these are 'major' events. This meant that both the lunar crust and outer mantle are relatively cool. In fact, the mantle might be barely plastic. As a result, there are no convection currents in the mantle, and hence no tectonism to deform the crust. The seismic data from Apollo 12 and Apollo 14 had suggested that the maria were 14 nmi thick, and rested on top of about 22 nmi of anorthositic gabbro. It was initially believed that this outer layer was basalt, but in fact there is a thin veneer of basalt on a thick layer of an intensely brecciated material.

The objective of the heat flow experiment was to measure the temperature and the thermal properties of the regolith, to determine the rate at which the interior of the Moon was leaking heat. Although the sensors could be emplaced at only half the desired depth of 8 to 10 ft, the data gave an astonishing insight into the current state of the lunar 'heat engine'. It was because this experiment was rated so highly that the drilling operation was permitted to supersede visiting the North Complex. The data indicated that in sunlight the temperature at the surface of the regolith at this site at mid-northern latitude rose to 380K. During the lunar 'day', there was a net heat-flow *into* the surface from solar irradiation. At sunset, the temperature fell to 100K, and at 'night' heat was lost to space by radiative cooling. The thermal conductivity of the regolith was highly temperature dependent. As the efficiency of the radiative transfer between the fine powdery particles of the surface regolith was proportional to the cube of the absolute temperature, the heat flowed more readily into the immediate subsurface during the 'day' than it leaked out at 'night', and at a depth of 1.5 ft the temperature maintained a constant 220K. Beneath the surface zone influenced by the 'diurnal' variation, the temperature of the regolith increased with depth as a result of the leakage of heat from the interior, and at a depth of 3 ft it was some 40K warmer. At 8 ft long, the hard-won 'deep core' was the deepest regolith sample yet. It was x-rayed to reveal the density of the soil and the concentration of the pebbles along its length. Although the top 18 in was completely 'gardened', the rest of the sample was finely stratified, with at least 42 and possibly as many as 58 distinct layers indicating that it was still as it had been deposited. This stratification greatly simplified the task of dating the levels. The discovery that the top of the unsorted material was 400 million years old indicated that activity on the plain at Hadley during this period was confined to the upper 1.5 ft of the regolith. On Earth during this interval, continents had split, drifted thousands of miles, and then erected mountain ranges when they collided again.

Summarizing the conclusions from the missions to date, the Imbrium basin formed about 3.8 billion years ago, the Tranquilitatis site was engulfed by lava 100 million years later, and Procellarum 500 million years after that. The history of the basalts on the rim of the rille indicates that the eastern Imbrium was still active at this late stage. This marked a great advance in the state of our knowledge of the Moon's early history compared to just a few years previously. Of course, much of this had been predicted; the problem was that so too had much else, and the challenge had been to sift the 'wheat' from the 'chaff'.

MISSION OBJECTIVES

Launch Vehicle Objectives

1. To launch on a 80–100°E of N flight azimuth and insert the S-IVB/instrument unit/spacecraft into the planned circular Earth parking orbit. *Achieved.*
2. To restart the S-IVB during either the second or third revolution and inject the S-IVB/instrument unit/spacecraft into the planned translunar trajectory. *Achieved.*
3. To provide the required attitude control for the S-IVB/instrument unit/spacecraft during transposition, docking, and ejection. *Achieved.*
4. To perform an evasive maneuver after ejection of the command and service module/lunar module from the S-IVB/instrument unit. *Achieved.*
5. To attempt to impact the S-IVB/instrument unit on the lunar surface within 189 nmi (350 km) of selenographic coordinates 3.65°S, 7.58°W. *Achieved.*
6. To determine actual impact point within 2.7 nmi (5.0 km) and time of impact within 1 sec. *Achieved.*
7. To vent and dump the remaining gases and propellants to safe the S-IVB/instrument unit. *Achieved.*

The Apollo 15 crew: Dave Scott, Al Worden and Jim Irwin.

The launch of Apollo 15.

The Apollo 15 objective was Hadley Rille at the foot of the Apennine mountains.

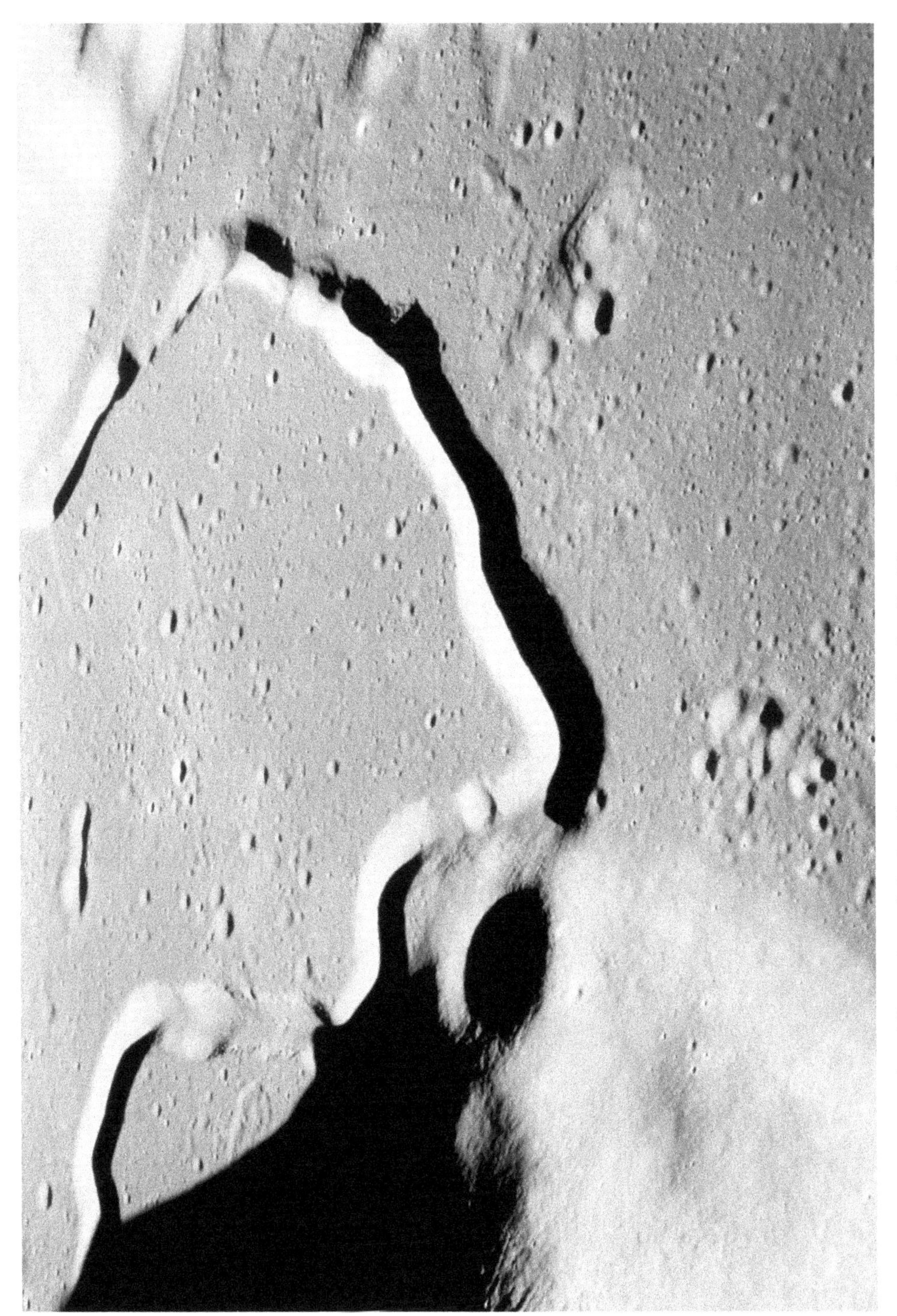

An oblique view of the landing site taken by Apollo 15 soon after entering lunar orbit.

An artist's depiction of the LRV traverses planned for Apollo 15. The first was to be southwest to the bend in Hadley Rille; the second, south past the craters of the South Complex and along the flank of Mount Hadley Delta; and the third, west to the rim of the rille, returning via the craters and mounds of the North Complex. (Courtesy NASA.)

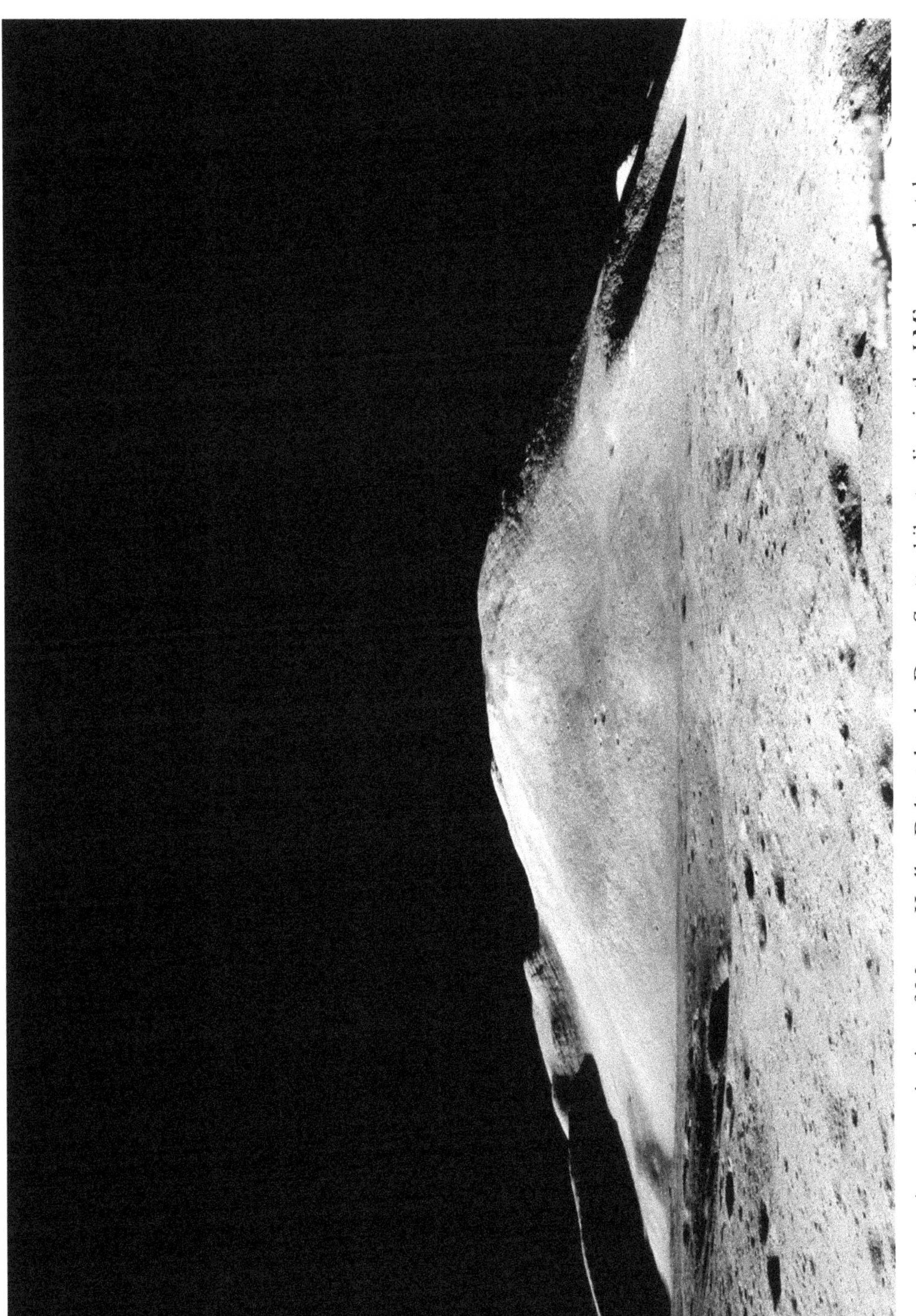

A panoramic view of Mount Hadley Delta taken by Dave Scott while standing in the LM's upper hatch.

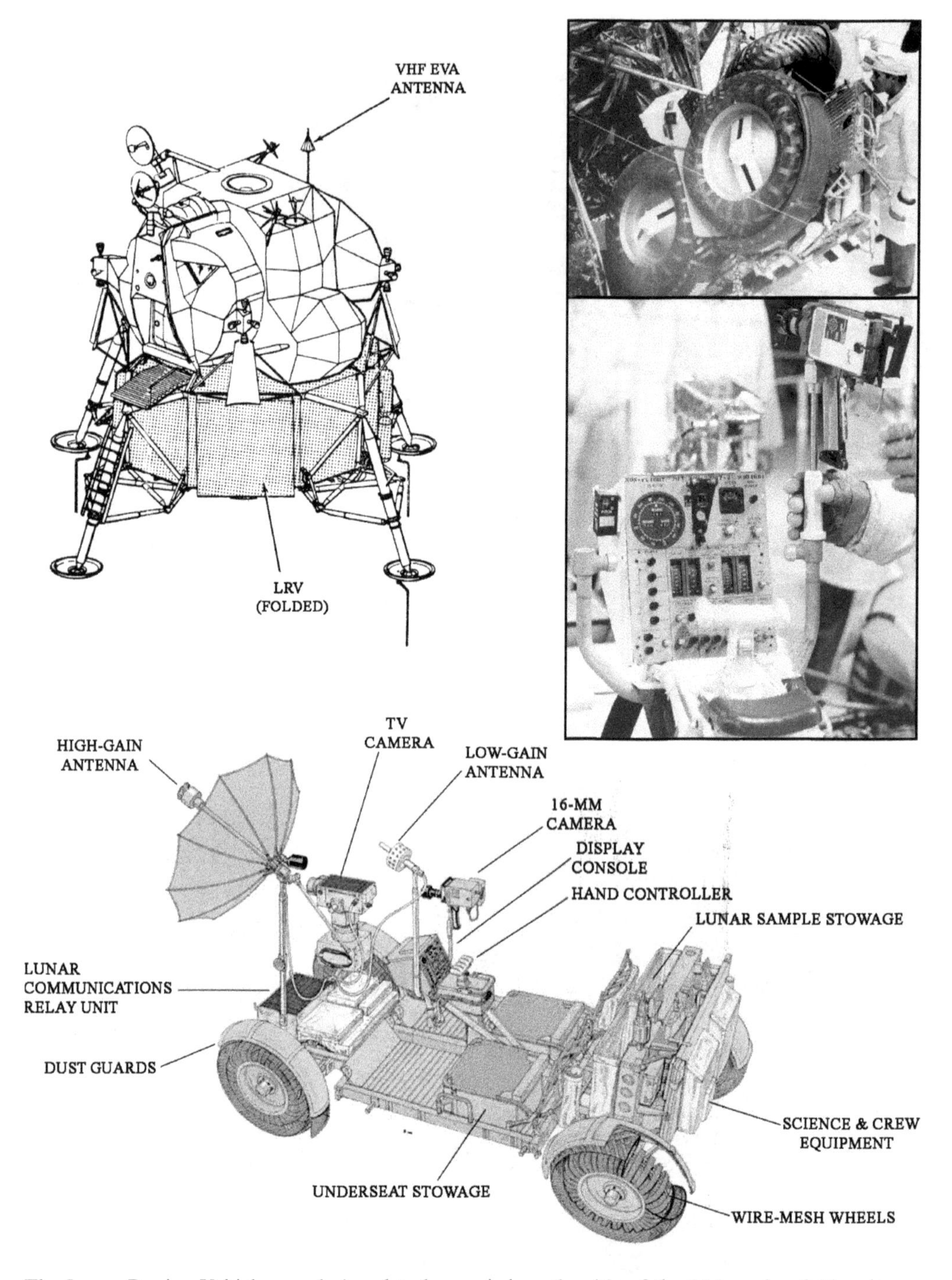

The Lunar Roving Vehicle was designed to be carried on the side of the LM, and unfold as it was lowered.

Dave Scott test drives the LRV.

Jim Irwin attends the LRV at the landing site, with Mount Hadley in the background.

Jim Irwin with the LRV at station 2, with Hadley Rille in the background.

A view from the flank of Mount Hadley Delta with the LM on the plain and the North Complex beyond.

Dave Scott with the LRV at station 9A, on the rim of Hadley Rille.

The far wall of Hadley Rille viewed from station 9A, with blocks on the near rim in the foreground.

At station 10 Dave Scott with a 500-mm-lens Hasselblad about to take additional pictures of the rille's far wall.

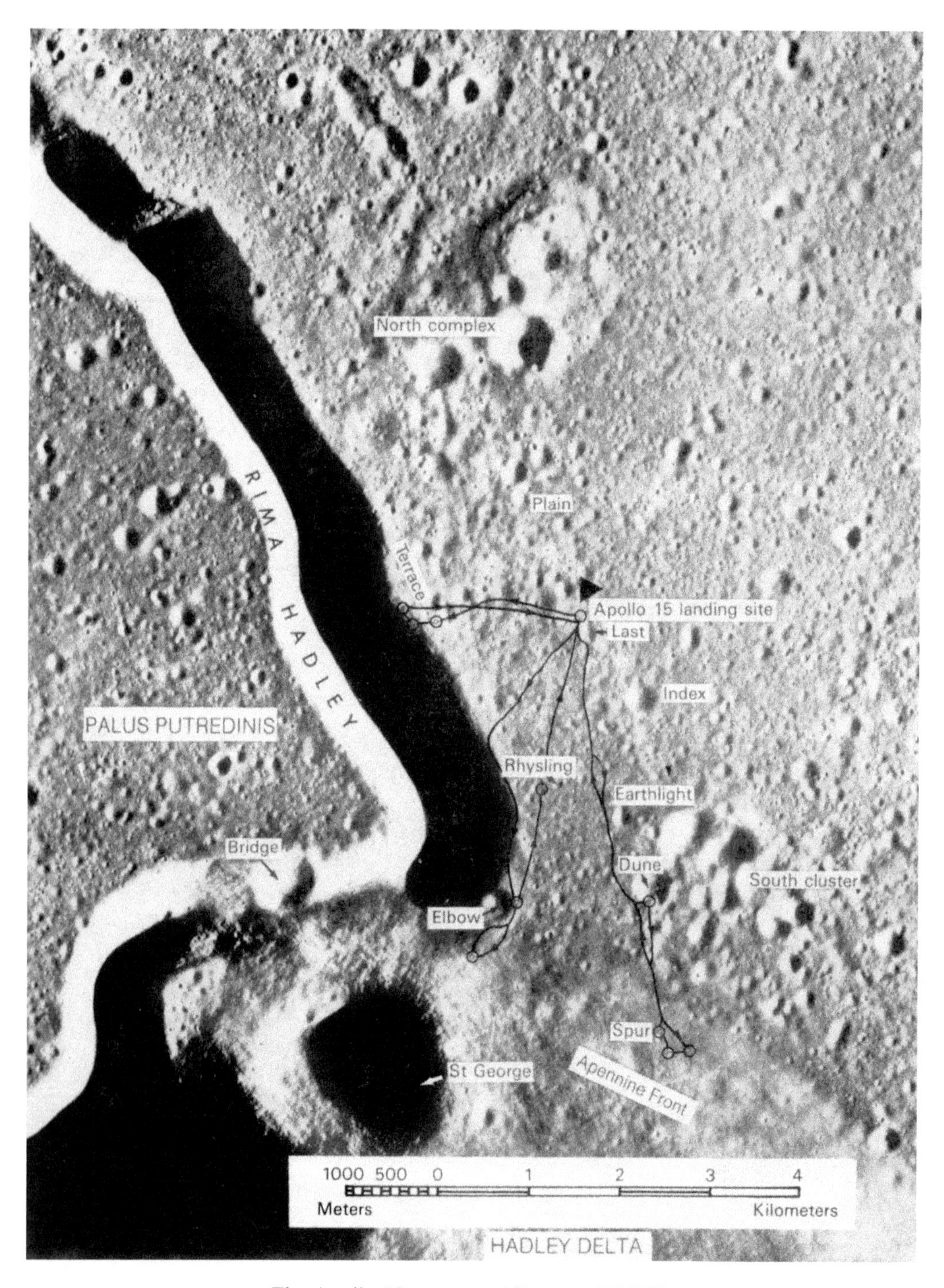

The Apollo 15 traverses. (Courtesy USGS.)

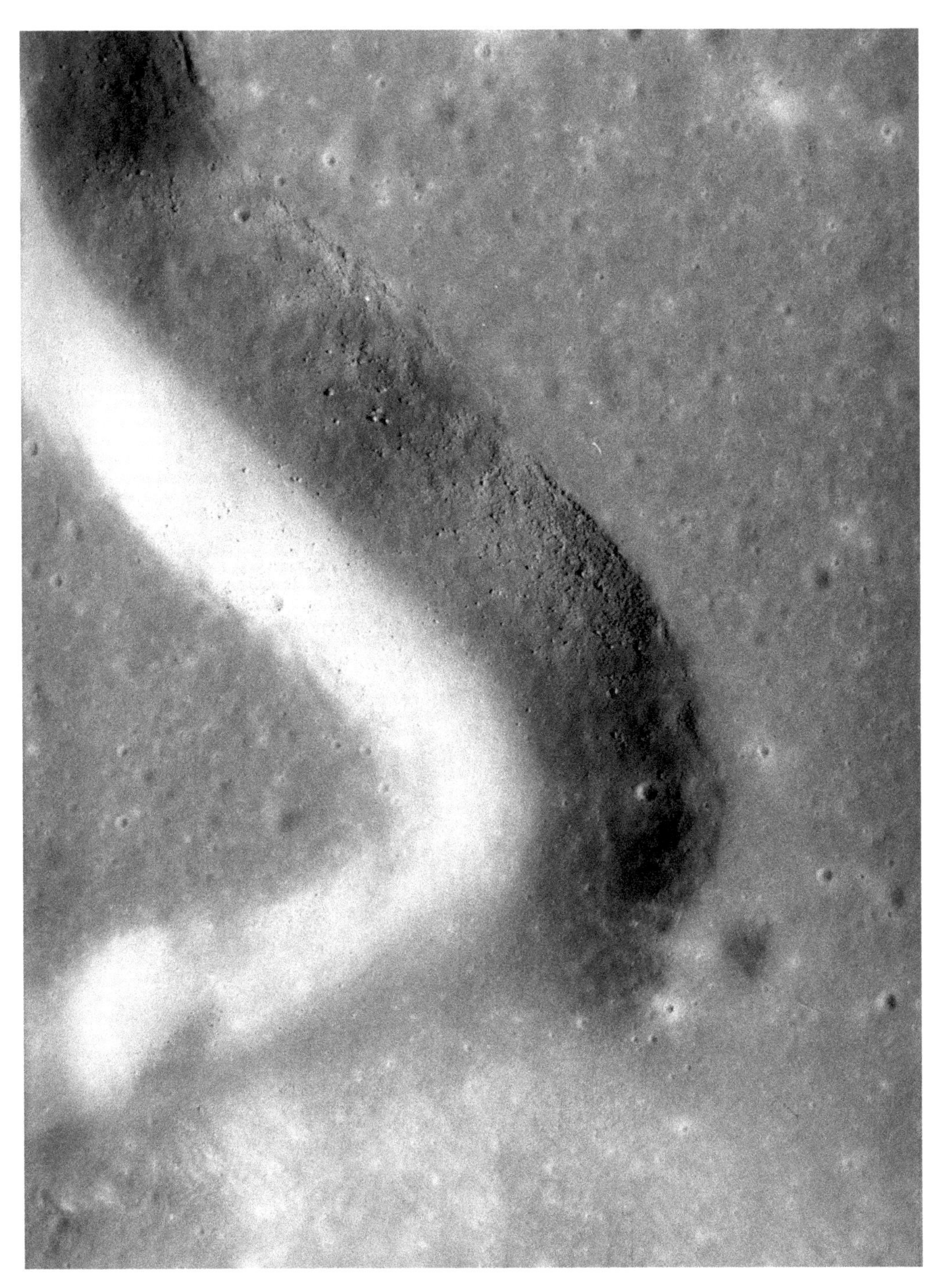

A view of Hadley Rille taken after the LM's return, with the eastern wall of the rille illuminated.

One of Apollo 15's parachutes became fouled.

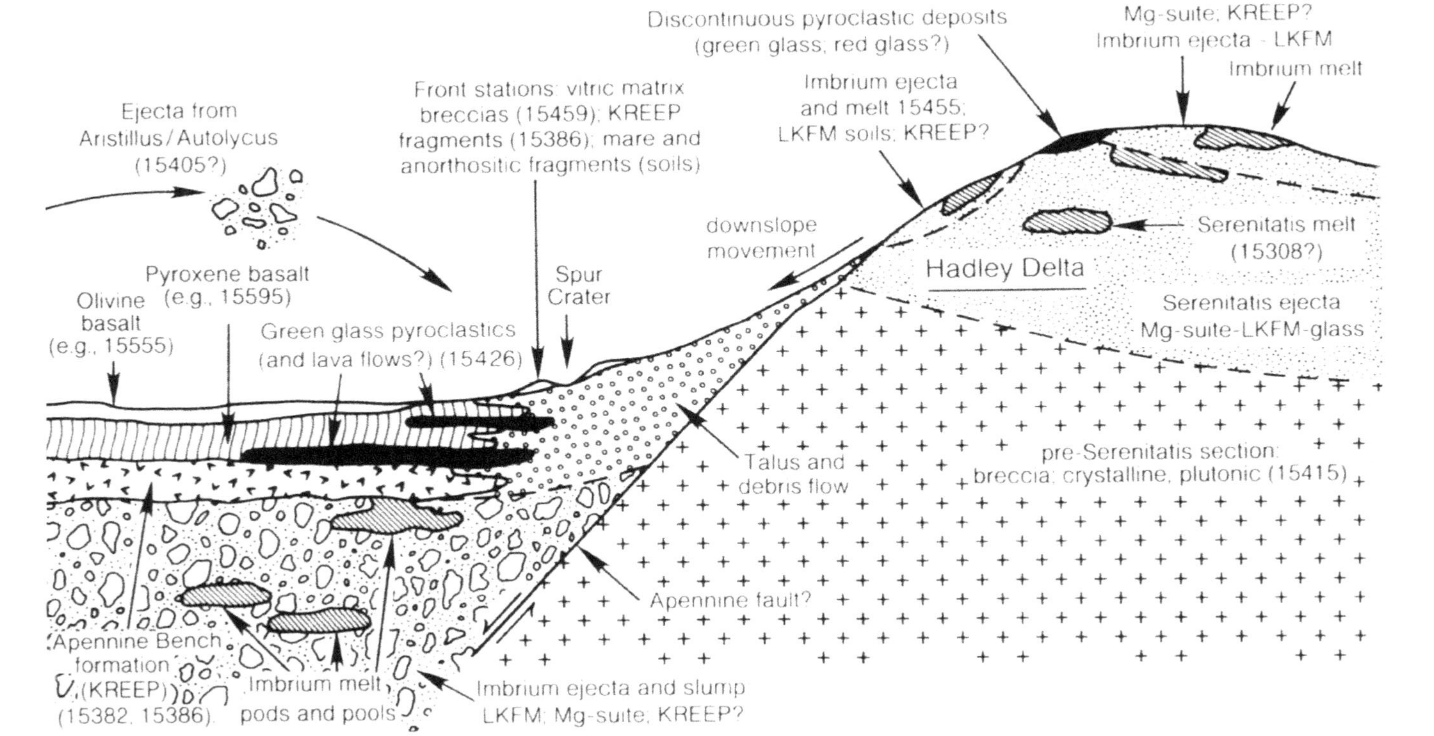

A schematic northwest–southeast cross-section through the Apollo 15 landing site showing the complex transition between mare and highlands (modified from P.D. Spudis and G. Ryder, 'Geology and petrology of the Apollo 15 landing site: past, present and future understanding', *EOS Transactions*, vol. 66, AGU, 1985; G.A. Swann *et al.*, 'Preliminary geologic investigations of the Apollo 15 landing site', in *Apollo 15 Preliminary Science Report*, SP-289, NASA, 1972; G.A. Swann, 'Some observations on the geology of the Apollo 15 landing site', in G. Ryder and P.D. Spudis (Eds.), *Workshop on the Geology and Petrology of the Apollo 15 Landing Site*, Lunar and Planetary Institute technical report no. 86-03, 1986). In the mare (left), post-Imbrian basalt lavas overlie a thick deposit of Imbrium ejecta. In the highland area (Mount Hadley Delta, right), older (pre-Imbrium) ejecta from the Serenitatis basin overlies the ancient crust. Numbers refer to specific collected samples representative of the various units inferred to be present. (Courtesy the Lunar and Planetary Institute and Cambridge University Press.)

Spacecraft Primary Objectives

1. To perform selenological inspection, survey, and sampling of materials and surface features in a preselected area of the Hadley–Apennine region. *Achieved.*
2. To emplace and activate surface experiments. *Achieved.*
3. To evaluate the capability of the Apollo equipment to provide extended lunar surface stay time, increased extravehicular operations, and surface mobility. *Achieved.*
4. To conduct in-flight experiments and photographic tasks from lunar orbit. *Achieved.*

Detailed Objectives

1. LRV evaluation. *Achieved.*
2. Extravehicular communications with the lunar communications relay unit and ground-controlled television assembly. *Achieved.*
3. Extravehicular mobility unit assessment on lunar surface. *Achieved.*
4. Lunar module landing effects evaluation. *Achieved.*
5. Service module orbital photographic tasks. *Achieved.*
6. Command module photographic tasks. *Achieved.*
7. Scientific instrument module thermal data. *Achieved.*
8. Scientific instrument module inspection during extravehicular activity. *Achieved.*
9. Scientific instrument module door jettison evaluation. *Achieved.*
10. Lunar module descent engine performance. *Achieved.*
11. Visual observations from lunar orbit. *Achieved.*
12. Visual light flash phenomenon. *Achieved.*

Experiments

1. Contingency sample collection. *Achieved.*
2. ALSEP V: Apollo Lunar Scientific Experiment Package.
   a. S-031: Passive seismic. *Achieved.*
   b. S-034: Lunar surface magnetometer. *Achieved.*
   c. S-035: Solar wind spectrometer. *Achieved.*
   d. S-036: Suprathermal ion detector. *Achieved.*
   e. S-037: Heat flow. *Partially achieved.*
   f. S-058: Cold cathode ion gauge. *Achieved.*
3. S-059: Lunar geology investigation. *Achieved.*
4. S-078: Laser-ranging retroreflector. *Achieved.*
5. S-080: Solar wind composition. *Achieved.*
6. S-160: Gamma-ray spectrometer. *Achieved.*
7. S-161: X-ray fluorescence. *Achieved.*
8. S-162: Alpha particle spectrometer. *Achieved.*
9. S-164: S-band transponder (command and service module and lunar module). *Achieved.*
10 S-165: Mass spectrometer. *Achieved.*
11. S-170: Downlink bistatic radar observations of the Moon. *Achieved.*
12. S-176: Apollo window meteoroid. *Achieved.*
13. S-177: Ultraviolet photography of the Earth and Moon. *Achieved.*
14. S-178: Gegenschein from lunar orbit. *Not achieved. The 14 35-mm photographs*

scheduled for this experiment were not obtained due to an error in the spacecraft photographic attitudes.

15. S-200: Soil mechanics. *Achieved.*
16. M-078: Bone mineral measurement. *Achieved.*
17. M-515: Lunar dust detector. *Achieved.*

Subsatellite Experiments

1. S-164: S-band transponder. *Achieved.*
2. S-173: Particle shadows/boundary layer. *Achieved.*
3. S-174: Magnetometer. *Achieved.*

Operational Tests

1. For Manned Spacecraft Center:
   (a) Lunar gravity measurement using the lunar module primary guidance system. *Achieved.*
   (b) Lunar module voice and data relay test. *Achieved.*
2. For Department of Defense/Kennedy Space Center:
   (a) Chapel Bell (classified Department of Defense test).
   (b) Radar skin tracking.
   (c) Ionospheric disturbance from missiles.
   (d Acoustic measurement of missile exhaust noise.
   (e) Army acoustic test.
   (f) Long-focal-length optical system.
   (g) Sonic boom measurement.

MISSION TIMELINE

Apollo 15 mission event	GET (h:m:s)	Date (GMT)	Time (h:m:s)
Terminal countdown started.	–028:00:00	24 Jul 1971	23:00:00
Scheduled 9-hr 34-min hold at T–9 hr.	–009:00:00	25 Jul 1971	18:00:00
Countdown resumed at T–9 hr.	–009:00:00	26 Jul 1971	03:34:00
Scheduled 1-hr hold at T–3 hr 30 min.	–003:30:00	26 Jul 1971	09:04:00
Countdown resumed at T–3 hr 30 min.	–003:30:00	26 Jul 1971	10:04:00
Guidance reference release.	–000:00:16.939	26 Jul 1971	13:33:43
S-IC engine start command.	–000:00:08.9	26 Jul 1971	13:33:51
S-IC engine ignition (#5).	–000:00:06.50	26 Jul 1971	13:33:53
All S-IC engines thrust OK.	–000:00:01.4	26 Jul 1971	13:33:58
Range zero.	000:00:00.00	26 Jul 1971	13:34:00
All holddown arms released (1st motion) (1.08 *g*).	000:00:00.3	26 Jul 1971	13:34:00
Liftoff (umbilical disconnected).	000:00:00.58	26 Jul 1971	13:34:00
Tower clearance yaw maneuver started.	000:00:01.68	26 Jul 1971	13:34:01
Yaw maneuver ended.	000:00:09.66	26 Jul 1971	13:34:09

Apollo 15 mission event – *continued*	GET (h:m:s)	Date (GMT)	Time (h:m:s)
Pitch and roll maneuver started.	000:00:12.21	26 Jul 1971	13:34:12
Roll maneuver ended.	000:00:23.02	26 Jul 1971	13:34:23
Mach 1 achieved.	000:01:05.0	26 Jul 1971	13:35:05
Maximum bending moment (80,000,000 lbf-in).	000:01:20.1	26 Jul 1971	13:35:20
Maximum dynamic pressure (768.58 lb/ft^2).	000:01:22.0	26 Jul 1971	13:35:22
S-IC center engine cutoff command.	000:02:15.96	26 Jul 1971	13:36:16
Pitch maneuver ended.	000:02:36.94	26 Jul 1971	13:36:36
S-IC outboard engine cutoff. Maximum total inertial acceleration (3.97 *g*).	000:02:39.56	26 Jul 1971	13:36:39
S-IC maximum Earth-fixed velocity.	000:02:40.00	26 Jul 1971	13:36:40
S-IC/S-II separation command.	000:02:41.2	26 Jul 1971	13:36:41
S-II engine start command.	000:02:41.9	26 Jul 1971	13:36:41
S-II ignition.	000:02:43.0	26 Jul 1971	13:36:43
S-II aft interstage jettisoned.	000:03:11.2	26 Jul 1971	13:37:11
Launch escape tower jettisoned.	000:03:15.9	26 Jul 1971	13:37:15
Iterative guidance mode initiated.	000:03:22.62	26 Jul 1971	13:37:22
S-IC apex.	000:04:37.562	26 Jul 1971	13:38:37
S-II maximum total inertial acceleration (1.79 *g*). S-II center engine cutoff.	000:07:39.56	26 Jul 1971	13:41:39
S-II outboard engine cutoff.	000:09:09.06	26 Jul 1971	13:43:09
S-II maximum Earth-fixed velocity.	000:09:10.00	26 Jul 1971	13:43:10
S-II/S-IVB separation command.	000:09:10.1	26 Jul 1971	13:43:10
S-IVB 1st burn start command.	000:09:10.20	26 Jul 1971	13:43:10
S-IVB 1st burn ignition.	000:09:13.20	26 Jul 1971	13:43:13
S-II apex.	000:09:13.225	26 Jul 1971	13:43:13
S-IC impact (theoretical).	000:09:20.839	26 Jul 1971	13:43:20
S-IVB ullage case jettisoned.	000:09:21.8	26 Jul 1971	13:43:21
S-IVB 1st burn maximum total inertial acceleration (0.65 *g*). S-IVB 1st burn cutoff.	000:11:34.67	26 Jul 1971	13:45:34
Earth orbit insertion. S-IVB 1st burn maximum Earth-fixed velocity.	000:11:44.67	26 Jul 1971	13:45:44
Orbital navigation started.	000:11:56.3	26 Jul 1971	13:45:56
Maneuver to local horizontal attitude started.	000:13:15.7	26 Jul 1971	13:47:15
S-II impact (theoretical).	000:19:43.912	26 Jul 1971	13:53:43
S-IVB 2nd burn restart preparation.	002:40:24.80	26 Jul 1971	16:14:24
S-IVB 2nd burn restart command.	002:49:54.90	26 Jul 1971	16:23:54
S-IVB 2nd burn ignition.	002:50:02.90	26 Jul 1971	16:24:02
S-IVB 2nd burn cutoff and maximum total inertial acceleration (1.40 *g*).	002:55:53.61	26 Jul 1971	16:29:53
S-IVB 2nd burn maximum Earth-fixed velocity.	002:55:54.00	26 Jul 1971	16:29:54
Translunar injection.	002:56:03.61	26 Jul 1971	16:30:03
Orbital navigation started.	002:58:26.0	26 Jul 1971	16:32:26
Maneuver to local horizontal attitude started.	002:58:26.2	26 Jul 1971	16:32:26
Maneuver to transposition and docking attitude started.	003:10:54.6	26 Jul 1971	16:44:54
CSM separated from S-IVB.	003:22:27.2	26 Jul 1971	16:56:27

Apollo 15 mission event – *continued*	GET (h:m:s)	Date (GMT)	Time (h:m:s)
TV transmission started.	003:25	26 Jul 1971	16:34
CSM docked with LM/S-IVB.	003:33:49.5	26 Jul 1971	17:07:49
TV transmission ended.	003:50	26 Jul 1971	16:34
CSM/LM ejected from S-IVB.	004:18:01.2	26 Jul 1971	17:52:01
S-IVB APS evasive maneuver ignition.	004:40:01.8	26 Jul 1971	18:14:01
S-IVB APS evasive maneuver cutoff.	004:41:22.0	26 Jul 1971	18:15:22
Maneuver to S-IVB LOX dump attitude initiated.	004:49:41.8	26 Jul 1971	18:23:41
S-IVB lunar impact maneuver – CVS venting closed.	004:56:40.6	26 Jul 1971	18:30:40
S-IVB lunar impact maneuver – LOX dump. Start of unplanned velocity increment due to J-2 engine control helium dump.	005:01:20.6	26 Jul 1971	18:35:20
S-IVB lunar impact maneuver – CVS vent opened.	005:01:40.6	26 Jul 1971	18:35:40
S-IVB lunar impact maneuver – LOX dump ended.	005:02:08.7	26 Jul 1971	18:36:08
S-IVB lunar impact maneuver – J-2 engine control helium dump ended.	005:18:51	26 Jul 1971	18:52:51
Maneuver to attitude required for final S-IVB APS burn initiated.	005:27:13.5	26 Jul 1971	19:01:13
S-IVB lunar impact maneuver – 1st APS ignition.	005:46:00.7	26 Jul 1971	19:20:00
S-IVB lunar impact maneuver – 1st APS cutoff.	005:52:01.9	26 Jul 1971	19:26:01
S-IVB lunar impact maneuver – Start of 1st unplanned velocity increment due to instrument unit thermal control system water valve operations and APS attitude engine reactions.	006:18:00	26 Jul 1971	19:52:00
S-IVB lunar impact maneuver – End of 1st velocity increment due to IU/TCS and APS effects.	006:23:00	26 Jul 1971	19:57:00
S-IVB lunar impact maneuver – Start of 2nd velocity increment due to IU/TCS and APS effects.	006:58:00	26 Jul 1971	20:32:00
S-IVB lunar impact maneuver – End of 2nd velocity increment due to IU/TCS and APS effects.	007:03:00	26 Jul 1971	20:37:00
S-IVB lunar impact maneuver – Start of 3rd velocity increment due to IU/TCS and APS effects.	007:38:00	26 Jul 1971	21:12:00
S-IVB lunar impact maneuver – End of 3rd velocity increment due to IU/TCS and APS effects.	007:43:00	26 Jul 1971	21:17:00
S-IVB lunar impact maneuver – Start of 4th velocity increment due to IU/TCS and APS effects.	008:18:00	26 Jul 1971	21:52:00
S-IVB lunar impact maneuver – End of 4th velocity increment due to IU/TCS and APS effects.	008:23:00	26 Jul 1971	21:57:00
S-IVB lunar impact maneuver – Start of 5th velocity increment due to IU/TCS and APS effects.	008:53:00	26 Jul 1971	22:27:00
S-IVB lunar impact maneuver – End of 5th velocity increment due to IU/TCS and APS effects.	008:58:00	26 Jul 1971	22:32:00
S-IVB lunar impact maneuver – Start of 6th velocity increment due to IU/TCS and APS effects.	009:28:00	26 Jul 1971	23:02:00
S-IVB lunar impact maneuver – End of 6th velocity increment due to IU/TCS and APS effects.	009:33:00	26 Jul 1971	23:07:00
S-IVB lunar impact maneuver – 2nd APS ignition.	010:00:01	26 Jul 1971	23:34:01
S-IVB lunar impact maneuver – 2nd APS cutoff.	010:01:12	26 Jul 1971	23:35:12
S-IVB 0.3 deg/sec solar heating avoidance roll command.	010:19:22	26 Jul 1971	23:53:22

Apollo 15 mission event – *continued*	GET (h:m:s)	Date (GMT)	Time (h:m:s)
Midcourse correction ignition.	028:40:22.0	27 Jul 1971	18:14:22
Midcourse correction cutoff.	028:40:22.8	27 Jul 1971	18:14:22
Sextant photography test started.	032:00	27 Jul 1971	21:34
Sextant photography test ended.	032:50	27 Jul 1971	22:24
Preparations for LM ingress.	033:25	27 Jul 1971	22:59
CDR and LMP entered LM for checkout.	033:56	27 Jul 1971	23:30
TV transmission of CM and LM interiors started.	034:55	28 Jul 1971	00:29
TV transmission of CM and LM interiors ended.	035:46	28 Jul 1971	01:20
CDR and LMP entered CM.	036:55	28 Jul 1971	02:29
Visual light flash phenomenon observations started.	051:37	28 Jul 1971	17:11
Visual light flash phenomenon observations ended.	052:33	28 Jul 1971	18:07
LM housekeeping.	056:26	28 Jul 1971	22:00
LM ingress and housekeeping.	056:26	28 Jul 1971	22:00
CDR and LMP entered LM for checkout.	057:00	28 Jul 1971	22:34
CDR and LMP entered CM.	058:00	28 Jul 1971	23:34
Equigravisphere.	063:55:20	29 Jul 1971	05:29:20
Midcourse correction ignition.	073:31:14.81	29 Jul 1971	15:05:14
Midcourse correction cutoff.	073:31:15.72	29 Jul 1971	15:05:15
Scientific instrument module door jettisoned.	074:06:47.1	29 Jul 1971	15:40:47
Lunar orbit insertion ignition (SPS).	078:31:46.70	29 Jul 1971	20:05:46
Lunar orbit insertion cutoff.	078:38:25.06	29 Jul 1971	20:12:25
S-IVB impact on lunar surface.	079:24:42.9	29 Jul 1971	20:58:42
Orbital science photography started.	080:35	29 Jul 1971	22:09
Orbital science photography ended.	080:50	29 Jul 1971	22:24
Terminator photography.	082:00	29 Jul 1971	23:34
Descent orbit insertion ignition (SPS).	082:39:49.09	30 Jul 1971	00:13:49
Descent orbit insertion cutoff.	082:40:13.62	30 Jul 1971	00:14:13
CSM landmark tracking.	083:45	30 Jul 1971	01:19
Terminator photography.	084:35	30 Jul 1971	02:09
TV transmission of landing site started.	095:00	30 Jul 1971	12:34
TV transmission of landing site ended.	095:10	30 Jul 1971	12:44
Descent orbit trim ignition (RCS).	095:56:44.70	30 Jul 1971	13:30:44
Descent orbit trim cutoff.	095:57:15.10	30 Jul 1971	13:31:15
CDR and LMP entered LM for activation, checkout, and platform alignment.	098:00	30 Jul 1971	15:34
CM/LM undocking failure due to loose CM/LM umbilical.	100:14	30 Jul 1971	17:48
LM undocking and separation.	100:39:16.2	30 Jul 1971	18:13:16
CSM orbit circularization ignition (SPS).	101:38:58.98	30 Jul 1971	19:12:59
CSM orbit circularization cutoff.	101:39:02.65	30 Jul 1971	19:13:02
CSM lunar surface landmark tracking.	102:35	30 Jul 1971	20:09
LM landing radar on.	104:25:13.0	30 Jul 1971	21:59:13
LM powered descent engine ignition.	104:30:09.4	30 Jul 1971	22:04:09
LM throttle to full-throttle position.	104:30:35.9	30 Jul 1971	22:04:35
LM manual target (landing site) update.	104:31:44.2	30 Jul 1971	22:05:44
LM pitchover started.	104:33:10.4	30 Jul 1971	22:07:10

Apollo 15 mission event – *continued*	GET (h:m:s)	Date (GMT)	Time (h:m:s)
LM landing radar range data good.	104:33:26.2	30 Jul 1971	22:07:26
LM landing radar altitude data good.	104:33:38.2	30 Jul 1971	22:07:38
LM landing radar updates enabled.	104:33:50.2	30 Jul 1971	22:07:50
LM throttle down.	104:37:31.1	30 Jul 1971	22:11:31
LM approach phase program selected.	104:39:32.2	30 Jul 1971	22:13:32
LM landing radar antenna to position 2.	104:39:39.0	30 Jul 1971	22:13:39
LM 1st landing point redesignation.	104:39:40.0	30 Jul 1971	22:13:40
LM landing radar switched to low scale.	104:40:13.0	30 Jul 1971	22:14:13
LM attitude hold mode selected.	104:41:08.7	30 Jul 1971	22:15:08
LM landing phase program selected.	104:41:10.2	30 Jul 1971	22:15:10
LM powered descent engine cutoff.	104:42:28.6	30 Jul 1971	22:16:28
LM lunar landing (right side and forward footpad contact).	104:42:29.3	30 Jul 1971	22:16:29
LM final settling.	104:42:31.1	30 Jul 1971	22:16:31
CSM orbital science photography.	106:00	30 Jul 1971	23:34
Stand-up EVA started (Scott).	106:42:49	31 Jul 1971	00:16:49
Stand-up EVA ended.	107:15:56	31 Jul 1971	00:49:56
CSM orbital science photography.	108:00	31 Jul 1971	01:34
CSM orbital science photography.	108:40	31 Jul 1971	02:14
CSM bistatic radar test.	110:00	31 Jul 1971	03:34
1st EVA started (LM cabin depressurized).	119:39:17	31 Jul 1971	13:13:17
TV transmission started for 1st EVA.	119:52:45	31 Jul 1971	13:26:45
Contingency sample collected.	120:00:05	31 Jul 1971	13:34:05
Lunar roving vehicle (LRV) offloaded.	120:18:31	31 Jul 1971	13:52:31
LRV deployed.	120:31:33	31 Jul 1971	14:05:33
LRV configured for traverse.	121:24:03	31 Jul 1971	14:58:03
Departed for station 1.	121:44:55	31 Jul 1971	15:18:55
Arrived at station 1. Performed radial sampling, gathered documented samples, and performed panoramic photography.	122:10:46	31 Jul 1971	15:44:46
Departed for station 2.	122:22:36	31 Jul 1971	15:56:36
Arrived at station 2. Gathered samples, obtained a double core tube sample and performed stereopanoramic and 500-mm photography.	122:34:44	31 Jul 1971	16:08:44
CSM deep space measurements.	122:40	31 Jul 1971	16:14
CSM sunrise solar corona photography.	123:05	31 Jul 1971	16:39
Departed for LM.	123:26:02	31 Jul 1971	17:02
Arrived at LM. Offloaded and deployed Apollo lunar surface experiment package (ALSEP), laser-ranging retroreflector, and solar wind composition experiment.	123:59:39	31 Jul 1971	17:33:39
CSM sunset solar corona photography.	124:30	31 Jul 1971	18:04
CSM lunar libration photography.	125:00	31 Jul 1971	18:34
1st ALSEP data received on Earth.	125:18:00	31 Jul 1971	18:52
TV transmission ended for 1st EVA.	125:55	31 Jul 1971	19:29
Cold cathode gage experiment turned on. CSM orbital science photography.	126:00	31 Jul 1971	19:34
1st EVA ended (cabin repressurized).	126:11:59	31 Jul 1971	19:45:59

Apollo 15 mission event – *continued*	GET (h:m:s)	Date (GMT)	Time (h:m:s)
Heat flow experiment turned on.	126:13	31 Jul 1971	19:47
CSM bistatic radar test.	131:40	01 Aug 1971	01:14
CSM orbital science photography.	142:00	01 Aug 1971	11:34
2nd EVA started (cabin depressurized).	142:14:48	01 Aug 1971	11:48:48
Equipment prepared for LRV traverse.	142:25:04	01 Aug 1971	11:59:04
TV transmission started for 2nd EVA.	142:35	01 Aug 1971	12:09
Departed for station 6.	143:10:43	01 Aug 1971	12:44:43
Arrived at station 6. Gathered samples, obtained a single core tube sample, obtained a special environmental sample from trench, and performed panoramic and 500-mm photography tasks.	143:53:46	01 Aug 1971	13:27:46
CSM Earthshine photography.	144:10:32	01 Aug 1971	13:44:32
Departed for station 6a.	144:58:49	01 Aug 1971	14:32:49
Arrived at station 6a. Gathered samples and performed panoramic photography tasks.	145:01:11	01 Aug 1971	14:35:11
Departed for station 7.	145:22:40	01 Aug 1971	14:56:40
Arrived at station 7. Gathered selected samples, a comprehensive soil sample, and performed panoramic photography.	145:26:25	01 Aug 1971	15:00:25
Departed for station 4.	146:16:09	01 Aug 1971	15:50:09
Arrived at station 4. Gathered samples and performed panoramic photography.	146:28:59	01 Aug 1971	16:02:59
CSM deep space measurements.	146:30	01 Aug 1971	16:04
Departed for LM.	146:45:44	01 Aug 1971	16:19:44
Arrived at LM. Offloaded samples and configured LRV for trip to station 8 (ALSEP site).	147:08:09	01 Aug 1971	16:42:09
Departed for station 8.	147:19:33	01 Aug 1971	16:53:33
CSM orbital science photography.	147:20	01 Aug 1971	16:54
Arrived at station 8. Gathered comprehensive geologic sample, gathered special environmental sample from trench, drilled second heat flow hole and emplaced probe, drilled deep core sample hole, and performed penetrometer experiments.	147:21:15	01 Aug 1971	16:55:15
Departed for LM.	148:31:08	01 Aug 1971	18:05:08
Arrived at LM. Deployed US flag and started EVA closeout.	148:32:17	01 Aug 1971	18:06:17
CSM zodiacal light photography.	148:40	01 Aug 1971	18:14
CSM orbital science photography.	149:10	01 Aug 1971	18:44
TV transmission ended for 2nd EVA.	149:20	01 Aug 1971	18:54
2nd EVA ended (cabin repressurized).	149:27:02	01 Aug 1971	19:01:02
3rd EVA started (LM cabin depressurized).	163:18:14	02 Aug 1971	08:52:14
TV transmission started for 3rd EVA.	163:45	02 Aug 1971	09:19
Departed for ALSEP site.	164:04:13	02 Aug 1971	09:38:13
Arrived at ALSEP site. Recovered deep core sample and photographed LRV operation.	164:09:00	02 Aug 1971	09:43:00
Departed for station 9.	164:48:05	02 Aug 1971	10:22:05

Apollo 15 mission event – *continued*	GET (h:m:s)	Date (GMT)	Time (h:m:s)
Arrived at station 9. Collected samples and performed panoramic photography tasks.	165:01:22	02 Aug 1971	10:35:22
CSM plane change ignition (SPS).	165:11:32.74	02 Aug 1971	10:45:32
CSM plane change cutoff.	165:11:51.05	02 Aug 1971	10:45:51
Departed for station 9a.	165:16:50	02 Aug 1971	10:50:50
Arrived at station 9a. Gathered extensive samples, obtained a double core tube and performed photographic tasks including 500-mm and stereoscopic panoramic photography.	165:19:26	02 Aug 1971	10:53:26
Departed for station 10.	166:14:25	02 Aug 1971	11:48:25
Arrived at station 10. Gathered samples and performed 500-mm and panoramic photography tasks.	166:16:45	02 Aug 1971	11:50:45
Departed for ALSEP site.	166:28:49	02 Aug 1971	12:02:49
Arrived at ALSEP site. Recovered drilled core sample and performed photographic tasks.	166:43:40	02 Aug 1971	12:17:40
Arrived at LM. EVA closeout procedures started.	166:45:45	02 Aug 1971	12:19:45
Solar wind composition experiment retrieved.	167:10	02 Aug 1971	12:44
Departed for final positioning of LRV to obtain television coverage of LM ascent.	167:32:18	02 Aug 1971	13:06:18
LRV parked in final position.	167:35:24	02 Aug 1971	13:09:24
3rd EVA ended (LM cabin repressurized).	168:08:04	02 Aug 1971	13:42:04
TV transmission ended for 3rd EVA.	168:20	02 Aug 1971	13:54
CSM Gegenschein photography.	168:30	02 Aug 1971	14:04
LM equipment jettisoned.	169:00	02 Aug 1971	14:34
CSM tracking of LM landing site.	169:30	02 Aug 1971	15:04
Surface television transmission started for lunar liftoff.	171:30	02 Aug 1971	17:04
LM lunar liftoff ignition (LM APS).	171:37:23.2	02 Aug 1971	17:11:23
Surface television transmission ended.	171:40	02 Aug 1971	17:14
Lunar ascent orbit cutoff.	171:44:34.2	02 Aug 1971	17:18:34
Terminal phase initiation ignition.	172:29:40.0	02 Aug 1971	18:03:40
Terminal phase initiation cutoff.	172:29:42.6	02 Aug 1971	18:03:42
TV transmission started.	173:05	02 Aug 1971	18:39
TV transmission ended.	173:10	02 Aug 1971	18:44
Terminal phase finalize.	173:11:07	02 Aug 1971	18:45:07
TV transmission started.	173:35	02 Aug 1971	19:09
CSM/LM docked.	173:36:25.5	02 Aug 1971	19:10:25
TV transmission ended. CDR and LMP prepared to transfer to CSM.	173:40	02 Aug 1971	19:14
Samples and equipment transferred to CSM.	175:00	02 Aug 1971	20:34
CDR and LMP entered CSM and hatch closed.	176:40	02 Aug 1971	22:14
LM ascent stage jettisoned.	179:30:01.4	03 Aug 1971	01:04:01
CSM separation maneuver from LM.	179:50	03 Aug 1971	01:24
LM ascent stage deorbit ignition.	181:04:19.8	03 Aug 1971	02:38:19
LM ascent stage fuel depletion.	181:05:42.8	03 Aug 1971	02:39:42
LM ascent stage impact on lunar surface.	181:29:37.0	03 Aug 1971	03:03:37
Deep space measurements and Gegenschein photography.	195:45	03 Aug 1971	17:19

Apollo 15 mission event – *continued*	GET (h:m:s)	Date (GMT)	Time (h:m:s)
Ultraviolet photography of lunar maria.	196:35	03 Aug 1971	18:09
Visual light flash phenomenon observations started.	197:00	03 Aug 1971	18:34
Visual observations from lunar orbit.	197:20	03 Aug 1971	18:54
Visual light flash phenomenon observations ended.	198:00	03 Aug 1971	19:34
Orbital science photography.	198:35	03 Aug 1971	20:09
Visual observations from lunar orbit.	199:00	03 Aug 1971	20:34
CSM lunar terminator photography.	199:30	03 Aug 1971	21:04
CSM lunar terminator photography.	200:30	03 Aug 1971	22:04
Orbital science photography.	200:50	03 Aug 1971	22:24
Ultraviolet photography of lunar surface.	201:00	03 Aug 1971	22:34
CSM lunar terminator photography.	201:40	03 Aug 1971	23:14
CSM boom photography.	202:20	03 Aug 1971	23:54
CSM lunar terminator photography.	214:05	04 Aug 1971	11:39
Orbital science photography.	214:35	04 Aug 1971	12:09
Deep space measurements.	215:40	04 Aug 1971	13:14
Sunrise solar corona photography.	216:00	04 Aug 1971	13:34
Orbital science photography.	217:00	04 Aug 1971	14:34
CSM lunar terminator photography.	217:20	04 Aug 1971	14:54
CSM lunar terminator photography.	219:20	04 Aug 1971	16:54
Orbit shaping maneuver ignition.	221:20:48.02	04 Aug 1971	18:54:48
Orbit shaping maneuver cutoff.	221:20:51.44	04 Aug 1971	18:54:51
Subsatellite deployed.	222:39:29.1	04 Aug 1971	20:13:29
Transearth injection ignition (SPS).	223:48:45.84	04 Aug 1971	21:22:45
Transearth injection cutoff.	223:51:06.74	04 Aug 1971	21:25:06
Moon and star field photography.	224:20	04 Aug 1971	21:54
Corona window calibration.	239:05	05 Aug 1971	12:39
Transearth EVA started (Worden).	241:57:12	05 Aug 1971	15:31:12
Transearth EVA – TV transmission started.	242:00	05 Aug 1971	15:34
Transearth EVA – TV and data acquisition cameras installed and adjusted.	242:02	05 Aug 1971	15:36
Transearth EVA – Camera cassette retrieved.	242:22	05 Aug 1971	15:56
Transearth EVA – TV transmission ended.	242:28	05 Aug 1971	16:02
Transearth EVA – Ingress and hatch closed.	242:33	05 Aug 1971	16:07
Transearth EVA ended.	242:36:19	05 Aug 1971	16:10:19
Visual light flash phenomenon observations started.	264:35	06 Aug 1971	14:09
Visual light flash phenomenon observations ended.	265:35	06 Aug 1971	15:09
Lunar eclipse photography.	269:00	06 Aug 1971	18:34
Sextant photography.	270:00	06 Aug 1971	19:34
Lunar eclipse photography.	271:00	06 Aug 1971	20:34
Contamination photography.	271:50	06 Aug 1971	21:24
Mass spectrometer boom retraction test.	272:45	06 Aug 1971	22:19
Midcourse correction ignition.	291:56:49.91	07 Aug 1971	17:30:49
Midcourse correction cutoff.	291:57:12.21	07 Aug 1971	17:31:12
CM/SM separation.	294:43:55.2	07 Aug 1971	20:17:55
Entry.	294:58:54.7	07 Aug 1971	20:32:54

Apollo 15 mission event – *continued*	GET (h:m:s)	Date (GMT)	Time (h:m:s)
Communication blackout ended.	295:02:31	07 Aug 1971	20:36:31
Radar contact with CM established by recovery ship.	295:03	07 Aug 1971	20:37
S-band contact with CM established by recovery aircraft.	295:04	07 Aug 1971	20:38
Forward heat shield jettisoned.	295:06:45	07 Aug 1971	20:40:45
Drogue parachute deployed.	295:06:46	07 Aug 1971	20:40:46
VHF recovery beacon contact with CM by recovery ship. Visual contact by recovery helicopters.	295:07	07 Aug 1971	20:41
Main parachute deployed.	295:07:34	07 Aug 1971	20:41:34
Voice contact with CM established.	295:09	07 Aug 1971	20:43
Splashdown (went to apex-up).	295:11:53.0	07 Aug 1971	20:45:53
Swimmers deployed to CM.	295:18	07 Aug 1971	20:52
Flotation collar installed and inflated on CM.	295:26	07 Aug 1971	21:00
Hatch opened for crew egress.	295:37	07 Aug 1971	21:11
Crew in egress raft.	295:38	07 Aug 1971	21:12
Crew aboard recovery helicopter.	295:46	07 Aug 1971	21:20
Crew aboard recovery ship.	295:51	07 Aug 1971	21:25
Communication blackout started.	295:59:13	07 Aug 1971	21:33:13
CM aboard recovery ship	296:46	07 Aug 1971	22:20
First sample flight departed recovery ship.	301:56	08 Aug 1971	03:30
First sample flight arrived in Hawaii.	303:46	08 Aug 1971	05:20
First sample flight departed Hawaii.	304:30	08 Aug 1971	06:04
First sample flight arrived in Houston, TX.	311:59	08 Aug 1971	13:33
Crew departed recovery ship.	315:36	08 Aug 1971	17:10
Crew arrived at Hickam AFB, Hawaii.	316:12	08 Aug 1971	17:46
CM arrived in Hawaii.	316:26	08 Aug 1971	18:00
Crew arrived at Ellington AFB, Houston.	324:45	09 Aug 1971	02:19
CM departed Hawaii for pyrotechnic safing at North Island Naval Air Station, San Diego, CA.	340:26	11 Aug 1971	18:00
CM arrived at North Island Naval Air Station.	467:26	17 Aug 1971	00:00
Pyrotechnic safing of CM completed.	474:06	17 Aug 1971	07:40
Fuel portion of CM reaction control system deactivated.	489:56	18 Aug 1971	23:30
Oxidizer portion of CM reaction control system deactivated.	513:46	19 Aug 1971	23:20
CM departed San Diego for contractor's facility in Downey, CA.	516:46	20 Aug 1971	02:20
CM arrived at contractor's facility.	603:41	20 Aug 1971	17:15
ALSEP central station turned off by ground command.	–	30 Sep 1977	–

Apollo 16

The tenth manned mission: the fifth lunar landing

16–27 April 1972

BACKGROUND

Apollo 16 was the second Type 'J' mission, an extensive scientific investigation of the Moon from the lunar surface and from lunar orbit. The vehicles and payload were similar to those of Apollo 15.

The primary objectives were:

- to perform selenological inspection, survey, and sampling of materials and surface features in a preselected area of the Moon;
- to emplace and activate surface experiments; and
- to conduct in-flight experiments and photographic tasks.

The crew members were Captain John Watts Young (USN), commander; Lieutenant Commander Thomas Kenneth 'Ken' Mattingly II (USN), command module pilot; and Lieutenant Colonel Charles Moss Duke Jr (USAF), lunar module pilot. Selected as an astronaut in 1962, Young was making his fourth spaceflight, and was only the second astronaut to achieve that distinction. He had been pilot of Gemini 3, command pilot of Gemini 10, and command module pilot of Apollo 10, the first test of the LM in lunar orbit and the dress rehearsal for the first manned landing on the Moon. Born on 24 September 1930 in San Francisco, California, Young was 41 years old at the time of the Apollo 16 mission. He received a BS in aeronautical engineering from the Georgia Institute of Technology in 1952. His backup for the mission was Fred Wallace Haise Jr. Mattingly, who had been removed from the Apollo 13 mission as a result of his susceptibility to German measles, was making his first spaceflight. Born on 17 March 1936 in Chicago, Illinois, Mattingly was 36 years old at the time of the Apollo 16 mission. He received a BS in aeronautical engineering from Auburn University in 1958, and was selected as an astronaut in 1966. His backup was Lieutenant Colonel Stuart Allen Roosa (USAF). Duke was making his first spaceflight. Born on 3 October 1935 in Charlotte, North Carolina, he was 36 years old at the time of the Apollo 16 mission. He received a BS in naval sciences from the US Naval Academy in 1957 and an MS in aeronautics and astronautics from the Massachusetts Institute of Technology in 1964. He was selected as an astronaut in 1966.

His backup was Captain Edgar Dean Mitchell (USN). The capsule communicators (CAPCOMs) for the mission were Major Donald Herod Peterson (USAF), Major Charles Gordon Fullerton (USAF), Colonel James Benson Irwin (USAF), Haise, Roosa, Mitchell, Major Henry Warren Hartsfield Jr (USAF), Anthony Wayne 'Tony' England PhD, and Lieutenant Colonel Robert Franklyn Overmyer (USMC). The support crew consisted of Peterson, England, Hartsfield, and Phillip Kenyon Chapman ScD. The flight directors were M.P. 'Pete' Frank and Philip C. Shaffer (first shift), Eugene F. Kranz and Donald R. Puddy (second shift), and Gerald D. Griffin, Neil B. Hutchinson, and Charles R. Lewis (third shift).

The Apollo 16 launch vehicle was a Saturn V, designated AS-511. The mission also carried the designation Eastern Test Range #1601. The CSM was designated CSM-113, and had the call-sign 'Casper'. The lunar module was designated LM-11, and had the call-sign 'Orion'.

LANDING SITE

With the results of four Apollo lunar landings, geologists were confident they understood the 500 million years since the creation of the Imbrium basin, and they were eager to send the next mission to a site in the highlands. Previously, the flight dynamics team had been reluctant to try for a highland site because the line of approach would be rough and there would be insufficient room for a large landing ellipse, but after Apollo 15 had flown over a range of mountains to land on a confined plain beyond, the 'highlands' did not seem so daunting. Propellant considerations limited the site selectors to the central highlands. In 1965 D.E. Wilhelms drew a distinction between the hummocky Fra Mauro and the rolling light plains in the highlands beyond, and produced a map in which the latter was labeled the Cayley Formation (this name derived from the fact that the first patch of it he mapped was in the vicinity of the 6-nmi-diameter crater Cayley). Whereas the maria were effusive eruptions of a dark basalt rich in mafic silicates that yielded a low-viscosity lava which spread out to form smooth plains, this 'highland basalt' was thought to be sufficiently enriched with silica as to make it semi-viscous, and after oozing from fissures it settled as isolated patches in low-lying areas. When Eggleton mapped the light-toned hills close by the crater Descartes in the central highlands, he listed them as an atypically patch of the Fra Mauro Formation. Lacking evidence to the contrary, it was natural to consider these domical hills to be extruded silica-rich rhyolite lava which, being viscous, had piled up as hills. Initially referred to as the Material of the Descartes Mountains, this hilly patch was renamed the Descartes Formation. Petrologists argued for Apollo 16 to set down on the Kant Plateau because it appeared to be a block of primitive crust that, although cratered, seemed not to have been masked by volcanism, but the geologists rejected it for precisely this reason; they were *seeking* volcanism. They pointed out that just beyond the western scarp of the plateau was the Descartes Formation, and a landing in one of valleys that was embayed by the Cayley Formation would enable a single mission to sample *both* types of terrain. Of course, by specifying only the contact between these formations, this criterion left the choice of landing site wide open. In the absence of a rock exposure, such as a convenient rille, the only way to investigate the vertical structure of the Cayley would be to perform a radial sample on a 'drill hole' crater. This not only narrowed the options, it also led the site selectors directly to a pair of fresh-looking ray craters, one each at the base of two mountains about 6 nmi apart on opposite sides of an east–west embayment. In view of its geological context, this site, some 45 nmi north of the crater Descartes, was named Descartes–Cayley. It was ideal for a three-traverse mission, because the first could explore the landing site in the middle of the plain, the second could explore the southern mountain and its ray crater, and the third could sample the northern area. Stu Roosa was to have photographed it using the Hiflex lunar terrain camera on Apollo 14, but after taking high-resolution pictures the Kant Plateau this mapping camera malfunctioned just as Descartes came into view; fortunately, he was able to hastily snap some pictures using a handheld Hasselblad with a 500-mm lens which were of sufficient resolution to select a landing site and plan traverse routes.

LAUNCH PREPARATIONS

The terminal countdown was picked up at T–28 hr at 03:54:00 GMT on 15 April 1972. Scheduled holds were initiated at T–9 hr for 9 hours and at T–3 hr 30 min for 1 hour. At launch time, the Cape Kennedy launch area was experiencing fair weather resulting from a ridge of high pressure extending westward, from the Atlantic Ocean through central Florida. Cumulus clouds covered 20% of the sky (base 3,000 ft), the temperature was 88.2°F, the relative humidity was 44%, and the barometric pressure was 14.769 psi. The winds, as measured by the anemometer on the light pole 60.0 ft above ground at the launch site were 12.2 kt at 269° from true north; the winds at 530 ft were 9.9 kt at 256° from true north.

ASCENT PHASE

Apollo 16 was launched from Pad A of Launch Complex 39 at the Kennedy Space Center at a Range Zero time of 17:54:00 GMT (12:54:00 EST) on 16 April 1972. The launch window extended to 21:43:00 GMT in order to take advantage of a Sun elevation angle on the lunar surface of 11.9 deg at the scheduled time of landing. Between T+12.7 and T+31.8 sec, the vehicle rolled from a pad azimuth of 90°E of N to a flight azimuth of 72.034°E of N. The maximum wind conditions encountered during ascent were 50.7 kt at 257° from true north at 38,880 ft, with a maximum wind shear of 0.0095/sec at 44,780 ft. The S-IC shut down at T+161.78, followed by S-IC/S-II separation, and S-II ignition. The S-II shut down at T+559.54 followed by separation from the S-IVB, which ignited at T+563.60. The first S-IVB cutoff occurred at T+706.21 with deviations from the planned trajectory of only +0.6 ft/sec in velocity; altitude was exactly as planned. At insertion, at T+716.21 (i.e. S-IVB cutoff plus 10 sec to account for engine tailoff and other transient effects), the parking orbit showed an apogee and perigee of 91.3 × 90.0 nmi, an inclination of 32.542 deg, a period of 87.85 min, and a velocity of 25,605.1 ft/sec. The apogee and perigee were based upon a spherical Earth with a radius of 3,443.934 nmi.

The COSPAR designation for the CSM upon achieving orbit was 1972-031A and the

Apollo 16 preparation event	Date
S-IVB-511 stage delivered to KSC.	01 Jul 1970
Spacecraft/LM adapter SLA-20 delivered to KSC.	17 Aug 1970
Saturn V instrument unit IU-511 delivered to KSC.	29 Sep 1970
S-II-11 stage delivered to KSC.	30 Sep 1970
Individual and combined CM and SM systems test completed at factory.	03 Dec 1970
LM-11 final engineering evaluation acceptance test at factory.	24 Feb 1971
Integrated CM and SM systems test completed at factory.	17 Mar 1971
Descent stage of LM-11 ready to ship from factory to KSC.	01 May 1971
Descent stage of LM-11 delivered to KSC.	05 May 1971
Ascent stage of LM-11 ready to ship from factory to KSC.	07 May 1971
Ascent stage of LM-11 delivered to KSC.	14 May 1971
CM-113 and SM-113 ready to ship from factory to KSC.	26 Jul 1971
CM-113 and SM-113 delivered to KSC.	29 Jul 1971
CM-113 and SM-113 mated.	02 Aug 1971
LRV-2 delivered to KSC.	01 Sep 1971
CSM-113 combined systems test completed.	13 Sep 1971
S-IC-11 stage delivered to KSC.	17 Sep 1971
S-IC-11 stage erected on MLP-3	21 Sep 1971
Saturn V instrument unit IU-511 delivered to KSC.	29 Sep 1971
S-II-11 stage erected.	01 Oct 1971
S-IVB-511 stage erected.	05 Oct 1971
IU-511 erected.	06 Oct 1971
Launch vehicle electrical systems test completed.	15 Oct 1971
LM-11 altitude tests completed.	19 Oct 1971
CSM-113 altitude tests completed.	21 Oct 1971
Launch vehicle propellant dispersion/malfunction overall test completed.	08 Nov 1971
LRV-2 installed.	16 Nov 1971
Launch vehicle service arm overall test completed.	18 Nov 1971
CSM-113 moved to VAB.	07 Dec 1971
Spacecraft erected.	08 Dec 1971
Space vehicle and MLP-3 transferred to Launch Complex 39A.	13 Dec 1971
CSM-113 integrated systems test completed.	03 Jan 1972
LM-11 combined systems test completed.	04 Jan 1972
Space vehicle and MLP-3 returned to VAB.	27 Jan 1972
Space vehicle and MLP-3 returned to Launch Complex 39A.	09 Feb 1972
CSM-113 integrated systems test repeated.	14 Feb 1972
CSM-113 electrically mated to launch vehicle.	21 Feb 1972
Space vehicle overall test #1 (plugs in) completed.	23 Feb 1972
LM-11 flight readiness test completed.	24 Feb 1972
Space vehicle flight readiness test completed.	02 Mar 1972
S-IC-11 stage RP-1 fuel loading completed.	20 Mar 1972
Space vehicle countdown demonstration test (wet) completed.	30 Mar 1972
Space vehicle countdown demonstration test (dry) completed.	31 Mar 1972

Apollo 16 ascent phase event	GET (h:m:s)	Altitude (nmi)	Range (nmi)	Earth-fixed velocity (ft/sec)	Space-fixed velocity (ft/sec)	Event duration (sec)	Geocentric latitude (°N)	Longitude (°E)	Space-fixed flight path angle (deg)	Space-fixed heading angle (°E of N)
Liftoff	000:00:00.59	0.060	0.000	0.0	1,340.7	–	28.4470	–80.6041	0.05	90.00
Mach 1 achieved	000:01:07.5	4.282	1.358	1,076.4	2,075.5	–	28.4539	–80.5797	26.79	84.51
Maximum dynamic pressure	000:01:26.0	7.755	3.800	1,759.6	2,785.9	–	28.4670	–80.5359	29.12	81.64
S-IC center engine cutoff*	000:02:17.85	24.548	26.821	5,488.2	6,658.8	144.55	28.5847	–80.1207	23.105	76.125
S-IC outboard engine cutoff	000:02:41.78	35.698	49.927	7,753.0	8,961.7	168.5	28.7009	–79.7028	19.914	75.328
S-IC/S-II separation*	000:02:43.5	36.560	51.929	7,767.8	8,979.2	–	28.7109	–79.6666	19.643	75.339
S-II center engine cutoff	000:07:41.77	92.441	592.660	17,039.0	18,357.7	296.57	30.9376	–69.6064	0.116	79.535
S-II outboard engine cutoff	000:09:19.54	93.445	894.079	21,539.3	22,858.7	394.34	31.7737	–63.8100	0.367	82.585
S-II/S-IVB separation*	000:09:20.5	93.468	897.389	21,550.4	22,869.8	–	31.7812	–63.7457	0.358	82.622
S-IVB 1st burn cutoff	000:11:46.21	93.374	1,430.142	24,280.1	25,600.0	142.61	32.5109	–53.2983	0.001	88.496
Earth orbit insertion	000:11:56.21	93.377	1,469.052	24,286.1	25,605.0	–	32.5262	–52.5300	0.001	88.932

*Only the commanded time is available for this event.

S-IVB was designated 1972-031B. After undocking at the Moon, the LM ascent stage would be designated 1972-031C, the descent stage 1972-031E, and the particles and fields subsatellite 1972-031D.

EARTH ORBIT PHASE

After in-flight systems checks, the 341.92-sec translunar injection maneuver (second S-IVB firing) was performed at 002:33:36.50. The S-IVB engine shut down at 002:39:18.42 and translunar injection occurred 10 sec later, at a velocity of 35,566.1 ft/sec after 1.5 Earth orbits lasting 2 hr 37 min 32.21 sec.

Apollo 16 earth orbit phase event	GET (h:m:s)	Space-fixed velocity (ft/sec)	Event duration (sec)	Velocity change (ft/sec)	Apogee (nmi)	Perigee (nmi)	Period (min)	Inclination (deg)
Earth orbit insertion	000:11:56.21	25,605.1	–	–	91.3	90.0	87.85	32.542
S-IVB 2nd burn ignition	002:33:36.50	25,598.1	–	–	–	–	–	–
S-IVB 2nd burn cutoff	002:39:18.42	35,590.2	341.92	10,389.6	–	–	–	32.511

TRANSLUNAR PHASE

At 003:04:59.0, the CSM was separated from the S-IVB stage, transposed, and docked at 003:21:53.4. The docked spacecraft were ejected from the S-IVB at 003:59:15.1, and an 80.2-sec separation maneuver was made at 004:18:08.3. Color television was transmitted for 18 min during the transposition and docking. At 005:40:07.2, a 54.2-sec propulsive force from the S-IVB auxiliary propulsion system targeted the S-IVB for impact on the Moon near the Apollo 12 landing site. As on previous missions, the impact was desired to produce calibrated seismic vibrations to study the nature of the lunar interior structure. Although launch vehicle systems malfunctions precluded a planned trajectory refinement, the impact point was within the desired area. A transponder failure at 027:09:59 caused loss of S-IVB telemetry and prevented determination of a precise impact time, making interpretation of seismic data uncertain, but it is estimated to have impacted at 075:08:04 at selenographic coordinates 1.3°N, 23.8°W, some 173 nmi from the aim point, 71 nmi from the Apollo 12 seismometer, 131 nmi from the Apollo 14 seismometer, and 593 nmi from the Apollo 15 seismometer. At impact, the S-IVB weighed 30,805 lb and was traveling 8,202 ft/sec.

During the CSM/LM docking, light-colored particles were seen coming from the LM area. These were unexplained. At 007:18, the crew reported a stream of particles emitting from the LM in the vicinity of aluminum closeout panel 51, located below the docking target on the +Z face of the LM ascent stage, which covered the Mylar insulation over reaction control system A. To determine the systems status, the crew entered the LM at

008:17 and powered up. All systems were normal and the LM was powered down at 008:52. The CM television was turned on at 008:45 to give the Mission Control Center a view of the particle emission. In order to point the high-gain antenna, panel 51 was rotated out of sunlight and a marked decrease was then noted in the quantity of particles. On the television picture, the source of the particles appeared to be a growth of grass-like particles at the base of the panel. The television was turned off at 009:06. Results of the investigation found that the particles were shredded thermal paint, and that the degraded thermal protection due to the paint shredding would have no effect on subsequent LM operations. The 45-min in-flight electrophoresis demonstration commenced on schedule at 025:05 and was successful. Ultraviolet photography of the Earth from 58,000 and 117,000 nmi was accomplished as planned. The only required midcourse correction was made at 030:39:00.66. It lasted 2.01 sec and was to ensure proper lunar orbit insertion.

At 038:18:56, the command module computer received an indication that an inertial measurement unit gimbal lock had occurred. The computer correctly downmoded the IMU to 'coarse align' mode and set the appropriate alarms. As the large number of LM panel particles floating near the spacecraft blocked the command module pilot's vision of the stars, realignment of the platform was accomplished using the Sun and Moon. It was suspected that the gimbal lock indication was an electrical transient caused by actuation of the thrust vector control enable relay when exiting the IMU alignment program. A program was uplinked to the crew and entered into the computer. The program would allow the computer to ignore gimbal lock indication during critical periods. The visual light flash phenomena experiment started at 049:10. Numerous flashes were reported by the crew prior to terminating the experiment at 050:16. The crew also reported the flashes left no afterglow, were instantaneous, and were white. The second LM housekeeping commenced about 053:30 and was completed at 055:11; all LM systems were nominal. The SM scientific instrumentation module door was jettisoned at 069:59:01.

At 074:28:27.87, at an altitude of 93.9 nmi above the Moon, the service propulsion system was fired for 374.90 sec to insert the spacecraft into a lunar orbit of 170.3 × 58.1 nmi. The translunar coast had lasted 71 hr 55 min 14.35 sec.

Apollo 16 translunar phase event	GET (h:m:s)	Altitude (nmi)	Space-fixed velocity (ft/sec)	Event duration (sec)	Velocity change (ft/sec)	Space-fixed flight path angle (deg)	Space-fixed heading angle (°E of N)
Translunar injection	002:39:28.42	171.243	35,566.1	–	–	7.461	59.524
CSM separated from S-IVB	003:04:59.0	3,870.361	24,824.8	–	–	45.397	69.807
CSM/LM ejected from S-IVB	003:59:15.1	12,492.7	16,533.5	–	–	61.07	88.39
Midcourse correction ignition	030:39:00.66	119,343.8	4,514.8	–	–	76.86	111.56
Midcourse correction cutoff	030:39:02.67	119,345.3	4,508.1	2.01	12.5	76.72	111.50

LUNAR ORBIT AND LUNAR SURFACE PHASE

At 078:33:45.04, a 24.35-sec service propulsion system maneuver was performed to enter the descent orbit of 58.5 × 10.9 nmi for undocking. LM activation started at 093:34, about 11 min early. The LM was powered up and all systems were nominal. Undocking and initial separation occurred at 096:13:31. The CSM was to have executed an orbit circularization maneuver on the 13th lunar revolution at 097:41:44, but oscillations were detected in a secondary system that controlled the direction of the gimbal of the service propulsion system engine. While flight controllers evaluated the problem, the CSM maneuvered into a stationkeeping mode with the LM and prepared to either redock or continue the mission. After 5 hr 45 min, tests and analyses showed that the system was still usable and safe; therefore, the vehicles were separated again and the mission continued on a revised timeline. A separation maneuver occurred at 102:30:00, and the platforms were realigned at 102:40. At 103:21:43.08, the service propulsion system was fired for 4.66 sec to place the CSM into a lunar orbit of 68.0 × 53.1 nmi in preparation for the acquisition of scientific data. The 731-sec powered descent was initiated at 104:17:25 at an altitude of 10.994 nmi. At 104:29:35 (02:23:35 GMT on 21 April) the LM landed at selenographic coordinates 8.97301°S, 15.50019°E, on the Descartes–Cayley some 668 ft north and 197 ft west of the planned point. At engine cutoff, approximately 102 sec of firing time remained. Since the LM had remained in lunar orbit 6 hr longer than planned, it was powered down to conserve electrical power, and the first extravehicular activity was delayed in order to provide the crew with a well-deserved sleep period.

The LM cabin was depressurized at 118:53:38 for the first extravehicular activity period. Television coverage was delayed until the LRV systems were activated because the LM steerable antenna, normally used for initial lunar surface television transmission, remained locked in one axis and could not be used. The Apollo lunar surface experiments package was successfully deployed, but the commander accidentally tripped over an electronics cable, breaking it, and rendering the heat flow experiment inoperative. After completing their activities at the experiments site, the crew drove the lunar roving vehicle (LRV-2) west to Flag Crater where they made visual observations, photographed items of interest, and collected lunar samples. The inbound traverse route was slightly south of the outbound route, and the next stop was Spook Crater. The crew then returned by way of the ALSEP site to the LM, at which time they deployed the solar wind composition experiment. Several LRV problems occurred during EVA-1. While ascending ridges and traversing very rocky terrain, there was no response from the rear wheels when full throttle was applied; the vehicle continued to move, but the front wheels were digging into the surface. The crew entered the LM and the cabin was repressurized at 126:04:40. The first extravehicular activity lasted 7 hr 11 min 2 sec. The distance traveled in the lunar roving vehicle was 2.3 nmi, the vehicle drive time was 43 min, the vehicle was parked for 3 hr 39 min, and 65.92 lb of samples were collected.

After 16 hr 30 min in the LM, the crew depressurized the cabin at 142:39:35 to begin the second extravehicular period. After preparing the LRV, they headed south-southeast to a sampling station near the Cinco Craters on the north slope of Stone Mountain. They then drove in a northwesterly direction, making stops near the craters Stubby and Wreck, where, at station 8, a rear-drive troubleshooting procedure was implemented, and a mismatch of power mode switching was identified as the cause of the problem. After a change in the switch configuration, the LRV was working properly. At station 9 the LRV range, bearing, and distance were reported to be inoperative. However, navigation heading was working. When the crew reset the power switches, the navigation system began operating nominally. After the crew arrived at station 10 (LM and ALSEP site), the surface activity was extended about 20 min because the crew's consumables usage was lower than predicted. The lunar module pilot then examined the damaged heat flow experiment. Visual inspection revealed that the cable separated at the connector. Results of troubleshooting a model of the experiment at Mission Control indicated that a fix could be accomplished. However, it was not attempted because the time required could affect the third EVA. The period ended with ingress and repressurization of the LM cabin at 150:02:44. During ingress, a 2-in portion of the commander's antenna was broken off, which produced a 15- to 18-dB drop in signal strength. Because the commander's backpack radio relayed the lunar module pilot's information to the LM and the lunar communications relay unit for transmission to ground stations, a decision was made later to have the commander use the lunar module pilot's oxygen purge system, which supported the antenna. The second extravehicular activity lasted 7 hr 23 min 9 sec. The distance traveled in the LRV was 6.1 nmi, the vehicle drive time was 1 hr 31 min, park time was 3 hr 56 min, and 63.93 lb of samples were collected.

The third extravehicular period began 30 min early when the cabin was depressurized at 165:31:28, but four stations were deleted because of time limitations. They first drove to the south rim of North Ray Crater where photographs were taken and samples gathered, then walked some way around the rim in order to sample House Rock, the largest single rock seen during the extravehicular activities. They then drove southeast to the final sampling area, Shadow Rock. On completing activities there, the crew drove the vehicle back to the LM, retracing the outbound route. They re-entered the LM, and the cabin was repressurized at 171:11:31, thereby ending the fifth human exploration of the Moon. The third extravehicular activity lasted 5 hr 40 min 3 sec. The distance traveled by the LRV was 6.2 nmi, the vehicle drive time was 1 hr 12 min, the vehicle was parked for 2 hr 26 min, and 78.04 lb of samples were collected.

For the mission, the total time spent outside the LM was 20 hr 14 min 14 sec, the total distance traveled by the LRV was 14.5 nmi, the vehicle drive time was 3 hr 26 min, the vehicle was parked during extravehicular activities for 10 hr 1 min, and the collected samples totaled 211.00 lb (95.71 kg; the official metric total as determined by the Lunar Receiving Laboratory in Houston). The farthest point traveled from the LM was 15,092 ft.

While the crew was on the lunar surface, the command module pilot had obtained photographs, measured physical properties of the Moon, and made visual observations. A 7.14-sec CSM plane change maneuver was made at 169:05:52.14 and adjusted the orbit to 64.6 × 55.0 nmi.

Ignition of the ascent stage engine for lunar liftoff occurred at 175:31:47.9 (01:25:47 GMT on 24 April) and was televised. The LM had been on the lunar surface for 71 hr 2 min 13 sec. The 427.8-sec firing of the ascent engine placed the vehicle into a 40.2 × 7.9 nmi orbit. Several rendezvous sequence maneuvers were required before docking could occur 2 hr later. First, a vernier adjustment was made at 175:42:18 at an altitude of 11.2

nmi. Then the terminal phase was initiated with a 2.5-sec maneuver at 176:26:05. This maneuver brought the ascent stage to an orbit of 64.2 × 40.1 nmi. Following a nominal rendezvous sequence, the ascent stage docked with the CM at 177:41:18 at an altitude of 65.6 nmi, after being undocked for 81 hr 27 min 47 sec. After the crew transferred the samples, film, and equipment to the CSM, the ascent stage was jettisoned at 195:00:12 at an altitude of 59.2 nmi. After jettison, the LM lost stability and began to tumble at a rate of about 3 deg/sec. This may have been due to a guidance circuit breaker inadvertently being left open. A maneuver was made at 195:03:13 to separate the CSM from the ascent stage. Because no deorbit burn maneuver was possible, the ascent stage remained in lunar orbit for approximately one year.[6] The deployment boom of the mass spectrometer in the scientific instrument bay stalled in a retract cycle and so this was jettisoned at 195:23:12.

Before the CSM was maneuvered from lunar orbit, a particles and fields subsatellite, similar to that deployed from Apollo 15, was deployed at 196:02:09 during the 62nd revolution into an orbit of 66 × 52 nmi at an inclination of $\simeq$10 deg to the lunar equator. The subsatellite was instrumented to measure plasma and energetic-particle fluxes, vector magnetic fields, and subsatellite velocity from which lunar gravitational anomalies could be determined. It was scheduled to be released during the 73rd revolution into an orbit of 170 × 58 nmi. However, as a result of the engine gimbal anomaly earlier in the mission, a planned CSM orbit-shaping maneuver was not performed before ejection of the subsatellite and it was placed into an orbit with a much shorter lifetime than planned. Due to communications interference resulting from the failure of the ascent stage to deorbit, it was not possible to activate the subsatellite until 20 hr after deployment. Loss of all tracking and telemetry data occurred at 20:31 GMT on 29 May 1972. Reacquisition of the signal was expected at 22:00 GMT on that day, but was not achieved. It is believed that the subsatellite struck the far side of the lunar surface during the 425th revolution at 110°E. The lower-than-desired orbit contributed to the short orbital life because the 'mascons' were located near the subsatellite's ground track.

The second CSM plane-change maneuver and some orbital science photography were deleted so that transearth injection could be performed 24 hours earlier than originally planned. This decision was made due to the engine problem experienced during the lunar orbit circularization maneuver. Following a 162.29-sec maneuver at 200:21:33.07 at an altitude of 52.2 nmi, transearth injection was achieved at 200:24:15.36 at a velocity of 8,663.0 ft/sec after 64 lunar orbits lasting 125 hr 49 min 32.59 sec.

[6] Later analysis indicated that the ascent stage struck the lunar surface before Apollo 17 commenced, but no data were available for substantiation.

Apollo 16 lunar orbit phase event	GET (h:m:s)	Altitude (nmi)	Space-fixed velocity (ft/sec)	Event duration (sec)	Velocity change (ft/sec)	Apolune (nmi)	Perilune (nmi)
Lunar orbit insertion ignition	074:28:27.87	93.9	8,105.4	–	–	–	–
Lunar orbit insertion cutoff	074:34:42.77	75.3	5,399.2	374.90	2,802	170.3	58.1
Descent orbit insertion ignition	078:33:45.04	58.5	5,486.3	–	–	–	–
Descent orbit insertion cutoff	078:34:09.39	58.4	5,281.9	24.35	209.5	58.5	10.9
LM undocking and separation	096:13:31	33.8	5,417.2	–	–	–	–
CSM orbit circularization ignition	103:21:43.08	59.2	5,277.8	–	–	–	–
CSM orbit circularization cutoff	103:21:47.74	59.1	5,348.7	4.66	81.6	68.0	53.1
LM powered descent initiation	104:17:25	10.944	5,548.8	–	–	–	–
LM powered descent cutoff	104:29:36	–	–	731	6,703	–	–
CSM plane change ignition	169:05:52.14	58.6	5,349.8	–	–	–	–
CSM plane change cutoff	169:05:59.28	58.6	5,349.9	7.14	124	64.6	55.0
LM lunar liftoff ignition	175:31:47.9	–	–	–	–	–	–
LM lunar ascent orbit cutoff	175:38:55.7	9.9	5,523.3	427.8	6,054.2	40.2	7.9
LM vernier adjustment	175:42:18	11.2	5,515.2	–	–	–	–
LM terminal phase initiation ignition	176:26:05	40.2	5,351.6	–	–	–	–
LM terminal phase initiation cutoff	176:26:07.5	–	–	2.5	78.0	–	–
LM terminal phase finalize	177:08:42	–	–	–	–	64.2	40.1
CSM/LM docked	177:41:18	65.6	5,313.7	–	–	–	–
LM ascent stage jettisoned	195:00:12	59.2	5,347.9	–	–	–	–
CSM separation maneuver	195:03:13	–	–	–	2.0	–	–
Subsatellite deployed	196:02:09	58.4	5,349.4	–	–	66	52

TRANSEARTH PHASE

Between 202:57 and 203:12, good-quality television pictures were transmitted from inside the CM. From 203:29 to 204:12, pictures were broadcast from the LRV camera on the lunar surface. The first of two midcourse corrections, a 22.6-sec 3.4-ft/sec maneuver, was made at 214:35:02.8 to achieve the desired entry interface conditions with Earth. At 218:39:46, the command module pilot began a transearth coast EVA. Television coverage was provided for the 1-hr 23-min 42-sec period, during which he retrieved film cassettes from the scientific instrument module cameras, visually inspected the equipment, and exposed an experiment for 10 min to provide data on microbial response to the space environment. This brought the total extravehicular activity for the mission to 22 hr 17 min 36 sec. A scheduled television press conference started at 243:35 and lasted for 18 min. During the conference, the crew gave a brief description of the far side of the Moon. An item of particular interest was the crew's description of Guyot Crater, which appeared to be full of material. The material seemed to have overflowed and spilled down the side of the crater. The crew compared their observations with similar geological formations in Hawaii. Additional activities during transearth coast included photography for a Skylab program study of the behavior and effects of particles emanating from the spacecraft, and the second light-flash observation session. The second midcourse correction, a 6.4-sec maneuver of 1.4 ft/sec, was made at 262:37:20.7.

Apollo 16 transearth phase event	GET (h:m:s)	Altitude (nmi)	Space-fixed velocity (ft/sec)	Event duration (sec)	Velocity change (ft/sec)	Space-fixed flight path angle (deg)	Space-fixed heading angle (°E of N)
Transearth injection ignition	200:21:33.07	52.2	5,383.6	–	–	0.15	–85.80
Transearth injection cutoff	200:24:15.36	59.7	8,663.0	162.29	3,370.9	5.12	–82.37
Midcourse correction ignition	214:35:02.8	183,668.0	3,806.8			–75.08	165.08
Midcourse correction cutoff	214:35:25.4	183,664.8	3,807.9	22.6	3.4	–80.35	164.99
Midcourse correction ignition	262:37:20.7	25,312.9	12,256.5	–	–	–69.02	157.11
Midcourse correction cutoff	262:37:27.1	25,305.2	12,258.3	6.4	1.4	–69.02	157.10

RECOVERY

The service module was jettisoned at 265:22:33, and the CM entry followed a normal profile. The command module re-entered Earth's atmosphere (at the 400,000-ft altitude of the 'entry interface') at 265:37:31 at 36,196.1 ft/sec, after a transearth coast of 65 hr 13 min 16 sec. While on the drogue parachutes, the CM was viewed on television, and continuous coverage was possible through crew recovery. The parachute system effected splashdown of the CM in the Pacific Ocean at 19:45:05 GMT on 27 April. The mission duration was 265:51:05. The impact point, estimated to be 0.70°S and 156.22°W, was about 3.0 nmi from the target point and 2.7 nmi from the recovery ship USS *Ticonderoga*. The CM assumed an apex-down flotation attitude, but was returned to the normal flotation attitude in 4 min 25 sec by the inflatable-bag-uprighting system. The crew was retrieved by helicopter and was on the recovery ship 37 min after splashdown. The CM was recovered 62 min later. The estimated CM weight at splashdown was 11,995 lb, and the estimated distance traveled for the mission was 1,208,746 nmi. The crew remained on board the *Ticonderoga* until 17:30 GMT on 29 April, when they were flown to Hickam Air Force Base, Hawaii, where they arrived at 19:21 GMT. They departed by C-141 aircraft for Ellington Air Force Base, Houston, at 20:07 GMT and arrived at 03:40 GMT on 30 April. The CM arrived in Hawaii at 03:30 GMT on 30 April. At 18:00 GMT on 1 May, it departed for North Island Naval Air Station, San Diego, for deactivation, arriving at 00:00 GMT on 6 May. As propellants were being removed from the CM the following day, a tank cart exploded because of overpressurization. Forty-six persons suspected of inhaling toxic fumes were hospitalized, but examination established no symptoms of inhalation. The CM was not damaged. An investigation board reported that the ratio of neutralizer to oxidizer being detanked had been too low because of the extra oxidizer retained in the CM tanks as a result of the parachute anomaly. To prevent future overpressurization, changes were ordered in the ground support equipment and detanking procedures. Deactivation was completed at 00:00 GMT on 11 May. The CM left North Island at 03:00 GMT on 12 May, and was transferred to the North American Rockwell Space Division facility at Downey, California, for post-flight analysis, arriving at 10:30 GMT that same day.

CONCLUSIONS

The overall performance of the mission was excellent, with all of the primary objectives and most of the detailed objectives being achieved, although the mission was terminated one day earlier than planned. Experiment data were gathered during lunar orbit, from the lunar surface, and during both the translunar and transearth coast phases for all detailed objectives and experiments except subsatellite tracking for autonomous navigation and the surface heat flow experiment. Especially significant scientific findings included the first photography obtained of the geocorona in the hydrogen (Lyman alpha) wavelength from outside Earth's atmosphere, and the discovery of two new auroral belts around Earth. The following conclusions were made from an analysis of post-mission data:

1. Lunar dust and soil continued to cause problems with some equipment, although procedural measures were taken and equipment changes and additions were made to control the condition.
2. Loss of the heat flow experiment emphasized that all hardware should be designed for loads accidentally induced by crew movements because of vision and mobility constraints while wearing the pressurized suits.
3. The capability of the S-band omnidirectional antenna system to support the overall Lunar Module mission operations was demonstrated after the failure experienced with the S-band steerable antenna.
4. The performance of the Apollo 16 particles and fields subsatellite showed that the lunar gravitational model was not sufficiently accurate for the orbital conditions that existed to accurately predict the time of impact.
5. The absence of cardiac arrhythmias on this mission was, in part, attributed to a better physiological balance of electrolytes and body fluids resulting from an augmented dietary intake of potassium and a better rest–work cycle that effectively improved the crew's sleep.
6. The ability of the crew and the capability of the spacecraft to land safely in the rough terrain of a lunar highlands region without having high resolution photography prior to the mission was demonstrated. Further, the capability of the lunar roving vehicle to operate under these conditions and on slopes up to 20 deg was demonstrated.

The 'ground truth' from Apollo 16 overturned the scientific rationale for the selection of its landing site. The Cayley Formation was certainly not a volcanic plain. The status of the Descartes Formation was uncertain, as it was not established that they had managed to sample it. If Stone Mountain was of volcanic origin, it was masked by ejecta from South Ray. Although Smoky Mountain had not been sampled, nothing was seen at North Ray to imply that it was volcanic. Although all this came as a considerable shock, it did not take long for another theory to emerge: the Cayley was 'fluidized fragmental ejecta' from a basin-forming event that had 'sloshed' into the highlands, turning valleys between peaks into fairly level plains. When the Cayley rocks were studied, there were found to be three types of breccia: regolith breccias, fragmental breccias and impact-melt breccias. The chemical composition of the impact-melt breccias resembled such rocks at Fra Mauro and Hadley–Apennine. Most of the Cayley rocks were breccias, many of which had suffered shock-melting. Some crystalline impact-melts were so enriched with alumina that a new category was coined: very-high alumina (VHA). The thermally processed breccias, which were relatively deficient in alumina, were otherwise chemically like the KREEPy basalts of the western maria, and may have originated there. Having been assigned the site with the most convincing evidence for 'highland basalt', Apollo 16 had discovered that it did not exist. Whilst the VHA matrix-rich clast-poor crystallized impact-melt resembled basalt, it had not been extruded, and thus was not evidence of volcanism. Basalt *was* recovered in the form of fragments in the regolith, but it undoubtedly originated elsewhere and was thrown in. The Cayley was sampled at widely spaced sites on a north–south line running across the valley. Instead of the predicted series of lava flows inter-bedded with ancient regolith, it was found to be made of four types of fragmental breccia and, although their populations varied across the plain, there was no significant difference in their chemical composition. In general, the Cayley breccias were derived from plutonic anorthosites and feldspathic gabbros, indicating that they are fragments of the crust that formed on the ancient magma ocean. The only evidence of stratification in the Cayley was in the walls of the ray craters, which showed distinct populations derived from layers of light-matrix and dark-matrix breccias. The dark breccias excavated by North Ray, on the edge of the plain, showed the Cayley to be several hundred feet thick. Thus the Cayley, having been carefully isolated from the Fra Mauro during mapping because it 'looked different', was, in fact, closely related. One lesson for the geologists was that having been so sure that the Cayley was volcanic, they had settled upon a single 'working hypothesis' and therefore, as a community, had not been receptive to contradictory evidence. Another realization was the over-reliance on terrestrial analogues – the Moon does not share the Earth's endogenic geological history. Apollo 16 suggested that there are probably few, if any, volcanically constructed mountains on the Moon. Mattingly was the first CMP to take a pair of binoculars to assist his orbital survey. Circling the Moon every 2 hr, he methodically recorded the difference in character of the near and far sides and was struck by the extent of the Imbrium sculpture. From his perspective, there was nothing distinctive about the so-called Descartes Formation – it fit right into the Imbrium structure. In fact, even as his surface colleagues were finding that the Cayley was not what they had been led to expect, Mattingly observed that it looked like "a pool of unconsolidated material which has been shaken until the surface is relatively flat."

The Apollo 16 crew: Ken Mattingly (left), John Young and Charlie Duke.

The launch of Apollo 16.

Apollo 16 was to land where the Cayley plain embayed the Descartes hills.

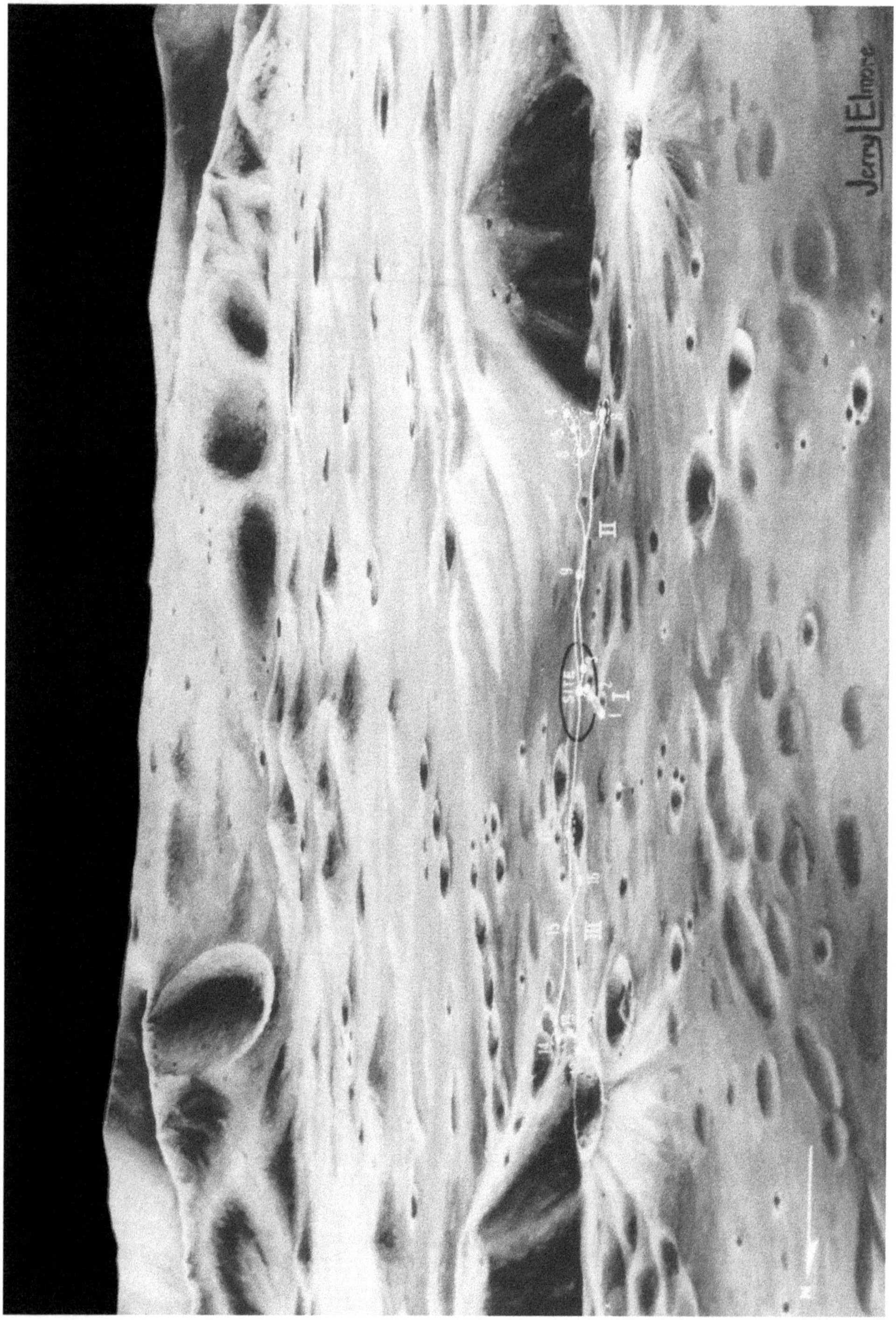

An artist's depiction of the three LRV traverses planned for Apollo 16. (Courtesy NASA.)

John Young sampling alongside the LRV.

A view from the ALSEP site, with the hummocky terrain masking the descent stage of the LM.

A view from station 4 on the flank of Stone mountain across the plain to Smoky mountain.

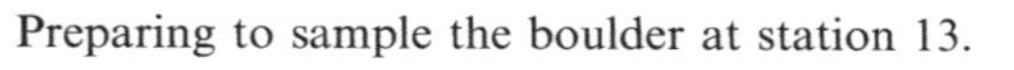

Preparing to sample the boulder at station 13.

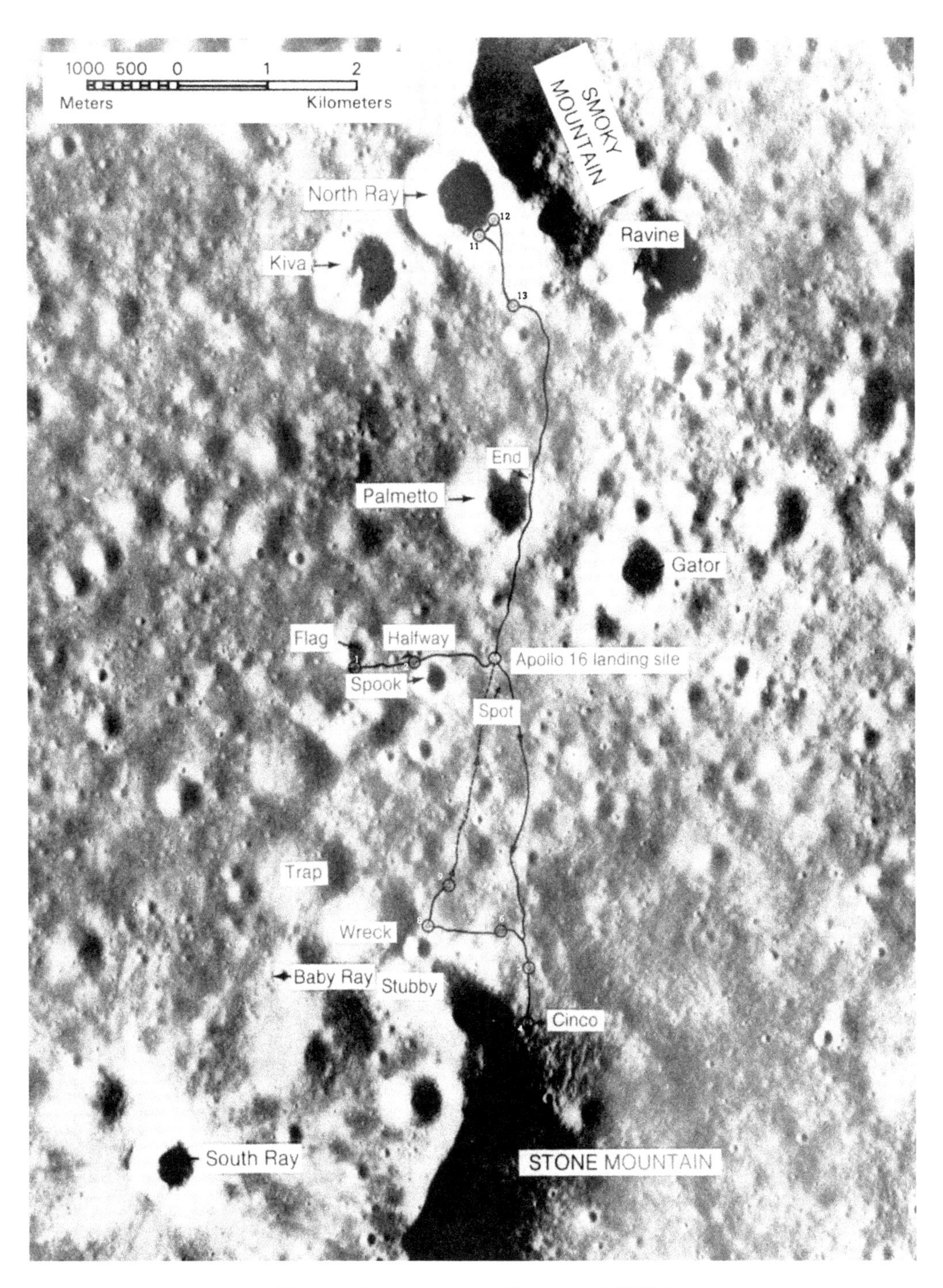

The Apollo 16 traverses. (Courtesy USGS.)

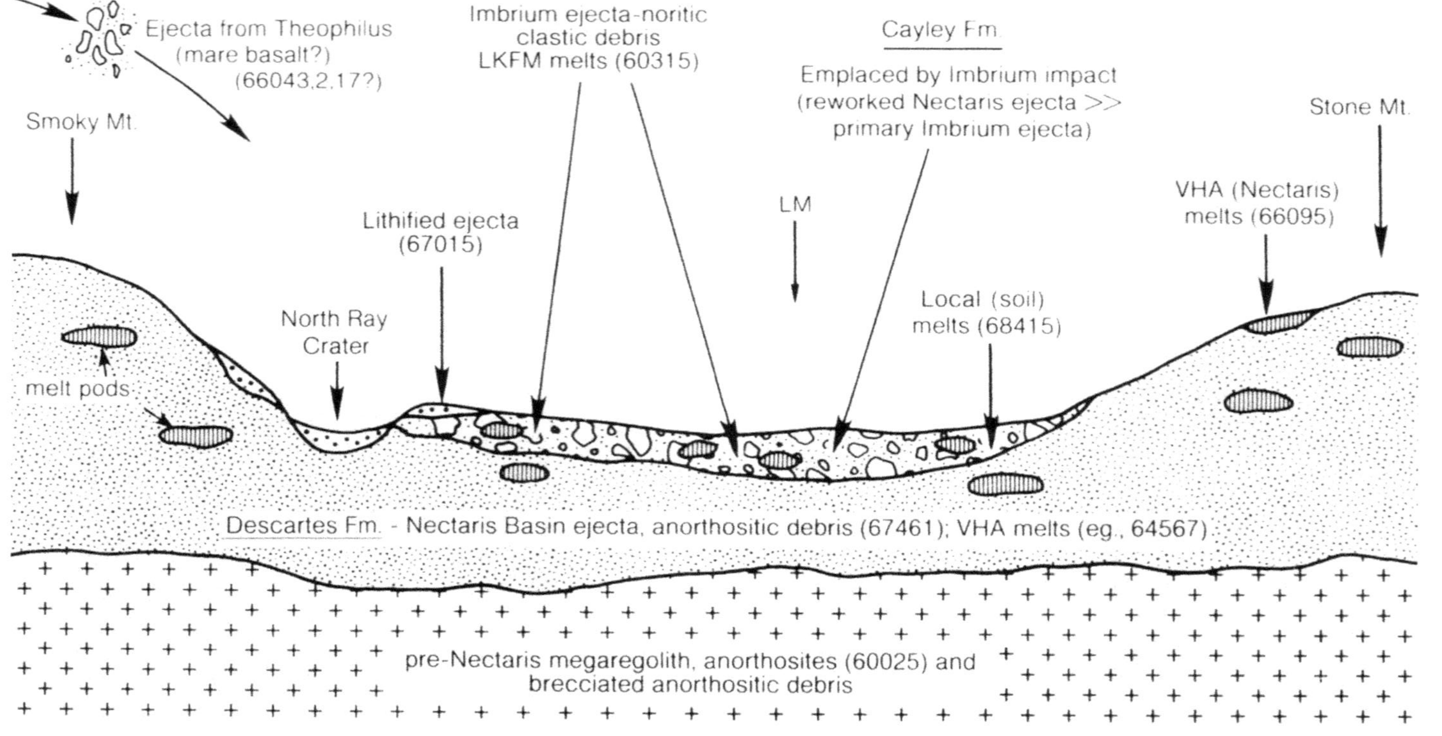

A schematic cross-section north–south through the Apollo 16 landing site showing the complex interrelationships of different units of impact-produced ejecta excavated from large basins (modified after P.D. Spudis, 'Apollo 16 site geology and impact melts: implications for the geologic history of the lunar highlands', *J. Geophys. Res.* vol. 89, 1984; D. Stoffler *et al.*, 'Composition and evolution of the lunar crust in the Descartes highlands, Apollo 16', *J. Geophys. Res.* vol. 89, 1984). The Descartes Formation is mostly ejecta from the older Nectaris basin. The overlying Cayley Formation, emplaced by a younger impact event (possibly the formation of the Imbrium basin) comprises mostly reworked materials from the Descartes Formation with fluidized ejecta from the new event. Both formations are inferred to rest on an older (pre-Nectaris) megaregolith composed of a thick layer of complex debris from many older impact events. Numbers refer to specific collected samples representative of the various units inferred to be present. 'LM' shows where the Lunar Module landed. (Courtesy the Lunar and Planetary Institute and Cambridge University Press.)

MISSION OBJECTIVES

Launch Vehicle Objectives

1. To launch on a 72–100°E of N flight azimuth and insert the S-IVB/instrument unit/spacecraft into the planned circular Earth parking orbit. *Achieved.*
2. To restart the S-IVB during either the second or third revolution and inject the S-IVB/instrument unit/spacecraft into the planned translunar trajectory. *Achieved.*
3. To provide the required attitude control for the S-IVB/instrument unit/spacecraft during transposition, docking, and ejection. *Achieved.*
4. To perform an evasive maneuver after ejection of the command and service module/lunar module from the S-IVB/instrument unit. *Achieved.*
5. To target the S-IVB/instrument stages for impact on the lunar surface at selenographic coordinates 2.3°S, 31.7°W. *Achieved.*
6. To determine actual impact point within 2.7 nmi (5.0 km) and time of impact within 1 sec. *Not achieved. The desired accuracy was not achieved.*
7. To vent and dump the remaining gases and propellants to safe the S-IVB/instrument unit. *Achieved.*

Spacecraft Primary Objectives

1. To perform selenological inspection, survey, and sampling of materials and surface features in a preselected area of the Descartes–Cayley region. *Achieved.*
2. To emplace and activate surface experiments. *Achieved.*
3. To conduct in-flight experiments and photographic tasks. *Achieved.*

Detailed Objectives

1. Service module orbital photographic tasks. *Achieved.*
2. Visual light flash phenomenon. *Achieved.*
3. Command module photographic tasks. *Partially Achieved. Timeline changes caused data loss.*
4. Visual observations from lunar orbit. *Achieved.*
5. Skylab contamination study. *Partially Achieved. Timeline changes caused data loss.*
6. Improved gas/water separator. *Not Achieved. Separator failed before it could be evaluated.*
7. Body fluid balance analysis. *Achieved.*
8. Subsatellite tracking for autonomous navigation. *Not achieved. Timeline changes caused data loss.*
9. Improved fecal collection bag. *Achieved.*
10. Skylab food package. *Achieved.*
11. LRV evaluation. *Achieved.*

Crew Participation Experiments

1. S-031: Passive seismic. *Partially achieved. No lunar module ascent stage impact.*
2. S-033: Active seismic. *Partially achieved. The fourth mortar was not fired.*
3. S-034: Lunar surface magnetometer. *Achieved.*
4. S-037: Heat flow. *Not achieved. Electronics package cable broken.*

5. S-059: Lunar geology investigation. *Achieved.*
6. S-080: Solar wind composition. *Achieved.*
7. S-152: Cosmic-ray detector (sheets). *Partially achieved. Partial deployment of panel #4.*
8. S-160: Gamma-ray spectrometer. *Achieved.*
9. S-161: X-ray fluorescence. *Achieved.*
10. S-162: Alpha particle spectrometer. *Achieved.*
11. S-164: S-band transponder (command and service module/lunar module). *Achieved.*
12. S-165: Mass spectrometer. *Achieved.*
13. S-170: Downlink bistatic radar observations of the Moon. *Achieved.*
14. S-177: Ultraviolet photography of Earth and Moon. *Partially achieved. Timeline changes caused data loss.*
15. S-178: Gegenschein from lunar orbit. *Achieved.*
16. S-198: Portable magnetometer. *Achieved.*
17. S-200: Soil mechanics. *Partially achieved. No trench was dug due to time constraints.*
18. S-201: Far ultraviolet camera/spectroscope. *Achieved.*
19. M-191: Microbial response in space environment. *Achieved.*

Passive Experiments
1. M-078: Bone mineral measurement. *Achieved.*
2. M-211: Biostack. *Achieved.*
3. S-176: Apollo window meteoroid. *Achieved.*

Operational Test
1. Lunar module voice and data relay. *Achieved.*

In-flight Demonstration
1. Fluid electrophoresis in space. *Achieved.*

Subsatellite Experiments
1. S-164: S-band transponder. *Achieved.*
2. S-173: Particle shadows/boundary layer. *Achieved.*
3. S-174: Magnetometer. *Achieved.*

Operational Tests – Manned Spacecraft Center and US Department of Defense
1. Chapel Bell (classified, Department of Defense test).
2. Radar skin tracking.
3. Ionospheric disturbance from missiles.
4. Acoustic measurement of missile exhaust noise.
5. Army acoustic test.
6. Long-focal-length optical system.
7. Sonic boom measurement.

MISSION TIMELINE

Apollo 16 mission event	GET (h:m:s)	Date (GMT)	Time (h:m:s)
Terminal countdown started.	–028:00:00	15 Apr 1972	03:54:00
Scheduled 9-hr hold at T–9 hr.	–009:00:00	15 Apr 1972	22:54:00
Countdown resumed at T–9 hr.	–009:00:00	16 Apr 1972	07:54:00
Scheduled 1-hr hold at T–3 hr 30 min.	–003:30:00	16 Apr 1972	13:24:00
Countdown resumed at T–3 hr 30 min.	–003:30:00	16 Apr 1972	14:24:00
Guidance reference release.	–000:00:16.963	16 Apr 1972	17:53:43
S-IC engine start command.	–000:00:08.9	16 Apr 1972	17:53:51
S-IC engine ignition (#5).	–000:00:06.7	16 Apr 1972	17:53:53
All S-IC engines thrust OK.	–000:00:01.9	16 Apr 1972	17:53:58
Range zero.	000:00:00.00	16 Apr 1972	17:54:00
All holddown arms released (1st motion) (1.08 *g*).	000:00:00.3	16 Apr 1972	17:54:00
Liftoff (umbilical disconnected).	000:00:00.59	16 Apr 1972	17:54:00
Tower clearance yaw maneuver started.	000:00:01.7	16 Apr 1972	17:54:01
Yaw maneuver ended.	000:00:10.9	16 Apr 1972	17:54:10
Pitch and roll maneuver started.	000:00:12.7	16 Apr 1972	17:54:12
Roll maneuver ended.	000:00:31.8	16 Apr 1972	17:54:31
Mach 1 achieved.	000:01:07.5	16 Apr 1972	17:55:07
Maximum dynamic pressure (726.81 lb/ft^2).	000:01:26.0	16 Apr 1972	17:55:26
Maximum bending moment (71,000,000 lbf-in).	000:01:26.5	16 Apr 1972	17:55:26
S-IC center engine cutoff command.	000:02:17.85	16 Apr 1972	17:56:17
Pitch maneuver ended.	000:02:38.9	16 Apr 1972	17:56:38
S-IC outboard engine cutoff. Maximum total inertial acceleration (3.82 *g*).	000:02:41.78	16 Apr 1972	17:56:41
S-IC maximum Earth-fixed velocity.	000:02:42.5	16 Apr 1972	17:56:42
S-IC/S-II separation command.	000:02:43.5	16 Apr 1972	17:56:43
S-II engine start command.	000:02:44.2	16 Apr 1972	17:56:44
S-II ignition.	000:02:45.2	16 Apr 1972	17:56:45
S-II aft interstage jettisoned.	000:03:13.5	16 Apr 1972	17:57:13
Launch escape tower jettisoned.	000:03:19.8	16 Apr 1972	17:57:19
Iterative guidance mode initiated.	000:03:24.5	16 Apr 1972	17:57:24
S-IC apex.	000:04:30.973	16 Apr 1972	17:58:31
S-II center engine cutoff. S-II maximum total inertial acceleration (1.74 *g*).	000:07:41.77	16 Apr 1972	18:01:41
S-IC impact (theoretical).	000:09:07.136	16 Apr 1972	18:03:07
S-II outboard engine cutoff.	000:09:19.54	16 Apr 1972	18:03:19
S-II maximum Earth-fixed velocity.	000:09:20.0	16 Apr 1972	18:03:20
S-II/S-IVB separation command.	000:09:20.5	16 Apr 1972	18:03:20
S-IVB 1st burn start command.	000:09:20.60	16 Apr 1972	18:03:20
S-IVB 1st burn ignition.	000:09:23.60	16 Apr 1972	18:03:23
S-IVB ullage case jettisoned.	000:09:32.3	16 Apr 1972	18:03:32
S-II apex.	000:09:44.122	16 Apr 1972	18:03:44
S-IVB 1st burn cutoff and maximum total inertial acceleration (0.67 *g*).	000:11:46.21	16 Apr 1972	18:05:46

Apollo 16 mission event – *continued*	GET (h:m:s)	Date (GMT)	Time (h:m:s)
Earth orbit insertion. S-IVB 1st burn maximum Earth-fixed velocity.	000:11:56.21	16 Apr 1972	18:05:56
Maneuver to local horizontal attitude started.	000:12:07.8	16 Apr 1972	18:06:07
Orbital navigation started.	000:13:26.1	16 Apr 1972	18:07:26
S-II impact (theoretical).	000:20:02.390	16 Apr 1972	18:14:02
S-IVB 2nd burn restart preparation.	002:23:58.60	16 Apr 1972	20:17:58
S-IVB 2nd burn restart command.	002:33:28.50	16 Apr 1972	20:27:28
S-IVB 2nd burn ignition.	002:33:36.50	16 Apr 1972	20:27:36
S-IVB 2nd burn cutoff and maximum total inertial acceleration (1.42 *g*).	002:39:18.42	16 Apr 1972	20:33:18
S-IVB safing procedures started.	002:39:19.1	16 Apr 1972	20:33:19
S-IVB 2nd burn maximum Earth-fixed velocity.	002:39:20.0	16 Apr 1972	20:33:20
Translunar injection.	002:39:28.42	16 Apr 1972	20:33:28
Maneuver to local horizontal attitude and orbital navigation started.	002:41:50.3	16 Apr 1972	20:35:50
Maneuver to transposition and docking attitude started.	002:54:19.3	16 Apr 1972	20:48:19
CSM separated from S-IVB.	003:04:59.0	16 Apr 1972	20:58:59
TV transmission started.	003:10	16 Apr 1972	21:04
CSM docked with LM/S-IVB.	003:21:53.4	16 Apr 1972	21:15:53
TV transmission ended.	003:28	16 Apr 1972	21:22
CSM/LM ejected from S-IVB.	003:59:15.1	16 Apr 1972	21:53:15
TV transmission started.	004:10	16 Apr 1972	22:04
S-IVB yaw maneuver to attain attitude for evasive maneuver.	004:10:01	16 Apr 1972	22:04:01
S-IVB APS evasive maneuver ignition.	004:18:08.3	16 Apr 1972	22:12:08
S-IVB APS evasive maneuver cutoff.	004:19:28.5	16 Apr 1972	22:13:28
TV transmission ended.	004:20	16 Apr 1972	22:14
Maneuver to S-IVB LOX dump attitude initiated.	004:27:48.4	16 Apr 1972	22:21:48
Alternate (second) maneuver to LOX dump attitude.	004:31:09	16 Apr 1972	22:25:09
S-IVB lunar impact maneuver – CVS vent opened.	004:34:47.1	16 Apr 1972	22:28:47
S-IVB lunar impact maneuver – LOX dump started.	004:39:27.1	16 Apr 1972	22:33:27
S-IVB lunar impact maneuver – CVS vent closed.	004:39:47.1	16 Apr 1972	22:33:47
S-IVB lunar impact maneuver – LOX dump ended.	004:40:15.1	16 Apr 1972	22:34:15
Maneuver to attitude required for final S-IVB APS burn initiated.	005:30:37.2	16 Apr 1972	23:24:37
S-IVB lunar impact maneuver – APS ignition.	005:40:07.2	16 Apr 1972	23:34:07
S-IVB lunar impact maneuver – APS cutoff.	005:41:01.4	16 Apr 1972	23:35:01
S-IVB lunar impact maneuver – 3-axis tumble command initiated.	005:55:06.2	16 Apr 1972	23:49:06
Command to inhibit instrument unit flight control computer to leave S-IVB in 3-axis tumble mode.	005:55:37	16 Apr 1972	23:49:37
Crew reported stream of particles coming from LM.	007:18	17 Apr 1972	01:12
Unscheduled crew transfer to LM for system checks.	008:17	17 Apr 1972	02:11
TV transmission to give Mission Control a view of the particle emissions started.	008:45	17 Apr 1972	02:39
LM powered down.	008:52	17 Apr 1972	02:46
TV transmission ended.	009:06	17 Apr 1972	03:00

Apollo 16 mission event – *continued*	GET (h:m:s)	Date (GMT)	Time (h:m:s)
Electrophoresis demonstration started.	025:05	17 Apr 1972	18:59
Electrophoresis demonstration ended.	025:50	17 Apr 1972	19:44
Loss of S-IVB tracking data precluded exact determination of impact time and location within mission objectives.	027:09:59	17 Apr 1972	21:03:59
Midcourse correction ignition (SPS).	030:39:00.66	18 Apr 1972	00:33
Midcourse correction cutoff.	030:39:02.67	18 Apr 1972	00:33:02
LM pressurized.	032:30	18 Apr 1972	02:24
CDR and LMP entered LM for housekeeping and communication checkout.	033:00	18 Apr 1972	02:54
CDR and LMP entered CM.	035:00	18 Apr 1972	04:54
False gimbal lock indication.	038:18:56	18 Apr 1972	08:12:56
Visual light flash phenomenon observations started.	049:10	18 Apr 1972	19:04
Visual light flash phenomenon observations ended.	050:16	18 Apr 1972	20:10
CDR and LMP entered LM for housekeeping.	053:30	18 Apr 1972	23:24
CDR and LMP entered CM.	055:11	19 Apr 1972	01:05
Skylab food test.	056:30	19 Apr 1972	02:24
Equigravisphere.	059:19:45	19 Apr 1972	05:13:45
Scientific instrument module door jettisoned.	069:59:01	19 Apr 1972	15:53:01
Lunar orbit insertion ignition (SPS).	074:28:27.87	19 Apr 1972	20:22:27
Lunar orbit insertion cutoff.	074:34:42.77	19 Apr 1972	20:28:42
S-IVB impact on lunar surface.	075:08:04	19 Apr 1972	21:02:04
Descent orbit insertion ignition (SPS).	078:33:45.04	20 Apr 1972	00:27:45
Descent orbit insertion cutoff.	078:34:09.39	20 Apr 1972	00:28:09
CSM landmark tracking.	079:30	20 Apr 1972	01:24
Solar monitor door/tie-down release.	080:10	20 Apr 1972	02:04
CDR and LMP entered LM.	092:50	20 Apr 1972	14:44
LM activation and system checks.	093:34	20 Apr 1972	15:28
Terminator photography.	094:40	20 Apr 1972	16:34
LM undocking and separation.	096:13:31	20 Apr 1972	18:07:31
CSM landmark tracking.	096:40	20 Apr 1972	18:34
CSM checkout indicated no rate feedback and SPS engine gimbal Position indicator showed yaw oscillations. Planned circularization maneuver at 097:41:44 not performed.	097:40	20 Apr 1972	19:34
Rendezvous (CSM active).	100:00	20 Apr 1972	21:54
LM separation from CSM.	102:30:00	21 Apr 1972	00:24
CSM and LM platforms realigned.	102:40	21 Apr 1972	00:34
CSM orbit circularization ignition (SPS).	103:21:43.08	21 Apr 1972	01:15:43
CSM orbit circularization cutoff.	103:21:47.74	21 Apr 1972	01:15:47
LM powered descent engine ignition (LM DPS).	104:17:25	21 Apr 1972	02:11:25
LM throttle to full-throttle position.	104:17:53	21 Apr 1972	02:11:53
LM manual target (landing site) update.	104:19:16	21 Apr 1972	02:13:16
CSM landmark tracking.	104:20	21 Apr 1972	02:14:20
LM landing radar velocity data good.	104:20:38	21 Apr 1972	02:14:38
LM landing radar range data good.	104:21:24	21 Apr 1972	02:15:24
LM landing radar updates enabled.	104:21:54	21 Apr 1972	02:15:54

Apollo 16 mission event – *continued*	GET (h:m:s)	Date (GMT)	Time (h:m:s)
LM landing point redesignation phase entered.	104:24:14	21 Apr 1972	02:18:14
LM throttle down.	104:24:54	21 Apr 1972	02:18:54
LM landing radar antenna to position 2.	104:26:50	21 Apr 1972	02:20:50
LM approach phase program selected and pitchover.	104:26:52	21 Apr 1972	02:20:52
LM 1st landing point redesignation.	104:27:20	21 Apr 1972	02:21:20
LM landing radar switched to low scale.	104:27:32	21 Apr 1972	02:21:32
LM attitude hold mode selected.	104:28:37	21 Apr 1972	02:22:37
LM landing phase program selected.	104:28:42	21 Apr 1972	02:22:42
LM lunar landing.	104:29:35	21 Apr 1972	02:23:35
LM powered descent engine cutoff.	104:29:36	21 Apr 1972	02:23:36
Mission clock updated (000:11:48.00 added).	118:06:31	21 Apr 1972	16:00:31
CSM terminator photography.	118:20	21 Apr 1972	16:14
1st EVA started (LM cabin depressurized).	118:53:38	21 Apr 1972	16:47:38
Lunar roving vehicle (LRV) offloaded.	119:25:29	21 Apr 1972	17:19:29
LRV deployed.	119:32:44	21 Apr 1972	17:26:44
Far ultraviolet camera/spectroscope deployed.	119:54:01	21 Apr 1972	17:48:01
TV transmission started for 1st EVA.	120:05:40	21 Apr 1972	17:59:40
US flag deployed.	120:15	21 Apr 1972	18:09
Apollo lunar surface experiments package (ALSEP) offloaded.	120:21:35	21 Apr 1972	18:15:35
CSM terminator photography.	120:30	21 Apr 1972	18:24
CSM Gum nebula photography.	121:20	21 Apr 1972	19:14
ALSEP deployed, deep core sample gathered, and LRV configured for traverse.	122:55:23	21 Apr 1972	20:49:23
Departed for station 1.	122:58:02	21 Apr 1972	20:52:02
CSM zodiacal photography.	123:00	21 Apr 1972	20:54
CDR reported bright flash on lunar surface.	123:09:40	21 Apr 1972	21:03:40
Arrived at station 1. Performed radial sampling, gathered rake and documented samples, and performed panoramic and stereographic photography.	123:23:54	21 Apr 1972	21:17:54
Departed for station 2.	124:14:32	21 Apr 1972	22:08:32
Arrived at station 2. Performed a lunar portable magnetometer measurement, gathered samples and performed panoramic and 500-mm photography.	124:21:10	21 Apr 1972	22:15:10
Departed for ALSEP site (station 3/10).	124:48:07	21 Apr 1972	22:42:07
Arrived at station 3/10. Performed the 'Lunar Grand Prix' with LRV, retrieved core sample, armed the active seismic experiment mortar package, and departed for LM.	124:54:14	21 Apr 1972	22:48:14
Arrived at LM. Deployed solar wind composition experiment, gathered samples, performed photography, and started EVA closeout.	125:05:09	21 Apr 1972	22:59:09
Solar wind composition experiment deployed.	125:07:00	21 Apr 1972	23:01
TV transmission ended for 1st EVA.	125:35	21 Apr 1972	23:29
1st EVA ended (LM cabin repressurized).	126:04:40	21 Apr 1972	23:58:40
CSM ultraviolet photography.	126:20	22 Apr 1972	00:14
CSM Gegenschein calibration.	127:00	22 Apr 1972	00:54

Apollo 16 mission event – *continued*	GET (h:m:s)	Date (GMT)	Time (h:m:s)
CSM orbital science visual observations.	128:00	22 Apr 1972	01:54
LM crew debriefing.	128:20	22 Apr 1972	02:14
CSM terminator photography.	128:30	22 Apr 1972	02:24
CSM orbital science visual observations.	129:25	22 Apr 1972	03:19
CSM orbital science photography.	130:00	22 Apr 1972	03:54
CSM terminator photography.	131:20	22 Apr 1972	05:14
CSM Gegenschein photography.	142:30	22 Apr 1972	16:24
2nd EVA started (LM cabin depressurized).	142:39:35	22 Apr 1972	16:33:35
LRV prepared for traverse.	142:49:29	22 Apr 1972	16:43:29
TV transmission started for 2nd EVA.	142:55	22 Apr 1972	16:49
Departed for station 4.	143:31:40	22 Apr 1972	17:25:40
Arrived at station 4. Performed penetrometer measurements, gathered samples, obtained a double core tube sample, gathered a soil trench sample, and performed 500-mm and panoramic photography.	144:07:26	22 Apr 1972	18:01:26
CSM deep space measurement.	144:45	22 Apr 1972	18:39
Departed for station 5.	145:05:16	22 Apr 1972	18:59:16
Arrived at station 5. Gathered samples, performed lunar portable magnetometer measurement, and performed panoramic photography.	145:10:05	22 Apr 1972	19:04:05
CSM orbital science photography.	145:35	22 Apr 1972	19:29
Departed for station 6.	145:58:40	22 Apr 1972	19:52:40
CSM orbital science visual observations.	146:05	22 Apr 1972	19:59
Arrived at station 6. Gathered samples and performed panoramic photography.	146:06:37	22 Apr 1972	20:00:37
Departed for station 8 (station 7 deleted).	146:29:18	22 Apr 1972	20:23:18
Arrived at station 8. Gathered samples, obtained a double core tube sample, and performed panoramic photography.	146:40:19	22 Apr 1972	20:34:19
CSM terminator photography.	147:15	22 Apr 1972	21:09
Departed for station 9.	147:48:15	22 Apr 1972	21:42:15
Arrived at station 9. Gathered samples, obtained single core tube sample, and performed panoramic photography.	147:53:12	22 Apr 1972	21:47:12
Departed for station 10.	148:29:45	22 Apr 1972	22:23:45
Arrived at station 10. Gathered samples, performed penetrometer measurements, obtained a double core tube sample, and performed panoramic photography.	148:54:16	22 Apr 1972	22:48:16
CSM solar corona photography.	149:05	22 Apr 1972	22:59
Departed for LM.	149:21:17	22 Apr 1972	23:15:17
Arrived at LM and started EVA activity closeout.	149:23:24	22 Apr 1972	23:17:24
TV transmission ended for 2nd EVA.	149:40	22 Apr 1972	23:34
2nd EVA ended (LM cabin repressurized).	150:02:44	22 Apr 1972	23:56:44
CSM photography of mass spectrometer boom.	153:05	23 Apr 1972	02:59
CSM orbital science visual observations.	153:40	23 Apr 1972	03:34
CSM terminator photography.	154:20	23 Apr 1972	04:14
CSM orbital science photography.	155:05	23 Apr 1972	04:59

Apollo 16 mission event – *continued*	GET (h:m:s)	Date (GMT)	Time (h:m:s)
CSM bistatic radar test started.	155:20	23 Apr 1972	05:14
CSM bistatic radar test ended.	156:00	23 Apr 1972	05:54
CSM mass spectrometer retraction test started.	165:30	23 Apr 1972	15:24
3rd EVA started (LM cabin depressurized).	165:31:28	23 Apr 1972	15:25:28
TV transmission started for 3rd EVA.	165:40	23 Apr 1972	15:34
LRV prepared for traverse.	165:43:29	23 Apr 1972	15:37:29
CSM mass spectrometer retraction test ended.	166:00	23 Apr 1972	15:54
Departed for station 11.	166:09:13	23 Apr 1972	16:03:13
Arrived at station 11. Gathered samples, performed 500-mm and panoramic photography.	166:44:50	23 Apr 1972	16:38:50
CSM solar camera photography.	166:50	23 Apr 1972	16:44
CSM orbital science visual observations.	167:50	23 Apr 1972	17:44
Departed for station 13.	168:09:46	23 Apr 1972	18:03:46
Arrived at station 13. Gathered samples, performed lunar portable magnetometer measurement and performed panoramic photography.	168:17:39	23 Apr 1972	18:11:39
Departed for station 10 prime.	168:46:33	23 Apr 1972	18:40:33
CSM plane change ignition (SPS).	169:05:52.14	23 Apr 1972	18:59:52
CSM plane change cutoff.	169:05:59.28	23 Apr 1972	18:59:59
Arrived at station 10 prime. Gathered samples, obtained a double core tube sample, and performed 500-mm and panoramic photography.	169:15:38	23 Apr 1972	19:09:38
LRV driven to LM. Samples gathered. EVA closeout started.	169:51:48	23 Apr 1972	19:45:48
Solar wind composition experiment retrieved.	170:12:00	23 Apr 1972	20:06
Departed for LRV final parking area.	170:23:06	23 Apr 1972	20:17:06
Arrived at final parking area. Performed two lunar portable magnetometer measurements, gathered samples and continued EVA closeout.	170:27:09	23 Apr 1972	20:21:09
CSM Gegenschein photography.	171:00	23 Apr 1972	20:54
Film from far ultraviolet camera/spectroscope retrieved.	171:01:42	23 Apr 1972	20:55:42
TV transmission ended for 3rd EVA. CSM deep space measurement.	171:10	23 Apr 1972	21:04
3rd EVA ended (LM cabin repressurized).	171:11:31	23 Apr 1972	21:05:31
LM equipment jettisoned.	172:15	23 Apr 1972	22:09
TV transmission started.	175:15	24 Apr 1972	01:09
LM lunar liftoff ignition (LM APS).	175:31:47.9	24 Apr 1972	01:25:47
Lunar ascent orbit cutoff.	175:38:55.7	24 Apr 1972	01:32:55
TV transmission ended.	175:40	24 Apr 1972	01:34
Vernier adjustment.	175:42:18	24 Apr 1972	01:36:18
TV transmission started.	176:18	24 Apr 1972	02:12
TV transmission ended.	176:25	24 Apr 1972	02:19
Terminal phase initiation ignition (LM APS).	176:26:05	24 Apr 1972	02:20:05
Terminal phase initiation cutoff.	176:26:07.5	24 Apr 1972	02:20:07
LM 1st midcourse correction.	176:35	24 Apr 1972	02:29
LM 2nd midcourse correction.	176:50	24 Apr 1972	02:44

Apollo 16 mission event – *continued*	GET (h:m:s)	Date (GMT)	Time (h:m:s)
Terminal phase finalize.	177:08:42	24 Apr 1972	03:02:42
CSM/LM docked.	177:41:18	24 Apr 1972	03:35:18
Transfer and stowing of equipment and samples started.	178:15	24 Apr 1972	04:09
Mass spectrometer deployed.	178:40	24 Apr 1972	04:34
Transfer and stowing of equipment and samples ended.	180:00	24 Apr 1972	05:54
Transfer of items to LM ascent stage started.	192:00	24 Apr 1972	17:54
Transfer of items to LM ascent stage ended. LM ascent stage activated.	192:30	24 Apr 1972	18:24
Maneuver to LM jettison attitude.	192:55	24 Apr 1972	18:49
Hatch closeout.	194:30	24 Apr 1972	20:24
LM prepared for jettison.	194:35	24 Apr 1972	20:29
LM ascent stage jettisoned.	195:00:12	24 Apr 1972	20:54:12
CSM separation maneuver.	195:03:13	24 Apr 1972	20:57:13
Mass spectrometer boom jettisoned.	195:23:12	24 Apr 1972	21:17:12
Subsatellite deployed.	196:02:09	24 Apr 1972	21:56:09
Sunrise solar corona photography.	196:40	24 Apr 1972	22:34
Transearth injection ignition (SPS).	200:21:33.07	25 Apr 1972	02:15:33
Transearth injection cutoff.	200:24:15.36	25 Apr 1972	02:18:15
X-ray spectrometer – Scorpius X-1 observation started.	201:31	25 Apr 1972	03:25
X-ray spectrometer – Scorpius X-1 observation ended.	202:11	24 Apr 1972	04:05
Mission clock updated (024:34:12 added).	202:18:12	25 Apr 1972	04:12:12
TV transmission from CM started.	202:57	25 Apr 1972	04:51
TV transmission from CM ended.	203:12	25 Apr 1972	05:06
TV transmission from lunar surface (LRV camera) started.	203:29	25 Apr 1972	05:23
TV transmission from lunar surface ended.	204:12	25 Apr 1972	06:06
Midcourse correction ignition.	214:35:02.8	25 Apr 1972	16:29:02
Midcourse correction cutoff.	214:35:25.4	25 Apr 1972	16:29:25
Transearth EVA started (Mattingly).	218:39:46	25 Apr 1972	20:33:46
TV transmission started for transearth EVA.	218:40	25 Apr 1972	20:34
Installation of television camera and data acquisition cameras started.	218:50	25 Apr 1972	20:44
Camera cassette retrieval and scientific instrument module inspection.	219:10	25 Apr 1972	21:04
Microbial response in space environment experiment.	219:30	25 Apr 1972	21:24
TV transmission ended for transearth EVA.	219:49	25 Apr 1972	21:43
Ingress and hatch closing started.	219:50	25 Apr 1972	21:44
Transearth EVA ended.	220:03:28	25 Apr 1972	21:57:28
X-ray spectrometer – Cygnus X-1 observation started.	221:01	25 Apr 1972	22:55
X-ray spectrometer – Cygnus X-1 observation ended.	224:01	26 Apr 1972	01:55
X-ray spectrometer – Scorpius X-1 observation started.	224:21	26 Apr 1972	02:15
Contamination control.	226:10	26 Apr 1972	04:04
Apollo 15 subsatellite reactivated.	226:50	26 Apr 1972	04:44
X-ray spectrometer – Scorpius X-1 observation ended.	226:51	26 Apr 1972	04:45
Visual light flash phenomenon observations started.	238:00	26 Apr 1972	15:54
Visual light flash phenomenon observations ended.	239:00	26 Apr 1972	16:54

Apollo 16 mission event – *continued*	GET (h:m:s)	Date (GMT)	Time (h:m:s)
X-ray spectrometer – Scorpius X-1 observation started.	242:21	26 Apr 1972	20:15
Televised press conference started.	243:35	26 Apr 1972	21:29
Televised press conference ended.	243:53	26 Apr 1972	21:47
Jet firing test.	245:00	26 Apr 1972	22:54
Skylab contamination photography started.	245:30	26 Apr 1972	23:24
X-ray spectrometer – Scorpius X-1 observation ended.	245:51	26 Apr 1972	23:45
Skylab contamination photography ended.	247:00	27 Apr 1972	00:54
X-ray spectrometer – Cygnus X-1 observation started.	248:51	27 Apr 1972	02:45
X-ray spectrometer – Cygnus X-1 observation ended.	251:51	27 Apr 1972	05:45
Midcourse correction ignition.	262:37:20.7	27 Apr 1972	16:31:20
Midcourse correction cutoff.	262:37:27.1	27 Apr 1972	16:31:27
Earth ultraviolet photography.	263:00	27 Apr 1972	16:54
CM/SM separation.	265:22:23	27 Apr 1972	19:16:23
Entry.	265:37:31	27 Apr 1972	19:31:31
Communication blackout started.	265:37:47	27 Apr 1972	19:31:47
Radar contact with CM by recovery ship.	265:40	27 Apr 1972	19:34
Communication blackout ended.	265:41:01	27 Apr 1972	19:35:01
Visual contact with CM established by recovery forces.	265:45	27 Apr 1972	19:39
Forward heat shield jettisoned.	265:45:25	27 Apr 1972	19:39:25
Drogue parachute deployed	265:45:26	27 Apr 1972	19:39:26
VHF recovery beacon contact with CM established by recovery ship.	265:46	27 Apr 1972	19:40
Main parachute deployed.	265:46:16	27 Apr 1972	19:40:16
Voice contact with CM established.	265:47	27 Apr 1972	19:41
Splashdown (went to apex-down).	265:51:05	27 Apr 1972	19:45:05
CM returned to apex-up position.	265:55:30	27 Apr 1972	19:49:30
Swimmers deployed to CM.	265:56	27 Apr 1972	19:50
Flotation collar inflated.	266:06	27 Apr 1972	20:00
Hatch opened for crew egress.	266:10	27 Apr 1972	20:04
Crew aboard recovery helicopter.	266:22	27 Apr 1972	20:16
Crew aboard recovery ship.	266:28	27 Apr 1972	20:22
CM aboard recovery ship.	267:30	27 Apr 1972	21:24
1st sample flight departed recovery ship.	305:51	29 Apr 1972	11:45
1st sample flight arrived in Hawaii.	308:20	29 Apr 1972	14:14
1st sample flight departed Hawaii.	309:09	29 Apr 1972	15:03
Crew departed recovery ship.	311:36	29 Apr 1972	17:30
Crew arrived in at Hickam AFB, Hawaii.	313:27	29 Apr 1972	19:21
Crew departed Hickam AFB.	314:13	29 Apr 1972	20:07
1st sample flight arrived in Houston, TX.	316:38	29 Apr 1972	22:32
CM arrived in Hawaii.	321:36	30 Apr 1972	03:30
Crew arrived at Ellington AFB, Houston.	321:46	30 Apr 1972	03:40
CM departed Hawaii.	360:06	01 May 1972	18:00
CM arrived at North Island Naval Air Station, San Diego, CA.	462:06	06 May 1972	00:00

Apollo 16 mission event – *continued*	GET (h:m:s)	Date (GMT)	Time (h:m:s)
Explosive failure of ground support equipment decontamination unit tank during deactivation of nitrogen tetroxide portion of CM RCS.		07 May 1972	
CM deactivated.	606:06	11 May 1972	00:00
CM departed North Island Naval Air Station for contractor's facility in Downey, CA.	609:06	12 May 1972	03:00
CM arrived at contractor's facility.	616:36	12 May 1972	10:30
Final telemetry from subsatellite (just before lunar surface impact).	1,034:37	29 May 1972	20:31
ALSEP central station turned off by ground command.	–	30 Sep 1977	–

Apollo 17

The eleventh manned mission: the sixth lunar landing

7–19 December 1972

BACKGROUND

Apollo 17 was the third Type 'J' mission, an extensive scientific investigation of the Moon on the lunar surface and from lunar orbit. Although the spacecraft and launch vehicle were similar to those for Apollo 15 and 16, some experiments were unique to this mission. It was also the final manned lunar landing mission of the Apollo program.

The primary objectives were:

- to perform selenological inspection, survey, and sampling of materials and surface features in a preselected area of the Moon;
- to emplace and activate surface experiments; and
- to conduct in-flight experiments and photographic tasks.

The crew members were Captain Eugene Andrew 'Gene' Cernan (USN), commander; Commander Ronald Ellwin Evans (USN), command module pilot; and Harrison Hagan 'Jack' Schmitt PhD, lunar module pilot. Selected as an astronaut in 1963, Cernan was making his third spaceflight. He had been pilot of Gemini 9-A and lunar module pilot of Apollo 10, the first test of the LM in lunar orbit and the dress rehearsal for the first manned landing on the Moon. Born on 14 March 1934 in Chicago, Illinois, Cernan was 38 years old at the time of the Apollo 17 mission. He received a BS in electrical engineering from Purdue University in 1956 and an MS in aeronautical engineering from the US Naval Postgraduate School in 1963. His backup for the mission was Captain John Watts Young (USN). Evans and Schmitt were making their first spaceflights. Born on 10 November 1933 in St Francis, Kansas, Evans was 39 years old at the time of the mission. He received a BS in electrical engineering from the University of Kansas in 1956 and a MS in aeronautical engineering from the US Naval Postgraduate School in 1964, and he was selected as an astronaut in 1966.[7] His backup was Lieutenant Colonel Stuart Allen Roosa (USAF). A geologist, Schmitt was the first true scientist to explore the Moon. Born on 3

[7] Evans died of a heart attack on 7 April 1990 in Scottsdale, Arizona.

July 1935 in Santa Rita, New Mexico, he was 37 years old at the time of the Apollo 17 mission. Schmitt received a BS in science from the California Institute of Technology in 1957 and a doctorate in geology from Harvard University in 1964. He was selected as an astronaut in 1965. His backup was Colonel Charles Moss Duke Jr (USAF). The capsule communicators (CAPCOMs) for the mission were Major Charles Gordon Fullerton (USAF), Lieutenant Colonel Robert Franklyn Overmyer (USMC), Robert Alan Ridley Parker PhD, Joseph Percival Allen IV PhD, Captain Alan Bartlett Shepard Jr (USN), Commander Thomas Kenneth 'Ken' Mattingly II (USN), Duke, Roosa, and Young. The support crew were Overmyer, Parker, and Fullerton. The flight directors were Gerald D. Griffin (first shift), Eugene F. Kranz and Neil B. Hutchinson (second shift), and M.P. 'Pete' Frank and Charles R. Lewis (third shift).

The Apollo 17 launch vehicle was a Saturn V, designated AS-512. The mission also carried the designation Eastern Test Range #1701. The CSM was designated CSM-114, and had the call-sign 'America'. The lunar module was designated LM-12, and had the call-sign 'Challenger'.

LANDING SITE

The Apollo 17 site was chosen very much with the fact that it would be the final mission in mind. As the selection occurred before Apollo 16 flew, it was influenced by the degree to which the 'crucial events' in lunar history had already been resolved, or were likely to be addressed by Apollo 16. The target of Apollo 17 was therefore hotly contested. Tycho in the southern highlands and Tsiolkovski on the far side were sites of 'special interest', but were impractical operationally. The Humorum basin, south of Procellarum, is only partly flooded, and the crater Gassendi resides between its rim and the northern shore of its internal mare. Being 50 nmi in diameter and inside the basin rim, Gassendi's floor was intensely fractured as it rose isostatically. A landing in this crater would sample its central peak, date the crater and likely shed light on the enclosing basin. However, the selection was driven by the imperative to refine the timescale over which the lunar 'heat engine' had been active. As the formation of the Imbrium basin was now well understood, as was the up-welling of lavas that had flooded most of the basins over the ensuing 500 million years, the goal was *late* volcanism. The Marius Hills and Davy Rille were not worthy of a 'J'-mission, so the choice came down to Gassendi, the 60-nmi-diameter crater Alphonsus, which appeared to have volcanoes on its floor, and the 'dark mantle' on the eastern rim of the Serenitatis basin. The sculpture on the Haemus Mountains showed that the Serenitatis basin was formed before the Imbrium basin. Although Serenitatis was not flooded by lava until some time later, this is believed to have started prior to the onset of up-welling in Imbrium. Evidently, Serenitatis was not flooded all at once. There is dark material around the southeastern rim, and the general view was that it was considerably younger than the lighter-toned lava in the middle. A visit to this area might be able to sample both ancient and very young terrain. Prior to being reassigned to Fra Mauro, Apollo 14 had been set to sample the dark mantle at a network of rilles and compressional 'wrinkle ridges' situated 33 nmi west of the 17-nmi-diameter crater Littrow, which was just outside the eastern rim of the basin. This was well suited to a single-traverse 'H'-mission, but unworthy of a 'J'-

mission. The selectors therefore examined a variety of sites all along the eastern arc of the rim, where it was embayed by the dark mantle. Al Worden surveyed the area on Apollo 15, while Scott and Irwin explored Hadley–Apennine, and noted a cluster of small 'dark halo' craters in a valley in the Taurus Mountains, south of Littrow. In addition, he observed dark streaks across the flanks of the mountains, which suggested the presence of fire fountains. This put Taurus–Littrow at the top of the list, and in February 1972 Apollo 17 was formally assigned the task of seeking the 'holy grail' of lunar geology in the form of unambiguous proof that the 'heat engine' was still active; it would be a fitting finale to the program.

Like the Apennines, the Taurus massifs are up-thrust crustal blocks. Although they appear to be half as tall, this is in part because they have been more thoroughly embayed. The landing site was just east of a cluster of 2,000-ft-diameter craters on the flat floor of the 6-nmi-wide valley. The massifs had been assigned obvious names: North, South and East Massifs. The lower domical hills east of the North Massif, running off to the east by the northern flank of the East Massif, appeared so 'etched' as to be named the Sculptured Hills. The evidence from Apollo 15, in the shape of orbital and surface photography, was that the summits of the Apennines had shed most of their regolith cover by mass-wastage, and although Scott and Irwin had hoped to locate boulders on the lower flank of Mount Hadley Delta in order to sample a higher elevation than could have been reached directly, the flank had been surprisingly clean. Intriguingly, Worden had spotted boulder tracks on the massifs defining this Taurus valley, and Cernan and Schmitt hoped to sample rocks whose tracks led back to specific outcrops. Although the search for recent volcanism was the motivation for selecting Taurus–Littrow, the primary objective was to characterize the massifs. The fact that the dark mantle followed the topography so well hinted that it was a lava flow, but the streaks on the massifs implied fire fountains, and the dark halo craters strengthened this by suggesting vents. Prior to Apollo 16's discovery that the Cayley was not volcanic, the consensus view was that the remarkably flat floor of the Taurus–Littrow valley was ancient basin in-fill recently mantled by dark pyroclastic. Having learned their lesson, the geologists openly embraced a range of working hypotheses. In fact, to cover themselves, they included the possibility that the floor of the valley might be fluidized brecciated ejecta like the Cayley. Accordingly, the geological maps described the valley floor morphologically, as the 'subfloor unit'. Determining its actual nature by inspecting any stratigraphy exposed in the walls of the large craters on the valley floor and sampling their ejecta was made the second objective. There was evidence that the valley was within a 'ray' of ejecta from Tycho, and it was possible that some of the craters on the valley floor were side effects of this event; if so then the fact that their rims were dark implied that the putative fire fountains were younger. Although proving this masterful photo-geological interpretation would be the significant result, the evidence was so tenuous that it had been subordinated to characterizing the massifs and the subfloor. The geologists were further favored with a fan of light-toned material at the base of the South Massif which looked *precisely* like an avalanche. As the slope was not particularly steep, certainly not beyond the angle of repose, a massive shock must have dislodged the material. The flanks of the South Massif were generally free of craters, but there was a cluster of them on its summit. The avalanche may have been caused by the shock of their formation. It was a long shot, but these craters were in-line between Tycho and the Central Cluster on the valley floor, and might all be related; if so then the 'exposure age' of the rock in the avalanche would match that of the rock excavated on the valley floor. If this turned out to be the case, it would be a clear determination of the age of Tycho, the most prominent ray crater in the southern highlands, which would be a welcome bonus. In order not to prejudge the issue, the putative avalanche fan was named the 'light mantle'. As if this was not enough, there was a scarp across the valley floor, with the eastern part of the valley on the lower side. Although 250 ft tall, the scarp looked to be a gentle slope, and it was hoped that Cernan and Schmitt would be able to drive up it in order to reach boulders that had rolled down the South Massif as part of the light mantle and engulfed the crater Nansen.

With such a rich variety of potentially highly significant features, Schmitt lobbied for an extended surface stay to allow a fourth excursion, but this would have involved trading some of the scientific experiments for extra consumables, which the geophysicists were unwilling to do. Three traverses were therefore developed to address the multiplicity of geological objectives, which were, in order of priority:

1) to characterize the rim of the Serenitatis basin by sampling boulders on the North and South Massifs whose context could be related by tracks to specific outcrops;
2) to determine the nature of the subfloor unit by sampling the craters of the Central Cluster;
3) to demonstrate that the Moon had been volcanically active within the last 100 million years by sampling the dark halo craters Shorty and Van Serg;
4) to sample the light mantle;
5) to investigate the relationship of the Sculptured Hills to the basin rim; and
6) to investigate the nature of the scarp.

The first traverse, south to take a look at some of the Central Cluster craters, would be contingent on deploying the ALSEP on schedule. The second would involve a long drive west onto the high side of the scarp in order to inspect boulders associated with the light mantle and sample Shorty. The third would be north, to sample boulders on the flank of the North Massif and investigate the Sculptured Hills. It was an *extremely* ambitious program.

LAUNCH PREPARATIONS

The terminal countdown was picked up at T–28 hr on at 12:53:00 GMT on 5 December 1972. Scheduled holds were initiated at T–9 hr for 9 hours and at T–3 hr 30 min for 1 hour. The countdown proceeded smoothly until 2 min 47 sec before the scheduled time of launch, when the Terminal Countdown Sequencer failed to issue the S-IVB LOX tank pressurization command. As a result, an automatic hold command was issued at T–30 sec which lasted 1 hour 5 min 11 sec. The countdown was recycled to T–22 min, but was held again at T–8 min to resolve the sequencer corrective action. This hold lasted 1 hour 13 min 19 sec. The countdown was then resumed at T–8 min and proceeded smoothly to launch. The delays totaled 2 hr 40 min. During the night launch of Apollo 17, the Cape Kennedy area was experiencing mild temperatures with gentle surface winds. These conditions resulted from a warm moist air mass covering most of Florida. This warm air was separated from an extremely cold air mass over the rest of the south by a cold front oriented northeast–southwest and passing through the Florida panhandle. Surface winds in the Cape Kennedy area were light and northwesterly. The maximum wind belt was located north of Florida, giving less intense wind flow aloft over the Cape Kennedy area. At launch time, stratocumulus clouds covered 20% of the sky (base 2,600 ft) and cirrus clouds covered 50% (base 26,000 ft), the temperature was 70.0°F, the relative humidity was 93%, and the barometric pressure was 14.795 psi. The winds, as measured by the anemometer on the light pole 60.0 ft above ground at the launch site were 8.0 kt at 5° from true north; the winds at 530 ft were 10.5 kt at 335° from true north.

ASCENT PHASE

Apollo 17 was launched from Pad A of Launch Complex 39 at the Kennedy Space Center at a Range Zero time of 05:33:00 GMT (00:33:00 EST) on 7 December 1972. The launch window extended from 02:53:00 GMT to 06:31:00 GMT in order to take advantage of a Sun elevation angle on the lunar surface of 13.3 deg at the scheduled time of landing. Between T + 12.9 and T + 14.3 sec, the vehicle rolled from a pad azimuth of 90°E of N to a flight azimuth of 91.503°E of N. The maximum wind conditions encountered during ascent were 87.6 kt at 311° from true north at 38,945 ft, with a maximum wind shear of 0.0177/sec at 26,164 ft. The S-IC shut down at T + 161.20, followed by S-IC/S-II separation, and S-II ignition. The S-II shut down at T + 559.66 followed by separation from the S-IVB, which ignited at T + 563.80. The first S-IVB cutoff occurred at T + 702.65, with deviations from the planned trajectory of only +1.0 ft/sec in velocity and only –0.1 nmi in altitude. At insertion, at T + 712.65 (i.e. S-IVB cutoff plus 10 sec to account for engine tailoff and other transient effects), the parking orbit showed an apogee and perigee of 90.3 × 90.0 nmi, an inclination of 28.526 deg, a period of 87.83 min, and a velocity of 25,603.9 ft/sec. The apogee and perigee were based upon a spherical Earth with a radius of 3,443.934 nmi.

The COSPAR designation for the CSM upon achieving orbit was 1972-096A and the S-IVB was designated 1972-096B. After undocking at the Moon, the LM ascent stage would be designated 1972-096C and the descent stage 1972-096D.

Apollo 17 preparation event	Date
S-II-12 stage delivered to KSC.	27 Oct 1970
S-IVB-512 stage delivered to KSC.	21 Dec 1970
Individual and combined CM and SM systems test completed at factory.	08 May 1971
LM-12 final engineering evaluation acceptance test at factory.	23 May 1971
LM-12 integrated test at factory.	23 May 1971
Ascent stage of LM-12 ready to ship from factory to KSC.	14 Jun 1971
Descent stage of LM-12 ready to ship from factory to KSC.	14 Jun 1971
Ascent stage of LM-12 delivered to KSC.	16 Jun 1971
Descent stage of LM-12 delivered to KSC.	17 Jun 1971
Integrated CM and SM systems test completed at factory.	02 Aug 1971
CM-114 and SM-114 ready to ship from factory to KSC.	17 Mar 1972
CM-114 and SM-114 delivered to KSC.	24 Mar 1972
Spacecraft/LM adapter SLA-21 delivered to KSC.	24 Mar 1972
CM-114 and SM-114 mated.	28 Mar 1972
CSM-114 combined systems test completed.	09 May 1972
S-IC-12 stage delivered to KSC.	11 May 1972
S-IC-12 stage erected on MLP-3.	15 May 1972
LM-12 stages mated.	18 May 1972
S-II-12 stage erected.	19 May 1972
LRV-3 delivered to KSC.	02 Jun 1972
LM-12 combined systems test completed.	07 Jun 1972
Saturn V instrument unit IU-512 delivered to KSC.	07 Jun 1972
CSM-114 altitude tests completed.	19 Jun 1972
IU-512 erected.	20 Jun 1972
S-IVB-512 stage erected.	23 Jun 1972
Launch vehicle electrical systems test completed.	12 Jul 1972
LM-12 altitude tests completed.	25 Jul 1972
Launch vehicle propellant dispersion/malfunction overall test completed.	01 Aug 1972
Launch vehicle service arm overall test completed.	11 Aug 1972
LRV-3 installed.	13 Aug 1972
CSM-114 moved to VAB.	22 Aug 1972
Spacecraft erected.	23 Aug 1972
Spacecraft moved to VAB.	24 Aug 1972
Space vehicle and MLP-3 transferred to Launch Complex 39A.	28 Aug 1972
LM-12 combined systems test completed.	06 Sep 1972
CSM-114 integrated systems test completed.	11 Sep 1972
LM-12 flight readiness test completed.	04 Oct 1972
CSM-114 electrically mated to launch vehicle.	11 Oct 1972
Space vehicle overall test #1 (plugs in) completed.	12 Oct 1972
Space vehicle overall test completed.	17 Oct 1972
Space vehicle flight readiness test completed.	20 Oct 1972
S-IC-12 stage RP-1 loading completed.	10 Nov 1972
Space vehicle countdown demonstration test (wet) completed.	20 Nov 1972
Space vehicle countdown demonstration test (dry) completed.	21 Nov 1972

Apollo 17 ascent phase event	GET (h:m:s)	Altitude (nmi)	Range (nmi)	Earth-fixed velocity (ft/sec)	Space-fixed velocity (ft/sec)	Event duration (sec)	Geocentric latitude (°N)	Longitude (°E)	Space-fixed flight path angle (deg)	Space-fixed heading angle (°E of N)
Liftoff	000:00:00.63	0.060	0.000	1.1	1,340.6	–	28.4470	–80.6041	0.05	90.00
Mach 1 achieved	000:01:07.5	4.315	1.265	1,076.7	2,085.8	–	28.4465	–80.5082	26.91	90.29
Maximum dynamic pressure	000:01:22.5	6.992	3.071	1,611.1	2,650.5	–	28.4457	–80.5460	28.89	91.04
S-IC center engine cutoff*	000:02:19.30	25.388	27.795	5,646.8	6,862.7	146.2	28.4329	–80.0781	23.199	91.355
S-IC outboard engine cutoff	000:02:41.20	35.900	49.145	7,757.4	9,012.1	168.1	28.4211	–79.6741	20.4285	91.718
S-IC/S-II separation*	000:02:42.9	36.776	51.112	7,778.4	9,036.1	–	28.4200	–79.6369	20.151	91.741
S-II center engine cutoff	000:07:41.21	93.420	591.254	17,064.6	18,439.6	296.61	27.5754	–69.4919	–0.058	97.647
S-II outboard engine cutoff	000:09:19.66	93.182	895.010	21,559.1	22,933.5	395.06	26.7251	–63.8908	0.254	100.395
S-II/S-IVB separation*	000:09:20.6	93.195	898.234	21,567.7	22,942.1	–	26.7147	–63.8314	0.244	100.424
S-IVB 1st burn cutoff	000:11:42.65	92.082	1,417.476	24,225.0	25,598.0	138.85	24.7139	–54.4952	0.00118	104.718
Earth orbit insertion	000:11:52.65	92.057	1,456.314	24,230.9	25,603.9	–	24.5384	–53.8107	0.0003	105.021

* Only the commanded time is available for this event.

EARTH ORBIT PHASE

After in-flight systems checks, the 351.04-sec translunar injection maneuver (second S-IVB firing) was performed at 003:12:36.60. The S-IVB engine shut down at 003:18:27.64 and translunar injection occurred 10 sec later at a velocity of 35,555.3 ft/sec after two Earth orbits lasting 3 hr 6 min 44.99 sec.

Apollo 17 earth orbit phase event	GET (h:m:s)	Space-fixed velocity (ft/sec)	Event duration (sec)	Velocity change (ft/sec)	Apogee (nmi)	Perigee (nmi)	Period (min)	Inclination (deg)
Earth orbit insertion	000:11:52.65	25,603.9	–	–	90.3	90.0	87.83	28.526
S-IVB 2nd burn ignition	003:12:36.60	22,589.4	–	–	–	–	–	–
S-IVB 2nd burn cutoff	003:18:27.64	35,579.5	351.04	10,376	–	–	–	28.466

TRANSLUNAR PHASE

At 003:42:27.6, the CSM was separated from the S-IVB stage, transposed, and docked at 003:57:10.7. During docking, there were indications of a ring latch malfunction. The LM was pressurized, the hatch removed, and troubleshooting revealed that the handles for latches 7, 9, and 10 were not locked. All were manually set and the docked spacecraft were ejected from the S-IVB at 004:45:02.3. A 79.9-sec separation maneuver was performed at 005:03:01.1. The S-IVB tanks were vented at 006:09:59.8, and the auxiliary propulsion system was fired for 98.2 sec to target the S-IVB for a lunar impact. A second, 102.2-sec maneuver was performed at 011:14:59.8. The stage impacted the lunar surface at 086:59:42.3 at selenographic coordinates 4.21°S, 12.31°W, some 84 nmi from the aim point, 183 nmi from the Apollo 12 seismometer, 85 nmi from the Apollo 14 seismometer, 557 nmi from the Apollo 15 seismometer, and 459 nmi from the Apollo 16 seismometer. The impact was recorded by all four instruments. At impact, the S-IVB weighed 30,712 lb and was traveling 8,366 ft/sec.

The 2-hr 40-min launch delay prompted ground controllers to modify Apollo 17's trajectory to arrange for it to arrive at the Moon at the originally scheduled time. They shortened the translunar coast time by having the crew make a 1.73-sec 10.5-ft/sec midcourse correction at 035:29:59.91. The commander and lunar module pilot transferred to the LM at 040:10. At ingress, it was discovered that docking latch 4 was not properly latched. The command module pilot moved the latch handle between 30 and 45 deg, disengaging the hook from the docking ring. After discussion with ground control, it was decided to curtail further action on the latch until the next LM activation. The remainder of the LM housekeeping was nominal and the LM was closed out at 042:11. The heat flow and convection demonstrations were conducted as planned. The first demonstration began at 042:55 and was performed with the spacecraft in attitude hold while the second run was accomplished with the spacecraft in the passive thermal control mode. The demonstrations produced satisfactory results, and were concluded at 046:00. The second LM housekeeping session commenced at 059:59 and was completed at 062:16. All LM systems checks were nominal. During the LM housekeeping period, the command module pilot performed troubleshooting on the docking latch 4 problem experienced during the first session. Following instructions from the ground controllers, he stroked the latch handle and succeeded in cocking the latch, which was left in the cocked position for the CSM/LM rendezvous. At 065:39, a 1-hr visual light flash phenomenon observation was conducted by the crew. They reported seeing light flashes ranging from bright to dull. The SM scientific instrument module bay door was jettisoned at 081:32:40. At 086:14:22.60, at an altitude of 76.8 nmi above the Moon, the service propulsion system was fired for 393.16 sec to insert the spacecraft into a lunar orbit of 170.0 × 52.6 nmi. The translunar coast had lasted 83 hr 2 min 18.11 sec.

Apollo 17 translunar phase event	GET (h:m:s)	Altitude (nmi)	Space-fixed velocity (ft/sec)	Event duration (sec)	Velocity change (ft/sec)	Space-fixed flight path angle (deg)	Space-fixed heading angle (°E of N)
Translunar injection	003:18:37.64	169.401	35,555.3	–	–	7.379	118.110
CSM separated from S-IVB	003:42:27.6	3,566.842	25,344.9	–	–	44.177	102.769
CSM/LM ejected from S-IVB	004:45:02.3	13,393.6	16,012.8	–	–	61.80	83.485
Midcourse correction ignition	035:29:59.91	128,217.7	4,058.1	–	–	76.40	66.71
Midcourse correction cutoff	035:30:01.64	128,246.9	4,066.8	1.7	10.5	76.48	66.84

LUNAR ORBIT AND LUNAR SURFACE PHASE

At 090:31:37.43, a 22.27-sec service propulsion system burn was performed and lowered the spacecraft to the descent orbit of 59.0 × 14.5 nmi in preparation for undocking of the LM. The CSM/LM combination was retained in this orbit 17 hours before the spacecraft were undocked and separated by a 3.4-sec maneuver at 107:47:56 at an altitude of 47.2 nmi, while in an orbit of 61.5 × 11.5 nmi. After undocking, a 3.80-sec maneuver at 109:17:28.92 circularized the CSM orbit to 70 × 54 nmi. The second LM descent orbit insertion maneuver, performed for 21.5 sec at 109:22:42, lowered the orbit to 59.6 × 6.2 nmi. The 725-sec powered descent maneuver was initiated from this orbit at 110:09:53 at an altitude of 8.7 nmi. The landing was at 110:21:58 (19:54:58 GMT on 11 December) at selenographic coordinates 20.19080°N, 30.77168°E, some 656 ft east of the planned point. At engine cutoff, approximately 117 sec of firing time remained.

The first extravehicular activity began at 114:21:49 with the depressurization of the LM cabin. The commander dedicated his first step onto the lunar surface to "all those who made it possible". No sooner had Schmitt emerged than he ruminated on the nature of the dark mantle: it was not a surface layer of pyroclastic, but very fine crystals mixed into the 'gardened' lithic fragments that were derived from a light-grey pyroxene whose coarse crystals and vesicles identified it as a gabbro that solidified at a shallow depth. The lunar roving vehicle (LRV-3) was offloaded at 114:51:10. By design, the television coverage did not start until the LRV system was activated. While working on the LRV, the commander inadvertently broke off the extension of the right-rear fender with the hammer in his shin pocket, but was able to secure it using tape. After an LRV test drive around the LM, the crew gathered some samples and took panoramic photography, then at 115:40:58 they deployed the US flag that had been on display at Mission Control throughout the Apollo program. At 115:58:30 retrieved the ALSEP from the LM, then deployed this 600 ft west–northwest of the landing site. At 119:11:02 they began the first traverse, arriving at station 1 at 119:24:02. It had been intended to inspect Steno, a 2,000-ft-diameter crater 0.4 nmi to the south that had excavated the subfloor, and then the similarly sized Emory, 0.55 nmi beyond, but the ALSEP deployment had run late and the visit to Emory had to be deleted, which was a pity because this crater was situated on the margin of the dark mantle. With time running short, and Steno's location proving elusive, they opted not to seek the designated sampling point, and instead accepted a 60-ft crater later identified as being within 500 ft of Steno. The fact that the blocks on the rim were the same as at the landing site showed that the subfloor was horizontally homogeneous. At 119:56:47 they departed. On the way back, the fender extension came off, and a 'rooster tail' showered the vehicle and crew with dust. After a brief stop to deploy a seismic profiling explosive charge, they reached a site on the opposite side of the LM from the ALSEP at 120:11:02 where they were to deploy the surface electrical properties experiment. They returned to the LM at 120:36:15 and entered the LM. The cabin was repressurized at 121:33:42. The first EVA lasted 7 hr 11 min 53 sec. The distance traveled in the LRV was 1.8 nmi, the vehicle drive time was 33 min, and 31.53 lb of samples were collected.

The second extravehicular activity began 80 min late, with cabin depressurization at 137:55:06. Prior to starting the EVA traverse, ground controllers relayed instructions for improvising a replacement for the lost fender extension: four maps, taped together and held in position by two clamps from portable utility lights, made an excellent substitute. The crew loaded the LRV and departed for the surface electrical properties experiment site at 138:44:02, and at 138:51:43 began the westerly traverse. On the way, they drove past Camelot, which, at 2,200 ft in diameter, was one of the largest craters in the Central Cluster, in order to verify that its southern rim was littered with large blocks. Next was Horatio, which, although not blocky, had layering in its wall that suggested the subfloor was overlain by a blanket of regolith 60–100 ft thick. Continuing west, they crossed onto the light mantle, which, at its periphery, was just a thin veneer of finely grained material. The scarp proved to be a smooth slope, and although the LRV had to 'tack' left and right to ascend the 30-deg slope, it readily reached the crest. Once on the elevated section of the valley floor, they headed for Nansen, a lenticular depression at the base of the South Massif, arriving at station 2, located some 4 nmi from the LM, at 140:01:30. The massif was generally tan-grey on its flank, but where the slope shallowed near the summit there were blue-grey outcrops from which boulders had rolled, many ending up in Nansen, the southern part of which was filled by talus. The sample site was among another cluster of rocks, some 300 ft east of Nansen. All the rocks proved to be breccias of various types. They left station 2 at 141:07:25, and at 141:48:38 arrived at station 3, which was a 35-ft crater on the light mantle at the base of the scarp, which they left at 142:25:56. Since the traverse was running late, most of the activities for station 4 at the 330-ft-diameter 'dark halo' crater Shorty that was suspected of being a volcanic vent, had been deleted, but on arriving at 142:42:57 they discovered an intriguing patch of 'orange soil' near a block on the rim, and they were given time to sample it with a core tube. The tentative conclusion was that it was chemical alteration by oxidation in a fumarole, which supported the case for ongoing volcanism. After leaving Shorty at 143:19:03, they raced to Camelot, where at 143:45:15 they established station 5 among the large blocks on its southern rim, which confirmed the subfloor to be pyroxene gabbro. They left at 144:15:58, and arrived back at the LM at 144:32:24. During the traverse, they had deployed three explosive packages in support of the lunar seismic profiling experiment, performed seven traverse gravimeter measurements, gathered numerous samples, and completed their 500-mm and panoramic photographic tasks. They entered the LM, and the cabin was repressurized at 145:32:02. The second extravehicular activity lasted 7 hr 36 min 56 sec. The distance traveled in the LRV was 11.0 nmi, the vehicle drive time was 2 hr 25 min, and 75.18 lb of samples were collected.

After a 15-hr 30-min period in the LM, the cabin was depressurized at 160:52:48 for the third EVA, about 50 min later than planned. At 161:42:36 they set off to explore the northeastern part of the valley. The first leg was north, aiming for a 'house sized' rock at the base of the North Massif, named Turning Point Rock, where they swung east and ran diagonally up the 20-deg slope, arriving at station 6 at 162:11:24 at an even larger block at an elevation of 250 ft above the valley floor. In rolling down from a high outcrop, this had left a prominent track, and on coming to rest had broken. For once, the schedule was flexible, and Schmitt was able to perform a thorough inspection of the entire train, which was 80 ft in length. Continuing east at 163:22:10, they drove back downslope to another boulder at station 7, arriving at 163:29:05. They left at 163:51:09 for station 8 at the base of the Sculptured Hills, arriving at 164:07:40 and departing at 164:55:33. They arrived at station 9, the crater Van Serg, at 165:13:10. This proved to be littered not with rock but with regolith breccia, where an impact had hit a thick patch of regolith and the shock had formed a great deal of 'instant rock'. Leaving at 166:09:25, they arrived back at the LM at 166:37:51. The cosmic ray experiment had been retrieved shortly before initiating the traverse. While traveling, 500-mm and panoramic photography was secured, nine traverse gravimeter measurements were made, and several more seismic profiling charges were deployed. While at station 9, the surface electrical properties experiment was terminated when it was noted that the receiver temperature had risen to a level that could affect the data tape; the recorder was removed. On returning to the LM, the lunar neutron probe experiment was retrieved. With Schmitt back in the LM, Cernan paused to reflect on their achievement, and of the future. "I'd like to just say what I believe history will record: that America's challenge of today has forged Man's destiny of tomorrow. And as we leave the Moon at Taurus–Littrow, we leave as we came and, God willing, as we shall return, with peace and hope for all Mankind." With that, he headed up the ladder. Following equipment jettison, the cabin was repressurized at 168:07:56, thus ending the sixth human exploration of the Moon. The third EVA lasted 7 hr 15 min 8 sec. The distance traveled in the LRV was 6.5 nmi, the vehicle drive time was 1 hr 31 min, and 136.69 lb of samples were collected.

For the mission, the total time spent outside the LM was 22 hr 3 min 57 sec, the total distance traveled in the LRV was 19.3 nmi, the vehicle drive time was 4 hr 29 min, and the collected samples totaled 243.65 lb (110.52 kg; the official metric total as determined by the Lunar Receiving Laboratory in Houston). The farthest point traveled from the LM was 25,029 ft. Good-quality television transmissions were received during all three EVAs.

Numerous science activities were conducted in lunar orbit while the surface was being explored. In addition to the panoramic camera, the mapping camera, and the laser altimeter (which were used on previous missions), three new experiments were included in the service module. An ultraviolet spectrometer measured lunar atmospheric density and composition, an infrared radiometer mapped the thermal characteristics of the Moon, and a lunar sounder acquired data on the subsurface structure. The orbit of the CSM did not decay as predicted while the LM was on the Moon. Consequently, a 37.50-sec orbital trim maneuver was performed at 178:54:05.45 to lower the orbit to 67.3 × 62.5 nmi, and the planned 20.05-sec plane change maneuver at 179:53:53.83 resulted in an orbit of 62.8 × 62.5 nmi.

Ignition of the ascent stage engine for lunar liftoff occurred at 185:21:37 (22:54:37 GMT on 14 December). The LM had been on the lunar surface for 74 hr 59 min 40 sec. The 441-sec maneuver was made to achieve the initial lunar orbit of 48.5 × 9.1 nmi. Several rendezvous sequence maneuvers were required before docking could occur 2 hours later. A 10-sec vernier adjustment maneuver at 185:32:12 adjusted the orbit to 48.5 × 9.4 nmi. Finally, the 3.2-sec terminal phase initiation at 186:15:58 brought the ascent stage to an orbit of 64.7 × 48.5 nmi. The ascent stage and the CSM docked at 187:37:15 at an altitude of 60.6 nmi. The two spacecraft had been undocked for 79 hr 49 min 19 sec. After transfer of the crew and samples to the CSM, the ascent stage was jettisoned at 191:18:31, and the CSM was prepared for transearth injection. The ascent stage was then maneuvered by remote control to strike the lunar surface. A 12-sec maneuver was made at 191:23:31 to separate the CSM from the ascent stage, and resulted in an orbit of 63.9 × 61.2 nmi. A 116-sec deorbit maneuver at 60.5 nmi altitude depleted the ascent stage's propellants by 193:00:10. Impact was at selenographic coordinates 19.96°N, 30.50°E at 193:17:20.8, 0.7 nmi from the aim point and 4.7 nmi southwest of the Apollo 17 landing site. The impact was recorded by the Apollo 12, 14, 15, and 16 seismic stations. The first two of eight explosive packages placed by the crew on the surface were detonated at 210:15:14.56 and 212:44:57.11. Both events were picked up by the lunar seismic profiling geophones, and the resulting flash and dust from the second explosion were seen on television. A third lunar surface explosive package was detonated at 229:35:34.67. The television assembly and lunar communications relay unit failed to operate when attempts were made to command the camera on at 218:20, 235:04, and 235:13. It was later determined that the relay unit experienced an over-temperature failure.

Following a 143.69-sec service propulsion system maneuver at 234:02:09.18 at an altitude of 62.1 nmi, transearth injection was achieved at 234:04:32.87, at a velocity of 8,374.3 ft/sec, after 75 lunar orbits lasting 147 hr 43 min 37.11 sec. The crew had spent an additional day in lunar orbit performing scientific experiments.

Apollo 17 lunar orbit phase event	GET (h:m:s)	Altitude (nmi)	Space-fixed velocity (ft/sec)	Event duration (sec)	Velocity change (ft/sec)	Apolune (nmi)	Perilune (nmi)
Lunar orbit insertion ignition	086:14:22.60	76.8	8,110.2	–	–	–	–
Lunar orbit insertion cutoff	086:20:55.76	51.2	5,512.1	393.16	2,988	170.0	52.6
1st descent orbit insertion ignition	090:31:37.43	51.1	5,512.7	–	–	–	–
1st descent orbit insertion cutoff	090:31:59.70	50.9	5,322.1	22.27	197	59.0	14.5
CSM/LM separation initiated	107:47:56	47.2	5,342.8	–	–	–	–
CSM/LM separation cutoff	107:47:59.4	–	–	3.4	1	61.5	11.5
CSM orbit circularization ignition	109:17:28.92	58.6	5,279.9	–	–	–	–
CSM orbit circularization cutoff	109:17:32.72	58.8	5,349.9	3.80	70.5	70	54
LM 2nd descent orbit insertion ignition	109:22:42	59.6	5,274.5	–	–	–	–
LM 2nd descent orbit insertion cutoff	109:23:03.5	59.6	5,267.0	21.5	7.5	59.6	6.2
LM powered descent initiation	110:09:53	8.7	5,550.3	–	–	–	–
LM powered descent cutoff	110:21:58	–	–	725	6,698	–	–
CSM orbital trim ignition	178:54:05.45	64.9	5,315.1	–	–	–	–
CSM orbital trim cutoff	178:54:42.95	–	–	37.50	9.2	67.3	62.5
CSM plane change ignition.	179:53:53.83	–	–	–	–	–	–
CSM plane change cutoff	179:54:13.88	60.5	5,341.1	20.05	366	62.8	62.5
LM lunar liftoff ignition	185:21:37	–	–	–	–	–	–
LM ascent orbit cutoff	185:28:58	8	5,542.3	441	6,075.7	48.5	9.1
LM vernier adjustment initiated	185:32:12	9.4	5,534.7	–	–	–	–
LM vernier adjustment cutoff	185:32:22	–	–	10	10.0	48.5	9.4
LM terminal phase initiation ignition	186:15:58	44.6	5,333.3	–	–	–	–
LM terminal phase initiation cutoff	186:16:01.2	–	–	3.2	53.8	64.7	48.5
CSM/LM docked	187:37:15	60.6	5,341.7	–	–	–	–
LM ascent stage jettisoned	191:18:31	60.6	5,343.4	–	–	–	–
CSM separation ignition.	191:23:31	–	–	–	–	–	–
CSM separation cutoff	191:23:43	–	–	12	2.0	63.9	61.2
LM ascent stage deorbit ignition	192:58:14	60.5	5,343.7	–	–	–	–
LM ascent stage deorbit cutoff	193:00:10	58.9	5,130.1	116	286.0	–	–

TRANSEARTH PHASE

The fourth and fifth lunar surface explosive packages were detonated at 235:09:36.79 and 238:12:46.08 and the geophones received strong signals. At 254:54:40, the command module pilot began a 1-hr 5-min 44-sec transearth coast extravehicular activity, televised to Earth, during which he retrieved the lunar sounder film, panoramic camera, and mapping camera cassettes in three trips to the scientific instrument module bay. This brought the total extravehicular activity for the mission to 23 hr 9 min 41 sec. Three final explosive packages were detonated at 257:43:41.06, 259:11:56.82, and 261:44:28.28, and were detected by the lunar surface geophones. During the remainder of transearth flight, the crew performed another light-flash experiment, and operated the infrared radiometer and ultraviolet spectrometer. One midcourse correction was required, a 9-sec 2.1-ft/sec maneuver at 298:38:01.

Apollo 17 transearth phase event	GET (h:m:s)	Altitude (nmi)	Space-fixed velocity (ft/sec)	Event duration (sec)	Velocity change (ft/sec)	Space-fixed flight path angle (deg)	Space-fixed heading angle (°E of N)
Transearth injection ignition	234:02:09.18	62.1	5,337.1	–	–	–0.18	257.32
Transearth injection cutoff	234:04:32.87	63.1	8,374.3	143.69	3,046.3	2.46	259.47
Midcourse correction ignition	298:38:01	25,016.3	12,021.1	–	–	–68.43	34.63
Midcourse correction cutoff	298:38:10	24,999.7	12,025.8	9	2.1	–68.42	34.63

RECOVERY

The service module was jettisoned at 301:23:49, and the CM entry followed a normal profile. The command module re-entered Earth's atmosphere (at the 400,000-ft altitude of the 'entry interface') at 301:38:38 at 36.090.3 ft/sec, after a transearth coast of 67 hr 34 min 05 sec. The parachute system effected splashdown of the CM in the Pacific Ocean at 19:24:59 GMT on 19 December. The mission duration was 301:51:59. The impact point, estimated to be 17.88°S and 166.11°W, was about 1.0 nmi from the target, and 3.5 nmi from the recovery ship USS *Ticonderoga.* The CM assumed an apex-up flotation attitude. The crew was retrieved by helicopter and was on the ship 52 min after splashdown. The CM was recovered 71 min later. The estimated CM weight at splashdown was 12,120 lb, and the estimated distance traveled for the mission was 1,291,299 nmi. The crew departed the *Ticonderoga* at 00:38 GMT on 21 December and arrived in Houston at 15:50 GMT. The CM was sent for deactivation to North Island Naval Air Station, San Diego, where it arrived at 19:30 GMT on 27 December, this work being completed at 22:00 GMT on 30 December. The CM left North Island at 19:00 GMT on 2 January, and three hours later was delivered to the North American Rockwell Space Division facility in Downey, California, for post-flight analysis.

CONCLUSIONS

Due to the experienced personnel and excellent performance of equipment, all facets of the final Apollo lunar landing mission were conducted with skill, precision, and relative ease. The following conclusions were made from an analysis of post-mission data:

1. The Apollo 17 mission was the most productive and trouble-free manned mission, and represented the culmination of continual advancements in hardware, procedures, training, planning, operations, and scientific experiments.
2. The Apollo 17 mission demonstrated the practicality of training scientists to become qualified astronauts while retaining their expertise and scientific knowledge.
3. Stars and the horizon were not visible during night launches, therefore out-of-the-window alignment techniques could not be used for attitude reference.
4. The dynamic environment within the cabin during the early phases of the launch made system troubleshooting or corrective actions by the crew impractical. Therefore, either the ground control or automation should be relied upon for system troubleshooting and, in some cases, corrective actions.
5. As a result of problems on this and other missions, further research was needed to increase the dependability of mechanisms used to extend and retract equipment repeatedly in the space environment.

The subfloor was found to be a lava flow. When the seismic charges that had been deployed on the valley floor were remotely detonated, they revealed the thickness of this material to exceed 1 nmi. Before the surface electrical properties experiment overheated, it recorded data consistent with this finding. The underlying pre-Serenitatis basement lies deep beneath the reach of an impact. Evidently, prior to this embayment, the massifs rose almost 3 nmi above the original valley floor, and most of the 'intra-massif' ejecta within the Taurus Mountains is now submerged. The fragments in the regolith on the dark valley floor (with the exception of the immediate vicinity of Van Serg) were derived from the subfloor. The fragments in the light mantle were various types of breccias, which derived from the South Massif. It is a characteristic of terrestrial avalanches that the larger debris settles at the bottom of the deposit. The fines fill in the cavities, and then accumulate on top, making a smooth fan. This seemed to match the light mantle at the base of the South Massif. The surface was all fines, the fragment size and population increased with depth, and only the largest craters had excavated large rocks. The solar wind 'exposure age' of the light mantle material indicated it to have been deposited about 100 million years ago. Given the logic set out prior to the mission, this was tentatively interpreted as dating the crater Tycho.

During 30 min of studying the large fragmented boulder at station 6, Schmitt 'read' its complex story. Laboratory analysis of the samples confirmed his observation that the blue-grey rock was a breccia. However, the tan-grey rock was not gabbro, it was impact-melt, and therefore a fine-matrix breccia. This misidentification was excusable, because it was difficult to distinguish an igneous rock from a fragment-poor melt in a laboratory, let alone 'in the field'. One thing was certain though: if they had simply taken a chip off one end and moved on, they would have completely missed the story the rock had to tell. In retrospect, it was clear that both kinds of breccia were also present in the talus at the base of the South Massif, but the tan-grey was not vesicular. The first vesicular type they saw was Turning Point Rock. Because its track led to an outcrop one-third of the way up the massif, the station 6 rock's context was known. A 3-ft-wide metamorphic transition between the two types of breccia reflected the enormity of the event that produced it, and because the story of this rock was the story of the massif, this extraordinary contact was a measure of shock associated with the impact that created the Serenitatis basin. By sending a professional geologist and granting him the time to really study a rock *in situ*, station 6 was the closest that Apollo ever came to reproducing the spirit of a terrestrial geological field trip.

Schmitt resolved the nature of the dark mantle while still on the Moon by noting that there were fine crystals mixed in with the regolith. These proved to have a homogeneous basaltic composition. The 'orange soil' at Shorty was found to be chemically identical but in the form of microscopic beads, which implied that it was gas-rich magma spewed out by a fire fountain, and trace amounts of sulphur, zinc, lead and other volatile elements on the surface of the beads proved it. Both were pyroclastics. The valley of Taurus–Littrow had been the site of intense volcanism, but not recently: at 3.64 billion years old, the 'orange soil' was almost as old as the subfloor (3.72 billion years). On emerging from the fire fountain, if the ilmenite-rich magma droplets had cooled slowly they had crystallized, making them black, but if they had cooled instantly, they had vitrified. The glassy beads were orange because of the iron-to-titanium ratio (just as the high magnesium content of the glass found at Hadley–Apennine had made it green). Furthermore, it was evident from the block on Shorty's rim that the crater had excavated bedrock. The extremely vesicular form of this particular piece of gabbro hinted that it was off the very top of the subfloor. Evidently, after the main slab of the subfloor unit had gained a solid crust it had received a thick pyroclastic deposit, with the two types of material forming distinct layers. Before the pyroclastic could be disturbed by gardening, it was covered by a late, fairly thin, lava flow that was probably from the vent that issued the fire fountain. When Shorty punched through this veneer, it excavated the compacted deposits beneath. Some of this material was built into the rim as if it were solid rock, and the rest was scattered on the flank. This resolved the issue of the dark halo. In view of this, interest in finding 'recent' volcanism waned, but this undoubtedly also reflected the fact that there would be no opportunities to seek the 'ground truth' required to test a hypothesis. There are, in fact, a number of convincing candidates. The flooded crater Flamsteed in Procellarum where Surveyor 1 landed was considered for Apollo 12, but the 'pin-point' landing demonstration was made by visiting Surveyor 3 at a site that was more conveniently located but less geologically significant. A site never considered for Apollo was the 10-nmi crater Lichtenberg in northwestern Procellarum. This is partly flooded, but has bright rays and therefore cannot be very old. And of course, also in this region, there are the Marius Hills which would have justified an 'H'-mission but wasted a 'J'-mission. If Apollo 13 had not aborted, Apollo 14 would have visited the dark mantle near a network of ridges and rilles in eastern Serenitatis, and the discovery that this was ancient would have enabled the final mission to be sent to a site where some form of volcanism may indeed still be ongoing.

MISSION OBJECTIVES

Launch Vehicle Objectives

1. To launch on a 72–100°E of N flight azimuth and insert the S-IVB/instrument unit/ spacecraft into the planned circular Earth parking orbit. *Achieved.*

The night launch of Apollo 17.

Prior to its being reassigned to the Fra Mauro Formation, Apollo 14 was to have landed by a system of rilles and ridges due west of the crater Littrow, to test the hypothesis that the darker section of the southeastern fringe of Mare Serenitatis was 'young' volcanism. In sending Apollo 17 to make this test, the investigation was broadened to include the Taurus mountains, forming the southeastern rim of the Serenitatis basin. The lower picture is a mosaic of images of the target valley and the approach route taken shortly prior to Apollo 17's landing.

This picture of Taurus–Littrow was taken from the LM shortly after separation (the CSM is at the center of the frame, set against the base of the South Massif). Contrast this confined landing site with the open plain selected for Apollo 11.

An artist's depiction of the three LRV traverses planned for Apollo 17, from the perspective of the most westerly Sculptured Hill. The first traverse was to be south to the crater Emory, the second west to the base of the South Massif, and the third to sample boulders on the North Massif and the Sculptured Hills. (Courtesy NASA.)

Gene Cernan with the LRV at station 1, with the most westerly of the Sculptured Hills in the background.

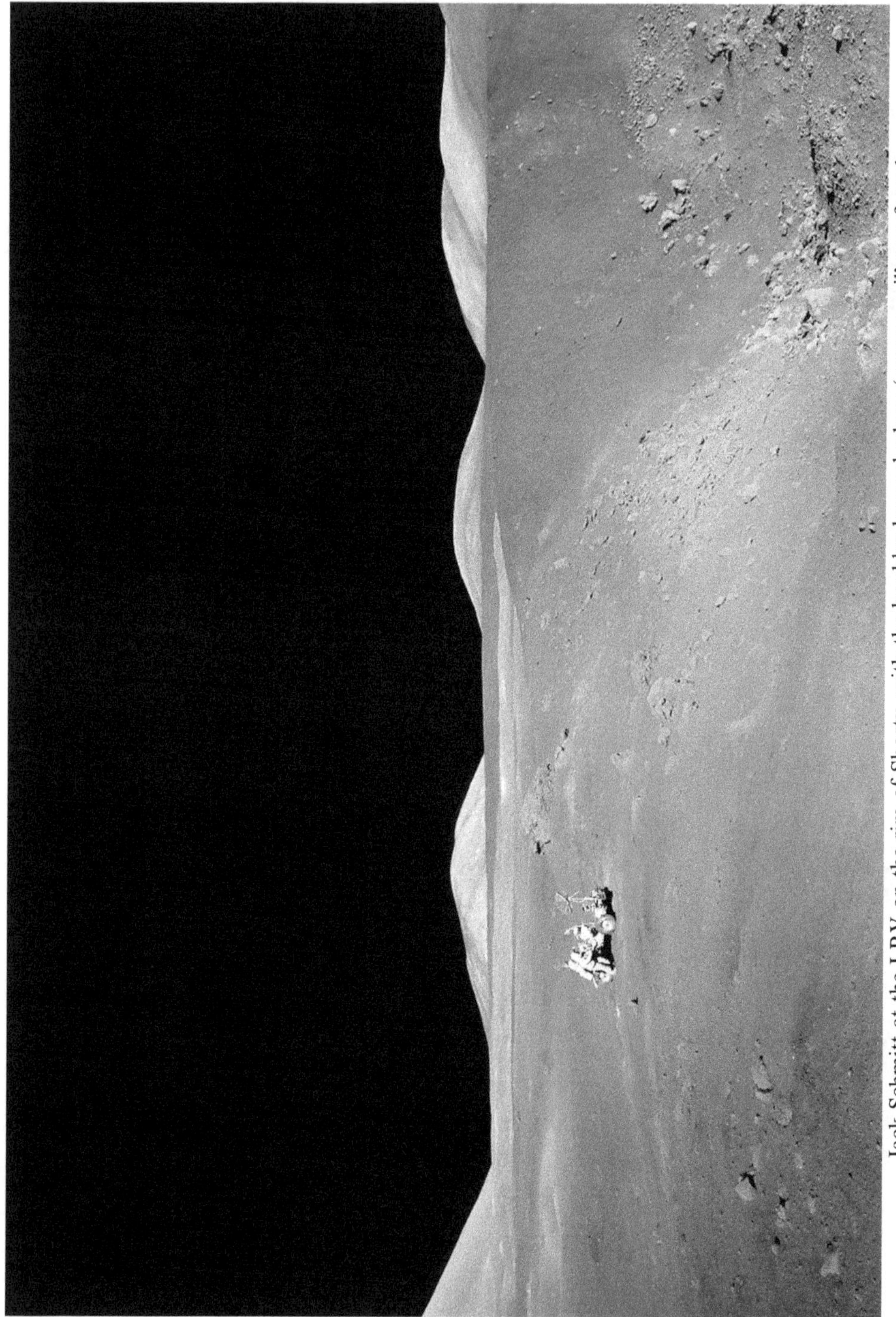

Jack Schmitt at the LRV on the rim of Shorty, with the boulder beyond where 'orange soil' was found.

At station 5 Jack Schmitt runs around the boulders on Camelot's rim.

Gene Cernan attends the LRV alongside the split boulder at station 6.

Jack Schmitt alongside the split boulder at station 6.

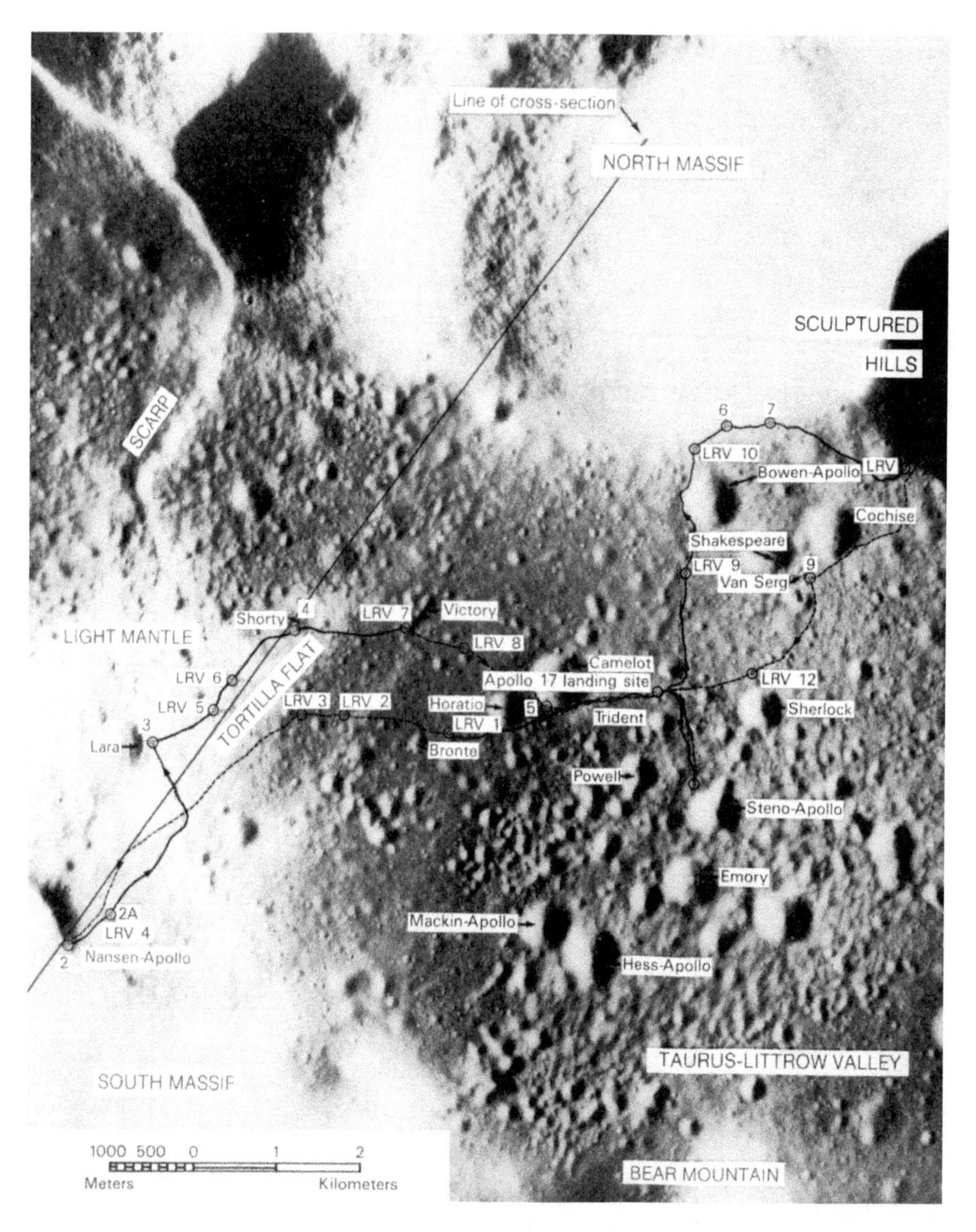

The Apollo 17 traverses. The diagonal line shows the inferred cross-section (see below). Note that although the crater near station 6 is listed as 'Bowen–Apollo', the crew actually referred to it as 'Henry the Navigator'. (Courtesy USGS.)

The Apollo 17 crew, Ron Evans, Jack Schmitt and Gene Cernan, on the recovery carrier.

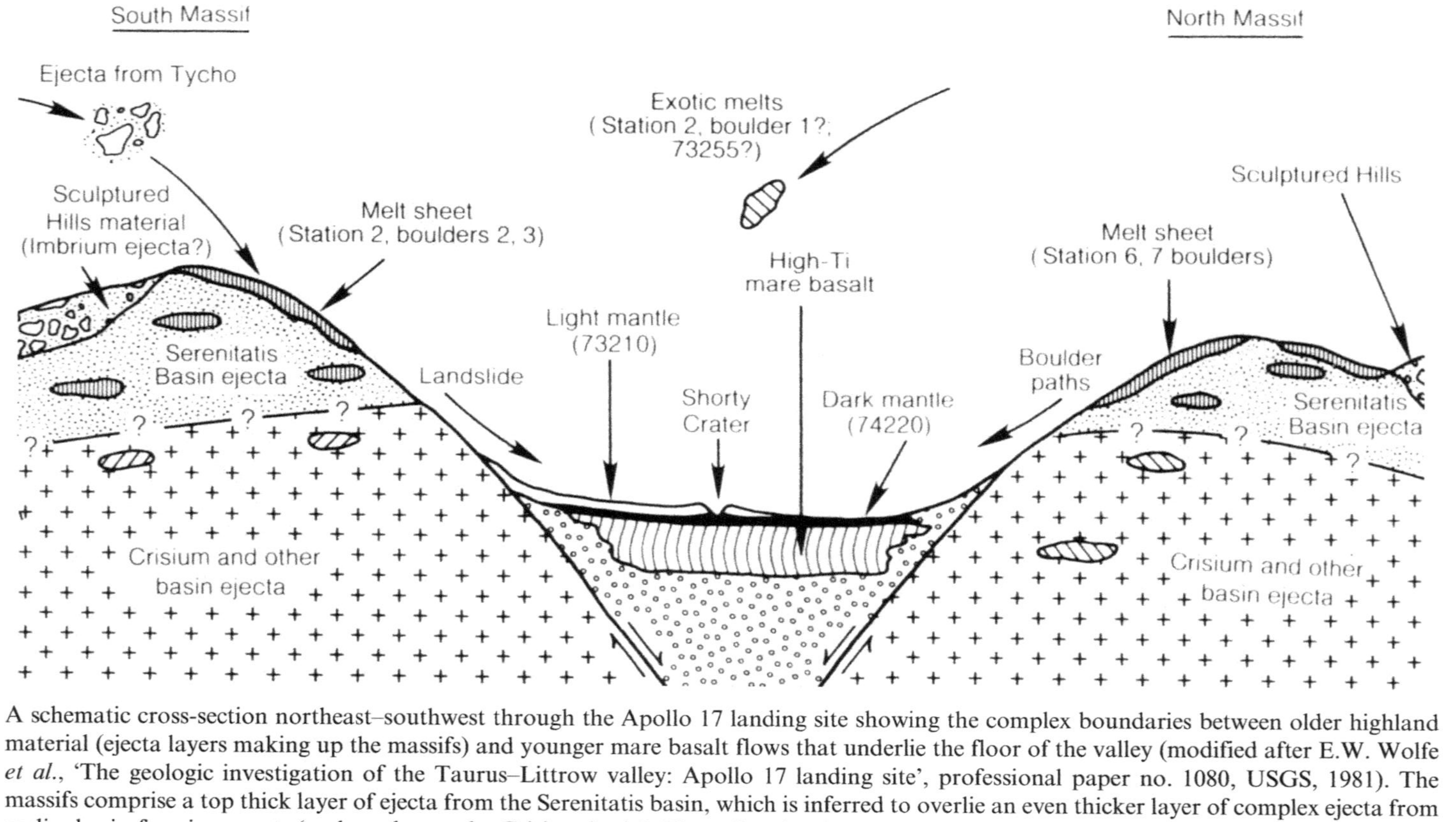

A schematic cross-section northeast–southwest through the Apollo 17 landing site showing the complex boundaries between older highland material (ejecta layers making up the massifs) and younger mare basalt flows that underlie the floor of the valley (modified after E.W. Wolfe *et al.*, 'The geologic investigation of the Taurus–Littrow valley: Apollo 17 landing site', professional paper no. 1080, USGS, 1981). The massifs comprise a top thick layer of ejecta from the Serenitatis basin, which is inferred to overlie an even thicker layer of complex ejecta from earlier basin-forming events (such as the nearby Crisium basin). The valley developed as a down-dropped wedge (graben) between two steep fractures (faults) that may have formed at the time of the Serenitatis impact event. It was then filled, first with fragmental debris that slumped off the massifs, then by basalt lava flows and pyroclastic deposits. More recent activity includes landslides and boulder falls from outcrops high on the massifs down onto the lower slopes and valley floor. Numbers refer to specific collected samples representative of the various units inferred to be present. (Courtesy the Lunar and Planetary Institute and Cambridge University Press.)

2. To restart the S-IVB during either the first or second opportunity over the Atlantic and inject the S-IVB/instrument unit/spacecraft into the planned translunar trajectory. *Achieved.*
3. To provide the required attitude control for the S-IVB/instrument unit/spacecraft during transposition, docking, and ejection. *Achieved.*
4. To perform an evasive maneuver after ejection of the command and service module/lunar module from the S-IVB/instrument unit. *Achieved.*
5. To attempt to impact the S-IVB/instrument unit on the lunar surface within 189 nmi (350 km) of selenographic latitude 7°S, longitude 8°W. *Achieved.*
6. To determine actual impact point within 2.7 nmi (5.0 km) and time of impact within 1 sec. *Achieved.*
7. To vent and dump the remaining gases and propellants to safe the S-IVB/instrument unit. *Achieved.*

Spacecraft Primary Objectives
1. To perform selenological inspection, survey, and sampling of materials and surface features in a preselected area of the Taurus–Littrow region. *Achieved.*
2. To emplace and activate surface experiments. *Achieved.*
3. To conduct in-flight experiments and photographic tasks. *Achieved.*

Detailed Objectives
1. To obtain (service module) lunar surface photographs and altitude data from lunar orbit. *Achieved.*
2. To obtain data on the visual light flash phenomenon. *Achieved.*
3. To obtain (command module) photographs of lunar surface features of scientific interest and photographs of low brightness astronomical and terrestrial sources. *Achieved.*
4. To record visual observations (from lunar orbit) of particular lunar surface features and processes. *Achieved.*
5. To obtain data on Apollo spacecraft-induced contamination (Skylab contamination study). *Achieved.*
6. To obtain data on whole body metabolic gains or losses, together with associated endocrinological controls (food compatibility assessment). *Achieved.*
7. To obtain data on the use of the protective pressure garment. *Achieved.*

Experiments
1. ALSEP V: Apollo Lunar Surface Experiments Package.
   (a) S-037: Heat flow experiment. *Achieved.*
   (b) S-202: Lunar ejecta and meteorites experiment. *Partially achieved. Operation was restricted during lunar day due to overheating.*
   (c) S-203: Lunar seismic profiling experiment. *Achieved.*
   (d) S-205: Lunar atmospheric composition experiment. *Achieved.*
   (e) S-207: Lunar surface gravimeter experiment. *Partially achieved. Data obtained in the seismic and free oscillation channels only.*
2. Collect and document samples and study lunar surface geology. *Achieved.*

3. S-152: Cosmic-ray detector (sheets) experiment. *Achieved.*
4. S-164: S-band transponder experiment (command and service module/lunar module). *Achieved.*
5. S-169: Far ultraviolet spectrometer experiment. *Achieved.*
6. S-171: Infrared scanning radiometer experiment. *Achieved.*
7. S-199: Traverse gravimeter experiment. *Achieved.*
8. S-204: Surface electrical properties experiment. *Achieved.*
9. S-209: Lunar sounder experiment. *Achieved.*
10. S-229: Lunar neutron probe experiment. *Achieved.*

In-flight Demonstrations

1. Heat flow and convection. *Achieved.*

Passive Objectives

1. Long-term lunar surface exposure. *Achieved.*
2. S-160: Gamma-ray spectrometer. *Achieved.*
3. S-176: Apollo window meteoroid. *Achieved.*
4. S-200: Soil mechanics. *Achieved.*
5. M-211: Biostack IIA. *Achieved.*
6. M-212: BIOCORE *Achieved.*

Operational Tests for Manned Spacecraft Center/Department of Defense.

1. Chapel Bell (classified Department of Defense test).
2. Radar skin tracking.
3. Ionospheric disturbance from missiles.
4. Acoustic measurement of missile exhaust noise.
5. Army acoustic test.
6. Long-focal-length optical system.
7. Sonic boom measurement.
8. Skylab Medical Mobile Laboratory.

MISSION TIMELINE

Apollo 17 mission event	GET (h:m:s)	Date (GMT)	Time (h:m:s)
Terminal countdown started.	–028:00:00	05 Dec 1972	12:53:00
Scheduled 9-hr hold at T–9 hr.	–009:00:00	06 Dec 1972	07:53:00
Countdown resumed at T–9 hr.	–009:00:00	06 Dec 1972	16:53:00
Scheduled 1-hr hold at T–3 hr 30 min.	–003:30:00	06 Dec 1972	22:23:00
Countdown resumed at T–3 hr 30 min.	–003:30:00	06 Dec 1972	23:23:00
Terminal Countdown Sequencer (TCS) failed to issue the S-IVB LOX pressurization command.	–000:02:47	07 Dec 1972	02:50:13

Apollo 17 mission event – *continued*	GET (h:m:s)	Date (GMT)	Time (h:m:s)
Unscheduled but automatic 1-hr 5-min 11-sec hold at T–30 sec due to TCS failure.	–000:00:30	07 Dec 1972	02:52:30
Countdown recycled to T–22 min.	–000:22:00	07 Dec 1972	03:57:41
Unscheduled 1-hr 13-min 19-sec hold at T–8 min to resolve TCS corrective action.	–000:08:00	07 Dec 1972	04:11:41
Countdown resumed at T–8 min.	–000:08:00	07 Dec 1972	05:25:00
Guidance reference release.	–000:00:16.960	07 Dec 1972	05:32:43
S-IC engine start command.	–000:00:08.9	07 Dec 1972	05:32:51
S-IC engine ignition (#5).	–000:00:06.9	07 Dec 1972	05:32:53
All S-IC engines thrust OK.	–000:00:01.6	07 Dec 1972	05:32:58
Range zero.	000:00:00.00	07 Dec 1972	05:33:00
All holddown arms released (1st motion) (1.08 *g*).	000:00:00.24	07 Dec 1972	05:33:00
Liftoff (umbilical disconnected).	000:00:00.63	07 Dec 1972	05:33:00
Tower clearance yaw maneuver started.	000:00:01.7	07 Dec 1972	05:33:01
Yaw maneuver ended.	000:00:09.7	07 Dec 1972	05:33:09
Pitch and roll maneuver started.	000:00:12.9	07 Dec 1972	05:33:12
Roll maneuver ended.	000:00:14.3	07 Dec 1972	05:33:14
Mach 1 achieved.	000:01:07.5	07 Dec 1972	05:34:07
Maximum bending moment (96,000,000 lbf-in).	000:01:19	07 Dec 1972	05:34:19
Maximum dynamic pressure (701.75 lb/ft^2).	000:01:22.5	07 Dec 1972	05:34:22
S-IC center engine cutoff command.	000:02:19.30	07 Dec 1972	05:35:19
Pitch maneuver ended.	000:02:40.1	07 Dec 1972	05:35:40
S-IC outboard engine cutoff. Maximum total inertial acceleration (3.87 *g*).	000:02:41.20	07 Dec 1972	05:35:41
S-IC maximum Earth-fixed velocity.	000:02:42.0	07 Dec 1972	05:35:42
S-IC/S-II separation command.	000:02:42.9	07 Dec 1972	05:35:42
S-II engine start command.	000:02:43.6	07 Dec 1972	05:35:43
S-II ignition.	000:02:44.6	07 Dec 1972	05:35:44
S-II aft interstage jettisoned.	000:03:12.9	07 Dec 1972	05:36:12
Launch escape tower jettisoned (planned time, actual time not recorded).	000:03:19	07 Dec 1972	05:36:19
Iterative guidance mode initiated.	000:03:24.1	07 Dec 1972	05:36:24
S-IC apex.	000:04:33.689	07 Dec 1972	05:37:33
S-II center engine cutoff. Maximum total inertial acceleration (1.74 *g*).	000:07:41.21	07 Dec 1972	05:40:41
S-IC impact (theoretical).	000:09:11.708	07 Dec 1972	05:42:11
S-II outboard engine cutoff.	000:09:19.66	07 Dec 1972	05:42:19
S-II/S-IVB separation command. S-II maximum Earth-fixed velocity.	000:09:20.6	07 Dec 1972	05:42:20
S-IVB 1st burn start command.	000:09:20.70	07 Dec 1972	05:42:20
S-IVB 1st burn ignition.	000:09:23.80	07 Dec 1972	05:42:23
S-IVB ullage case jettisoned.	000:09:32.4	07 Dec 1972	05:42:32
S-II apex.	000:09:34.527	07 Dec 1972	05:42:34
S-IVB 1st burn cutoff and maximum total inertial acceleration (0.67 *g*).	000:11:42.65	07 Dec 1972	05:44:42

Apollo 17 mission event – *continued*	GET (h:m:s)	Date (GMT)	Time (h:m:s)
Earth orbit insertion.	000:11:52.65	07 Dec 1972	05:44:52
S-IVB 1st burn maximum Earth-fixed velocity.	000:11:52.7	07 Dec 1972	05:44:52
Maneuver to local horizontal attitude started.	000:12:04.4	07 Dec 1972	05:45:04
S-II impact (theoretical).	000:19:56.947	07 Dec 1972	05:52:56
S-IVB 2nd burn restart preparation.	003:02:58.60	07 Dec 1972	08:35:58
S-IVB 2nd burn restart command.	003:12:28.60	07 Dec 1972	08:45:28
S-IVB 2nd burn ignition.	003:12:36.60	07 Dec 1972	08:45:36
S-IVB 2nd burn cutoff and maximum total inertial acceleration (1.41 *g*).	003:18:27.64	07 Dec 1972	08:51:27
S-IVB safing procedure – CVS opened.	003:18:28.3	07 Dec 1972	08:51:28
S-IVB 2nd burn maximum Earth-fixed velocity.	003:18:28.5	07 Dec 1972	08:51:28
Translunar injection.	003:18:37.64	07 Dec 1972	08:51:37
Maneuver to local horizontal attitude and orbital navigation started.	003:20:59.6	07 Dec 1972	08:53:59
Maneuver to transposition and docking attitude started.	003:33:28.9	07 Dec 1972	09:06:28
CSM separated from S-IVB.	003:42:27.6	07 Dec 1972	09:15:27
TV transmission started.	003:50	07 Dec 1972	09:23
CSM docked with LM/S-IVB.	003:57:10.7	07 Dec 1972	09:30:10
TV transmission ended.	004:10	07 Dec 1972	09:43
CSM/LM ejected from S-IVB.	004:45:02.3	07 Dec 1972	10:18:02
S-IVB APS evasive maneuver ignition.	005:03:01.1	07 Dec 1972	10:36:01
S-IVB APS evasive maneuver cutoff (estimated).	005:04:21.0	07 Dec 1972	10:37:21
S-IVB lunar impact maneuver – CVS opened.	005:19:39.8	07 Dec 1972	10:52:39
S-IVB lunar impact maneuver – LOX dump started.	005:24:20.2	07 Dec 1972	10:57:20
S-IVB lunar impact maneuver – CVS closed.	005:24:40.0	07 Dec 1972	10:57:40
S-IVB lunar impact maneuver – LOX dump ended.	005:25:07.9	07 Dec 1972	10:58:07
Maneuver to attitude for 1st S-IVB APS lunar impact burn.	006:02:15	07 Dec 1972	11:35:15
S-IVB lunar impact maneuver – 1st APS ignition command.	006:09:59.8	07 Dec 1972	11:42:59
S-IVB lunar impact maneuver – 1st APS cutoff command.	006:11:38.0	07 Dec 1972	11:44:38
Maneuver to S-IVB solar heating attitude.	006:17:44	07 Dec 1972	11:50:44
Maneuver to attitude for 2nd S-IVB APS lunar impact burn.	011:02:40	07 Dec 1972	16:35:40
S-IVB lunar impact maneuver – 2nd APS ignition command.	011:14:59.8	07 Dec 1972	16:47:59
S-IVB lunar impact maneuver – 2nd APS cutoff command.	011:16:42.0	07 Dec 1972	16:49:42
S-IVB 3-axis tumble mode initiated.	011:31:42	07 Dec 1972	17:04:42
S-IVB passive thermal control maneuver.	011:31:50	07 Dec 1972	17:04:50
Command to inhibit instrument unit flight control computer to leave the S-IVB in 3-axis tumble mode.	011:32:12.5	07 Dec 1972	17:05:12
Midcourse correction ignition (SPS).	035:29:59.91	08 Dec 1972	17:02:59
Midcourse correction cutoff.	035:30:01.64	08 Dec 1972	17:03:01
Maneuver to LM checkout attitude.	039:05	08 Dec 1972	20:38
Preparations for intravehicular transfer.	039:20	08 Dec 1972	20:53
LM pressurization started.	039:30	08 Dec 1972	21:03
CDR and LMP entered LM for housekeeping and communications check.	040:10	08 Dec 1972	21:43
LM closeout.	042:11	08 Dec 1972	23:44

Apollo 17 mission event – *continued*	GET (h:m:s)	Date (GMT)	Time (h:m:s)
Heat flow and convection demonstration started.	042:55	09 Dec 1972	00:28
Heat flow and convection demonstration ended.	043:45	09 Dec 1972	01:18
Heat flow and convection demonstration started.	045:20	09 Dec 1972	02:53
Heat flow and convection demonstration ended.	046:00	09 Dec 1972	03:33
LM pressurization started.	059:30	09 Dec 1972	17:03
CDR and LMP entered LM for telemetry checkout.	059:59	09 Dec 1972	17:32
CDR and LMP entered CM.	062:16	09 Dec 1972	19:49
Mission clock updated (002:40:00 added).	065:00	09 Dec 1972	22:33
Apollo light flash phenomenon experiment started.	065:39	09 Dec 1972	23:12
Apollo light flash phenomenon experiment ended.	066:39	10 Dec 1972	00:12
Equigravisphere.	070:37:45	10 Dec 1972	04:10:45
Scientific instrument module door jettisoned.	081:32:40	10 Dec 1972	15:05:40
In-flight science phase of mission initiated with turn-on of Far Ultraviolet Spectrometer.	083:26	10 Dec 1972	16:59
Ultraviolet photography of dark Moon.	084:50	10 Dec 1972	18:23
Lunar orbit insertion ignition (SPS).	086:14:22.60	10 Dec 1972	19:47:22
Lunar orbit insertion cutoff.	086:20:55.76	10 Dec 1972	19:53:55
S-IVB impact on lunar surface.	086:59:42.3	10 Dec 1972	20:32:42
Terminator photography.	087:05	10 Dec 1972	20:38
Orbital science visual observations.	087:15	10 Dec 1972	20:48
Orbital science photography.	088:00	10 Dec 1972	21:33
1st descent orbit insertion ignition (SPS).	090:31:37.43	11 Dec 1972	00:04:37
1st descent orbit insertion cutoff.	090:31:59.70	11 Dec 1972	00:04:59
Landmark observations.	090:50	11 Dec 1972	00:23
CDR and LMP entered LM.	105:02	11 Dec 1972	14:35
CSM/LM separation maneuver initiated (RCS).	107:47:56	11 Dec 1972	17:20:56
CSM/LM separation maneuver cutoff.	107:47:59.4	11 Dec 1972	17:20:59
CSM orbit circularization ignition (SPS).	109:17:28.92	11 Dec 1972	18:50:28
CSM orbit circularization cutoff.	109:17:32.72	11 Dec 1972	18:50:32
2nd descent orbit insertion ignition (LM RCS).	109:22:42	11 Dec 1972	18:55:42
2nd descent orbit insertion cutoff.	109:23:03.5	11 Dec 1972	18:56:03
CSM landmark tracking started.	109:40	11 Dec 1972	19:13
LM powered descent engine ignition (DPS).	110:09:53	11 Dec 1972	19:42:53
LM throttle to full-throttle position.	110:10:21	11 Dec 1972	19:43:21
LM manual target (landing site) update.	110:11:25	11 Dec 1972	19:44:25
LM landing radar velocity data good.	110:13:28	11 Dec 1972	19:46:28
LM landing radar range data good.	110:14:06	11 Dec 1972	19:47:06
LM landing radar updates enabled.	110:14:32	11 Dec 1972	19:47:32
LM throttle down.	110:17:19	11 Dec 1972	19:50:19
LM approach phase program selected.	110:19:15	11 Dec 1972	19:52:15
LM landing radar antenna to position 2.	110:19:16	11 Dec 1972	19:52:16
LM 1st landing point redesignation.	110:19:26	11 Dec 1972	19:52:26
LM landing radar switched to low scale.	110:19:54	11 Dec 1972	19:52:54
LM landing phase program selected.	110:20:51	11 Dec 1972	19:53:51
LM lunar landing and powered descent engine cutoff.	110:21:58	11 Dec 1972	19:54:58

Apollo 17 mission event – *continued*	GET (h:m:s)	Date (GMT)	Time (h:m:s)
CSM landmark tracking ended.	111:20	11 Dec 1972	20:53
1st EVA started (LM cabin depressurized).	114:21:49	11 Dec 1972	23:54:49
CSM orbital science visual observations.	114:45	12 Dec 1972	00:18
Lunar roving vehicle (LRV) offloaded.	114:51:10	12 Dec 1972	00:24:10
LRV deployed, test drive performed and documented with photography, gathered samples and performed 500-mm and panoramic photography.	115:13:50	12 Dec 1972	00:46:50
CSM orbital science photography.	115:15	12 Dec 1972	00:48
US flag deployed and documented with photographs and stereo photography.	115:40:58	12 Dec 1972	01:13:58
Traverse gravimeter experiment reading obtained.	115:50:51	12 Dec 1972	01:23:51
Cosmic ray experiment deployed.	115:54:40	12 Dec 1972	01:27:40
Apollo lunar surface experiment (ALSEP) package offloaded.	115:58:30	12 Dec 1972	01:31:30
Traverse gravimeter experiment reading obtained.	116:06:01	12 Dec 1972	01:39:01
Traverse gravimeter experiment reading obtained.	116:11:54	12 Dec 1972	01:44:54
Traverse gravimeter experiment reading obtained.	116:46:17	12 Dec 1972	02:19:17
CSM orbital science photography.	117:10	12 Dec 1972	02:43
1st ALSEP data received on Earth.	117:21:00	12 Dec 1972	02:54
Heat flow experiment turned on.	117:29	12 Dec 1972	03:02
ALSEP deployment completed and documented with photographs and panoramic photography.	118:07:43	12 Dec 1972	03:40:43
CSM terminator photography.	118:10	12 Dec 1972	03:43
Lunar seismic profiling experiment (S-203) turned on. CSM Earthshine photography started.	118:25	12 Dec 1972	03:58
Deep core sample obtained and lunar neutron probe experiment deployed.	118:35:27	12 Dec 1972	04:08:27
Traverse gravimeter experiment reading obtained.	118:43:08	12 Dec 1972	04:16:08
CSM Earthshine photography ended.	118:50	12 Dec 1972	04:23
Departed for station 1.	119:11:02	12 Dec 1972	04:44:02
Arrived at station 1 and deployed seismic profiling experiment explosive charge 6, obtained traverse gravimeter experiment reading and documented rake samples and performed panoramic photography.	119:24:02	12 Dec 1972	04:57:02
Lunar surface gravimeter experiment (S-207) activated.	119:50	12 Dec 1972	05:23
Departed for surface electrical properties experiment site with a stop to deploy seismic profiling experiment explosive charge 7, and performed panoramic photography.	119:56:47	12 Dec 1972	05:29:47
Arrived at surface electrical properties experiment site. Deployed antennas and the transmitter, gathered samples and performed documentary and panoramic photograph traverse gravimeter experiment reading obtained.	120:11:02	12 Dec 1972	05:44:02
Departed for LM.	120:33:39	12 Dec 1972	06:06:39
Arrived at LM and started EVA activity closeout.	120:36:15	12 Dec 1972	06:09:15
Traverse gravimeter experiment reading obtained.	121:16:37	12 Dec 1972	06:49:37
Traverse gravimeter experiment reading obtained.	121:21:11	12 Dec 1972	06:54:11

Apollo 17 mission event – *continued*	GET (h:m:s)	Date (GMT)	Time (h:m:s)
1st EVA ended (LM cabin repressurized).	121:33:42	12 Dec 1972	07:06:42
CSM zodiacal light photography.	130:35	12 Dec 1972	16:08
CSM orbital science photography.	134:00	12 Dec 1972	19:33
CSM solar corona photography.	134:50	12 Dec 1972	20:23
CSM orbital science visual observations.	137:00	12 Dec 1972	22:33
2nd EVA started (LM cabin depressurized).	137:55:06	12 Dec 1972	23:28:06
Traverse gravimeter experiment reading obtained.	138:04:08	12 Dec 1972	23:37:08
TV transmission started for 2nd EVA.	138:05	12 Dec 1972	23:38
LRV loaded for traverse and a traverse gravimeter experiment reading obtained.	138:39:00	13 Dec 1972	00:12
Departed for surface electrical properties experiment site.	138:44:02	13 Dec 1972	00:17:02
Arrived at surface electrical properties experiment site. Activated experiment, gathered samples, and performed panoramic photography.	138:47:05	13 Dec 1972	00:20:05
Departed for station 2 with four short stops – one to deploy seismic profiling experiment explosive charge 4, and three to gather en route samples.	138:51:43	13 Dec 1972	00:24:43
CSM orbital science photography and visual observations.	139:45	13 Dec 1972	01:18
Arrived at station 2. Traverse gravimeter experiment reading obtained, gathered samples including a rake sample, and performed documentary and panoramic photography.	140:01:30	13 Dec 1972	01:34:30
Departed for station 3 with one stop to obtain a traverse gravimeter experiment reading, gather samples and perform panoramic and 500-mm photography.	141:07:25	13 Dec 1972	02:40:25
Arrived at station 3. Traverse gravimeter experiment reading obtained, gathered samples including a double core-tube sample and a rake sample, and performed panoramic and 500-mm photography.	141:48:38	13 Dec 1972	03:21:38
CSM terminator photography.	142:05	13 Dec 1972	03:38
Departed for station 4 with two short stops to gather en route samples.	142:25:56	13 Dec 1972	03:58:56
Arrived at station 4. Traverse gravimeter experiment reading obtained, gathered samples including a trench sample and a double core-tube sample, and performed documentary and panoramic photography.	142:42:57	13 Dec 1972	04:15:57
Departed for station 5 with one stop to deploy seismic profiling experiment explosive charge, gather samples, and perform panoramic photography.	143:19:03	13 Dec 1972	04:52:03
Arrived at station 5. Traverse gravimeter experiment reading obtained, gathered samples, and performed documentary and panoramic photography.	143:45:15	13 Dec 1972	05:18:15
Departed for the LM with a short stop to deploy seismic profiling experiment explosive charge 8 documented with photographs, and a stop at the ALSEP site to allow the LMP to relevel the lunar surface gravimeter experiment.	144:15:58	13 Dec 1972	05:48:58

Apollo 17 mission event – *continued*	GET (h:m:s)	Date (GMT)	Time (h:m:s)
Arrived at the LM and started EVA closeout.	144:32:24	13 Dec 1972	06:05:24
TV transmission ended for 2nd EVA.	144:55	13 Dec 1972	06:28
Traverse gravimeter experiment reading obtained.	145:19:24	13 Dec 1972	06:52:24
2nd EVA ended (LM cabin repressurized).	145:32:02	13 Dec 1972	07:05:02
CSM orbital science photography.	154:40	13 Dec 1972	16:13
CSM terminator photography.	156:50	13 Dec 1972	18:23
3rd EVA started (LM cabin depressurized).	160:52:48	13 Dec 1972	22:25:48
Zodiacal light photography.	160:55	13 Dec 1972	22:28
Traverse gravimeter experiment reading obtained.	161:02:40	13 Dec 1972	22:35:40
TV transmission started for 3rd EVA.	161:15	13 Dec 1972	22:48
LRV loaded for traverse, and panoramic and 500-mm photography performed.	161:16:15	13 Dec 1972	22:49:15
Traverse gravimeter experiment reading obtained.	161:19:45	13 Dec 1972	22:52:45
Cosmic ray experiment retrieved.	161:20:17	13 Dec 1972	22:53:17
Departed for surface electrical properties experiment site.	161:36:31	13 Dec 1972	23:09:31
Arrived at surface electrical properties experiment site. Activated the experiment, gathered samples, and performed documentary photography.	161:39:07	13 Dec 1972	23:12:07
Departed for station 6 with two short stops to gather en route samples.	161:42:36	13 Dec 1972	23:15:36
CSM orbital science photography.	161:50	13 Dec 1972	23:23
Arrived at station 6. Traverse gravimeter experiment reading obtained, gathered samples including a single core-tube sample, a rake sample, and performed documentary, panoramic, and 500-mm photography.	162:11:24	13 Dec 1972	23:44:24
Departed for station 7.	163:22:10	14 Dec 1972	00:55:10
Arrived at station 7. Gathered samples and performed documentary and panoramic photography.	163:29:05	14 Dec 1972	01:02:05
CSM orbital science visual observations.	163:30	14 Dec 1972	01:03
Departed for station 8 with one short stop to gather en route samples.	163:51:09	14 Dec 1972	01:24:09
Arrived at station 8. Two traverse gravimeter experiment readings obtained, gathered samples including rake and trench samples, and performed documentary and panoramic photography.	164:07:40	14 Dec 1972	01:40:40
Departed for station 9.	164:55:33	14 Dec 1972	02:28:33
Arrived at station 9. Seismic profiling experiment explosive charge 5 deployed, two traverse gravimeter readings obtained, gathered samples including a trench sample and a double core-tube sample, and performed documentary, panoramic and 500-mm photography. Removed data storage electronics assembly from surface electrical properties receiver.	165:13:10	14 Dec 1972	02:46:10

Apollo 17 mission event – *continued*	GET (h:m:s)	Date (GMT)	Time (h:m:s)
Departed for the LM with two short stops – one to gather en route samples and the other to deploy seismic profiling experiment explosive charge 2 and perform documentary and panoramic photography.	166:09:25	14 Dec 1972	03:42:25
Arrived at LM and started EVA closeout.	166:37:51	14 Dec 1972	04:10:51
Traverse gravimeter experiment reading obtained.	166:55:09	14 Dec 1972	04:28:09
Final traverse gravimeter experiment reading obtained.	167:11:11	14 Dec 1972	04:44:11
ALSEP photography completed.	167:33:58	14 Dec 1972	05:06:58
Lunar neutron probe experiment retrieved.	167:36:43	14 Dec 1972	05:09:43
LRV positioned to monitor LM ascent.	167:39:57	14 Dec 1972	05:12:57
Seismic profiling experiment explosive charge 3 deployed.	167:44:41	14 Dec 1972	05:17:41
Equipment jettisoned.	167:45	14 Dec 1972	05:18
3rd EVA ended (LM cabin repressurized).	168:07:56	14 Dec 1972	05:40:56
Orbital trim maneuver ignition (RCS).	178:54:05.45	14 Dec 1972	16:27:05
Orbital trim maneuver cutoff.	178:54:42.95	14 Dec 1972	16:27:42
CSM plane change ignition (RCS).	179:53:53.83	14 Dec 1972	17:26:53
CSM plane change cutoff.	179:54:13.88	14 Dec 1972	17:27:13
1st equipment jettison from LM.	180:15	14 Dec 1972	17:48
CSM zodiacal light photography.	182:20	14 Dec 1972	19:53
CSM landmark tracking.	182:40	14 Dec 1972	20:13
CSM landmark tracking.	183:00	14 Dec 1972	20:33
2nd equipment jettison from LM.	183:24	14 Dec 1972	20:57
LM lunar liftoff ignition (LM APS).	185:21:37	14 Dec 1972	22:54:37
LM ascent orbit cutoff.	185:28:58	14 Dec 1972	23:01:58
Vernier adjustment maneuver initiated (LM RCS).	185:32:12	14 Dec 1972	23:05:12
Vernier adjustment maneuver cutoff.	185:32:22	14 Dec 1972	23:05:22
Terminal phase initiation ignition (LM APS).	186:15:58	14 Dec 1972	23:48:58
Terminal phase initiation cutoff.	186:16:01.2	14 Dec 1972	23:49:01
LM midcourse corrections.	186:30	14 Dec 1972	00:03
CSM/LM docked.	187:37:15	15 Dec 1972	01:10:15
Transfer, stowing of equipment and samples started.	188:00	15 Dec 1972	01:33
Transfer, stowing of equipment and samples ended.	190:05	15 Dec 1972	03:38
CDR and LMP entered CM.	190:10	15 Dec 1972	03:43
LM closeout.	190:30	15 Dec 1972	04:03
LM ascent stage jettisoned.	191:18:31	15 Dec 1972	04:51:31
Separation maneuver initiated.	191:23:31	15 Dec 1972	04:56:31
Separation maneuver cutoff.	191:23:43	15 Dec 1972	04:56:43
LM ascent stage deorbit ignition.	192:58:14	15 Dec 1972	06:31:14
LM ascent stage deorbit cutoff.	193:00:10	15 Dec 1972	06:33:10
LM ascent stage impact on lunar surface.	193:17:20.8	15 Dec 1972	06:50:20
Terminator photography.	206:20	15 Dec 1972	19:53
Orbital science visual observations.	206:40	15 Dec 1972	20:13
Orbital science visual observations.	207:10	15 Dec 1972	20:43
Explosive package #6 detonated on lunar surface.	210:15:14.56	15 Dec 1972	23:48:14
Explosive package #7 detonated on lunar surface.	212:44:57.11	16 Dec 1972	02:17:57

Apollo 17 mission event – *continued*	GET (h:m:s)	Date (GMT)	Time (h:m:s)
Orbital science photography.	213:10	16 Dec 1972	02:43
Terminator photography.	215:20	16 Dec 1972	04:53
Explosive package #4 detonated on lunar surface.	229:35:34.67	16 Dec 1972	19:08:34
Terminator photography.	231:20	16 Dec 1972	20:53
Transearth injection ignition (SPS).	234:02:09.18	16 Dec 1972	23:35:09
Transearth injection cutoff.	234:04:32.87	16 Dec 1972	23:37:32
TV transmission started.	234:10	16 Dec 1972	23:43
TV transmission ended.	234:35	17 Dec 1972	00:08
Ultraviolet spectrometer of Lyman Alpha region started.	235:00	17 Dec 1972	00:33
Explosive package #1 detonated on lunar surface.	235:09:36.79	17 Dec 1972	00:42:36
Ultraviolet spectrometer of Lyman Alpha region ended.	236:00	17 Dec 1972	01:33
Ultraviolet spectrometer of Earth.	236:05	17 Dec 1972	01:38
Ultraviolet spectrometer of Moon.	237:15	17 Dec 1972	02:48
Ultraviolet spectrometer of Moon off.	238:00	17 Dec 1972	03:33
Explosive package #8 detonated on lunar surface.	238:12:46.08	17 Dec 1972	03:45:46
Ultraviolet spectrometer on for passive thermal control galactic scan.	239:40	17 Dec 1972	05:13
Transearth EVA started (Evans).	254:54:40	17 Dec 1972	20:27:40
Installation of television camera and data acquisition camera started.	255:00	17 Dec 1972	20:33
Panoramic film cassette retrieved.	255:23	17 Dec 1972	20:56
Mapping camera film cassette retrieved.	255:36	17 Dec 1972	21:09
CM hatch closed.	255:40	17 Dec 1972	21:13
TV transmission ended for transearth EVA.	255:42	17 Dec 1972	21:15
Transearth EVA ended.	256:00:24	17 Dec 1972	21:33:24
Ultraviolet coma cluster observation.	257:00	17 Dec 1972	22:33
Explosive package #5 detonated on lunar surface.	257:43:41.06	17 Dec 1972	23:16:41
Explosive package #2 detonated on lunar surface.	259:11:56.82	18 Dec 1972	01:44:56
Ultraviolet Alpha ERI measurements.	260:30	18 Dec 1972	02:03
Ultraviolet passive thermal control measurements for Alpha ERI and Alpha GRU.	261:20	18 Dec 1972	02:53
Explosive package #3 detonated on lunar surface.	261:44:22.28	18 Dec 1972	03:07:22
Ultraviolet passive thermal control for galactic scan.	262:30	18 Dec 1972	04:03
Ultraviolet dark north observation.	274:30	18 Dec 1972	16:03
Apollo light flash observation and investigation.	277:10	18 Dec 1972	18:43
Ultraviolet spectrometer of Virgo cluster.	279:10	18 Dec 1972	20:43
Ultraviolet spectrometer viewing dark south.	280:50	18 Dec 1972	22:23
TV transmission for in-flight press conference started.	281:20	18 Dec 1972	22:53
TV transmission ended.	281:47	18 Dec 1972	23:20
Ultraviolet spectrometer of Spica.	283:45	19 Dec 1972	01:18
Ultraviolet passive thermal control for galactic scan.	285:30	19 Dec 1972	03:03
Midcourse correction ignition (RCS).	298:38:01	19 Dec 1972	16:11:01
Midcourse correction cutoff.	298:38:10	19 Dec 1972	16:11:10
In-flight science phase of mission ended with turn-off of Far Ultraviolet Spectrometer.	299:20	19 Dec 1972	16:53

Apollo 17 mission event – *continued*	GET (h:m:s)	Date (GMT)	Time (h:m:s)
CM/SM separation.	301:23:49	19 Dec 1972	18:56:49
Entry.	301:38:38	19 Dec 1972	19:11:38
Communication blackout started.	301:38:55	19 Dec 1972	19:11:55
Radar contact with CM established by recovery ship.	301:41	19 Dec 1972	19:14
Communication blackout ended.	301:42:15	19 Dec 1972	19:15:15
Forward heat shield jettisoned.	301:46:20	19 Dec 1972	19:19:20
Drogue parachute deployed	301:46:22	19 Dec 1972	19:19:22
Visual contact with CM established by recovery ship and photo helicopter.	301:47	19 Dec 1972	19:20
Main parachute deployed.	301:47:13	19 Dec 1972	19:20:13
VHF recovery beacon contact with CM established by recovery ship.	301:48	19 Dec 1972	19:21
Voice contact with CM established by recovery ship.	301:49	19 Dec 1972	19:22
Splashdown (went to apex-up).	301:51:59	19 Dec 1972	19:24:59
Swimmers deployed to CM.	302:02	19 Dec 1972	19:35
Flotation collar inflated.	302:08	19 Dec 1972	19:41
Hatch opened for crew egress.	302:21	19 Dec 1972	19:54
Crew on board recovery helicopter.	302:33	19 Dec 1972	20:06
Crew on board recovery ship.	302:44	19 Dec 1972	20:17
CM on board recovery ship.	303:55	19 Dec 1972	21:28
1st sample flight departed recovery ship.	323:52	20 Dec 1972	17:25
1st sample flight arrived in Hawaii.	330:27	21 Dec 1972	00:00
Crew departed recovery ship.	331:05	21 Dec 1972	00:38
1st sample flight departed Hawaii.	333:37	21 Dec 1972	03:10
1st sample flight arrived in Houston, TX.	340:43	21 Dec 1972	10:16
Crew arrived at Ellington AFB, Houston.	346:17	21 Dec 1972	15:50
CM arrived at North Island Naval Air Station, San Diego, CA.	493:57	27 Dec 1972	19:30
CM deactivated.	568:27	30 Dec 1972	22:00
CM departed North Island Naval Air Station for contractor's facility in Downey, CA.	637:27	02 Jan 1973	19:00
CM arrived at contractor's facility.	640:27	02 Jan 1973	22:00
ALSEP central station turned off by ground command.	–	30 Sep 1977	–

In retrospect

CANCELLED MISSIONS

Developing the technology to achieve the challenge of a manned lunar landing by the end of the 1960s was costly, but a large stock of Saturn V launchers were built in case it was necessary to undertake a lengthy series of test flights. After the Apollo 11 landing, which was primarily a Cold War statement, the stock of hardware enabled NASA to switch the focus to exploration. By increasing the capacity of its transportation system, it was able to fly advanced missions which were, first and foremost, science missions.

Lunar surface mission	Type	Traverse distance (km)	Experiments (kg)	EVA time (h:m)	Lunar samples (kg)
Apollo 11	'G'	0.25	102	2:24	21
Apollo 12	'H'	2.0	166	7:29	34
Apollo 14	'H'	3.3	209	9:23	43
Apollo 15	'J'	27.9	550	18:33	77
Apollo 16	'J'	27.0	563	20:12	94
Apollo 17	'J'	35.0	514	22:05	110
TOTAL	–	95.45	2,104	80	379

The three 'J'-missions contributed 76% of the surface activity time, 77% of the equipment landed, 74% of the samples returned, and, with the benefit of LRV, 94% of the traverse range. If Apollo 18 and Apollo 19 had matched Apollo 17, the ratios would have increased 87% of the experiments, surface time and samples, and 96% of the traverse distance. Because the hardware had already been built, flying these two missions would have significantly increased the scientific output of the program for a trivial extra cost. Instead, these magnificent vehicles now adorn the parking lots of NASA centers and languish in museums.

NASA'S ASSESSMENT

The 'Top Ten' scientific discoveries of Apollo's exploration of the Moon, as detailed by NASA soon after the program was completed,[8] were:

1. *The Moon is not a primordial object; it is an evolved terrestrial planet with internal zoning similar to that of Earth.* Before Apollo, the state of the Moon was a subject of almost unlimited speculation. We now know that the Moon is made of rocky material that has been variously melted, erupted through volcanoes, and crushed by meteorite impacts. The Moon possesses a thick crust (60 km), a fairly uniform lithosphere (60–1,000 km), and a partly liquid asthenosphere (1,000–1,740 km); a small iron core at the bottom of the asthenosphere is possible but unconfirmed. Some rocks give hints for ancient magnetic fields although no planetary field exists today.
2. *The Moon is ancient and still preserves an early history (the first billion years) that must be common to all terrestrial planets.* The extensive record of meteorite craters on the Moon, when calibrated using absolute ages of rock samples, provides a key for unraveling time scales for the geologic evolution of Mercury, Venus, and Mars based on their individual crater records. Photogeologic interpretation of other planets is based largely on lessons learned from the Moon. Before Apollo, however, the origin of lunar impact craters was not fully understood and the origin of similar craters on Earth was highly debated.
3. *The youngest Moon rocks are virtually as old as the oldest Earth rocks. The earliest processes and events that probably affected both planetary bodies can now only be found on the Moon.* Moon rock ages range from about 3.2 billion years in the maria (dark, low basins) to nearly 4.6 billion years in the terrae (light, rugged highlands). Active geologic forces, including plate tectonics and erosion, continuously repave the oldest surfaces on Earth whereas old surfaces persist with little disturbance on the Moon.
4. *The Moon and Earth are genetically related and formed from different proportions of a common reservoir of materials.* The distinctively similar oxygen isotopic compositions of Moon rocks and Earth rocks clearly show common ancestry. Relative to Earth, however, the Moon was highly depleted in iron and in volatile elements that are needed to form atmospheric gases and water.
5. *The Moon is lifeless; it contains no living organisms, fossils, or native organic compounds.* Extensive testing revealed no evidence for life, past or present, among the lunar samples. Even nonbiological organic compounds are amazingly absent; traces can be attributed to contamination by meteorites.
6. *All Moon rocks originated through high-temperature processes with little or no involvement with water. They are roughly divisible into three types: basalts, anorthosites, and breccias.* Basalts are dark lava rocks that fill mare basins; they generally resemble, but are much older than, lavas that comprise the oceanic crust of Earth. Anorthosites are light rocks that form the ancient highlands; they generally resemble, but are much older than, the most ancient rocks on Earth. Breccias are composite rocks formed

[8] http://www.nasm.si.edu/collections/imagery/apollo/apollotop10.htm

from all other rock types through crushing, mixing, and sintering during meteorite impacts. The Moon has no sandstone, shale, or limestone such as testify to the importance of water-borne processes on Earth.

7. *Early in its history, the Moon was melted to great depths to form a 'magma ocean'. The lunar highlands contain the remnants of early, low density rocks that floated to the surface of the magma ocean.* The lunar highlands were formed about 4.4–4.6 billion years ago by flotation of an early, feldspar-rich crust on a magma ocean that covered the Moon to a depth of many tens of kilometers or more. Innumerable meteorite impacts through geologic time reduced much of the ancient crust to arcuate mountain ranges between basins.
8. *The lunar magma ocean was followed by a series of huge asteroid impacts that created basins which were later filled by lava flows.* The large, dark basins such as Imbrium are gigantic impact craters, formed early in lunar history, that were later filled by lava flows about 3.2–3.9 billion years ago. Lunar volcanism occurred mostly as lava floods that spread horizontally; volcanic fire fountains produced deposits of black, orange and emerald-green glass beads.
9. *The Moon is slightly asymmetrical in bulk form, possibly as a consequence of its evolution under Earth's gravitational influence. Its crust is thicker on the far side, while most volcanic basins – and unusual mass concentrations – occur on the near side.* Mass is not distributed uniformly inside the Moon. Large mass concentrations ('mascons') lie beneath the surface of many large lunar basins and probably represent thick accumulations of dense lava. Relative to its geometric center, the Moon's center of mass is displaced toward Earth by several kilometers.
10. *The surface of the Moon is covered by a rubble pile of rock fragments and dust, called the lunar regolith, that contains a unique radiation history of the Sun which is of importance to understanding climate changes on Earth.* The regolith was produced by innumerable meteorite impacts through geologic time. Surface rocks and mineral grains are distinctively enriched in chemical elements and isotopes implanted by solar radiation. As such, the Moon has recorded four billion years of the Sun's history to a degree of completeness we are unlikely to find elsewhere.

THE OUTSTANDING MYSTERY

Significantly, the question of the origin of the Moon was not immediately resolved. When Apollo 11 set off for the Moon, there were three theories for how the Moon formed: (1) that it split away from Earth when our planet was new and still molten; (2) that it formed in the same part of the Solar System at the same time as Earth, and was captured by it early on; and (3) that it formed elsewhere in the Solar System, and was captured much later. To general consternation, the Apollo samples ruled out *all three* theories! It was not until 1986 that a consensus was arrived at in which the Moon formed by the accretion of debris left over from the collision of a Mars-sized body with Earth very soon after Earth formed in the solar nebula.

Appendices

Apollo Program Budget Appropriations ($000)[1]

	1960	1961	1962	1963	1964	1965	1966
Advanced Technical Development Studies	$100	$1,000	$0	$0	$0	$0	$0
Orbital Flight Tests	$0	$0	$63,900	$0	$0	$0	$0
Biomedical Flight Tests	$0	$0	$16,550	$0	$0	$0	$0
High-Speed Re-entry Tests	$0	$0	$27,550	$0	$0	$0	$0
Spacecraft Development	$0	$0	$52,000	$0	$0	$0	$0
Instrumentation & Scientific Equipment	$0	$0	$0	$11,500	$0	$0	$0
Operational Support	$0	$0	$0	$2,500	$0	$0	$0
Little Joe II Development	$0	$0	$0	$8,800	$0	$0	$0
Supporting Development	$0	$0	$0	$3,000	$0	$0	$0
Command and Service Modules	$0	$0	$0	$345,000	$545,874	$577,834	$615,000
Lunar Module	$0	$0	$0	$123,100	$135,000	$242,600	$310,800
Guidance & Navigation	$0	$0	$0	$32,400	$91,499	$81,038	$115,000
Integration, Reliability, & Checkout	$0	$0	$0	$0	$60,699	$24,763	$34,400
Spacecraft Support	$0	$0	$0	$0	$43,503	$83,663	$95,400
Saturn C-1	$0	$0	$0	$90,864	$0	$0	$0
Saturn I	$0	$0	$0	$0	$187,077	$40,265	$800
Saturn IB	$0	$0	$0	$0	$146,817	$262,690	$274,185
Saturn V	$0	$0	$0	$0	$763,382	$964,924	$1,177,320
Engine Development	$0	$0	$0	$0	$166,000	$166,300	$134,095
Apollo Mission Support	$0	$0	$0	$0	$133,101	$170,542	$210,385
Manned Space Flight Operations	$0	$0	$0	$0	$0	$0	$0
Advanced Development	$0	$0	$0	$0	$0	$0	$0
Flight Modules	$0	$0	$0	$0	$0	$0	$0
Science Payloads	$0	$0	$0	$0	$0	$0	$0
Ground Support	$0	$0	$0	$0	$0	$0	$0
Spacecraft	$0	$0	$0	$0	$0	$0	$0
Apollo Program	**$100**	**$1,000**	**$160,000**	**$617,164**	**$2,272,952**	**$2,614,619**	**$2,967,385**
NASA Total	**$523,575**	**$964,000**	**$1,671,750**	**$3,674,115**	**$3,974,979**	**$4,270,695**	**$4,511,644**
Apollo Share of Total Budget	**<1%**	**<1%**	**10%**	**17%**	**57%**	**61%**	**66%**

[1] *The Apollo Spacecraft: A Chronology,* volumes I through IV.

1967	1968	1969	1970	1971	1972	1973	Total
$0	$0	$0	$0	$0	$0	$0	$1,100
$0	$0	$0	$0	$0	$0	$0	$63,900
$0	$0	$0	$0	$0	$0	$0	$16,550
$0	$0	$0	$0	$0	$0	$0	$27,550
$0	$0	$0	$0	$0	$0	$0	$52,000
$0	$0	$0	$0	$0	$0	$0	$11,500
$0	$0	$0	$0	$0	$0	$0	$2,500
$0	$0	$0	$0	$0	$0	$0	$8,800
$0	$0	$0	$0	$0	$0	$0	$3,000
$560,400	$455,300	$346,000	$282,821	$0	$0	$0	$3,728,229
$472,500	$399,600	$326,000	$231,433	$0	$0	$0	$2,241,033
$76,654	$113,000	$43,900	$33,866	$0	$0	$0	$587,357
$29,975	$66,600	$65,100	$0	$0	$0	$0	$281,537
$110,771	$60,500	$121,800	$170,764	$0	$0	$0	$686,401
$0	$0	$0	$0	$0	$0	$0	$90,864
$0	$0	$0	$0	$0	$0	$0	$228,142
$236,600	$146,600	$41,347	$0	$0	$0	$0	$1,108,239
$1,135,600	$998,900	$534,453	$484,439	$189,059	$142,458	$26,300	$6,416,835
$49,800	$18,700	$0	$0	$0	$0	$0	$534,895
$243,900	$296,800	$0	$0	$0	$0	$0	$1,054,728
$0	$0	$546,400	$422,728	$314,963	$307,450	$0	$1,591,541
$0	$0	$0	$0	$11,500	$12,500	$0	$24,000
$0	$0	$0	$0	$245,542	$55,033	$0	$300,575
$0	$0	$0	$60,094	$106,194	$52,100	$0	$218,388
$0	$0	$0	$0	$46,411	$31,659	$0	$78,070
$0	$0	$0	$0	$0	$0	$50,400	$50,400
$2,916,200	**$2,556,000**	**$2,025,000**	**$1,686,145**	**$913,669**	**$601,200**	**$76,700**	**$19,408,134**
$4,175,100	**$3,970,000**	**$3,193,559**	**$3,113,765**	**$2,555,000**	**$2,517,700**	**$2,509,900**	**$56,661,332**
70%	**64%**	**63%**	**54%**	**36%**	**24%**	**3%**	**34%**

Ascent Data[1]

	Apollo 7	Apollo 8	Apollo 9	Apollo 10	Apollo 11	Apollo 12	Apollo 13	Apollo 14	Apollo 15	Apollo 16	Apollo 17
Pre-Staging											
Pad Azimuth (°E of N)	100	90.0	90.0	90.0	90.0	90.0	90.0	90.0	90.0	90.0	90.0
Flight Azimuth (°E of N)	72	72.124	72.0	72.028	72.058	72.029	72.043	75.558	80.088	72.034	91.503
Mach 1 – GET (sec)	62.15	61.48	68.2	66.8	66.3	66.1	68.4	68.0	65.0	67.5	67.5
Mach 1 Altitude (ft)	25,034	24,128	25,781	25,788	25,736	25,610	26,697	26,355	25,663	26,019	26,221
Maximum Bending Moment – GET (sec)	73.1	74.7	79.4	84.6	91.5	77.5	76	76	80.1	86.5	79
Maximum Bending Moment (lbf-in)	7,546,000	60,000,000	86,000,000	88,000,000	33,200,000	37,000,000	69,000,000	116,000,000	80,000,000	71,000,000	96,000,000
Maximum q – GET (sec)	75.5	78.9	85.5	82.6	83.0	81.1	81.3	81.0	82.0	86.0	82.5
Maximum q Altitude (ft)	39,903	44,062	45,138	43,366	44,512	42,133	40,876	40,398	44,971	47,122	42,847
Maximum q (lb/ft^2)	665.60	776.938	630.73	694.232	735.17	682.95	651.63	655.80	768.58	726.81	701.75
S-IC Stage Burn (S-IB Apollo 7)											
Duration (sec)	147.31	160.41	169.06	168.03	168.03	168.2	170.3	170.6	166.1	168.5	168.1
Maximum Total Inertial Acceleration – GET (sec)	140.10	153.92	162.84	161.71	161.71	161.82	163.70	164.18	159.56	161.78	161.20
Maximum Total Inertial Acceleration – (ft/sec^2)	137.76	127.46	123.75	126.21	126.67	125.79	123.36	122.90	127.85	122.90	124.51
Maximum Total Inertial Acceleration – (*g*)	4.28	3.96	3.85	3.92	3.94	3.91	3.83	3.82	3.97	3.82	3.87
Maximum Earth-Fixed Velocity – GET (sec)	144.6	154.47	163.45	161.96	162.30	162.18	164.10	164.59	160.00	162.5	162.0
Maximum Earth-Fixed Velocity (ft/sec)	6,490.1	7,727.36	7,837.89	7,835.76	7,882.9	7,852.0	7,820.9	7,774.9	7,387.6	7,779.5	7,790.0
Apex – GET (sec)	259.4	266.54	266.03	266.87	269.1	275.6	271.7	271.8	277.562	270.973	273.689
Apex – Altitude (nmi)	64.4	64.69	59.23	60.61	62.1	66.4	63.1	62.9	68.8	63.1	64.9
Apex – Range (nmi)	132.6	175.70	172.37	172.90	176.8	181.4	176.0	174.5	182.9	174.8	177.2
S-II Stage Burn											
Duration (sec)	–	367.85	371.06	388.59	384.22	389.14	426.64	392.55	386.06	394.34	395.06

Maximum Total Inertial Acceleration – GET (sec)	–	524.14	536.31	460.69	460.70	460.83	537.00	463.17	459.56	461.77	461.21
Maximum Total Inertial Acceleration – (ft/sec^2)	–	59.71	64.34	58.46	58.53	58.79	53.31	58.10	57.58	56.00	56.00
Maximum Total Inertial Acceleration – (*g*)	–	1.86	2.00	1.82	1.82	1.83	1.66	1.81	1.79	1.74	1.74
Maximum Earth-Fixed Velocity GET (sec)	–	524.90	536.45	553.50	549.00	553.20	593.50	560.07	550.00	560.0	560.6
Maximum Earth-Fixed Velocity (ft/sec)	–	21,068.14	21,441.11	21,317.81	21,377.0	21,517.8	21,301.6	21,574.5	21,601.4	21,550.9	21,567.6
Apex – GET (sec)	–	560.34	593.58	597.21	587.0	581.7	632.2	600.2	553.225	584.122	574.527
Apex – Altitude (nmi)	–	104.21	102.50	102.31	101.9	103.2	103.0	102.4	95.2	93.7	93.3
Apex – Range (nmi)	–	934.06	1,026.36	1,035.06	1,005.9	985.3	1,098.8	1,032.2	888.9	978.7	946.2
S-IVB First Burn											
Duration (sec)	469.79	156.69	123.84	146.95	147.13	137.31	152.93	137.16	141.47	142.61	138.85
Maximum Total Inertial Acceleration – GET (sec)	616.9	685.08	664.74	703.84	699.41	693.99	750.00	700.66	694.67	706.21	702.65
Maximum Total Inertial Acceleration (ft/sec^2)	82.22	23.10	25.72	22.60	22.08	22.21	21.85	21.62	21.00	21.59	21.46
Maximum Total Inertial Acceleration (*g*)	2.56	0.72	0.80	0.70	0.69	0.69	0.68	0.67	0.65	0.67	0.67
Maximum Earth-Fixed Velocity – GET (sec)	619.3	685.50	674.66	703.84	709.33	703.91	750.50	710.56	704.67	716.21	712.70
Maximum Earth-Fixed Velocity (ft/sec)	24,208.4	24,244.26	24,246.39	24,240.09	24,243.8	24,242.3	24,243.1	24,221.8	24,242.4	24,286.1	24,231.0
S-IVB Second Burn											
Duration (sec)	–	317.72	62.06	343.06	346.83	341.14	350.85	350.84	350.71	341.92	351.04
Maximum Total Inertial Acceleration – GET^2	–	002:55:55.61	004:46:57.68	002:39:10.66	002:50:03.11	002:53:04.02	002:41:37.23	002:34:23.34	002:55:53.61	002:39:18:42	003:18:27.64
Maximum Total Inertial Acceleration (ft/sec^2)	–	49.77	39.90	47.90	46.65	47.74	46.23	46.56	45.01	45.64	45.44
Maximum Total Inertial Acceleration (*g*)	–	1.55	1.24	1.49	1.45	1.48	1.44	1.45	1.40	1.42	1.41

	Apollo 7	Apollo 8	Apollo 9	Apollo 10	Apollo 11	Apollo 12	Apollo 13	Apollo 14	Apollo 15	Apollo 16	Apollo 17
Maximum Earth-Fixed Velocity											
– GET	–	002:55:56.00	004:46:58.20	002:39:11.30	002:50:03.50	002:53:04.32	002:41:37.80	002:34:23.67	002:55:54.00	002:39:20.0	003:18:28.5
Maximum Earth-Fixed Velocity (ft/sec)	–	34,178.74	26,432.58	34,251.67	34,230.3	34,063.0	34,231.0	34,194.9	34,236.9	34,269.0	34,202.4
S-IVB Third Burn											
Duration (sec)	–	–	242.06	–	–	–	–	–	–	–	–
Maximum Total Inertial Acceleration											
– GET	–	–	006:08:53.00	–	–	–	–	–	–	–	–
Maximum Total Inertial Acceleration											
(ft/sec^2)	–	–	54.40	–	–	–	–	–	–	–	–
Maximum Total Inertial Acceleration (*g*)	–	–	1.69	–	–	–	–	–	–	–	–
Maximum Earth-Fixed Velocity – GET	–	–	006:11:23.50	–	–	–	–	–	–	–	–
Maximum Earth-Fixed Velocity (ft/sec)	–	–	29,923.49	–	–	–	–	–	–	–	–

[1] Compiled from Saturn V launch vehicle flight evaluation reports, Apollo/Saturn V post-flight trajectory reports, and mission reports.
[2] GET is expressed as hours:minutes:seconds (h:m:s) for the S-IVB second and third burns.

Command Module Cabin Temperature History (°F)[1]

Mission	Apollo 7	Apollo 8	Apollo 9	Apollo 10	Apollo 11	Apollo 12	Apollo 13	Apollo 14	Apollo 15	Apollo 16	Apollo 17
Launch	70	65	65	75	70	70	70	70	70	70	70
Average	70	72	70	73	63	67	64	74	69	70	69
High	79	81	72	80	73	80	71	77	81	80	81
Low	64	61	65	64	55	58	58	60	59	57	61
Re-entry	65	61	67	58	55	60	75	59	59	57	6

[1] *Biomedical Results of Apollo*, SP-368, p. 133. All temperatures were measured at the heat exchanger inlet. During the Apollo 13 mission, the LM environmental control system provided a habitable environment for about 83 hours (57:45 to 141:05 GET). Cabin temperature remained low due to low electrical power levels. This caused crew discomfort during much of this period, with cabin temperatures ranging between 49°F and 55°F.

Call Signs

Mission	Command Module	Lunar Module
Apollo 7	"Apollo 7".	None.
Apollo 8	"Apollo 8".	None.
Apollo 9	"Gumdrop". Derived from the appearance of the spacecraft when transported on Earth. During shipment, it was wrapped in blue wrappings giving appearance of a wrapped gumdrop.	"Spider", derived from its bug-like configuration.
Apollo 10	"Charlie Brown", from a character in comic strip *Peanuts* © drawn by Charles L. Schulz. As in the comic, the CM "Charlie Brown" would be the guardian of the LM "Snoopy."	"Snoopy," after the beagle dog character in the same comic strip. The name referred to the fact that the LM would be "snooping" around the lunar surface in low orbit. Also, at the Manned Spacecraft Center, Snoopy was symbol of quality performance. Employees who did outstanding work were awarded a silver Snoopy pin.
Apollo 11	"Columbia", after "Columbiad", the canon used to launch Jules Verne's moonship (commonly thought to be the moonship itself which was referred to only as "the projectile"); also used because of the close relationship of the word to the United States' origins.	"Eagle," after the eagle selected for the mission insignia.
Apollo 12	"Yankee Clipper", selected from names submitted by employees of the command module prime contractor.	"Intrepid", selected from names submitted by employees of the lunar module prime contractor.
Apollo 13	"Odyssey," reminiscent of the long voyage of Odysseus of Greek mythology.	"Aquarius," after the Egyptian god Aquarius, the water carrier. Aquarius brought fertility and therefore life and knowledge to the Nile Valley, as the Apollo 13 crew hoped to bring knowledge from the Moon.
Apollo 14	"Kitty Hawk", the site of the Wright brothers' first flight in Kitty Hawk, NC.	"Antares", for the star on which the LM oriented itself for lunar landing.
Apollo 15	"Endeavor," for the ship which carried Captain James Cook on his 18th-century scientific voyages.	"Falcon," named for the USAF Academy mascot by Apollo 15's all-Air Force crew.
Apollo 16	"Casper", named for a cartoon character, "Casper the Friendly Ghost," because the white Teflon suits worn by the crew looked shapeless on television screens.	"Orion," for a constellation, because the crew would depend on star sightings to navigate in cislunar space.
Apollo 17	"America", as a tribute and a symbol of thanks to the American people who made the Apollo program possible.	"Challenger," indicative of the challenges of the future, beyond the Apollo program.

Excerpted and edited from *Astronaut Mission Patches and Spacecraft Callsigns*, by Dick Lattimer, unpublished draft in JSC History Office, later published as *All We Did Was Fly To The Moon: Astronaut Insignias and Callsigns; Space Patches From Mercury to the Space Shuttle*; and various NASA documents.

Accumulated Time in Space During Apollo Missions[1]

	Apollo 7	Apollo 8	Apollo 9	Apollo 10	Apollo 11	Apollo 12	Apollo 13	Apollo 14	Apollo 15	Apollo 16	Apollo 17	Flight time (sec)	Flight time (h:m:s)
Mission Duration (h:m:s)	260:09:03	147:00:42	241:00:54	192:03:23	195:18:35	244:36:25	142:54:41	216:01:58	295:11:53	265:51:05	301:51:59		
Mission Duration (sec)	936,543	529,242	867,654	691,403	703,115	880,585	514,481	777,718	1,062,713	957,065	1,086,719		
David Randolph Scott			867,654						1,062,713			1,930,367	536:12:47
Eugene Andrew Cernan				691,403							1,086,719	1,778,122	493:55:22
John Watts Young				691,403						957,065		1,648,468	457:54:28
Ronald Ellwin Evans											1,086,719	1,086,719	301:51:59
Harrison Hagan Schmitt											1,086,719	1,086,719	301:51:59
James Benson Irwin									1,062,713			1,062,713	295:11:53
Alfred Merrill Worden									1,062,713			1,062,713	295:11:53
James Arthur Lovell, Jr		529,242					514,481					1,043,723	289:55:23
Charles Moss Duke, Jr										957,065		957,065	265:51:05
Thomas Kenneth Mattingly II										957,065		957,065	265:51:05
Ronnie Walter Cunningham	936,543											936,543	260:09:03
Donn Fulton Eisele	936,543											936,543	260:09:03
Walter Marty Schirra, Jr	936,543											936,543	260:09:03
Alan LaVern Bean						880,585						880,585	244:36:25
Charles Conrad, Jr						880,585						880,585	244:36:25
Richard Francis Gordon, Jr						880,585						880,585	244:36:25
James Alton McDivitt			867,654									867,654	241:00:54
Russell Louis Schweickart			867,654									867,654	241:00:54
Edgar Dean Mitchell								777,718				777,718	216:01:58
Stuart Allen Roosa								777,718				777,718	216:01:58
Alan Bartlett Shepard, Jr								777,718				777,718	216:01:58
Edwin Eugene Aldrin, Jr					703,115							703,115	195:18:35
Neil Alden Armstrong					703,115							703,115	195:18:35
Michael Collins					703,115							703,115	195:18:35
Thomas Patten Stafford				691,403								691,403	192:03:23
William Alison Anders		529,242										529,242	147:00:42
Frank Frederick Borman II		529,242										529,242	147:00:42

Fred Wallace Haise, Jr							514,481					514,481	142:54:41
John Leonard Swigert, Jr							514,481					514,481	142:54:41
Total Man-Seconds From Liftoff	**2,809,629**	**1,587,726**	**2,602,962**	**2,074,209**	**2,109,345**	**2,641,755**	**1,543,443**	**2,333,154**	**3,188,139**	**2,871,195**	**3,260,157**	**27,021,714**	
Total Time In Space (h:m:s)	**780:27:09**	**441:02:06**	**723:02:42**	**576:10:09**	**585:55:45**	**733:49:15**	**428:44:03**	**648:05:54**	**885:35:39**	**797:33:15**	**905:35:57**	**7,506:01:54**	**7,506:01:54**

[1] Calculated

Apollo Space Vehicle Configuration

S-IB (Apollo 7)
- Reached 1.640 million lb of thrust at liftoff
- Accelerated total space vehicle to ~7,620 ft/sec
- Reached ~33 nmi in ~2.5 min

S-IC
- Reached to 7.650 million lb of thrust at liftoff
- Accelerated total space vehicle to ~7,880 ft/sec
- Reached ~58 nmi in ~ 2.5 min

S-II interstage
- Interfaced first and second stages
- Housed second stage engines
- Provided ullage for S-II engine start

S-II
- Accelerated vehicle from ~7,880 ft/sec to ~22,850 ft/sec in ~370 sec.
- Achieved altitude of ~101 nmi
- Housed S-II retro-rocket mounting

S-IVB Interstage
- Provided structural transition from diameter of S-II to S-IVB
- Housed S-IVB engine
- Had attitude control about 3 axes and +X ullage with APS, up to 505 sec of burn time

S-IVB
- Increased velocity from 7,620 ft/sec to 25,553 ft/sec in 470 sec to accomplish orbit (Apollo 7)
- Increased velocity from 22,850 ft/sec to 25,568 ft/sec in 154 sec to accomplish orbit (all other flights)
- Accelerated space vehicle to ~35,500 ft/sec for TLI (all except Apollo 7 and Apollo 9)

Instrument Unit
- Provided launch vehicle guidance; navigation; control signals; telemetry; command communications; tracking; EDS rates and display activation timing and stage functional sequencing

Spacecraft/Lunar Module Adapter
- Housed and supported the LM, aerodynamically enclosed, supported LM
- Provided the structural electrical interface between spacecraft and launch vehicle
- Provided diameter transition from S-IVB to CSM
- Allowed LM extraction

Lunar Module Descent Stage
- Provided velocity change for lunar deorbit and lunar landing (throttleable)
- Protected ascent stage from landing damage
- Provided ascent stage /descent stage staging
- Provided LM ascent stage launch pad
- Stowed lunar scientific equipment

Lunar Module Ascent Stage

- Provided mission life support for 2 crewmen
- Contained secondary command control and communications
- Computed and performed lunar landing abort, launch, rendezvous and docking with CSM
- Facilitated CM, LM ingress/egress inter- and extra-vehicular activities
- Maneuvered about and along 3 axed in the near-lunar environment

Service Module

- Provided velocity change for course correction, lunar orbit insertion, transearth injection and CSM aborts
- Provided attitude control and translation
- Supplemented environmental, electrical power and reaction control requirements of CM

Command Module

- Provided mission life support for 3 crewmen
- Provided inertial/space-fixed navigation
- Provided command control and communication center
- Provided attitude control about 3 axes
- Acted as a limited lifting body
- Provided CM-LM ingress/egress for inter- and extravehicular activity

Launch Escape System

- Transported CM away from space vehicle (and mainland) during launch abort
- Oriented CM attitude for launch abort descent
- Jettisoned safely as required
- Sensed flight dynamics
- Provided CM thermal protection

Apportionment of Training According to Mission Type[1]

Training Category	Missions Before 1st Lunar Landing (Apollo 7–10)		Early Lunar Landing Missions (Apollo 11–14)		Final Lunar Landing Missions (Apollo 15–17)	
	Hours	% of Total	Hours	% of Total	Hours	% of Total
Simulators	11,511	36	15,029	56	11,413	45
Special Purpose	4,023	13	5,379	20	9,246	36
Procedures	7,924	25	2,084	8	1,265	5
Briefings	5,894	18	3.070	11	2,142	9
Spacecraft Tests	2,576	8	1,260	5	1,255	5
Total	31,928	100	26,822	100	25,321	100

[1] *Apollo Program Summary Report* (JSC-09423), pp. 6–20 to 6–23. Includes participation of Mission Control Center personnel. Numbers in parentheses indicate simulations accomplished by follow-on or support crewmen.

Post-flight Medical Problems in Apollo Crews[1]

Diagnosis	Etiology	Cases
Barotitis media	Eustachian tube blockage	7
Folliculitis, right anterior chest	Bacterial	1
Gastroenteritis	Bacterial	1
Herpetic lesion, lip	Herpes virus	1
Influenza syndrome	Influenza B virus	1
	Undetermined	1
	Influenza A virus	1
Laceration of the forehead	Trauma	1
Rhinorrhea, mild	Fiberglass particle	1
Papular lesions, parasacral	Bacteria	1
Prostatitis	Undetermined	2
Pulpitis, tooth No. 7		1
Pustules, eyelids		1
Rhinitis	Viral	3
Acute maxillary sinusitis	Bacterial	1
Ligamentous strain, right shoulder		1
Urinary tract infection	Pseudomonas	1
Vestibular dysfunction, mild		1
Rhinitis and pharyngitis	Influenza B virus	1
Rhinitis and secondary bronchitis	Beta-streptococcus (not group A)	1
Contact dermatitis	Fiberglass	
	Beta cloth	1
	Micropore tape	6
Subungual hemorrhages, finger nails	Trauma	3

[1] *Biomedical Results of Apollo*, SP-368.

Apollo Medical Kits[1]

	Apollo 7	Apollo 8	Apollo 9	Apollo 10	Apollo 11	Apollo 12	Apollo 13	Apollo 14	Apollo 15	Apollo 16	Apollo 17
Command Module Medical Kit											
Methylcellulose eye drops (0.25%)	2/1	2/2	2/0	2/0	2/0	2/0	2/0	2/0	1/0	2/0	1/0
Tetrahydrozoline HCl (Visine)	–	–	–	–	–	–	–	–	–	–	1/1
Compress - bandage	2/0	2/0	2/0	2/0	2/0	2/0	2/0	2/0	2/0	2/0	2/0
Band-Aids	12/2	12/0	12/0	12/0	12/0	12/0	12/0	12/0	12/0	12/0	12/0
Antibiotic ointment	1/1	1/0	1/0	1/0	1/0	2/0	2/0	2/0	2/0	2/1	2/1
Skin cream	1/0	1/1	1/1	1/0	1/0	1/0	1/0	1/0	1/0	1/1	1/0
Demerol injectors (90 mg)	3/0	3/0	3/0	3/0	3/0	3/0	3/0	3/0	3/0	–	–
Marezine injectors	3/0	3/0	3/0	3/0	3/0	3/0	3/0	3/0	3/0	–	–
Marezine tablets (50 mg)	24/3	24/1	24/4	12/0	–	–	–	–	–	–	–
Dexedrine tablets (5 mg)	12/1	12/0	12/0	12/0	12/0	12/0	12/1	12/0	12/0	12/0	12/0
Darvon compound capsules (60 mg)	12/2	18/0	18/0	18/0	18/0	18/0	12/1	18/0	18/0	18/0	18/0
Actifed tablets (60 mg)	24/24	60/0	60/12	60/2	60/0	60/18	60/0	60/0	60/0	60/0	60/1
Lomotil tablets	24/8	24/3	24/1	24/13	24/2	24/0	24/1	24/0	24/0	24/0	48/5
Nasal emollient	1/0	2/1	2/1	1/0	1/0	1/0	1/0	1/0	1/0	1/0	1/0
Aspirin tablets (5 gr)	72/48	72/8	72/2	72/16	72/Unk	72/6	72/30	72/0	72/0	72/0	72/0
Tetracycline (250 mg)	24/02	24/0	24/0	15/0	–	–	–	–	60/0	60/0	60/0
Ampicillin	–	60/0	60/0	45/0	60/0	60/0	60/0	60/0	60/0	60/0	60/0
Seconal capsules (100 mg)	–	21/1	21/10	21/0	21/0	21/6	21/0	–	21/0	21/3	21/16
Seconal capsules (50 mg)	–	12/7	–	–	–	–	–	–	–	–	–
Nose drops (Afrin)	–	3/0	3/1	3/0	3/0	3/1	3/0	3/1	3/0	3/0	3/3
Benadryl (50 mg)	–	8/0	–	–	–	–	–	–	–	–	–
Tylenol (325 mg)	–	14/7	–	–	–	–	–	–	–	–	–
Bacitracin eye ointment	–	–	1/0	–	–	–	–	–	–	–	–
Scopolamine (0.3 mg) – Dexedrine (5-mg capsules)	–	–	–	–	12/6	12/0	12/2	12/0	12/0	12/0	12/1
Mylicon tablets	–	–	–	–	40/0	40/0	40/0	40/0	40/0	40/0	40/0
Opthaine	–	–	–	–	–	–	1/0	1/0	1/0	1/0	1/0
Multi-Vitamins	–	–	–	–	–	–	–	20/0	–	–	–
Auxiliary Medications											
Pronestyl	–	–	–	–	–	–	–	–	–	80/0	80/0
Lidocaine	–	–	–	–	–	–	–	–	–	12/0	12/0
Atropine	–	–	–	–	–	–	–	–	–	12/0	12/0
Demerol	–	–	–	–	–	–	–	–	–	6/0	6/0

Apollo Medical Accessories Kit											
Constant Wear Garment Harness											
Plug	–	–	–	–	—	–	–	–	3	3	3
ECG Sponge Packages	–	–	–	–	–	–	–	–	14	14	14
Electrode Bag	1	1	1	1	1	1	1	1	1	1	1
Electrode Attachment Assembly	12	12	12	12	20	20	20	20	100	100	100
Micropore Disc	12	12	12	12	20	20	20	20	50	50	50
Sternal Harness	1	1	1	1	3	3	3	3	3	3	3
Axillary Harness	1	1	1	1	1	1	1	1	1	1	1
Electrode Paste	1	1	1	1	1	1	1	1	1	1	1
Oral Thermometer	1	1	1	1	1	1	1	1	1	1	1
pH Paper	1	1	1	1	1	1	1	1	1	None	None
Urine Collection and Transfer Assembly Roll-On Cuffs	3	3	6	6	6	6	6	6	6	6	6
Lunar Module Medical Kit[2]											
Rucksack	–	–	–	–	–	1					
Stimulant Pill s (Dexedrine)	–	–	–	–	–	4					
Pain Pills (Darvon)	–	–	–	–	–	4					
Decongestant Pills (Actifed)	–	–	–	–	–	8					
Diarrhea Pills (Lomotil)	–	–	–	–	–	12					
Aspirin	–	–	–	–	–	12					
Band-Aids	–	–	–	–	–	6					
Compress Bandages	–	–	–	–	–	2					
Eye Drops (Methylcellulose)	–	–	–	–	–	1					
Antibiotic Ointment (Neosporin)	–	–	–	–	–	1					
Sleeping Pills (Seconal)	–	–	–	–	–	6					
Anesthetic Eye Drops	–	–	–	–	–	1					
Nose Drops (Afrin)	–	–	–	–	–	1					
Urine Collection and Transfer Assembly Roll-On Cuffs	–	–	–	–	–	6					
Pronestyl	–	–	–	–	–	12					
Injectable Drug Kit											
Injectable Drug Kit Rucksack	–	–	–	–	–	1					
Lidocaine (cardiac)	–	–	–	–	–	8					
Atropine (cardiac)	–	–	–	–	–	4					
Demerol (pain)	–	–	–	–	–	2					

[1] *Biomedical Results of Apollo*, SP-368, P. 33.

[2] Typical quantities and items; there was no "standard" lunar module medical kit. The adequacy of the kits was reviewed after each mission and appropriate modifications were made for the next mission.

Apollo Training Exercises[1]

Exercise	Apollo 7	Apollo 8	Apollo 9	Apollo 10	Apollo 11	Apollo 12	Apollo 13	Apollo 14	Apollo 15	Apollo 16	Apollo 17
Lunar Surface Activity Simulations (Sessions)											
Surface Operations	–	–	–	–	20	31	42	43	91	67	47
Operations Before/ After EVA	–	–	–	–	10	4	11	18	20	10	20
Total Per Mission	–	–	–	–	30	35	53	61	111	77	67
Geology Field Trips[2]	–	–	–	–	1	4	7	7	12	18	13
Integrated Crew/ Ground Mission Simulations (Days)											
Command Module Simulator	18	14	10	11	6 (1)	10	13	12 (3)	13 (6)	16 (5)	13 (2)
Lunar Module Simulator	0	0	2	0	4	3	5	5 (2)	5	7 (1)	6
Command Module and Lunar Module Simulators	0	0	8	7	7	12	9	12 (1)	7	10	9
Total Per Mission	18	14	20	18	17 (1)	25	27	29 (6)	25 (6)	33 (6)	28 (2)

[1] *Apollo Program Summary Report* (JSC-09423), pages 6–20 to 6–23. Includes participation of Mission Control Center personnel. Numbers in parentheses indicate simulations accomplished by follow-on or support crewmen.

[2] Each field trip lasted from one to seven days.

Baseline Apollo Food and Beverage List[1]

Abbreviations
RSB – rehydratable spoon bowl
RD – rehydratable drink
IM – intermediate moisture
D – dehydrated
T – thermostabilized
NS – natural state

Beverages
Cocoa (RD)
Coffee (RD)
Grape Drink (RD)
Grapefruit Drink (RD)
Grape Punch (RD)
Orange-Grapefruit Drink (RD)
Orange Juice (RD)
Pineapple-Grapefruit drink (RD)
Pineapple-Orange-Drink (RD)

Breakfast Items
Bacon squares [8] (IM)
Cinnamon Toasted Bread Cubes [4] (D)
Canadian Bacon and Applesauce (RSB)
Cornflakes (RSB)
Fruit cocktail (RSB)
Sausage Patties (RSB)
Scrambled Eggs (RSB)
Peaches (RSB)
Spiced Fruit Cereal (RSB)
Apricot (IM)
Peaches (IM)

Cubes and Candy
Brownies [4] (IM)
Caramel Candy (IM)
Chocolate Bar (IM)
Creamed Chicken Bites [6] (D)
Cheese Crackers (D)
Cheese Sandwiches [4] (D)
Beef Sandwiches [4] (D)
Jellied Fruit Candy (IM)
Beef Jerky (IM)
Peanut Cubes [4] (NS)
Pecans [6] (IM)
Pineapple Fruitcake (IM)
Sugar Cookies [4] (D)
Turkey Bites [4] (D)

Desserts
Applesauce (RSB)
Banana Pudding (RSB)
Butterscotch Pudding (RSB)
Chocolate Pudding (RSB)
Cranberry-Orange Sauce (RSB)
Peach Ambrosia (RSB)

Salads and Soups
Chicken and Rice Soup (RSB)
Lobster Bisque (RSB)
Pea Soup (RSB)
Potato Soup (RSB)
Shrimp Cocktail (RSB)
Tomato Soup (RSB)
Tuna Salad (RSB)

Sandwich Spreads and Bread
Bread [Slice] (NS)
Catsup (NS)
Cheddar Cheese [2 oz] (NS)
Chicken Salad [8 oz] (T)
Ham Salad [8 oz] (T)
Jelly (NS)
Mustard (NS)
Peanut Butter (NS)

Meats
Beef Pot Roast (RSB)
Beef and Vegetables (RS)
Beef Stew (RSB)
Chicken and Rice (RSB)
Chicken and Vegetables (RSB)
Chicken Stew (RSB)
Pork and Scalloped Potatoes (RSB)
Spaghetti, Meat Sauce (RSB)
Beef and Gravy (T)
Frankfurters (T)
Meatballs, Sauce (T)
Turkey and Gravy (T)

[1] Lyndon B. Johnson Space Center Flight Crew Support Division

Flight Directors[1]

Apollo 7	Directors
Shift #1	Glynn S. Lunney
Shift #2	Eugene F. Kranz
Shift #3	Gerald D. Griffin

Apollo 8	Directors
Shift #1	Clifford E. Charlesworth
Shift #2	Glynn S. Lunney
Shift #3	Milton L. Windler

Apollo 9	Directors
Shift #1	Eugene F. Kranz
Shift #2	Gerald D. Griffin
Shift #3	M. P. 'Pete' Frank III

Apollo 10	Directors
Shift #1	Glynn S. Lunney
Shift #2	Gerald D. Griffin
Shift #3	Milton L. Windler
	M. P. 'Pete' Frank III

Apollo 11	Directors
Shift #1	Clifford E. Charlesworth
	Gerald D. Griffin
Shift #2	Eugene F. Kranz
Shift #3	Glynn S. Lunney

Apollo 12	Directors
Shift #1	Gerald D. Griffin
Shift #2	M. P. 'Pete' Frank III
Shift #3	Clifford E. Charlesworth
Shift #4	Milton L. Windler

Apollo 13	Directors
Shift #1	Milton L. Windler
Shift #2	Gerald D. Griffin
Shift #3	Eugene F. Kranz
Shift #4	Glynn S. Lunney

Apollo 14	Directors
Shift #1	M. P. "Pete" Frank III
	Glynn S. Lunney
Shift #2	Milton L. Windler
Shift #3	Gerald D. Griffin
Shift #4	Glynn S. Lunney

Apollo 15	Directors
Shift #1	Gerald D. Griffin
Shift #2	Milton L. Windler
Shift #3	Glynn S. Lunney
	Eugene F. Kranz

Apollo 16	Directors
Shift #1	M. P. 'Pete' Frank III
	Philip C. Shaffer
Shift #2	Eugene F. Kranz
	Donald R. Puddy
Shift #3	Gerald D. Griffin
	Neil B. Hutchinson
	Charles R. Lewis

Apollo 17	Directors
Shift #1	Gerald D. Griffin
Shift #2	Eugene F. Kranz
	Neil B. Hutchinson
Shift #3	M. P. 'Pete' Frank III
	Charles R. Lewis

[1] Compiled from various documents and memoranda in the Rice University archives. According to Pete Frank, the initials "M.P." do not represent any names, per a telephone conversation between author Orloff and Mr. Frank.

Earth Orbit Data[1]

	Apollo 7	Apollo 8	Apollo 9	Apollo 10	Apollo 11	Apollo 12	Apollo 13	Apollo 14	Apollo 15	Apollo 16	Apollo 17
Earth Orbit Insertion											
Insertion – GET (sec)	626.76	694.98	674.66	713.76	709.33	703.91	759.83	710.56	704.67	716.21	712.65
Altitude (ft)	748,439	627,819	626,777	627,869	626,909	626,360	628,710	626,364	566,387	567,371	559,348
Surface Range (nmi)	1,121.743	1,430.363	1,335.515	1,469.790	1,460.697	1,438.608	1,572.300	1,444.989	1,445.652	1,469.052	1,456.314
Earth Fixed Velocity (ft/sec)	24,208.5	24,242.9	24,246.39	24,244.3	24,243.9	24,242.3	24,242.1	24,221.6	24,242.4	24,286.1	24,230.9
Space-Fixed Velocity (ft/sec)	25,553.2	25,567.06	25,569.78	25,567.88	25,567.9	25,565.9	25,566.1	25,565.8	25,602.6	25,605.0	25,603.9
Geocentric Latitude (°N)	31.4091	32.4741	32.4599	32.5303	32.5027	31.5128	32.5249	31.0806	29.2052	32.5262	24.5384
Geodetic Latitude (°N)	31.58	32.6487	32.629	32.700	32.672	32.6823	32.6945	31.2460	29.3650	32.6963	24.6805
Longitude (°E)	–61.2293	–53.2923	–55.1658	–52.5260	–52.6941	–53.1311	–50.4902	–52.9826	–53.0807	–52.5300	–53.8107
Space-Fixed Flight Path Angle (deg)	0.005	0.0006	–0.0058	–0.0049	0.012	–0.014	0.005	–0.003	0.015	0.001	0.003
Space-Fixed Heading Angle (°E of N)	86.32	88.532	87.412	89.933	88.848	88.580	90.148	91.656	95.531	88.932	105.021
Apogee (nmi)	152.34	99.99	100.74	100.32	100.4	100.1	100.3	100.1	91.5	91.3	90.3
Perigee (nmi)	123.03	99.57	99.68	99.71	98.9	97.8	99.3	98.9	89.6	90.0	90.0
Period (min)	89.55	88.19	88.20	88.20	88.18	88.16	88.19	88.18	87.84	87.85	87.83
Inclination (deg)	31.608	32.509	32.552	32.546	32.521	32.540	32.547	31.120	29.679	32.542	28.526
Descending Node (deg)		42.415	45.538	123.132	123.088	123.126	123.084	117.455	109.314	123.123	86.978
Eccentricity	0.0045	0.00006	0.000149	0.000086	0.00021	0.00032	0.0001	0.0002	0.0003	0.0002	0.0000
Earth Orbit – Revolutions	163.0	1.5	151.0	1.5	1.5	1.5	1.5	1.5	1.5	1.5	2.0
Earth Orbit Duration	259:42:59.24	002:44:30.53	240:32:55.54	002:27:26.82	002:38:23.70	002:41:30.03	002:28:27.32	002:22:42.68	002:44:18.94	002:27:32.21	003:06:44.99

[1] Compiled from Saturn V launch vehicle flight evaluation reports, Apollo/Saturn V post-flight trajectory reports and mission reports.

Crew Information – Earth Orbit and Lunar Orbit Missions[1]

	Apollo 7	Apollo 8	Apollo 9	Apollo 10
Commander	Walter Marty Schirra, Jr	Frank Frederick Borman II	James Alton McDivitt	Thomas Patten Stafford
Date of Birth	12 Mar 1923	14 Mar 1928	10 Jun 1929	17 Sep 1930
Place of Birth	Hackensack, NJ	Gary, IN	Chicago IL	Weatherford, OK
Age on Launch Date	45	40	39	38
Status	Captain	Colonel	Colonel	Colonel
	USN	USAF	USAF	USAF
Year Selected Astronaut	1959	1962	1962	1962
Prior Space Flights	MA-8, GT-6A	GT-7	GT-4	GT-6A, GT-9A
Backup	Thomas Patten Stafford	Neil Alden Armstrong	Charles Conrad, Jr	Leroy Gordon Cooper, Jr
Status	Colonel	Civilian	Commander	Colonel
	USAF	NASA	USN	USAF
Command Module Pilot	Donn Fulton Eisele	James Arthur Lovell, Jr	David Randolph Scott	John Watts Young
Date of Birth	23 Jun 1930	25 Mar 1928	06 Jun 1932	24 Sep 1930
Place of Birth	Columbus, OH	Cleveland, OH	San Antonio, TX	San Francisco, CA
Date of Death	01 Dec 1987	—	—	—
Place of Death	Tokyo, Japan	—	—	—
Age on Launch Date	38	40	36	38
Status	Major	Captain	Colonel	Commander
	USAF	USN	USAF	USN
Year Selected Astronaut	1963	1962	1963	1962
Prior Space Flights	None	GT-7, GT-12	GT-8	GT-3, GT-10
Backup	John Watts Young	Edwin Eugene Aldrin, Jr	Richard Francis Gordon, Jr	Donn Fulton Eisele
Status	Commander	Colonel	Commander	Lt. Colonel
	USN	USAF	USN	USAF

Lunar Module Pilot	Ronnie Walter Cunningham	William Alison Anders	Russell Louis Schweickart	Eugene Andrew Cernan
Date of Birth	16 Mar 1932	17 Oct 1933	25 Oct 1935	14 Mar 1934
Place of Birth	Creston, IA	Hong Kong	Neptune, NJ	Chicago, IL
Age on Launch Date	36	35	33	35
Status	Civilian	Major	Civilian	Commander
	—	USAF	—	USN
Year Selected Astronaut	1963	1963	1963	1963
Prior Space Flights	None	None	None	GT-9A
Backup	Eugene Andrew Cernan	Fred Wallace Haise, Jr	Alan LaVern Bean	Edgar Dean Mitchell
Status	Commander	Civilian	Commander	Commander
	USN	NASA	USN	USN

[1] Compiled from press kits and mission reports, and *Who's Who in Space* (Cassutt).

Crew Information – Lunar Landing Missions[1]

	Apollo 11	Apollo 12	Apollo 13	Apollo 14	Apollo 15	Apollo 16	Apollo 17
Commander	Neil Alden Armstrong	Charles Conrad, Jr	James Arthur Lovell, Jr	Alan Bartlett Shepard, Jr	David Randolph Scott	John Watts Young	Eugene Andrew Cernan
Date of Birth	05 Aug 1930	02 Jun 1930	25 Mar 1928	18 Nov 1923	06 Jun 1932	24 Sep 1930	14 Mar 1934
Place of Birth	Wapakoneta, OH	Philadelphia, PA	Cleveland, OH	East Derry, NH	San Antonio, TX	San Francisco, CA	Chicago, IL
Date of Death	–	08 Jul 1999	–	21 Jul 1998	–	–	–
Place of Death	–	Ojai, CA	–	Monterey, CA	–	–	–
Age on Launch Date	38	39	42	47	39	41	38
Status	Civilian	Commander	Captain	Captain	Colonel	Captain	Captain
	–	USN	USN	USN	USAF	USN	USN
Year Selected Astronaut	1962	1962	1962	1959	1963	1962	1963
Prior Space Flights	GT-8	GT-5, GT-11	GT-7, GT-12, Apollo 8	MR-3	GT-8, Apollo 9	GT-3, GT-10, Apollo 10	GT-9A, Apollo 10
Backup	James Arthur Lovell, Jr	David Randolph Scott	John Watts Young	Eugene Andrew Cernan	Richard Francis Gordon, Jr	Fred Wallace Haise, Jr	John Watts Young
Status	Captain	Colonel	Commander	Captain	Captain	Civilian	Captain
	USN	USAF	USN	USN	USN	NASA	USN

	Apollo 11	Apollo 12	Apollo 13	Apollo 14	Apollo 15	Apollo 16	Apollo 17
Command Module Pilot	Michael Collins	Richard Francis Gordon, Jr	John Leonard Swigert, Jr	Stuart Allen Roosa	Alfred Merrill Worden	Thomas Kenneth Mattingly II	Ronald Ellwin Evans
Date of Birth	31 Oct 1930	05 Oct 1929	30 Aug 1931	16 Aug 1933	07 Feb 1932	17 Mar 1936	10 Nov 1933
Place of Birth	Rome, Italy	Seattle, WA	Denver, CO	Durango, CO	Jackson, MI	Chicago, IL	St Francis, KS
Date of Death	–	–	27 Dec 1982	12 Dec 1994	–	–	07 Apr 1990
Place of Death	–	–	Washington, DC	Washington, DC	–	–	Scottsdale, AZ
Age On Launch Date	38	40	38	37	39	36	39
Status	Lt Colonel USAF	Commander USN	Civilian –	Major USAF	Major USAF	Lt Commander USN	Commander USN
Year Selected Astronaut	1963	1963	1966	1966	1966	1966	1966
Prior Space Flights	GT-10	GT-11	None	None	None	None	None
Backup	William Alison Anders	Alfred Merrill Worden	Thomas Kenneth Mattingly II	Ronald Ellwin Evans	Vance DeVoe Brand	Stuart Allen Roosa	Stuart Allen Roosa
Status	Lt Colonel USAF	Major USAF	Lt Commander USN	Commander USN	Civilian NASA	Lt Colonel USAF	Lt Colonel USAF
Lunar Module Pilot	Edwin Eugene Aldrin, Jr	Alan LaVern Bean	Fred Wallace Haise, Jr	Edgar Dean Mitchell	James Benson Irwin	Charles Moss Duke, Jr	Harrison Hagan Schmitt
Date of Birth	20 Jan 1930	15 Mar 1932	14 Nov 1933	17 Sep 1930	17 Mar 1930	03 Oct 1935	03 Jul 1935
Place of Birth	Montclair, NJ	Wheeler, TX	Biloxi, MS	Hereford, TX	Pittsburgh, PA	Charlotte, NC	Santa Rita, NM
Date of Death	–	–	–	–	08-Aug-91	–	–
Place of Death	–	–	–	–	Glenwood Springs, CO	–	–
Age on Launch Date	39	37	36	40	41	36	37
Status	Colonel, ScD USAF	Commander USN	Civilian –	Commander, ScD USN	Lt Colonel USAF	Lt Colonel USAF	Civilian, PhD –
Year Selected Astronaut	1963	1963	1966	1966	1966	1966	1965
Prior Space Flights	GT-12	None	None	None	None	None	None
Backup	Fred Wallace Haise, Jr	James Benson Irwin	Charles Moss Duke, Jr	Joe Henry Engle	Harrison Hagan Schmitt	Edgar Dean Mitchell	Charles Moss Duke, Jr
Status	Civilian NASA	Lt Colonel USAF	Major USAF	Lt Colonel USAF	Civilian NASA	Captain USN	Colonel USAF

[1] Compiled from press kits and mission reports, and *Who's Who in Space* (Cassutt).

Crew Weight History (kg)[1]

Mission	Crewman	30 Days Before Launch	30-Day Average	Launch	Recovery
Apollo 7	Schirra	87.1	87.8	88.0	86.1
	Eisele	69.4	69.5	71.2	66.7
	Cunningham	69.4	70.7	70.8	67.8
Apollo 8	Borman	76.2	76.6	76.6	72.8
	Lovell	76.4	76.8	78.0	74.4
	Anders	66.0	66.4	64.4	62.6
Apollo 9	McDivitt	73.5	73.0	72.1	69.6
	Scott	82.8	82.0	80.7	78.2
	Schweickart	74.7	74.3	71.2	69.4
Apollo 10	Stafford	80.1	79.6	77.6	76.4
	Young	76.6	76.8	74.8	72.3
	Cernan	79.4	79.4	78.5	73.9
Apollo 11	Armstrong	78.0	78.4	78.0	74.4
	Collins	74.4	75.6	75.3	72.1
	Aldrin	77.6	78.1	75.7	75.3
Apollo 12	Conrad	66.2	66.6	67.7	65.8
	Gordon	71.0	70.7	70.4	67.1
	Bean	69.4	69.9	69.1	63.5
Apollo 13	Lovell	79.8	78.7	80.5	74.2
	Swigert	89.1	89.4	89.3	84.4
	Haise	71.0	70.8	70.8	67.8
Apollo 14	Shepard	78.0	78.4	76.2	76.6
	Roosa	74.2	75.3	74.8	69.4
	Mitchell	83.5	83.2	79.8	80.3

Mission	Crewman	30 Days Before Launch	30-Day Average	Launch	Recovery
Apollo 15	Scott	80.5	81.1	80.2	78.9
	Worden	73.7	73.6	73.5	72.1
	Irwin	74.3	74.3	73.2	70.8
Apollo 16	Young	80.8	80.1	78.9	75.5
	Mattingly	63.2	62.6	61.5	58.5
	Duke	73.1	73.2	73.0	70.5
Apollo 17	Cernan	81.0	80.7	80.3	76.1
	Evans	78.2	77.3	75.7	74.6
	Schmitt	76.0	76.0	74.8	72.9

[1] *Biomedical Results of Apollo*, SP-368, pp. 76–77.

Designations[1]

	Apollo 7	Apollo 8	Apollo 9	Apollo 10	Apollo 11	Apollo 12	Apollo 13	Apollo 14	Apollo 15	Apollo 16	Apollo 17
Call-Signs											
Command Module	Apollo 7	Apollo 8	Gumdrop	Charlie Brown	Columbia	Yankee Clipper	Odyssey	Kitty Hawk	Endeavour	Casper	America
Lunar Module	–	–	Spider	Snoopy	Eagle	Intrepid	Aquarius	Antares	Falcon	Orion	Challenger
NASA/Contractor Designations											
Space Vehicle	AS-205	AS-503	AS-504	AS-505	AS-506	AS-507	AS-508	AS-509	AS-510	AS-511	AS-512
Launch Vehicle	SA-205	SA-503	SA-504	SA-505	SA-506	SA-507	SA-508	SA-509	SA-510	SA-511	SA-512
Launch Vehicle Type	Saturn IB	Saturn V	Saturn V	Saturn V	Saturn V	Saturn V	Saturn V	Saturn V	Saturn V	Saturn V	Saturn V
Launch Vehicle 1st Stage	S-IB-5	S-IC-3	S-IC-4	S-IC-5	S-IC-6	S-IC-7	S-IC-8	S-IC-9	S-IC-10	S-IC-11	S-IC-12
Launch Vehicle 2nd Stage	S-IVB-205	S-II-3	S-II-4	S-II-5	S-II-6	S-II-7	S-II-8	S-II-9	S-II-10	S-II-11	S-II-12
Launch Vehicle 3rd Stage	–	S-IVB-503	S-IVB-504	S-IVB-505	S-IVB-506	S-IVB-507	S-IVB-508	S-IVB-509	S-IVB-510	S-IVB-511	S-IVB-512
Instrument Unit	S-IU-205	S-IU-503	S-IU-504	S-IU-505	S-IU-506	S-IU-507	S-IU-508	S-IU-509	S-IU-510	S-IU-511	S-IU-512
Spacecraft/LM Adapter	SLA-5	SLA-11A	SLA-12A	SLA-13A	SLA-14	SLA-15	SLA-16	SLA-17	SLA-19	SLA-20	SLA-21

Command Module	CM-101	CM-103	CM-104	CM-106	CM-107	CM-108	CM-109	CM-110	CM-112	CM-113	CM-114
Service Module	SM-101	SM-103	SM-104	SM-106	SM-107	SM-108	SM-109	SM-110	SM-112	SM-113	SM-114
Lunar Module	–	Lunar Module Test Article (LTA-B)	LM-3	LM-4	LM-5	LM-6	LM-7	LM-8	LM-10	LM-11	LM-12
Lunar Roving Vehicle	–	–	–	–	–	–	–	–	LRV-1	LRV-2	LRV-3
VAB High Bay	–	1	3	2	1	3	1	3	3	3	3
Firing Room	–	1	2	3	1	2	1	2	1	1	1
Mobile Launcher Platform	–	MLP-1	MLP-2	MLP-3	MLP-1	MLP-2	MLP-3	MLP-2	MLP-3	MLP-3	MLP-3
Computer Programs – CSM	Sundisk	Colossus	Colossus	Colossus 2	Comanche 55	Colossus 2C	Comanche 72 Rev. 3	Colossus 2E	Colossus 3	Colossus 3	Colossus 3
– LM			Sundance	Luminary 1	Luminary 99	Luminary 116	Luminary 131 Rev. 9	Luminary 1D	Luminary 1E	Luminary 1E	Luminary 1G
Eastern Test Range Number	66	170	9025	920	5307	2793	3381	7194	7744	1601	1701
International Designations											
CSM	1968-089A	1968-118A	1969-018A	1969-043A	1969-059A	1969-099A	1970-029A	1971-008A	1971-063A	1972-031A	1972-096A
S-IVB Stage	1968-089B	1968-118B	1969-018B	1969-043B	1969-059B	1969-099B	1970-029B	1971-008B	1971-063B	1972-031B	1972-096B
LM Ascent Stage[2]	–	–	1969-018C	1969-043D	1969-059C	1969-099C	1970-029C	1971-008C	1971-063C	1972-031C	1972-096C
LM Descent Stage	–	–	1969-018D	1969-043C	1969-059D	1969-099D	1970-029C	1971-008D	1971-063E	1972-031E	1972-096D
Lunar Subsatellite	–	–	–	–	–	–	–	–	1971-063D	1972-031D	–
NORAD Designations											
CSM	03486	03626	03769	03941	04039	04225	04371	04900	05351	06000	06300
S-IVB Stage	03487	03627	03770	03943	04040	04226	04372	04904	05352	06001	06301
LM Ascent Stage	–	–	03771	03949	04041	04246	–	04905	05366	06005	06307
LM Descent Stage	–	–	03780	03948	–	–	–	–	–	–	–
Lunar Subsatellite	–	–	–	–	–	–	–	–	05377	06009	–

[1] Compiled from *RAE Table of Earth Satellites 1957-1986*; press kits; mission implementation plans; Saturn V flight evaluation reports; *Apollo Program Summary Report*; *Stages to Saturn: A Technological History of the Apollo/Saturn Launch Vehicles;* and other sources, with special thanks to David Baker for the software designations.

[2] Ascent and descent stages for Apollo 13 remained docked throughout the mission and were jettisoned together.

Geology and Soil Mechanics Tools and Equipment[1]

Item	Apollo 11	Apollo 12	Apollo 14	Apollo 15	Apollo 16	Apollo 17
Apollo Lunar Surface Hand Tools						
Hammer	1	1	1	1	1	1
Large Scoop	1	1	1	0	0	0
Adjustable Scoop	0	0	0	1	1	1
Extension Handle	1	1	1	1	2	2
Gnomon	1	1	1	1	1	1
Tongs	1	1	1	1	2	2
Adjustable Trenching Tool	0	0	1	0	0	0
Rake	0	0	0	1	1	1
Core Tubes	2	4	6	0	0	0
Core Tube Caps	2	1	0	0	0	0
Drive Tubes (Lower)	0	0	0	5	5	5
Drive Tubes (Upper)	0	0	0	4	4	4
Drive Tube Cape and Bracket Assembly	0	0	0	3	5	5
Drive Tube Tool Assembly	0	0	0	0	1	1
Spring Scale	1	1	0	0	0	0
Sample Scale	0	0	1	1	1	1
Tool Carrier	0	0	0	1	1	0
Sample Return Container	2	2	2	2	2	2
Bags and Special Containers						
Small Sample Bags	5	0	0	0	0	0
Documented Sample Bags (15-Bag Dispenser)	1	3	1	0	0	0
Documented Sample Bags (20-Bag Dispenser)	0	0	0	6	7	8
Documented Sample Bags (35-Bag Dispenser)	0	1	2	0	0	0
Round Documented Sample Bag	0	0	0	0	0	48
Protective Padded Sample Bag	0	0	0	0	2	0
Documented Sample Weigh Bag	2	4	4	0	0	0
Sample Collection Bag	0	0	0	2	2	2
Gas Analysis Sample Container	1	1	0	0	0	0

Core Sample Vacuum Container	0	1	3	3	1	1
Solar Wind Composition Bag	2	1	1	0	0	0
Magnetic Shield Sample Container	0	0	1	0	0	0
Extra Sample Collection Bags	0	0	0	4	6	6
Organic Control Sample	0	1	2	2	2	0
Lunar Surface Sampler (Beta Cloth)	0	0	0	0	1	0
Lunar Surface Sampler (Velvet)	0	0	0	0	1	0
Lunar Rover Vehicle Soil Sampler	0	0	0	0	0	1
Magnetic Sample Assembly	0	0	0	0	1	0
Tether Hook	1	1	1	0	0	0
Lunar Surface Drill	0	0	0	1	1	1
Core Stem With Bit	0	0	0	1	1	1
Core Stems Without Bit	0	0	0	5	5	5
Core Stem Cap and Retainer Assembly	0	0	0	2	2	2
Self-Recording Penetrometer	0	0	0	1	1	0

[1] JSC-09423, pp. 3–27.

LM Ascent and Ascent Stage Lunar Impact[1]

	Apollo 11	Apollo 12	Apollo 14	Apollo 15	Apollo 16[2]	Apollo 17
LM Ascent						
GET	124:22:00.79	142:03:47.78	141:45:40	171:37:23.2	175:31:47.9	185:21:37
KSC Date	21 Jul 1969	20 Nov 1969	06 Feb 1971	02 Aug 1971	23 Apr 1972	14 Dec 1972
GMT Date	21 Jul 1969	20 Nov 1969	06 Feb 1971	02 Aug 1971	24 Apr 1972	14 Dec 1972
KSC Time	01:54:00 p.m.	09:25:47 a.m.	01:48:42 p.m.	01:11:23 p.m.	08:25:47 p.m.	05:54:37 p.m.
KSC Time Zone	EDT	EST	EST	EDT	EST	EST
GMT Time	17:54:00	14:25:47	18:48:42	17:11:23	01:25:47	22:54:37
LM Ascent Stage Lunar Impact						
GET	–	149:55:17.7	147:42:23.7	181:29:37.0	–	193:17:20.8
KSC Date	–	20 Nov 1969	06 Feb 1971	02 Aug 1971	–	15 Dec 1972
GMT Date	–	20 Nov 1969	07 Feb 1971	03 Aug 1971	–	15 Dec 1972

	Apollo 11	Apollo 12	Apollo 14	Apollo 15	Apollo 16[2]	Apollo 17
KSC Time	–	05:17:17 p.m.	07:45:25 p.m.	11:03:37 p.m.	–	01:50:20 a.m.
Time Zone	–	EST	EST	EDT	–	EST
GMT Time	–	22:17:17.7	00:45:25.7	03:03:37.0	–	06:50:20.8
Selenocentric Latitude (°N)	–	–3.94	–3.42	26.36	–	19.96
Selenocentric Longitude (°E)	–	–21.20	–19.67	0.25	–	30.50
Selenocentric Latitude	–	3°56′ 24″ S	3°25′ 12″ S	26°21′ 21¤ N	–	19°57′ 58″ N
Selenocentric Longitude	–	21°12′ 00″ W	19°40′ 01″ W	0°15′ 00¤ E	–	30°29′ 23″ E
Velocity (ft/sec)	–	5,512	5,512	5,577	–	5,479
Mass (lbm)	–	5,254	5,077	5,258	–	4,982
LM Ascent Stage Lunar Impact Energy (ergs)	–	3.36×10^{16}	3.25×10^{16}	3.44×10^{16}	–	3.15×10^{16}
Angle From Horizontal (deg)	–	3.7	3.6	3.2	–	4.9
Heading Angle (deg)	–	305.85	282	284	–	283
Crater Diameter (calculated) (ft)	–	29.9	29.6	30.2	–	–
Distance To Target (nmi)	–	35	7	12	–	0.7
Distance to LM Descent Stage Landing Site (nmi)	–	41.0	36	50	–	4.7
Distance to Seismic Stations (nmi)						
Apollo 12	–	39	62	610	–	945
Apollo 14	–	–	36	566	–	863
Apollo 15	–	–	–	50	–	416
Apollo 16	–	–	–	–	–	532
Azimuth to Seismic Stations (deg)						
Apollo 12	–	112	096	036	–	064
Apollo 14	–	–	276	029	–	061
Apollo 15	–	–	–	276	–	098
Apollo 16	–	–	–	–	–	027

[1] Compiled from Saturn V launch vehicle flight evaluation report and mission report for each flight. Times are when signals received on Earth. Actual landing site coordinates based on International Astronomical Union (IAU) Mean Earth Polar Axis coordinate system as described in the *Journal of Geophysical Research*, vol. 105, pp. 20,227 to 20,280, 2000.

[2] Deorbit maneuver was not possible and LM ascent stage remained in lunar orbit for about one year. No impact information is available.

Capsule Communicators (Capcoms)[1]

Apollo 7
Col Thomas Patten Stafford, USAF
Lt Cdr Ronald Ellwin Evans, USN
Maj William Reid Pogue, USAF
John Leonard Swigert, Jr
Cdr John Watts Young, USN
Cdr Eugene Andrew Cernan, USN

Apollo 8
Lt Col Michael Collins, USAF
Lt Cdr Thomas Kenneth Mattingly II, USN
Maj Gerald Paul Carr, USMC
Neil Alden Armstrong
Col Edwin Eugene Aldrin, USAF/ScD
Vance DeVoe Brand
Fred Wallace Haise, Jr

Apollo 9
Maj Stuart Allen Roosa, USAF
Lt Cdr Ronald Ellwin Evans, USN
Maj Alfred Merrill Worden, USAF
Cdr Charles Conrad, Jr, USN
Cdr Richard Francis Gordon, Jr, USN
Cdr Alan LaVern Bean, USN

Apollo 10
Maj Charles Moss Duke, Jr, USAF
Maj Joe Henry Engle, USAF
Maj Jack Robert Lousma, USMC
Lt Cdr Bruce McCandless II, USN

Apollo 11
Maj Charles Moss Duke, Jr, USAF
Lt Cdr Ronald Ellwin Evans, USN
Lt Cdr Bruce McCandless II, USN
Capt James Arthur Lovell, Jr, USN
Lt Col William Alison Anders, USAF
Lt Cdr Thomas Kenneth Mattingly II, USN
Fred Wallace Haise, Jr
Don Leslie Lind, PhD
Owen Kay Garriott, Jr, PhD
Harrison Hagan Schmitt, PhD

Apollo 12
Lt Col Gerald Paul Carr, USMC
Edward George Gibson, PhD
Cdr Paul Joseph Weitz, USN
Don Leslie Lind, PhD
Col David Randolph Scott, USAF
Maj Alfred Merrill Worden, USAF
Lt Col James Benson Irwin, USAF

Civilian Backup Capcoms
Dickie K. Warren
James O. Rippey
James L. Lewis
Michael R. Wash

Apollo 13
Cdr Joseph Peter Kerwin, USN/MD/MC
Vance DeVoe Brand
Maj Jack Robert Lousma, USMC
Cdr John Watts Young, USN
Lt Cdr Thomas Kenneth Mattingly II, USN

Apollo 14
Maj Charles Gordon Fullerton, USAF
Lt Cdr Bruce McCandless II, USN
Fred Wallace Haise, Jr
Lt Cdr Ronald Ellwin Evans, USN

Apollo 15
Joseph Percival Allen, IV, PhD
Maj Charles Gordon Fullerton, USAF
Karl Gordon Henize, PhD
Cdr Edgar Dean Mitchell, USN/ScD
Robert Alan Ridley Parker, PhD
Harrison Hagan Schmitt, PhD
Capt Alan Bartlett Shepard, Jr, USN
Capt Richard Francis Gordon, Jr, USN
Vance DeVoe Brand

Apollo 16
Maj Donald Herod Peterson, USAF
Maj Charles Gordon Fullerton, USAF
Col James Benson Irwin, USAF
Fred Wallace Haise, Jr
Lt Col Stuart Allen Roosa, USAF
Cdr Edgar Dean Mitchell, USN
Maj Henry Warren Hartsfield, Jr, USAF
Anthony Wayne England, PhD
Lt Col Robert Franklyn Overmyer, USMC

Apollo 17
Maj Charles Gordon Fullerton, USAF
Maj Charles Gordon Fullerton, USAF
Robert Alan Ridley Parker, PhD
Joseph Percival Allen, IV PhD
Capt Alan Bartlett Shepard, Jr, USN
Cdr Thomas Kenneth Mattingly II, USN
Col Charles Moss Duke, Jr, USAF
Lt Col Stuart Allen Roosa, USAF
Capt John Watts Young, USN

[1] Derived from various documents and memoranda in Rice University archives. Military ranks for astronauts who are not also backups are implied from available information and B. Hello (Rockwell) memo, 10 December 1969.

Extravehicular Activity[1]

		Apollo 9	Apollo 11	Apollo 12	Apollo 14	Apollo 15	Apollo 16	Apollo 17
Earth Orbit EVA	1st EVA Participant	Scott	–	–	–	–	–	–
	1st EVA Duration	01:01	–	–	–	–	–	–
	2nd EVA Participant	Schweickart	–	–	–	–	–	–
	2nd EVA Duration	01:07:00	–	–	–	–	–	–
	2nd EVA Duration Outside LM	00:47:01	–	–	–	–	–	–
LM Stand-Up EVA	Participant					Scott		
	Duration					00:33:07		
First Surface EVA	Duration	–	02:31:40	03:56:03	04:47:50	06:32:42	07:11:02	07:11:53
	Total Distance Traveled	–	~3,300 ft	~3,300 ft	~3,300 ft	5.6 nmi	2.3 nmi	1.8 nmi
	LRV Ride Time	–	–	–	–	01:02	00:43	00:33
	LRV Park Time	–	–	–	–	01:14	03:39	–
	Total LRV Time	–	–	–	–	02:16	04:22	–
	Samples Collected (lbm)[2]	–	47.51	36.82	45.19	31.97	65.92	31.53
Second Surface EVA	Duration	–	–	03:49:15	04:34:41	07:12:14	07:23:09	07:36:56
	Total Distance Traveled	–	–	~4,300 ft	~9,800 ft	6.7 nmi	6.1 nmi	11.0 nmi
	LRV Ride Time	–	–	–	–	01:23	01:31	02:25
	LRV Park Time	–	–	–	–	02:34	03:56	–
	Total LRV Time	–	–	–	–	03:57	05:27	–
	Samples Collected (lbm)	–	–	38.80	49.16	76.94	63.93	75.18
Third Surface EVA	Duration	–	–	–	–	04:49:50	05:40:03	07:15:08
	Total Distance Traveled	–	–	–	–	2.7 nmi	6.2 nmi	6.5 nmi
	LRV Ride Time	–	–	–	–	00:35	01:12	01:31
	LRV Park Time	–	–	–	–	01:22	02:26	–
	Total LRV Time	–	–	–	–	01:57	03:38	–
	Samples Collected (lbm)	–	–	–	–	60.19	78.04	136.69
Total Lunar Surface EVA	Total Duration	–	02:31:40	07:45:18	09:22:31	18:34:46	20:14:14	22:03:57
	Total Distance Traveled	–	~3,300 ft	~7,600 ft	~13,000 ft	15.1 nmi	14.5 nmi	19.3 nmi
	Total Samples Collected (lbm)	–	47.51	75.73	93.21	170.44	211.00	243.65

	Total LRV Ride Time	–	–	–	–	3:00	03:26	04:29
	Total LRV Park Time	–	–	–	–	05:10	10:01	–
	Total LRV Time	–	–	–	–	08:10	13:27	–
	Maximum Distance Traveled From LM (ft)	–	200[3]	1,350[4]	4,770[5]	16,470	15,092[6]	25,029
Transearth EVA	Participant	–	–	–	–	Worden	Mattingly	Evans
	Duration	–	–	–	–	00:39:07	01:23:42	01:05:44

[1] From mission reports. Lunar surface EVAs are time outside LM. Others are based on start/stop times. Apollo 9 EVAs are from depressurization start to repressurization stop.

[2] Returned sample weights provided by Lunar Sample Curator, NASA Johnson Space Center.

[3] *Apollo 11 Preliminary Science Report* (SP-214), p. 44.

[4] *Apollo 12 Preliminary Science Report*, p. 26 (measured from map).

[5] *Skylab: A Chronology* (SP-4011), pages 420-421 for Apollo 14, Apollo 15 and Apollo 17.

[6] Measured from map in *Apollo 16 Preliminary Science Report* (SP-315).

General Background[1]

	Apollo 7	Apollo 8	Apollo 9	Apollo 10	Apollo 11	Apollo 12	Apollo 13	Apollo 14	Apollo 15	Apollo 16	Apollo 17
Mission Information											
Mission Type	C	C prime	D	F	G	H–1	H–2	H–3	J–1	J–2	J–3
Purpose	CSM piloted flight demonstration in Earth orbit.	CSM piloted flight demonstration in lunar orbit.	Lunar module piloted flight demonstration in Earth orbit.	Lunar module piloted flight demonstration in lunar orbit.	Manned lunar landing demonstration.	Precision piloted lunar landing demonstration and systematic lunar exploration.	Precision piloted lunar landing demonstration and systematic lunar exploration.	Precision piloted lunar landing demonstration and systematic lunar exploration.	Extensive scientific investigation of Moon on lunar surface and from lunar orbit.	Extensive scientific investigation of Moon on lunar surface and from lunar orbit.	Extensive scientific investigation of Moon on lunar surface and from lunar orbit.
Trajectory Type	Earth Orbital	Lunar Orbital	Earth Orbital	Lunar Orbital	Lunar Landing	Lunar Landing	Lunar Landing	Lunar Landing	Lunar Landing	Lunar Landing	Lunar Landing
Payload Description	Block II CSM, adapter, and LES.	Block II CSM, LM test article, adapter, and LES.	Block II CSM, LM, adapter, and LES.	Block II CSM, LM, adapter, and LES.	Block II CSM, LM, adapter, and LES.	Block II CSM, LM, adapter, and LES.	Block II CSM, LM, adapter, and LES.	Block II CSM, LM, adapter, and LES.	Block II CSM, LM, adapter, and LES.	Block II CSM, LM, adapter, and LES.	Block II CSM, LM, adapter, and LES.

	Apollo 7	Apollo 8	Apollo 9	Apollo 10	Apollo 11	Apollo 12	Apollo 13	Apollo 14	Apollo 15	Apollo 16	Apollo 17
Launch Information											
Launch Sites (Florida)	Cape Kennedy	Kennedy Space Center	Kennedy Space Center	Kennedy Space Center	Kennedy Space Center	Kennedy Space Center	Kennedy Space Center	Kennedy Space Center	Kennedy Space Center	Kennedy Space Center	Kennedy Space Center
Launch Complex	Complex 34	Complex 39A	Complex 39A	Complex 39B	Complex 39A	Complex 39A	Complex 39A	Complex 39A	Complex 39A	Complex 39A	Complex 39A
Geodetic Latitude (°N)	28.521963	28.608422	28.608422	28.627306	28.608422	28.608422	28.608422	28.608422	28.608422	28.608422	28.608422
Geocentric Latitude (°N)	28.3608	28.4470	28.4470	28.4658	28.4470	28.4470	28.4470	28.4470	28.4470	28.4470	28.4470
Longitude (°E)	–80.561141	–80.604133	–80.604133	–80.620869	–80.604133	–80.604133	–80.604133	–80.604133	–80.604133	–80.604133	–80.604133
Range Zero[2]											
KSC Date	11 Oct 1968	21 Dec 1968	03 Mar 1969	18 May 1969	16 Jul 1969	14 Nov 1969	11 Apr 1970	31 Jan 1971	26 Jul 1971	16 Apr 1972	07 Dec 1972
KSC Time	11:02:45 a.m.	07:51:00 a.m.	11:00:00 a.m.	12:49:00 p.m.	09:32:00 a.m.	11:22:00 a.m.	02:13:00 p.m.	04:03:02 p.m.	09:34:00 a.m.	12:54:00 p.m.	12:33:00 a.m.
KSC Time Zone	EDT	EST	EST	EDT	EDT	EST	EST	EST	EDT	EST	EST
GMT Date	11 Oct 1968	21 Dec 1968	03 Mar 1969	18 May 1969	16 Jul 1969	14 Nov 1969	11 Apr 1970	31 Jan 1971	26 Jul 1971	16 Apr 1972	07 Dec 1972
GMT Time	15:02:45	12:51:00	16:00:00	16:49:00	13:32:00	16:22:00	19:13:00	21:03:02	13:34:00	17:54:00	05:33:00
Actual GMT Liftoff Time	15:02:45.36	12:51:00.67	16:00:00.67	16:49:00.58	13:32:00.63	16:22:00.68	19:13:00.61	21:03:02.57	13:34:00.58	17:54:00.59	05:33:00.63
Selected Durations (h:m:s)											
Ascent to Orbit (sec)	626.76	694.98	674.66	713.76	709.33	703.91	759.83	710.56	704.67	716.21	712.65
Earth Orbit	259:42:59	002:44:30.53	240:32:55.5	002:27:26.82	002:38:23.70	002:41:30.03	002:28:07.32	002:22:42.68	002:44:18.94	002:27:32.21	003:06:44.99
Revolutions	163.0	1.5	151.0	1.5	1.5	1.5	1.5	1.5	1.5	1.5	2.0
Translunar Coast	–	066:16:21.8	–	073:22:29.5	073:05:34.87	080:38:01.67	–	079:28:18.30	075:42:21.45	071:55:14.35	083:02:18.12
Time on Lunar Surface	–	–	–	–	021:36:21	031:31:12	–	033:30:29	066:54:54	071:02:13	074:59:39
Lunar Orbit	–	020:10:13.0	–	061:43:23.6	059:30:25.79	088:58:11.52	–	066:35:39.99	145:12:41.68	125:49:32.59	147:43:37.11
Revolutions	–	10	–	31	30	45	–	34	74	64	75
CSM/LM Undocked	–	–	006:22:50	008:10:05	027:51:00.0	037:42:17.9	–	039:45:08.9	072:57:09.3	081:27:47	079:49:19
Transearth Coast	–	057:23:32.5	–	054:09:40.8	059:36:52.0	071:52:51.96	–	067:09:13.8	071:07:48	065:13:16	067:34:05
CM Earth Entry (sec)	937	869.2	1,003.8	868.5	929.3	845.9	835.3	852.8	778.3	814	801
Mission Duration	260:09:03	147:00:42.0	241:00:54	192:03:23	195:18:35	244:36:25	142:54:41	216:01:58.1	295:11:53.0	265:51:05	301:51:59

[1] Compiled from mission reports, launch vehicle reports, and other sources

[2] Range Zero was the integral second before liftoff.

Entry, Splashdown, and Recovery[1]

	Apollo 7	Apollo 8	Apollo 9	Apollo 10	Apollo 11	Apollo 12	Apollo 13	Apollo 14	Apollo 15	Apollo 16	Apollo 17[2]
Earth Entry											
GET	259:53:26	146:46:12.8	240:44:10.2	191:48:54.5	195:03:05.7	244:22:19.09	142:40:45.7	215:47:45	294:58:54.7	265:37:31	301:38:38
Velocity (ft/sec)	25,846.4	36,221.1	25,894	36,314	36,194.4	36,116.618	36,210.6	36,170.2	36,096.4	36,196.1	36,090.3
Maximum Entry Velocity (ft/sec)	25,955	36,303	25,989	36,397	36,277						
Maximum g	3.33	6.84	3.35	6.78	6.56	6.57	5.56	6.76	6.23	7.19	6.49
Range (nmi)	1,594	1,292	1,835	1,295	1,497	1,250	1,250	1,234	1,184	1,190	1,190
Geodetic Latitude (°N)	–29.92	20.83	33.52	–23.60	–3.19	–13.80	–28.23	–36.36	14.23	–19.87	0.71
Longitude (°E)	92.62	–179.89	–99.05	174.39	171.96	173.52	173.44	165.80	–175.02	–162.13	–173.34
Flight Path Angle (°E of N)	–2.0720	–6.50	–1.74	–6.54	–6.48	–6.48	–6.269	–6.370	–6.51	–6.55	–6.49
Heading Angle (deg)	87.47	121.57	99.26	71.89	50.18	98.16	77.21	70.84	52.06	21.08	156.53
Lift to Drag Ratio	–	0.300	–	0.305	0.300	0.309	0.291	0.280	0.290	0.286	0.290
Max. Heating Rate (BTU/ft^2/sec)	–	296	–	296	286	285	271	310	289	346	346
Total Heating Load (BTU/ft^2)	–	26,140	–	25,728	26,482	26,224	25,710	27,111	25,881	27,939	27,939
Duration (sec)	937.0	869.2	1,003.8	868.5	929.3	845.9	835.3	852.8	778.3	814.0	801.0
Avg. Radiation Skin Dose (Rads)[3]	0.16	0.16	0.20	0.48	0.18	0.58	0.24	1.14	0.30	0.51	0.55
Earth Splashdown											
GET	260:09:03	147:00:42.0	241:00:54	192:03:23	195:18:35	244:36:25	142:54:41	216:01:58.1	295:11:53.0	265:51:05	301:51:59
KSC Date	22 Oct 1968	27 Dec 1968	13 Mar 1969	26 May 1969	24 Jul 1969	24 Nov 1969	17 Apr 1970	09 Feb 1971	07 Aug 1971	27 Apr 1972	19 Dec 1972
GMT Date	22 Oct 1968	27 Dec 1968	13 Mar 1969	26 May 1969	24 Jul 1969	24 Nov 1969	17 Apr 1970	09 Feb 1971	07 Aug 1971	27 Apr 1972	19 Dec 1972
KSC Time	07:11:48 a.m.	10:51:42 a.m.	12:00:54 p.m.	12:52:23 a.m.	12:50:35 p.m.	03:58:25 p.m.	01:07:41 p.m.	04:05:00 p.m.	04:45:53 p.m.	02:45:05 p.m.	02:24:59 p.m.
Time Zone	EDT	EST	EST	EDT	EDT	EST	EST	EST	EDT	EST	EST
GMT Time	11:11:48	15:51:42	17:00:54	16:52:23	16:50:35	20:58:25	18:07:41	21:05:00	20:45:53	19:45:05	19:24:59
Splashdown Site (Ocean)	Atlantic	Pacific	Atlantic	Pacific	Pacific	Pacific	Pacific	Pacific	Pacific	Pacific	Pacific
Latitude (°N)	27.63	8.10	23.22	–15.07	13.30	–15.78	–21.63	–27.02	26.13	–0.70	–17.88
Longitude (°E)	–64.15	–165.00	–67.98	–164.65	–169.15	–165.15	–165.37	–172.67	–158.13	–156.22	–166.11
CM Weight (lb)	11,409	10,977	11,094	10,901	10,873.0	11,050.2	11,132.9	11,481.2	11,731	11,995	12,120

	Apollo 7	Apollo 8	Apollo 9	Apollo 10	Apollo 11	Apollo 12	Apollo 13	Apollo 14	Apollo 15	Apollo 16	Apollo 17[2]
Distance to Target (nmi)	1.9	1.4	2.7	1.3	1.7	2.0	1.0	0.6	1.0	3.0	1.0
Distance to Recovery Ship (nmi)	7	2.6	3	2.9	13	3.91	3.5	3.8	5	2.7	3.5
Distance Traveled (nmi)	3,953,842	504,006	3,664,820	721,250	828,743	828,134	541,103	1,000,279	1,107,945	1,208,746	1,291,299
Maximum Distance Traveled from Earth (nmi)	244.2	203,752.37	275.0	215,548	210,391	–	–	–	–	–	–
Splashdown Weather											
1st Level Cloud Type	Light rain showers	Scattered clouds	30%	10%	–	–	Broken	High Scattered	Scattered	Scattered	Scattered
1st Level Cloud Cover (ft)	600 (overcast)	2,000	2,000	2,000	–	–	2,000	2,000	2,000	2,000	3,000
2nd Level Cloud Type	–	Overcast	Broken	20%	–	–	–	–	–	–	–
2nd Level Cloud Cover (ft)	–	9,000	9,000	7,000	–	–	–	–	–	–	–
Visibility (n mi)	2	10	10	10	12	10	10	10	10	10	10
Wind Speed (ft/sec)	27	32	15	8	27						
Wind Speed (knots)	16	19	9	5	16	15	10	15	10	10	10
Wind Direction (deg from True N)	260	70	200	100		68				110	130
Air Temperature (°F)	74	–	79	–	–	–	–	–	–	–	–
Water Temperature (°F)	81	82	76	85	–	–	–	–	–	–	–
Wave Height (ft)	3	6	7	3	3	3, with 15 ft swells	4	4	3	4	2 to 3
Wave Direction (deg from True N)	260	110	340	–	–	–	–	–	–	–	–
Crew Recovery											
Minutes To Crew Pickup	56	88	49	39	63	60	45	48	39	37	52
Launch Site Pickup Time	08:08 a.m.	12:20 p.m.	12:49:33 p.m.	01:31 p.m.	01:53 p.m.	04:58 p.m.	01:53 p.m.	04:53 p.m.	05:25 p.m.	03:22 p.m.	03:17 p.m.
Time Zone	EDT	EST	EST	EDT	EDT	EST	EST	EST	EDT	EST	EST
GMT Pickup Time	12:08	17:20	17:49:33	17:31	17:53	21:58	18:53	21:53	21:25	20:22	20:17

Recovery Ship	Essex (CVS–9)	Yorktown (CVS–10)	Guadalcanal (LPH–7)	Princeton (LPH–5)	Hornet (CVS–12)	Hornet (CVS–12)	Iwo Jima (LPH–2)	New Orleans (LPH–11)	Okinawa (LPH–3)	Ticonderoga (CVS–14)	Ticonderoga (CVS–14)
Commanding Officer (Captain)	John A. Harkins	John G. Fifield	Roy M. Sudduth	Carl M. Cruise	Carl J. Seiberlich	Carl J. Seiberlich	Leland E. Kirkemo	Robert W. Carius	Andrew F. Huff	Frank T. Hamler	Frank T. Hamler
Spacecraft Recovery											
Flotation Attitude	Inverted	Inverted	Upright	Upright	Inverted	Inverted	Upright	Upright	Upright	Inverted	Upright
Minutes To Upright	12.0	6.0	0.0	0.0	7.6	4.5	0.0	0.0	0.0	4.5	0.0
Minutes To CM Pickup	111	148	132	96	188	108	88	124	94	99	123
Launch Site Pickup Time	09:03 a.m.	01:20 p.m.	02:13 p.m.	02:28 p.m.	03:58 p.m.	05:46 p.m.	02:36 p.m.	06:09 p.m.	06:20 p.m.	04:24 p.m.	04:28 p.m.
Time Zone	EDT	EST	EST	EDT	EDT	EST	EST	EST	EDT	EST	EST
GMT Pickup Time	13:03	18:20	19:13	18:28	19:58	22:46	19:36	23:09	22:20	21:24	21:28
Recovery Forces[4]											
Navy Ships Deployed	9	12	6	8	5	5	4	5	4	4	3
Atlantic	4	6	3	4	3	3	2	3	2	1	1
Pacific	5	6	3	4	2	2	2	2	2	3	2
Aircraft Deployed	31	43	29	30	31	26	22	19	17	17	15
Navy	8	21	7	10	13	9	8	5	5	6	5
Air Force	23	22	22	20	18	17	14	14	12	11	10

[1] Compiled from mission reports, USN Historical Office data, *Apollo Program Summary Report* (JSC-09423) and other sources.

[2] Some Apollo 17 entry phase data are pre-flight predictions because actual data were not obtained.

[3] *Space Physiology & Medicine*, SP-447.

[4] JSC–09423, pp. 7–18.

Ground Ignition Weights[1]

Weights In Pounds Mass	Apollo 7	Apollo 8	Apollo 9	Apollo 10	Apollo 11	Apollo 12	Apollo 13	Apollo 14	Apollo 15	Apollo 16	Apollo 17
Ground Ignition Time Relative to Range Zero (sec)	**–2.988**	**–6.585**	**–6.3**	**–6.4**	**–6.4**	**–6.5**	**–6.7**	**–6.5**	**–6.5**	**–6.7**	**–6.9**
S–IB stage, dry	84,530	–	–	–	–	–	–	–	–	–	–
S–IB stage, fuel	276,900	–	–	–	–	–	–	–	–	–	–
S–IB stage, oxidizer	631,300	–	–	–	–	–	–	–	–	–	–
S–IB stage, other	1,182	–	–	–	–	–	–	–	–	–	–
S–IB stage, total	**993,912**	–	–	–	–	–	–	–	–	–	–
S–IB/S–IVB interstage, dry	**5,543**	–	–	–	–	–	–	–	–	–	–
Retromotor Propellant	1,061	–	–	–	–	–	–	–	–	–	–
S–IC stage, dry	–	305,232	294,468	293,974	287,531	287,898	287,899	287,310	286,208	287,855	287,356
S–IC stage, fuel	–	1,357,634	1,431,678	1,423,254	1,424,889	1,424,287	1,431,384	1,428,561	1,410,798	1,439,894	1,431,921
S–IC stage, oxidizer	–	3,128,034	3,301,203	3,302,827	3,305,786	3,310,199	3,304,734	3,312,769	3,312,030	3,311,226	3,314,388
S–IC stage, other	–	6,226	5,508	5,491	5,442	5,442	5,401	5,194	4,283	5,396	5,395
S–IC stage, total	–	**4,797,126**	**5,032,857**	**5,025,546**	**5,023,648**	**5,027,826**	**5,029,418**	**5,033,834**	**5,013,319**	**5,044,371**	**5,039,060**
S–IC/S–II interstage, dry	–	**12,436**	**11,591**	**11,585**	**11,477**	**11,509**	**11,454**	**11,400**	**9,083**	**10,091**	**9,975**
S–II stage, dry	–	88,500	84,312	84,273	79,714	80,236	77,947	78,120	78,908	80,362	80,423
S–II stage, fuel	–	793,795	821,504	823,325	819,050	825,406	836,741	837,484	837,991	846,157	844,094
S–II stage, oxidizer	–	154,907	158,663	158,541	158,116	157,986	159,931	159,232	158,966	160,551	160,451
S–II stage, other	–	1,426	1,188	1,250	1,260	1,250	1,114	1,051	1,082	991	934
S–II stage, total	–	**1,038,628**	**1,065,667**	**1,067,389**	**1,058,140**	**1,064,878**	**1,075,733**	**1,075,887**	**1,076,947**	**1,088,061**	**1,085,902**
S–II/S–IVB interstage, dry	–	**8,731**	**7,998**	**8,045**	**8,076**	**8,021**	**8,081**	**8,060**	**8,029**	**8,055**	**8,019**

S–IVB stage, dry	21,852	25,926	25,089	25,680	24,852	25,064	25,097	25,030	25,198	25,099	25,040
S–IVB stage, fuel	39,909	43,395	43,709	43,388	43,608	43,663	43,657	43,546	43,674	43,727	43,752
S–IVB stage, oxidizer	193,330	192,840	189,686	192,089	192,497	190,587	191,890	190,473	195,788	195,372	195,636
S–IVB stage, other	1,432	1,626	1,667	1,684	1,656	1,873	1,673	1,687	1,655	1,643	1,658
S–IVB stage, total	**256,523**	**263,787**	**260,151**	**262,841**	**262,613**	**261,187**	**262,317**	**260,736**	**266,315**	**265,841**	**266,086**
Total Instrument Unit	**4,263**	**4,842**	**4,281**	**4,267**	**4,275**	**4,277**	**4,502**	**4,505**	**4,487**	**4,502**	**4,470**
Spacecraft/Lunar Module Adapter	3,943	3,951	4,012	3,969	3,951	3,960	3,947	3,962	3,964	3,961	3,961
LM (LTA Apollo 8)	–	19,900	32,034	30,735	33,278	33,562	33,493	33,685	36,238	36,237	36,262
Command and Service Module	32,495	63,531	59,116	63,560	63,507	63,559	63,795	64,448	66,925	66,949	66,942
Total Launch Escape System	8,874	8,890	8,869	8,936	8,910	8,963	8,991	9,027	9,108	9,167	9,104
Total Spacecraft (CSM)	**45,312**	**96,272**	**104,031**	**107,200**	**109,646**	**110,044**	**110,226**	**111,122**	**116,235**	**116,314**	**116,269**
Total Vehicle	**1,306,614**	**6,221,823**	**6,486,577**	**6,486,873**	**6,477,875**	**6,487,742**	**6,501,733**	**6,505,548**	**6,494,415**	**6,537,238**	**6,529,784**

[1] Actual weights at S–IC stage ignition, compiled from Saturn launch vehicle flight evaluation reports. Weights to do not add to vehicle totals due to truncated decimal data in report.

Lunar Surface Experiment Assignments[1]

Designation	Experiment	Apollo 11	Apollo 12	Apollo 14	Apollo 15	Apollo 16	Apollo 17
M-515	Lunar Dust Detector		X	X	X		
S-031	Passive Seismic Experiment	X	X	X	X	X	
S-033	Active Seismic Experiment			X		X	
S-034	Lunar Surface Magnetometer		X		X	X	
S-035	Solar Wind Spectrometer		X		X		
S-036	Suprathermal Ion Detector		X	X	X		
S-037	Heat Flow Experiment				X	X	X
S-038	Charged Particle Lunar Environment			X			
S-058	Cold Cathode Ion Gauge		X	X	X		
S-059	Lunar Field Geology	X	X	X	X	X	X
S-078	Laser Ranging Retroreflector	X		X	X		
S-080	Solar Wind Composition	X	X	X	X	X	
S-151	Cosmic-Ray Detection (helmets)	X					
S-152	Cosmic-Ray Detector (sheets)					X	X
S-184	Lunar Surface Close-up (photography)	X	X	X			
S-198	Portable Magnetometer			X		X	
S-199	Lunar Gravity Traverse						X
S-200	Soil Mechanics			X	X	X	X
S-201	Far-Ultraviolet Camera/Spectroscope					X	
S-202	Lunar Ejecta and Meteorites						X
S-203	Lunar Seismic Profiling						X
S-204	Surface Electrical Properties						X
S-205	Lunar Atmospheric Composition						X
S-207	Lunar Surface Gravimeter						X
S-229	Lunar Neutron Probe						X
–	Lunar Sample Analysis	X	X	X	X	X	X
–	Surveyor III Analysis		X				
–	Long-Term Lunar Surface Exposure						X

[1] *Project Apollo:* NASA Facts.

Lunar Surface Experiment Descriptions[1]

Central Station
The heart of the experiment package, provided the radio frequency link to Earth for telemetering data, command/control, and power distribution to the experiments.

Early Apollo Scientific Experiment Package (EASEP)
Flown on Apollo 11 only, this experiment package was powered by solar energy and contained an abbreviated set of experiments. It continued to return data for 71 days.

Active Seismic
Used an astronaut-activated thumper device and mortar firing explosive charges to generate seismic signals. This experiment used geophone seismic listening devices to determine lunar structure to depths of about 1,000 ft.

Heat Flow
Probes containing temperature sensors were implanted in holes to depths of 8 ft to measure the near-surface temperature gradient and thermal conductivity from which heat flow from the lunar interior could be determined.

Lunar Mass Spectrometer
Used a magnetic deflection mass spectrometer to identify lunar atmospheric components and their relative abundance.

Lunar Seismic Profiling
Flown on Apollo 17 only, this experiment was an advanced version of the Active Seismic Experiment. It used four geophones to detect seismic signals generated by eight explosive charges weighing from about 1/10 to 6.5 lb. The charges were deployed at distances up to 2 nmi from the Lunar Module and were detonated by timers after the Lunar Module departed. Lunar structure to depths of 1.5 nmi was measured. Used in a listening mode, the experiment continued to provide data on Moon/thermal quakes and meteoroid impacts beyond its planned lifetime

Solar Wind Spectrometer
Measured interaction between the Moon and the solar wind by sensing flow-direction and energies of both electrons and positive ions. Results showed that solar wind plasma measurements on the lunar surface are indistinguishable from simultaneous plasma measurements made by nearby satellites

Suprathermal Ion Detector
Provided information on the energy and mass spectra of positive ions near the lunar surface. Evidence of prompt ionization and acceleration of gases generated on the Moon was found in the return data.

Charged Particle Lunar Environment
Measured the fluxes of charged particles, both electrons and ions, having energies from 50 to 50,000 eV. The instrument measured plasma particles originating in the Sun and low-energy particle flux in the magnetic tail of the Earth.

Laser Ranging Retroreflector
The retroreflector bounced laser pulses back to Earth ground stations to provide data for precise measurements of the Earth–Moon distance to determine Earth wobble about its axis, continental drift, lunar librations, etc. Arrays of 100 retroreflecting corners were flown on Apollos 11 and 14, and an array of 300 corners was flown on Apollo 15.

Lunar Surface Magnetometer
Measured the intrinsic remnant lunar magnetic field and the magnetic response of the Moon to large-scale solar and terrestrial magnetic fields. The electrical conductivity of the lunar interior was also determined from measurements of the Moon's response to magnetic field step-transients. Three boom-mounted sensors measured mutually-orthogonal components of the field

Lunar Ejecta and Meteorites
Three separate detectors which measured energy, speed, and direction of dust particles. Oriented east, west, and up. The dust particles measured were meteorites, secondary ejecta from meteorites, and, possibly, lunar surface particles levitated and accelerated by lunar surface phenomena.

Cold Cathode Ion Gauge
A separate experiment combined in an integrated package with the Suprathermal Ion Detector. It determined the density of neutral gas particles in the lunar atmosphere.

Passive Seismic
Detected Moon-quakes and meteoroid impacts to enable scientists to determine the Moon's internal composition.

Radioisotope Thermoelectric Generator
Supplied about 70 watts of electrical power for continuous day–night operation.

Lunar Surface Gravimeter
Measured and sensed changes in the vertical component of lunar gravity, using a spring mass suspension. It also provided data on the lunar tides.

---

[1] *Apollo Lunar Surface Experiments Package (ALSEP): Five Years of Lunar Science and Still Going Strong*, Bendix Aerospace.

Mission Insignias[1]

Project Apollo
The Project Apollo insignia was a disk circumscribed by a band displaying the words "Apollo" and "NASA." The center disk bore a large letter "A" with the constellation Orion positioned so that its three central stars formed the bar of the letter. To the right was the Earth, with the Moon in the upper left of the center disc. The Moon's face represented the mythical god Apollo. A double trajectory passed behind both spheres and through the central stars.

Apollo 1
The insignia for the first piloted Apollo flight depicted an Apollo spacecraft in Earth orbit. In the background were the stars and stripes of the U.S. flag. The crew members' names appeared in the inner border. The Moon appeared at the right, reminding us of the project goal.

Apollo 7
Symbolizing the Earth-orbital nature of the mission, a CSM circled the globe trailing an ellipse of orange flame. The background was navy-blue, symbolizing the depth of space. In the center, the Earth, with North and South America appearing against light blue oceans. The crew's names appeared in an arc at the bottom. A Roman numeral VII appeared in the Pacific region of the globe.

Apollo 8
The shape of the insignia symbolized the Apollo CM. The red figure 8 circled the Earth and Moon, representing not only the number of the mission but the translunar and transearth trajectories.

Apollo 9
Orbiting near the CM, the LM symbolized the first piloted flight of the spacecraft that would take humans to the lunar surface. A Saturn V was at the left. The crew names appeared around the top of the insignia, and the mission name appeared along the bottom. The 'D' in McDivitt had a red interior identifying this as the "D" mission in the Apollo series.

Apollo 10
The shield-shaped insignia was based more on mechanics than on mission goals. The three-dimensional Roman numeral X identified the mission and gave the effect of sitting on the Moon. The CM circled the Moon as the LM made its low pass over the surface, with the Earth in the background. Although Apollo 10 did not land, the prominence of the X indicated the mission would make a significant contribution to the Apollo program.

Apollo 11
The American eagle, symbolic of the United States, was about to land on the Moon. In its talons, an olive branch indicated the crew "came in peace for all mankind." The Earth, the place from which the crew came and would return safely in order to fulfill President John F. Kennedy's challenge to the nation, rested on a field of black, representing the vast unknown of space.

Apollo 12
An American clipper ship and blue-and-gold motif signified an all-Navy crew and related the era of the clipper ship to the era of space flight. As the clipper brought foreign shores closer to the U.S., and marked our increased utilization of the seas, spacecraft opened the way to other planets. Apollo 12 marked the increased utilization of space based on knowledge gained in earlier missions. The portion of the Moon shown represented the Ocean of Storms area in which Apollo would land. The four stars represented the crew and C. C. Williams, original LMP who died in an air crash.

Apollo 13
Apollo, the sun god of Greek mythology, was represented as the sun, with three horses driving his chariot across the surface of the Moon, symbolizing how the Apollo flights have extended the light of knowledge to all mankind. The Latin phrase "Ex Luna, Scientia" means "From the Moon, Knowledge."

Apollo 14
The Apollo 14 insignia featured the astronaut insignia approaching the Moon and leaving a comet trail from the liftoff point on Earth. The mission name and crew name appeared in the border

Apollo 15
Three stylized birds, or symbols of flight, representing the Apollo 15 crew, were superimposed over an artist's concept of the landing site, next to the Hadley Rille at the foot of the Lunar Apennines. To the right of the symbols, was an "XV", signifying the mission number.

Apollo 16
Resting on a gray field representing the lunar surface, the American eagle and red, white and blue striped shield paid tribute to the people of the United States. Crossing the shield while orbiting the Moon was a gold NASA vector. Sixteen stars, representing the mission number, and the crew names, appeared on a blue border, outlined in gold.

Apollo 17
The insignia was dominated by the image of Apollo, the Greek sun god. Suspended in space behind the head of Apollo was an American eagle of contemporary design, the red bars of the eagle's wing represented the bars in the U.S. flag; the three white stars symbolized the three astronaut crewmen. The background was deep blue space and within it were the Moon, the planet Saturn and a spiral galaxy or nebula. The Moon was partially overlaid by the eagle's wing suggesting that this was a celestial body that man has visited and in that sense conquered. The thrust of the eagle and the gaze of Apollo to the right and toward Saturn and the galaxy was meant to imply that man's goals in space would someday include the planets and perhaps the stars. The colors of the emblem were red, white and blue, the colors of the U.S. flag; with the addition of gold, to symbolize the golden age of space flight that would begin with this Apollo 17 lunar landing. The Apollo image used in this insignia was the Apollo of Belvedere sculpture that was in the Vatican Gallery in Rome. This emblem was designed by artist Robert T. McCall in collaboration with the astronauts.

[1] Excerpted and edited from *Astronaut Mission Patches and Spacecraft Callsigns*, by Dick Lattimer, unpublished draft in JSC History Office; *Space Patches From Mercury to the Space Shuttle*; and various NASA documents.

Launch Vehicle Propellant Usage[1]

	Burn Start	Burn End	Burn Time	Burn Rate lb/sec	Burn Start	Burn End	Burn Time	Burn Rate lb/sec	Burn Start	Burn End	Burn Time	Burn Rate lb/sec	Burn Start	Burn End	Burn Time	Burn Rate lb/sec
	Apollo 7	Apollo 7	Apollo 7	Apollo 7	Apollo 8	Apollo 8	Apollo 8	Apollo 8	Apollo 9	Apollo 9	Apollo 9	Apollo 9	Apollo 10	Apollo 10	Apollo 10	Apollo 10
S-IB Burn (sec)	–2.988	144.32	147.31	–	–	–	–	–	–	–	–	–	–	–	–	–
Oxidizer (LOX), lb	631,300	3,231	628,069	4,263.6	–	–	–	–	–	–	–	–	–	–	–	–
Fuel (RP-1), lb	276,900	4,728	272,172	1,847.6	–	–	–	–	–	–	–	–	–	–	–	–
Total, lb	908,200	7,959	900,241	6,111.3	–	–	–	–	–	–	–	–	–	–	–	–
S-IC Burn (sec)	–	–	–	–	–6.585	153.82	160.41	–	–6.3	162.76	169.06	–	–6.4	161.63	168.03	–
Oxidizer (LOX), lb	–	–	–	–	3,128,034	46,065	3,081,969	19,213.7	3,301,203	45,230	3,255,973	19,259.3	3,302,827	40,592	3,262,235	19,414.6
Fuel (RP-1)	–	–	–	–	1,357,634	26,622	1,331,012	8,297.8	1,431,678	42,390	1,389,288	8,217.7	1,423,254	28,537	1,394,717	8,300.4
Total, lb	–	–	–	–	4,485,668	72,687	4,412,981	27,511.5	4,732,881	87,620	4,645,261	27,477.0	4,726,081	69,129	4,656,952	27,715.0
S-II Burn (sec)	–	–	–	–	156.19	524.04	367.85	–	165.16	536.22	371.06	–	164.05	552.64	388.59	388.59
Oxidizer (LOX), lb	–	–	–	–	793,795	5,169	788,626	2,143.9	821,504	3,230	818,274	2,205.2	823,325	3,536	819,789	2,109.7
Fuel (LH_2), lb	–	–	–	–	154,907	4,514	150,393	408.8	158,663	3,381	155,282	418.5	158,541	4,622	153,919	396.1
Total, lb	–	–	–	–	948,702	9,683	939,019	2,552.7	980,167	6,611	973,556	2,623.7	981,866	8,158	973,708	2,505.7
S-IVB 1st Burn (sec)	146.97	616.76	469.79		528.29	684.98	156.69	–	540.82	664.66	123.84	–	556.81	703.76	146.95	–
Oxidizer (LOX), lb	193,330	1,671	191,659	408.0	192,840	132,220	60,620	386.9	189,686	133,421	56,265	454.3	192,089	133,883	58,206	396.1
Fuel (LH_2), lb	39,909	2,502	37,407	79.6	43,395	30,678	12,717	81.2	43,709	32,999	10,710	86.5	43,388	31,564	11,824	80.5
Total, lb	233,239	4,173	229,066	487.6	236,235	162,898	73,337	468.0	233,395	166,420	66,975	540.8	235,477	165,447	70,030	476.6
S-IVB 2nd Burn (sec)	–	–	–	–	10,237.79	10,555.51	317.72	–	17,155.54	17,217.60	62.06	–	9,207.52	9,550.58	343.06	343.06
Oxidizer (LOX), lb	–	–	–	–	131,975	8,064	123,911	390.0	132,988	109,298	23,690	381.7	133,471	5,274	128,197	373.7
Fuel (LH_2), lb	–	–	–	–	28,358	2,759	25,599	80.6	29,369	24,476	4,893	78.8	29,116	2,177	26,939	78.5
Total, lb	–	–	–	–	160,333	10,823	149,510	470.6	162,357	133,774	28,583	460.6	162,587	7,451	155,136	452.2
S-IVB 3rd Burn (sec)	–	–	–	–	–	–	–	–	22,039.26	22,281.32	242.06	–	–	–	–	–
Oxidizer (LOX), lb	–	–	–	–	–	–	–	–	108,927	34,051	74,876	309.3	–	–	–	–
Fuel (LH_2), lb	–	–	–	–	–	–	–	–	23,520	8,951	14,569	60.2	–	–	–	–

Total, lb	–	–	–	–	–	–	–	–	132,447	43,002	89,445	369.5	–	–	–	–
Oxidizer–Fuel Ratio																
S-IB Stage	2.280	–	2.308	–	–	–	–	–	–	–	–	–	–	–	–	–
S-IC Stage	–	–	–	–	2.304	–	2.316	–	2.306	–	2.344	–	2.321	–	2.339	–
S-II Stage	–	–	–	–	5.124	–	5.244	–	5.178	–	5.270	–	5.193	–	5.326	–
S-IVB Stage 1st burn	4.844	–	5.124	–	4.444	–	4.767	–	4.340	–	5.254	–	4.427	–	4.923	–
S-IVB Stage 2nd burn	–	–	–	–	4.654	–	4.840	–	4.528	–	4.842	–	4.584	–	4.759	–
S-IVB Stage 3rd burn	–	–	–	–	–	–	–	–	4.631	–	5.139	–	–	–	–	–

	Apollo 11	Apollo 11	Apollo 11	Apollo 11	Apollo 12	Apollo 12	Apollo 12	Apollo 12	Apollo 13	Apollo 13	Apollo 13	Apollo 13	Apollo 14	Apollo 14	Apollo 14	Apollo 14
S-IC Burn (sec)	–6.4	161.63	168.03	–	–6.5	161.74	168.24	–	–6.7	163.60	170.30	–	–6.5	164.10	170.60	–
Oxidizer (LOX), lb	3,305,786	39,772	3,266,014	19,437.1	3,310,199	42,093	3,268,106	19,425.3	3,304,734	38,921	3,265,813	19,176.8	3,312,769	42,570	3,270,199	19,168.8
Fuel (RP–1), lb	1,424,889	30,763	1,394,126	8,296.9	1,424,287	36,309	1,387,978	8,250.0	1,431,384	27,573	1,403,811	8,243.2	1,428,561	32,312	1,396,249	8,184.3
Total, lb	4,730,675	70,535	4,660,140	27,734.0	4,734,486	78,402	4,656,084	27,675.2	4,736,118	66,494	4,669,624	27,420.0	4,741,330	74,882	4,666,448	27,353.2
S-II Burn (sec)	164.00	548.22	384.22	–	163.20	552.34	389.14	–	166.00	592.64	426.64	–	166.50	559.05	392.55	–
Oxidizer (LOX), lb	819,050	3,536	815,514	2,122.5	825,406	3,536	821,870	2,112.0	836,741	3,533	833,208	1,953.0	837,484	2,949	834,535	2,125.9
Fuel (LH_2), lb	158,116	10,818	147,298	383.4	157,986	4,610	153,376	394.1	159,931	4,532	155,399	364.2	159,232	3,232	156,000	397.4
Total, lb	977,166	14,354	962,812	2,505.9	983,392	8,146	975,246	2,506.2	996,672	8,065	988,607	2,317.2	996,716	6,181	990,535	2,523.3
S-IVB 1st Burn (sec)	552.20	699.33	147.13	–	556.60	693.91	137.31	–	596.90	749.83	152.93	–	563.40	700.56	137.16	–
Oxidizer (LOX), lb	192,497	135,144	57,353	389.8	190,587	135,909	54,678	398.2	191,890	132,768	59,122	386.6	190,473	136,815	53,658	391.2
Fuel (LH_2), lb	43,608	31,736	11,872	80.7	43,663	32,346	11,317	82.4	43,657	31,455	12,202	79.8	43,546	32,605	10,941	79.8
Total, lb	236,105	166,880	69,225	470.5	234,250	168,255	65,995	480.6	235,547	164,223	71,324	466.4	234,019	169,420	64,599	471.0
S-IVB 2nd Burn (sec)	9,856.20	10,203.03	346.83	–	10,042.80	10,383.94	341.14	–	9,346.30	9,697.15	350.85	–	8,912.40	9,263.24	350.84	–
Oxidizer (LOX), lb	134,817	5,350	129,467	373.3	135,617	4,659	130,958	383.9	132,525	3,832	128,693	366.8	136,551	5,812	130,739	372.6
Fuel (LH_2), lb	29,324	2,112	27,212	78.5	29,804	2,109	27,695	81.2	29,367	1,963	27,404	78.1	30,428	2,672	27,756	79.1
Total, lb	164,141	7,462	156,679	451.7	165,421	6,768	158,653	465.1	161,892	5,795	156,097	444.9	166,979	8,484	158,495	451.8
Oxidizer–Fuel Ratio																
S-IC Stage	2.320	–	2.343	–	2.324	–	2.355	–	2.309	–	2.326	–	2.319	–	2.342	–
S-II Stage	5.180	–	5.536	–	5.225	–	5.359	–	5.232	–	5.362	–	5.260	–	5.350	–
S-IVB Stage 1st burn	4.414	–	4.831	–	4.365	–	4.831	–	4.395	–	4.845	–	4.374	–	4.904	–
S-IVB Stage 2nd burn	4.597	–	4.758	–	4.550	–	4.729	–	4.513	–	4.696	–	4.488	–	4.710	–

	Burn Start	Burn End	Burn Time	Burn Rate lb/sec	Burn Start	Burn End	Burn Time	Burn Rate lb/sec	Burn Start	Burn End	Burn Time	Burn Rate lb/sec	Burn Start	Burn End	Burn Time	Burn Rate lb/sec
	Apollo 15	**Apollo 15**	**Apollo 15**	**Apollo 15**	**Apollo 16**	**Apollo 16**	**Apollo 16**	**Apollo 16**	**Apollo 17**	**Apollo 17**	**Apollo 17**	**Apollo 17**	**Program totals**	**Program totals**	**Program totals**	**Program totals**
S-IC Burn (sec)	–6.5	159.56	166.06	–	–6.7	161.78	168.48	–	–6.9	161.20	168.10	–	–	–	1,677.31	–
Oxidizer (LOX), lb	3,312,030	31,135	3,280,895	19,757.3	3,311,226	34,028	3,277,198	19,451.6	3,314,388	36,479	3,277,909	19,499.8	32,903,196	396,885	32,506,311	19,380.1
Fuel (RP-1), lb	1,410,798	27,142	1,383,656	8,332.3	1,439,894	31,601	1,408,293	8,358.8	1,431,921	26,305	1,405,616	8,361.8	14,204,300	309,554	13,894,746	8,284.0
Total, lb	4,722,828	58,277	4,664,551	28,089.6	4,751,120	65,629	4,685,491	27,810.4	4,746,309	62,784	4,683,525	27,861.5	47,107,496	706,439	46,401,057	27,664.1
S-II Burn (sec)	163.00	549.06	386.06	–	165.20	559.54	394.34	–	164.60	559.66	395.06	–	–	–	3,895.51	–
Oxidizer (LOX), lb	837,991	3,109	834,882	2,162.6	846,157	3,141	843,016	2,137.8	844,094	3,137	840,957	2,128.7	8,285,547	34,876	8,250,671	2,118.0
Fuel (LH_2), lb	158,966	4,022	154,944	401.3	160,551	2,884	157,667	399.8	160,451	3,024	157,427	398.5	1,587,344	45,639	1,541,705	395.8
Total, lb	996,957	7,131	989,826	2,563.9	1,006,708	6,025	1,000,683	2,537.6	1,004,545	6,161	998,384	2,527.2	9,872,891	80,515	9,792,376	2,513.8
S-IVB 1st Burn (sec)	553.20	694.67	141.47	–	563.60	706.21	142.61	–	563.80	702.65	138.85	–	–	–	1,424.94	–
Oxidizer (LOX), lb	195,788	140,293	55,495	392.3	195,372	138,937	56,435	395.7	195,636	140,047	55,589	400.4	1,926,858	1,359,437	567,421	398.2
Fuel (LH_2), lb	43,674	32,416	11,258	79.6	43,727	32,081	11,646	81.7	43,752	32,685	11,067	79.7	436,119	320,565	115,554	81.1
Total, lb	239,462	172,709	66,753	471.9	239,099	171,018	68,081	477.4	239,388	172,732	66,656	480.1	2,362,977	1,680,002	682,975	479.3
S-IVB 2nd Burn (sec)	10,202.90	10,553.61	350.71	–	9,216.50	9,558.42	341.92	–	11,556.60	11,907.64	351.04	–	–	–	3,156.17	–
Oxidizer (LOX), lb	139,665	4,273	135,392	386.1	138,532	3,869	134,663	393.8	139,879	4,219	135,660	386.5	1,356,020	154,650	1,201,370	380.6
Fuel (LH_2), lb	29,799	1,722	28,077	80.1	29,968	2,190	27,778	81.2	30,050	2,212	27,838	79.3	295,583	44,392	251,191	79.6
Total, lb	169,464	5,995	163,469	466.1	168,500	6,059	162,441	475.1	169,929	6,431	163,498	465.8	1,651,603	199,042	1,452,561	460.2
Oxidizer-Fuel Ratio																
S-IC Stage	2.348	–	2.371	–	2.300	–	2.327	–	2.315	–	2.332	–	2.316	–	2.339	–
S-II Stage	5.272	–	5.388	–	5.270	–	5.347	–	5.261	–	5.342	–	5.220	–	5.352	–
S-IVB Stage 1st burn	4.483	–	4.929	–	4.468	–	4.846	–	4.471	–	5.023	–	4.418	–	4.910	–
S-IVB Stage 2nd burn	4.687	–	4.822	–	4.623	–	4.848	–	4.655	–	4.873	–	4.588	–	4.783	–

[1] All times are referenced to Range Zero; all other values represent actual usage, in pounds mass. Sources are the Saturn V launch vehicle flight evaluation reports and *Results of the Fifth Saturn IB Vehicle Test Flight (Apollo 7)*.

Earth Orbit and Lunar Orbit Experiments[1]

Designation	Experiment	Apollo 8	Apollo 11	Apollo 12	Apollo 14	Apollo 15	Apollo 16	Apollo 17
S-151	Cosmic Ray Detector (Helmets)	X		X				
S-158	Multispectral Photography			X				
S-160	Gamma-Ray Spectrometer					X	X	X
S-161	X-Ray Fluorescence					X	X	
S-162	Alpha-Particle Spectrometer					X	X	
S-164	S-Band Transponder (CSM/LM)				X	X	X	X
S-164	S-Band Transponder (Subsatellite)					X	X	
S-165	Mass Spectrometer					X	X	
S-169	Far-Ultraviolet Spectrometer							X
S-170	Bistatic Radar				X	X	X	
S-171	Infrared Scanning Radiometer							X
S-173[2]	Particle Shadows/boundary Layer					X	X	
S-174	Magnetometer					X	X	
S-176	Command Module Window Meteoroid				X	X	X	X
S-177	Ultraviolet Photography, Earth and Moon					X	X	
S-178	Gegenschein from Lunar Orbit				X	X	X	
S-209	Lunar Sounder							X
–	Candidate Exploration Sites			X	X			
–	CM Orbital Science Photography				X			
–	CM Photographic Tasks					X	X	X
–	Dim Light Photography				X			
–	Lunar Mission Photography From CM	X		X	X			
–	Selenodetic Reference Point Update			X	X			
–	SM Orbital Photographic Tasks[3]					X	X	X
–	Transearth Lunar Photography				X			
–	Visual Observations From Lunar Orbit					X	X	X

[1] *Apollo Program Summary Report* (JSC-09423).

[2] Experiments S-173 and S-174 were Particles and Fields Subsatellite experiments.

[3] Included panoramic camera photography, mapping camera photography, and laser altimetry. Also supported geologic objectives.

Launch Vehicle/Spacecraft Key Facts[1]

	Apollo 7	Apollo 8	Apollo 9	Apollo 10	Apollo 11	Apollo 12	Apollo 13	Apollo 14	Apollo 15	Apollo 16	Apollo 17
First Stage (S-IB)											
Contractor	Chrysler	–	–	–	–	–	–	–	–	–	–
Diameter, base, ft	21.500	–	–	–	–	–	–	–	–	–	–
Diameter, top, ft	21.667	–	–	–	–	–	–	–	–	–	–
Height, ft	80.200	–	–	–	–	–	–	–	–	–	–
Engines, type/number	H-1/8	–	–	–	–	–	–	–	–	–	–
Fuel	RP-1	–	–	–	–	–	–	–	–	–	–
Oxidizer	LO_2										
Rated thrust each engine, lbf	200.000	–	–	–	–	–	–	–	–	–	–
Rated thrust total, lbf	1,600,000	–	–	–	–	–	–	–	–	–	–
Thrust at 35 to 38 sec, lbf	1,744,400	–	–	–	–	–	–	–	–	–	–
First Stage (S-IC)											
Contractor	–	Boeing	Boeing	Boeing	Boeing	Boeing	Boeing	Boeing	Boeing	Boeing	Boeing
Diameter, base, ft	–	33.000	33.000	33.000	33.000	33.000	33.000	33.000	33.000	33.000	33.000
Diameter, top, ft	–	33.000	33.000	33.000	33.000	33.000	33.000	33.000	33.000	33.000	33.000
Height, ft	–	138.030	138.030	138.030	138.030	138.030	138.030	138.030	138.030	138.030	138.030
Engines, type/number	–	F-1/5	F-1/5	F-1/5	F-1/5	F-1/5	F-1/5	F-1/5	F-1/5	F-1/5	F-1/5
Fuel	–	RP-1	RP-1	RP-1	RP-1	RP-1	RP-1	RP-1	RP-1	RP-1	RP-1
Oxidizer	–	LO_2	LO_2	LO_2	LO_2	LO_2	LO_2	LO_2	LO_2	LO_2	LO_2
Rated thrust each engine, lbf	–	1,500,000	1,522,000	1,522,000	1,522,000	1,522,000	1,522,000	1,522,000	1,522,000	1,522,000	1,522,000
Rated thrust total, lbf	–	7,500,000	7,610,000	7,610,000	7,610,000	7,610,000	7,610,000	7,610,000	7,610,000	7,610,000	7,610,000
Thrust at 35 to 38 sec, lbf	–	7,560,000	7,576,000	7,536,000	7,552,000	7,594,000	7,560,000	7,504,000	7,558,000	7,620,000	7,599,000
Second Stage (S-II)											
Contractor	–	North American Rockwell	North American Rockwell	North American Rockwell	North American Rockwell	North American Rockwell	North American Rockwell	North American Rockwell	North American Rockwell	North American Rockwell	North American Rockwell
Diameter, ft	–	33.000	33.000	33.000	33.000	33.000	33.000	33.000	33.000	33.000	33.000
Height, ft	–	81.500	81.500	81.500	81.500	81.500	81.500	81.500	81.500	81.500	81.500
Engines, type/number	–	J-2/5	J-2/5	J-2/5	J-2/5	J-2/5	J-2/5	J-2/5	J-2/5	J-2/5	J-2/5
Fuel	–	LH_2	LH_2	LH_2	LH_2	LH_2	LH_2	LH_2	LH_2	LH_2	LH_2

Oxidizer	–	LO_2	LO_2	LO_2	LO_2	LO_2	LO_2	LO_2	LO_2	LO_2	LO_2
Rated thrust each engine, lbf	–	225,000	230,000	230,000	230,000	230,000	230,000	230,000	230,000	230,000	230,000
Rated thrust total, lbf	–	1,125,000	1,150,000	1,150,000	1,150,000	1,150,000	1,150,000	1,150,000	1,150,000	1,150,000	1,150,000
Thrust, engine start command +61 sec, lbf	–	1,143,578	1,155,611	1,159,477	1,155,859	1,161,534	1,160,767	1,164,464	1,169,662	1,163,534	1,156,694
Thrust, outboard engine cutoff, lbf	–	865,302	730,000	642,068	625,751	611,266	635,725	580,478	548,783	787,380	787,009
Third Stage (S-IVB) (2nd stage for Apollo 7)											
Contractor	McDonnell Douglas	McDonnell Douglas	McDonnell Douglas	McDonnell Douglas	McDonnell Douglas	McDonnell Douglas	McDonnell Douglas	McDonnell Douglas	McDonnell Douglas	McDonnell Douglas	McDonnell Dougla
Diameter, ft (base)	33.000	33.000	33.000	33.000	33.000	33.000	33.000	33.000	33.000	33.000	33.000
Diameter, ft (top)	21.667	21.667	21.667	21.667	21.667	21.667	21.667	21.667	21.667	21.667	21.667
Height, ft	58.400	58.630	58.630	58.630	58.630	58.630	58.630	58.630	58.630	58.630	58.630
Engines, type/number	J-2/1	J-2/1	J-2/1	J-2/1	J-2/1	J-2/1	J-2/1	J-2/1	J-2/1	J-2/1	J-2/1
Fuel	LH_2	LH_2	LH_2	LH_2	LH_2	LH_2	LH_2	LH_2	LH_2	LH_2	LH_2
Oxidizer	LO_2	LO_2	LO_2	LO_2	LO_2	LO_2	LO_2	LO_2	LO_2	LO_2	LO_2
Rated thrust total, lbf	200,000	230,000	230,000	230,000	230,000	230,000	230,000	230,000	230,000	230,000	230,000
Thrust, lbf – 1st burn	207,802	202,678	232,366	204,965	202,603	206,956	199,577	201,572	202,965	206,439	205,797
Thrust, lbf – 2nd burn	–	201,777	203,568	204,712	201,061	207,688	198,536	201,738	203,111	206,807	205,608
Thrust, lbf – 3rd burn	–	–	199,516	–	–	–	–	–	–	–	–
Instrument Unit (IU)											
Contractor	IBM	IBM	IBM	IBM	IBM	IBM	IBM	IBM	IBM	IBM	IBM
Diameter, ft	21.667	21.667	21.667	21.667	21.667	21.667	21.667	21.667	21.667	21.667	21.667
Height, ft	3.000	3.000	3.000	3.000	3.000	3.000	3.000	3.000	3.000	3.000	3.000
Service Module (SM)											
Contractor	North American Rockwell	North American Rockwell	North American Rockwell	North American Rockwell	North American Rockwell	North American Rockwell	North American Rockwell	North American Rockwell	North American Rockwell	North American Rockwell	North American Rockwell
Diameter, ft	12.833	12.833	12.833	12.833	12.833	12.833	12.833	12.833	12.833	12.833	12.833
Height (with engine bell), ft	24.583	24.583	24.583	24.583	24.583	24.583	24.583	24.583	24.583	24.583	24.583
Height (engine bell), ft	9.750	9.750	9.750	9.750	9.750	9.750	9.750	9.750	9.750	9.750	9.750
Fairing, ft	24.583	24.583	24.583	24.583	24.583	24.583	24.583	24.583	24.583	24.583	24.583
Main structure, ft	1.917	1.917	1.917	1.917	1.917	1.917	1.917	1.917	1.917	1.917	1.917

	Apollo 7	Apollo 8	Apollo 9	Apollo 10	Apollo 11	Apollo 12	Apollo 13	Apollo 14	Apollo 15	Apollo 16	Apollo 17
SPS nozzle structure	12.917	12.917	12.917	12.917	12.917	12.917	12.917	12.917	12.917	12.917	12.917
Weight, lb	19,730	51,258	36,159	51,371	51,243	51,105	51,105	51,744	54,063	54,044	54,044
Weight, dry, lb									13,470	13,450	13,450
Propellant, lb									40,593	40,594	40,594
Rated Thrust, SPS engine, lbf	20,500	20,500	20,500	20,500	20,500	20,500	20,500	20,500	20,500	20,500	20,500
Spacecraft/LM Adapter											
Contractor	Grumman	Grumman	Grumman	Grumman	Grumman	Grumman	Grumman	Grumman	Grumman	Grumman	Grumman
Minimum diameter, ft	12.833	12.833	12.833	12.833	12.833	12.833	12.833	12.833	12.833	12.833	12.833
Maximum diameter, ft	21.667	21.667	21.667	21.667	21.667	21.667	21.667	21.667	21.667	21.667	21.667
Height, ft	28.000	27.999	27.999	27.999	27.999	27.999	27.999	27.999	27.999	27.999	27.999
Upper jettisonable panels, ft	21.129	21.208	21.208	21.208	21.208	21.208	21.208	21.208	21.208	21.208	21.208
Lower fixed panels, ft	6.871	6.791	6.791	6.791	6.791	6.791	6.791	6.791	6.791	6.791	6.791
Lunar Module (LM)											
Contractor	Grumman	Grumman	Grumman	Grumman	Grumman	Grumman	Grumman	Grumman	Grumman	Grumman	Grumman
Overall											
Width, ft	–	–	31.000	31.000	31.000	31.000	31.000	31.000	31.000	31.000	31.000
Height, ft	–	–	22.917	22.917	22.917	22.917	22.917	22.917	22.917	22.917	22.917
Footpad diameter, ft	–	–	3.083	3.083	3.083	3.083	3.083	3.083	3.083	3.083	3.083
Sensing probe length, ft	–	–	5.667	5.667	5.667	5.667	5.667	5.667	5.667	5.667	5.667
Weight (lb)	–	(LTA) 19,900	32,034	30,735	33,278	33,562	33,493	33,685	36,238	36,237	36,262
LM Descent Stage	–	–									
Diameter, ft	–	–	14.083	14.083	14.083	14.083	14.083	14.083	14.083	14.083	14.083
Height, ft	–	–	10.583	10.583	10.583	10.583	10.583	10.583	10.583	10.583	10.583
Weight, dry, lb	–	–	4,265	4,703	4,483	4,875	4,650	4,716	6,179	6,083	6,155
Maximum rated thrust, lb	–	–	9,870	9,870	9,870	9,870	9,870	9,870	9,870	9,870	9,870
LM Ascent Stage	–	–									
Diameter, ft	–	–	14.083	14.083	14.083	14.083	14.083	14.083	14.083	14.083	14.083
Height, ft	–	–	12.333	12.333	12.333	12.333	12.333	12.333	12.333	12.333	12.333
Cabin volume, cu ft	–	–	235	235	235	235	235	235	235	235	235
Habitable volume, cu ft	–	–	160	160	160	160	160	160	160	160	160
Crew compartment height, ft	–	–	7.667	7.667	7.667	7.667	7.667	7.667	7.833	7.833	7.833
Crew compartment depth, ft	–	–	3.500	3.500	3.500	3.500	3.500	3.500	3.500	3.500	3.500

Weight, dry, lb	–	–	5,071	4,781	4,804	4,760	4,668	4,691	4,690	4,704	4,729
Maximum rated thrust, lb	–	–	2,524	1,650	3,218	3,224	N/A	3,218.2	3,225.6	3,224.7	3,234.8
Lunar Rover Vehicle (LRV)											
Contractor	–	–	–	–	–	–	–	–	Boeing	Boeing	Boeing
Length, ft	–	–	–	–	–	–	–	–	10.167	10.167	10.167
Width, ft	–	–	–	–	–	–	–	–	6.000	6.000	6.000
Wheel base, ft	–	–	–	–	–	–	–	–	7.500	7.500	7.500
Weight, lb	–	–	–	–	–	–	–	–	462	462	462
Payload capacity, lb	–	–	–	–	–	–	–	–	1,080	1,080	1,080
Command Module (CM)											
Contractor	North American Rockwell	North American Rockwell	North American Rockwell	North American Rockwell	North American Rockwell	North American Rockwell	North American Rockwell	North American Rockwell	North American Rockwell	North American Rockwell	North American Rockwell
Diameter, ft	12.833	12.833	12.833	12.833	12.833	12.833	12.833	12.833	12.833	12.833	12.833
Height, ft	11.417	11.417	11.417	11.417	11.417	11.417	11.417	11.417	11.417	11.417	11.417
Docking probe cone, ft	2.583	2.583	2.583	2.583	2.583	2.583	2.583	2.583	2.583	2.583	2.583
Main structure, ft	6.750	6.750	6.750	6.750	6.750	6.750	6.750	6.750	6.750	6.750	6.750
Aft/heat shield, ft	2.083	2.083	2.083	2.083	2.083	2.083	2.083	2.083	2.083	2.083	2.083
Weight, lb	12,659	12,392	12,405	12,277	12,250	12,365	12,365	12,831	12,831	12,874	12,874
Habitable volume, cu ft	210	210	210	210	210	210	210	210	210	210	210
Launch Escape System (LES)											
Contractor	North American Rockwell	North American Rockwell	North American Rockwell	North American Rockwell	North American Rockwell	North American Rockwell	North American Rockwell	North American Rockwell	North American Rockwell	North American Rockwell	North American Rockwell
Diameter, ft	4.000	4.000	4.000	4.000	4.000	4.000	4.000	4.000	4.000	4.000	4.000
Height, ft	33.460	33.460	33.460	33.460	33.460	33.460	33.460	33.460	33.460	33.460	33.460
Rocket motors (1 each)											
Thrust, LES, lb	155,000	147,000	147,000	147,000	147,000	147,000	147,000	147,000	147,000	147,000	147,000
Thrust, pitch control motor, lb	3,000	2,400	2,400	2,400	2,400	2,400	2,400	2,400	2,400	2,400	2,400
Thrust tower jettison motor, lb	33,000	31,500	31,500	31,500	31,500	31,500	31,500	31,500	31,500	31,500	31,500
Total Vehicle Height (ft)	**223.488**	**363.013**	**363.013**	**363.013**	**363.013**	**363.013**	**363.013**	**363.013**	**363.013**	**363.013**	**363.013**

[1] Compiled from Saturn launch vehicle flight evaluation reports. Thrust for S-IC stage is at sea level and for the S-II and S-IVB stages is at altitude. Thrust listed at "35 to 38 sec", "Engine Start Command +61 seconds", and at "Outboard Engine Cutoff" is actual thrust as flown. LM ascent and descent stages, LRV and CM dry weights are as published in mission press kits. All other weights are "as flown."

Launch Weather[1]

	Apollo 7	Apollo 8	Apollo 9	Apollo 10	Apollo 11	Apollo 12	Apollo 13	Apollo 14	Apollo 15	Apollo 16	Apollo 17
Surface Observations											
Pressure (lb/in)	14.765	14.804	14.642	14.779	14.798	14.621	14.676	14.652	14.788	14.769	14.795
Temperature (°F)	82.9	59.0	67.3	80.1	84.9	68.0	75.9	71.1	85.6	88.2	70.0
Relative Humidity	65%	88%	61%	75%	73%	92%	57%	86%	68%	44%	93%
Dew Point (°F)	70	56	53	72	75	65	60	67	74	62.6	68.0
Visibility (smi)	11.5	9.9	9.9	11.2	9.9	3.7	9.9	9.9	9.9	9.9	6.8
Surface Wind Conditions											
1st Level Wind Site (ft)	64.0	60.0	60.0	60.0	60.0	60.0	60.0	60.0	60.0	60.0	60.0
1st Level Wind Speed (ft/sec)	33.5	18.7	22.6	32.2	10.8	22.3	20.7	16.4	16.7	20.7	13.5
1st Level Wind Direction (deg)	090	348	160	142	175	280	105	255	156	269	005
2nd Level Wind Site (ft)	N/R[2]	N/R	N/R	N/R	N/R	N/R	N/R	530.0	530.0	530.0	530.0
2nd Level Wind Speed (ft/sec)	N/R	N/R	N/R	N/R	N/R	N/R	N/R	27.9	17.7	16.7	17.7
2nd Level Wind Direction (deg)	N/R	N/R	N/R	N/R	N/R	N/R	N/R	275	158	256	335
Cloud Coverage											
1st Level Cover	30%	40%	70%	40%	10%	100%/rain	40%	70%	70%	20%	20%
1st Level Type	Cumulo-nimbus	Cirrus	Strato-cumulus	Cumulus	Cumulus	Strato-cumulus	Alto-cumulus	Cumulus	Cirrus	Cumulus	Strato-cumulus
1st Level Altitude (ft)	2,100	N/R	3,500	2,200	2,400	2,100	19,000	4,000	25,000	3,000	26,000
2nd Level Cover	–	–	100%	20%	20%	–	100%	20%	–	–	50%
2nd Level Type	–	–	Altostratus	Altocumulus	Altocumulus	–	Cirrostratus	Altocumulus	–	–	Cirrus
2nd Level Altitude (ft)	–	–	9,000	11,000	15,000	–	26,000	8,000	–	–	26,000
3rd Level Cover	–	–	–	100%	90%	–	–	–	–	–	–
3rd Level Type	–	–	–	Cirrus	Cirrostratus	–	–	–	–	–	–
3rd Level Altitude (ft)	–	–	–	Unknown	Unknown	–	–	–	–	–	–
Maximum Wind Speed/Ascent											
Speed (ft/sec)	136.2	150.9	250.0	154	203	256	246	207	249.3	85.6	252.6
Altitude (ft)	172,000	108,300	38,480	295,276	183,727	180,446	256,562	193,570	182,900	38,880	145,996

Maximum Dynamic Pressure											
Ground Elapsed Time (sec)	75.5	78.9	85.5	82.6	83.0	81.1	81.3	81.0	82.0	86.0	82.5
Max q (lb/ft^2)	665.60	776.938	630.73	694.232	735.17	682.95	651.63	655.8	768.58	726.81	701.75
Altitude (ft)	39,903	44,062	45,138	43,366	44,512	42,133	40,876	40,398	44,971	47,122	42,847
Maximum Wind Conditions in High Dynamic Pressure Region											
Altitude (ft)	44,500	49,900	38,480	46,520	37,400	46,670	44,540	43,270	45,110	38,880	39,945
Wind Speed (ft/sec)	51.1	114.1	250.0	139.4	31.6	156.1	182.5	173.2	61.1	85.6	147.9
Wind Direction (deg)	309	284	264	270	297	245	252	255	063	257	311
Maximum Wind Components											
Pitch Plane – Pitch (ft/sec)	51.8	102.4	244.4	133.9	24.9	154.9	182.4	173.2	-58.4	85.3	114.2
Pitch Plane – Altitude (ft)	36,800	49,500	38,390	45,280	36,680	46,670	44,540	43,720	45,030	38,880	39,945
Yaw Plane – Yaw (ft/sec)	51.5	74.1	71.2	61.4	23.3	-64.0	49.2	81.7	24.0	41.0	95.8
Yaw Plane – Altitude (ft)	47,500	51,800	37,500	48,720	39,530	44,780	42,750	33,460	44,040	50,850	37,237
Maximum Shear Values (D h=1000 m)											
Pitch Plane Shear (sec^{-1})	0.0113	0.0103	0.0248	0.0203	0.0077	0.0183	0.0166	0.0201	0.0110	0.0095	0.0177
Pitch Plane Altitude (ft)	48,100	52,500	49,700	50,200	48,490	46,750	50,610	43,720	36,830	44,780	26,164
Yaw Plane Shear (sec^{-1})	0.0085	0.0157	0.0254	0.0125	0.0056	0.0178	0.0178	0.0251	0.0071	0.0114	0.0148
Yaw Plane Altitude (ft)	46,500	57,800	48,160	50,950	33,790	47,820	45,850	38,880	47,330	50,850	34,940
Maximum % Density Deviations											
Negative Deviation From PRA63[3]	–0.1	–0.7	–6.1	–1.0	–0.2	–7.6	–2.8	-5.0	None	-0.8	-0.0
Altitude (nmi)	4.32	4.32	7.56	4.32	4.45	8.50	7.69	7.69	None	4.86	0.00
Positive Deviation from PRA63	+1.3	+3.3	None	+3.3	+4.4	+1.2	+0.5	None	+4.2	+4.0	+1.7
Altitude (nmi)	5.80	8.50	None	7.56	7.69	5.67	8.64	None	7.56	8.64	7.02

[1] Compiled from Saturn launch vehicle reports, trajectory reconstruction reports, and *Summary of Atmospheric Data Observations For 155 Flights of MSFC/ABMA Related Aerospace Vehicles.*

[2] This measurement not used or not recorded at launch time.

[3] Patrick Air Force Base Reference Atmosphere, 1963

LM Lunar Landing[1]

	Apollo 10[2]	Apollo 11	Apollo 12	Apollo 13[3]	Apollo 14	Apollo 15	Apollo 16	Apollo 17
LM Lunar Landing Conditions								
PDI Burn Duration (sec)	–	756.39	717.0	–	764.61	739.2	734	721
Hover Time Remaining (sec)	–	45	103	–	68	103	102	117
Landing Site	Sea of Tranquility	Sea of Tranquility	Ocean of Storms	Fra Mauro	Fra Mauro	Hadley–Apennine	Plain of Descartes	Taurus–Littrow
Targeted Latitude (°N)	0.7333°	0.6833°	–2.9833°	–3.6167°	–3.6719°	26.0816°	–9.0002°	20.1639°
Targeted Longitude (°E)	23.6500°	23.7167°	–23.4000°	–17.5500°	–17.4627	3.6583°	15.5164°	30.7495°
Actual Landing Latitude (°N)	–	0.67408	–3.01239	–	–3.64530	26.13222	–8.97301	20.19080
Actual Landing Longitude (°E)	–	23.47297	–23.42157	–	–17.47136	3.63386	15.50019	30.77168
GET	–	102:45:39.9	110:32:36.2	–	108:15:11.40	104:42:29.3	104:29:35	110:21:58
KSC Date		20 Jul 1969	19 Nov 1969		05 Feb 1971	30 Jul 1971	20 Apr 1972	11 Dec 1972
GMT Date		20 Jul 1969	19 Nov 1969		05 Feb 1971	30 Jul 1971	21 Apr 1972	11 Dec 1972
KSC Time	–	04:17:39 p.m.	01:54:36 a.m.	–	04:18:13 a.m.	06:16:29 p.m.	09:23:35 p.m.	02:54:58 p.m.
Time Zone	–	EDT	EST	–	EST	EDT	EST	EST
GMT Time	–	20:17:39	06:54:36	–	09:18:13	22:16:29	02:23:35	19:54:58
Sun Angle (deg)	11.0	10.8	5.1	18.5	10.3	12.2	11.9	13.0
LM Surface Angle (deg)	–	4.5°tilt east; yaw 13°south	3°pitch up, 3.8°roll left	–	1°pitch down; 6.9°roll right; 1.4°yaw left in tilt of 11° from horizontal	6.9°pitch up; 8.6°roll left resulting	0°roll, 2.5° pitch up, slight yaw south	4 to 5° pitch up, 0°roll, near 0°yaw
LM Distance to Target (ft)		22,500 ft W of landing ellipse center	535 ft NW of Surveyor III	–	55 ft N; 165 ft E	1,800 ft NW	668 ft N; 197 ft W	656 ft

Distance to Seismic Stations (nmi)							
Apollo 12	–	–	–	98	641	641	–
Apollo 14	–	98	–	–	591	544	–
Apollo 15	–	641	–	591	–	604	–
Apollo 16	–	641	–	544	604	–	–
Azimuth to Seismic Stations (deg)							
Apollo 12	–	–	–	96	40	100	–
Apollo 14	–	276	–	–	33	101	–
Apollo 15	–	226	–	218	–	160	–
Apollo 16	–	276	–	277	342	–	–

[1] Compiled from mission reports and summary science reports. Actual landing site coordinates based on International Astronomical Union (IAU) Mean Earth Polar Axis coordinate system as described in the Journal of Geophysical Research, vol. 105, pages 20,227 to 20,280, 2000.

[2] Although not planned as a lunar landing mission, Apollo 10 flew over the area to be targeted by the first lunar landing mission.

[3] Data is for intended landing site; mission aborted.

Lunar Surface Experiments Package Arrays and Status[1]

Experiment	Principal Investigator	Apollo 11	Apollo 12	Apollo 14	Apollo 15	Apollo 16	Apollo 17
Array		EASEP	ALSEP A	ALSEP C	ALSEP A–2	ALSEP D	ALSEP E
Deploy Site Latitude		0.6735	–3.00942	–3.64398	26.13407	–8.97537	20.19209
Deploy Site Longitude		23.4730	–23.42458	–17.47748	–3.62981	15.49812	30.76492
Design Life (days)		14	365	365	365	365	730
Uplink Frequency (MHz)		2119.0	2119.0	2119.0	2119.0	2119.0	2119.0
Downlink Frequency (MHz)		2276.5	2278.5	2279.5	2278.0	2276.0	2275.5
Date Commanded On		21 Jul 1969	19 Nov 1969	05 Feb 1971	31 Jul 1971	21 Apr 1972	12 Dec 1972
Time Commanded On		04:40:39 GMT	14:21 GMT	17:23 GMT	18:37 GMT	19:38 GMT	02:53 GMT
Date Commanded Off		27 Aug 1969	30 Sep 1977	Failed Jan 1976	30 Sep 1977	30 Sep 1977	30 Sep 1977
Passive Seismic Experiment	Gary Latham, University of Texas	X	X	X	X	X	
Laser Ranging Retroreflector	J.E. Faller, Wesleyan University	100 corner		100 corner	300 corner		
Deploy Site Latitude		0.67337		–3.64421	26.13333		
Deploy Site Longitude		23.47293		–17.47880	3.62837		

Experiment	Principal Investigator	Apollo 11	Apollo 12	Apollo 14	Apollo 15	Apollo 16	Apollo 17
Lunar Surface Magnetometer	Palmer Dyal, Ames Research Center Charles Sonett, University of Arizona		Commanded off 14 Jun 1974	X	Commanded off 14 Jun 1974		
Solar Wind Composition (Exposure)	Conway W. Snyder, Jet Propulsion Laboratory	1 hr 17 min[2]	18 hr 42 min	21 hr 0 min	41 hr 8 min	45 hr 5 min	
Suprathermal Ion Detector Experiment	John Freeman, Rice University		X	X	X		
Heat Flow Experiment	Mark Langseth, Lamont-Doherty Geological Observatory, Columbia University				X	X	X
Charged Particle Lunar Environment Experiment	D. Reasoner, Rice University			X			
Cold Cathode Ion Gauge Experiment	Francis Johnson, University of Texas		X	X	X		
Active Seismic Experiment	Robert Kovach, Stanford University			X		X	
Lunar Seismic Profiling Experiment	Robert Kovach, Stanford University						X
Lunar Surface Gravimeter	Joseph Weber, University of Maryland						X
Lunar Mass Spectrometer	John H. Hoffman, University of Texas						X
Lunar Ejecta Meteoroid Experiment	Otto Berg, Goddard Space Flight Center						X
Dust Detector	James Bates, Manned Spacecraft Center	X	X	X	X		

[1] *Apollo Lunar Surface Experiments Package (ALSEP): Five Years of Lunar Science and Still Going Strong*, Bendix Aerospace. Coordinates based on IAU Mean Earth Polar Axis coordinate system, as described in Davies and Colvin, *Journal of Geophysical Research*, vol. 105, pages 20,227 – 20,280, 2000. Command dates and times and uplink/downlink frequencies provided by National Space Science Data Center (NSSDC) at the Goddard Space Flight Center (all other missions). Apollo 11 central station no longer accepted commands as of 27 Aug 1969.

[2] JSC-09423, pp. 3–54.

Launch Windows[1]

	Apollo 7	Apollo 8	Apollo 9	Apollo 10	Apollo 11	Apollo 12	Apollo 13	Apollo 14	Apollo 15	Apollo 16	Apollo 17
Launch Window Opening											
KSC Date	11 Oct 1968	21 Dec 1968	03 Mar 1969	18 May 1969	16 Jul 1969	14 Nov 1969	11 Apr 1970	31 Jan 1971	26 Jul 1971	16 Apr 1972	06 Dec 1972
KSC Time	11:00:00 am	07:50:22 am	11:00:00 am	12:49:00 pm	09:32:00 am	11:22:00 am	02:13:00 pm	03:23:00 pm	09:34:00 am	12:54:00 pm	09:53:00 pm
Time Zone	EDT	EST	EST	EDT	EDT	EST	EST	EST	EDT	EST	EST
GMT Date	11 Oct 1968	21 Dec 1968	03 Mar 1969	18 May 1969	16 Jul 1969	14 Nov 1969	11 Apr 1970	31 Jan 1971	26 Jul 1971	16 Apr 1972	07 Dec 1972
GMT Time	16:00:00	12:50:22	16:00:00	16:49:00	13:32:00	16:22:00	19:13:00	20:23:00	13:34:00	17:54:00	02:53:00
Launch Window Closing											
KSC Date	11 Oct 1968	21 Dec 1968	03 Mar 1969	18 May 1969	16 Jul 1969	14 Nov 1969	11 Apr 1970	31 Jan 1971	26 Jul 1971	16 Apr 1972	07 Dec 1972
KSC Time	03:00:00 pm	12:31:40 pm	02:15:00 pm	05:09:00 pm	01:54:00 pm	02:28:00 pm	05:36:00 pm	07:12:00 pm	12:11:00 pm	04:43:00 pm	01:31:00 am
Time Zone	EDT	EST	EST	EDT	EDT	EST	EST	EST	EDT	EST	EST
GMT Date	11 Oct 1968	21 Dec 1968	03 Mar 1969	18 May 1969	16 Jul 1969	14 Nov 1969	11 Apr 1970	01 Feb 1971	26 Jul 1971	16 Apr 1972	07 Dec 1972
GMT Time	20:00:00	17:31:40	19:15:00	21:09:00	17:54:00	19:28:00	22:36:00	00:12:00	16:11:00	21:43:00	06:31:00
Window Duration											
H:M:S	4:00:00	4:41:18	3:15:00	4:20:00	4:22:00	3:06:00	3:23:00	3:49:00	3:37:00	3:49:00	3:38:00
Minutes	240	281	195	260	262	186	203	229	217	229	218
Targeted Lunar Sun Elevation Angle (deg)	—	6.74	—	11.0	10.8	5.1	10.0	10.3	12.0	11.9	13.3

[1]Compiled from press kits, mission implementation plans, and mission reports.

Selected Mission Weights (lbs)[1]

	Apollo 7	Apollo 8	Apollo 9	Apollo 10	Apollo 11	Apollo 12	Apollo 13	Apollo 14	Apollo 15	Apollo 16	Apollo 17
CSM/LM at EOI	36,419	87,382	95,231	98,273	100,756.4	101,126.9	101,261.2	102,083.6	107,142	107,226	107,161
CSM/LM at Separation	–	–	–	94,063	96,566.6	–	–	–	–	–	–
CSM/LM at Transposition & Docking	–	–	91,055	94,243	96,767.5	97,119.8	97,219.4	98,037.2	103,105	103,175	103,167
CSM at Transposition & Docking	–	–	58,925	63,560	63,473.0	63,535.6	63,720.3	64,388.0	66,885	66,923	66,893
LM at Transposition & Docking	–	–	32,130	30,683	33,294.5	33,584.2	33,499.1	33,649.2	36,220	36,252	36,274
CSM/LM at 1st MCC Ignition	–	63,307	–	93,889	96,418.2	96,870.6	97,081.5	97,901.5	–	–	–
CSM/LM at 1st MCC Cutoff	–	–	–	93,413	96,204.2	96,401.2	96,851.1	–	–	–	–
CSM/LM Before Cryogenic Tank Anomaly	–	–	–	–	–	–	96,646.9	–	–	–	–
CSM/LM After Cryogenic Tank Anomaly	–	–	–	–	–	–	96,038.7	–	–	–	–
CSM/LM at 2nd MCC Ignition	–	62,845	–	–	–	–	95,959.9	97,104.1	–	–	–
CSM/LM at 2nd MCC Cutoff	–	–	–	–	–	–	95,647.1	–	–	–	–
CSM at TEI Ignition	–	45,931	–	37,254	36,965.7	34,130.6	95,424.0	34,554.4	35,899	38,697	36,394
CSM at TEI Cutoff	–	–	–	26,172	26,792.7	25,724.5	87,456.0	–	–	–	–
CSM at 3rd MCC Ignition	–	32,008	–	–	–	–	87,325.3	24,631.9	–	–	–
CSM at 3rd MCC Cutoff	–	–	–	–	–	–	87,263.3	–	–	–	–
CSM/LM at LOI Ignition	–	62,827	–	93,319	96,061.6	96,261.1	–	97,033.1	102,589	102,642	102,639
CSM/LM at LOI Cutoff	–	46,743	–	69,429	72,037.6	72,335.6	–	71,823.0	76,329	77,647	76,540
CSM/LM at Circularization Ignition	–	46,716	–	69,385	72,019.9	72,243.7	–	–	–	–	–
CSM/LM at Circularization Cutoff	–	–	–	68,455	70,905.9	71,028.4	–	–	–	–	–
CSM/LM at Descent Orbit Insertion	–	–	–	–	–	–	–	71,768.8	76,278	77,595	76,354
CSM/LM at Separation for Lunar Landing	–	–	–	68,238	70,760.3	70,897.3	–	70,162.3	74,460	76,590	74,762
CSM at Separation for Lunar Landing	–	–	–	37,072	37,076.8	36,911.8	–	36,036.4	37,742	39,847	37,991
LM at Separation for Lunar Landing	–	–	–	31,166	33,683.5	33,985.5	–	34,125.9	36,718	36,743	36,771
LM at Powered Descent Initiation	–	–	–	–	–	–	–	34,067.8	36,634	36,617	36,686
LM at Descent Orbit Insertion Ignition	–	–	–	31,137	33,669.6	33,971.8	–	–	–	–	–
LM at Descent Orbit Insertion Cutoff	–	–	–	30,903	33,401.6	33,719.3	–	–	–	–	–
LM at Lunar Landing	–	–	–	–	16,153.2	16,564.2	–	16,371.7	18,175	18,208	18,305
CSM at Plane Burn Time	–	–	–	–	–	–	–	35,610.4	37,219	38,994	37,464

CSM at Circularization Ignition	–	–	–	–	–	–	–	35,996.3	37,716	39,595	37,960
LM at Phasing Ignition	–	–	–	30,824	–	–	–	–	–	–	–
LM at Phasing Cutoff	–	–	–	30,283	–	–	–	–	–	–	–
LM at Fuel Depletion	–	–	5,616	5,243	–	–	–	–	–	–	–
CSM/LM Ascent Stage at Docking	–	–	36,828	44,930	42,585.4	41,071.8	–	39,906.8	41,754	44,318	41,914
CSM at Docking	–	–	26,895	36,995	36,847.4	35,306.2	–	34,125.5	35,928	38,452	36,036
LM Ascent Stage at Lunar Liftoff	–	–	–	–	10,776.6	10,749.6	–	10,779.8	10,915	10,949	10,997
LM Ascent Stage at Orbit Insertion for Docking	–	–	–	8,077	5,928.6	5,965.6	–	5,917.8	5,985	6,001	6,042
LM Ascent Stage at Terminal Phase Initiation	–	–	–	–	–	–	–	5,880.1	5,965	5,972	5,970
LM Ascent Stage After Staging	–	–	–	8,273	–	–	–	–	–	–	–
LM Ascent Stage at Coelliptic Sequence Initiation	–	–	–	8,052	5,881.5	5,885.9	–	–	–	–	–
LM Ascent Stage at Docking	–	–	9,933	7,935	5,738.0	5,765.6	–	5,781.3	5,826	5,866	5,878
CSM at After Post-Docking Jettison	–	–	27,139	–	37,100.5	35,622.9	–	34,596.3	36,407	38,992	36,619
LM Ascent Stage After Post-Docking Jettison	–	–	–	7,663	5,462.5	5,436.5	–	5,307.6	5,325	5,306	5,277
CSM (CSM/LM) at Subsatellite Jettison	–	–	–	–	–	–	–	–	36,019	38,830	–
CSM at 4th MCC Ignition	–	–	–	–	–	–	87,132.1	–	–	–	–
CSM at 4th MCC Cutoff	–	–	–	–	–	–	87,101.5	–	–	–	–
CSM at Pre-Entry Separation	23,435	31,768	24,183	25,095	26,656.5	25,444.2	–	24,375.0	26,323	27,225	26,659
CSM/LM Before CSM/LM Separation	–	–	–	–	–	–	87,057.3	–	–	–	–
CM/LM After CSM/LM Separation	–	–	–	–	–	–	37,109.7	–	–	–	–
SM After Pre-Entry Separation	11,071	19,589	11,924	12,957	14,549.1	13,160.7	–	11,659.9	13,358	14,199	13,507
CM After Pre-Entry Separation	12,364	12,179	12,259	12,138	12,107.4	12,283.5	12,367.6	12,715.1	12,965	13,026	13,152
CM at Entry	12,356	12,171	12,257	12,137	12,095.5	12,275.5	12,361.4	12,703.5	12,953	13,015	13,140
CM at Drogue Deployment	11,936	11,712	11,839	11,639	11,603.7	11,785.7	11,869.4	–	–	–	–
CM at Main Parachute Deployment	11,855	11,631	11,758	11,558	11,318.9	11,496.1	11,579.8	12,130.8	12,381	12,442	12,567
CM at Landing	11,409	10,977	11,094	10,901	10,873.0	11,050.2	11,132.9	11,481.2	11,731	11,995	12,120

[1] Compiled from mission reports. Apollo 7 and Apollo 8 did not have a LM. Apollo 13 includes CSM and LM until separation before Earth entry.

Saturn Stage Earth Impact[1]

	Apollo 7	Apollo 8	Apollo 9	Apollo 10	Apollo 11	Apollo 12	Apollo 13	Apollo 14	Apollo 15	Apollo 16	Apollo 17
S-IB Impact											
GET (sec)	560.2	–	–	–	–	–	–	–	–	–	–
Surface Range (nmi)	265.002	–	–	–	–	–	–	–	–	–	–
Geodetic Latitude (°N)	29.7605	–	–	–	–	–	–	–	–	–	–
Longitude (°E)	–75.7183	–	–	–	–	–	–	–	–	–	–
S-IC Impact											
GET (sec)	–	540.410	536.436	539.12	543.7	554.5	546.9	546.2	560.389	547.136	551.708
Surface Range (nmi)	–	353.462	346.635	348.800	357.1	365.200	355.300	351.700	368.800	351.600	356.6
Geodetic Latitude (°N)	–	30.2040	30.1830	30.188	30.212	30.273	30.177	29.835	29.4200	30.207	28.219
Longitude (°E)	–	–74.1090	–74.238	–74.207	–74.038	–73.895	–74.0650	–74.0420	–73.6530	–74.147	–73.8780
S-II Impact											
GET (sec)	–	1,145.106	1,205.346	1,217.89	1,213.7	1,221.6	1,258.1	1,246.3	1,143.912	1,202.390	1,146.947
Surface Range (nmi)	–	2,245.913	2,413.198	2,389.290	2,371.8	2,404.4	2,452.600	2,462.100	2,261.3	2,312.000	2292.800
Geodetic Latitude (°N)	–	31.8338	31.4618	31.522	31.535	31.465	31.320	29.049	26.975	31.726	20.056
Longitude (°E)	–	–37.2774	–34.0408	–34.512	–34.844	–34.214	–33.2890	–33.567	–37.924	–35.990	–39.6040
S-IVB Earth Impact											
GET	162:27:15	–	–	–	–	–	–	–	–	–	–
KSC Date	18 Oct 1968	–	–	–	–	–	–	–	–	–	–
GMT Date	18 Oct 1968	–	–	–	–	–	–	–	–	–	–
KSC Time	05:30:00 am	–	–	–	–	–	–	–	–	–	–
Time Zone	EDT	–	–	–	–	–	–	–	–	–	–
GMT Time	09:30:00 GMT	–	–	–	–	–	–	–	–	–	–
Latitude (°N)	–8.90	–	–	–	–	–	–	–	–	–	–
Longitude (°E)	081.6	–	–	–	–	–	–	–	–	–	–

[1] Theoretical impacts compiled from Saturn V launch vehicle flight evaluation reports, and Apollo/Saturn V post-flight trajectory reports. Impact date is same as launch date except for S-IVB stage, as indicated.

Translunar Injection[1]

	Apollo 8	Apollo 10	Apollo 11	Apollo 12	Apollo 13	Apollo 14	Apollo 15	Apollo 16	Apollo 17
GET	002:56:05.51	002:39:20.58	002:50:13.03	002:53:13.94	002:41:47.15	002:34:33.24	002:56:03.61	002:39:28.42	003:18:37.64
KSC Date	21 Dec 1968	18 May 1969	16 Jul 1969	14 Nov 1969	11 Apr 1970	31 Jan 1971	26 Jul 1971	16 Apr 1972	07 Dec 1972
GMT Date	21 Dec 1968	18 May 1969	16 Jul 1969	14 Nov 1969	11 Apr 1970	31 Jan 1971	26 Jul 1971	16 Apr 1972	07 Dec 1972
KSC Time	10:47:05 am	03:28:20 pm	12:22:13 pm	02:15:13 pm	04:54:47 pm	06:37:35 pm	12:30:03 pm	03:33:28 pm	03:51:37 am
Time Zone	EST	EDT	EDT	EST	EST	EST	EDT	EST	EST
GMT Time	15:47:05	19:28:20	16:22:13	19:15:13	21:54:47	23:37:35	16:30:03	20:33:28	08:51:37
Altitude (ft)	1,137,577	1,093,217	1,097,229	1,209,284	1,108,555	1,090,930	1,055,296	1,040,493	1,029,299
Altitude (nmi)	187.221	179.920	180.581	199.023	182.445	179.544	173.679	171.243	169.401
Earth Fixed Velocity (ft/sec)	34,140.1	34,217.2	34,195.6	34,020.5	34,195.3	34,151.5	34,202.2	34,236.6	34,168.3
Space-Fixed Velocity (ft/sec)	35,505.41	35,562.96	35,545.6	35,389.8	35,538.4	35,511.6	35,579.1	35,566.1	35,555.3
Geocentric Latitude (°N)	21.3460	–13.5435	9.9204	16.0791	–3.8635	–19.4388	24.8341	–11.9117	4.6824
Geodetic Latitude (°N)	21.477	–13.627	9.983	16.176	–3.8602	–19.554	24.9700	–11.9881	4.7100
Longitude (°E)	–143.9242	159.9201	–164.8373	–154.2798	167.2074	141.7312	–142.1295	162.4820	–53.1190
Flight Path Angle (deg)[2]	7.897	7.379	7.367	8.584	7.635	7.480	7.430	7.461	7.379
Heading Angle (°E of N)	67.494	61.065	60.073	63.902	59.318	65.583	73.173	59.524	118.110
Inclination (deg)	30.636	31.698	31.383	30.555	31.817	30.834	29.696	32.511	28.466
Descending Node (deg)	38.983	123.515	121.847	120.388	122.997	117.394	108.439	122.463	86.042
Eccentricity	0.97553	0.97834	0.97696	0.96966	0.9772	0.9722	0.9760	0.9741	0.9722
C3 (ft^2/sec^2)	–15,918,930	–14,084,265	–14,979,133	–19,745,586	–14,814,090	–18,096,135	–15,643,934	–16,881,439	–18,152,226

[1] Compiled from Saturn V launch vehicle flight evaluation reports and mission reports.

[2] Flight path angle and heading angle are 'space–fixed' for these measurements.

In-flight Medical Problems in Apollo Crews[1]

Symptom/Finding	Etiology	Cases
Barotitis	Barotrauma	1
Cardiac arrhythmia	Undetermined, possibly linked with potassium deficit	2
Dehydration	Reduced water intake during emergency	2
Dysbarism (bends)[2]	Undetermined	1
Excoriation, urethral meatus	Prolonged wearing of urine collection device	2
Eye irritation	Spacecraft atmosphere	4
	Fiberglass	1
Flatulence	Undetermined	3
Genito-urinary infection with prostatic congestion	Pseudomonas aeruginosa	1
Head cold	Undetermined	3
Headache	Spacecraft environment	1
Nasal stuffiness	Zero gravity	2
Nausea, vomiting	Labyrinthine	1
	Undetermined (possibly virus-related)	1
Pharyngitis	Undetermined	1
Rash, facial, recurrent inguinal	Contact dermatitis	1
	Prolonged wearing of urine collection device	11
Respiratory irritation	Fiberglass	1
Rhinitis	Oxygen, low relative humidity	2
Seborrhea	Activated by spacecraft environment	2
Shoulder strain	Lunar core drilling	1
Skin irritation	Biosensor sites	11
	Fiberglass	2
	Undetermined	1
Stomach awareness	Labyrinthine	6
Stomatitis	Aphthous ulcers	1
Subungual hemorrhages	Glove fit	5
Urinary tract infection	Undetermined	1

[1] *Biomedical Results of Apollo*, SP-368.

[2] Also occurred during Gemini 10; later incidences were reported by the same crewman five years after his Apollo mission.

LM Ascent Stage Propellant Status[1]

Weight (lbm)	Apollo 9	Apollo 10	Apollo 11	Apollo 12	Apollo 14	Apollo 15	Apollo 16	Apollo 17
Loaded								
Fuel	1,626	981	2,020	2,012	2,007.0	2,011.4	2,017.8	2,026.9
Oxidizer	2,524	1,650	3,218	3,224	3,218.2	3,225.6	3,224.7	3,234.8
Total	4,150	2,631	5,238	5,236	5,225.2	5,237.0	5,242.5	5,261.7
Transferred from RCS								
Fuel	–	–	–	–	–	–	16.0	–
Oxidizer	–	–	–	–	–	–	44.0	–
Total	–	–	–	–	–	–	60.0	–
Consumed by RCS								
Fuel	22	13.9	23	31	–	–	–	–
Oxidizer	44	28.0	46	62	–	–	–	–
Total	66	41.9	69	93	–	–	–	–
Consumed by APS Prior to Jettison								
Fuel	31	67	1,833	1,831	–	–	–	–
Oxidizer	59	108	2,934	2,943	–	–	–	–
Total	90	175	4,767	4,774	–	–	–	–
Remaining at Jettison								
Fuel	–	–	164	150	128.0	118.0	164.0	108.9
Oxidizer	–	–	238	219	204.2	173.0	257.7	175.6
Total	–	–	402	369	332.2	291.0	421.7	284.5
Consumed at Fuel Depletion								
Fuel	–	13	–	–	–	–	–	–
Oxidizer	–	106	–	–	–	–	–	–
Total	–	119	–	–	–	–	–	–
Consumed at Oxidizer Depletion								
Fuel	68	–	–	–	–	–	–	–
Oxidizer	0	–	–	–	–	–	–	–
Total	68	–	–	–	–	–	–	–
Total Consumed								
Fuel	1,558	887	1,856	1,862	1,879.0	1,893.4	1,869.8	1,918.0
Oxidizer	2,524	1,408	2,980	3,005	3,014.0	3,052.6	3,011.0	3,059.2
Total	4,082	2,295	4,836	4,867	4,893.0	4,946.0	4,880.8	4,977.2

[1] Compiled from mission reports.

LM Descent Stage Propellant Status[1]

Weight (lbm)	Apollo 9	Apollo 10	Apollo 11	Apollo 12	Apollo 13	Apollo 14	Apollo 15	Apollo 16	Apollo 17
Loaded									
Fuel	6,977	7,009.5	6,975	7,079	7,083.6	7,072.8	7,537.6	7,530.4	7,521.7
Oxidizer	11,063	11,209.2	11,209	11,350	11,350.9	11,344.4	12,023.9	12,028.9	12,042.5
Total	18,040	18,218.7	18,184	18,429	18,434.5	18,417.2	19,561.5	19,559.3	19,564.2
Consumed									
Fuel	4,127	295.0	6,724	6,658	3,225.5	6,812.8	7,058.3	7,105.4	7,041.3
Oxidizer	6,524	470.0	10,690	10,596	5,117.4	10,810.4	11,315.0	11,221.9	11,207.6
Total	10,651	765.0	17,414	17,254	8,342.9	17,623.2	18,373.3	18,327.3	18,248.9
Remaining at Cutoff									
Fuel	–	–	251	421	–	260.0	479	425	480.0
Oxidizer	–	–	519	754	–	534.0	709	807	835.0
Total	–	–	770	1,175	–	794.0	1,188	1,232	1,315.0
Usable at Cutoff									
Fuel	–	–	216	386	–	228.0	433	396	455.0
Oxidizer	–	–	458	693	–	400.0	622	732	770.0
Total	–	–	674	1,079	–	628.0	1,055	1,128	1,225.0
Remaining at Cutoff (No Landing)									
Fuel	2,850	6,714.5	–	–	3,858.1	–	–	–	–
Oxidizer	4,539	10,739.2	–	–	6,233.5	–	–	–	–
Total	7,389	17,453.7	–	–	10,091.6	–	–	–	–

[1]Compiled from mission reports.

Lunar Subsatellites[1]

	Apollo 15	Apollo 16
Designations		
International	1971 063D	1972 031D
NORAD	05377	06009
Deploy Conditions		
GET	222:39:29.1	196:02:09
KSC Date	04 Aug 1971	24 Apr 1972
GMT Date	04 Aug 1971	24 Apr 1972
KSC Time	04:13:29 p.m.	04:56:09 p.m.
KSC Time Zone	EDT	EST
GMT Time	20:13:29	21:56:09
CM Revolution at Deploy	74	62
Weight (lbm)	78.5	90
Apolune (nmi)	76.3	66
Perilune (nmi)	55.1	52
Inclination (deg)	151.28	169.2810
Period (min)	119.75	119
Flight Path Angle (deg)	–0.60	–0.41
Heading Angle (deg)	–41.78	–79.43
Eccentricity	0.00935	0.0108
Weight (lbm)	79	93
Status	Selenocentric orbit, 1984	Impacted lunar surface
GET (h:m)	[Unknown]	1,034:37
KSC Date	[Unknown]	29 May 1972
GMT Date	22 January 1973	
(ground support terminated)	29 May 1972	
KSC Time	[Unknown]	03:31 p.m. EDT
GMT Time	[Unknown]	20:31
Revolutions	[Unknown]	425
Lunar Impact Latitude (°N)	[Unknown]	[Unknown]
Lunar Impact Longitude (°E)	[Unknown]	110

[1] Compiled from *Apollo 15 Preliminary Science Report* (SP-289); National Space Science Data Center (NSSDC) at the Goddard Space Flight Center, MD; *Apollo 16 Preliminary Science Report* (SP-315); and mission reports.

S-IVB Lunar Impact[1]

	Apollo 13	Apollo 14	Apollo 15	Apollo 16	Apollo 17
S-IVB Lunar Impact					
GET	077:56:40.0	082:37:53.4	079:24:42.9	075:08:04	086:59:42.3
KSC Date	14 Apr 1970	04 Feb 1971	29 Jul 1971	19 Apr 1972	10 Dec 1972
GMT Date	15 Apr 1970	04 Feb 1971	29 Jul 1971	19 Apr 1972	10 Dec 1972
KSC Time	08:09:40 p.m.	02:40:55 a.m.	04:58:42 p.m.	04:02:04 p.m.	03:32:42 p.m.
Time Zone	EST	EST	EDT	EST	EST
GMT Time	01:09:40.0	07:40:55.4	20:58:42.9	21:02:04	20:32:42.3
Weight (lb)	29,599	30,836	30,880	30,805	30,712
Velocity (ft/sec)	8,465	8,333	8,465	8,202	8,366
Energy (erg)	4.63×10^{17}	4.52×10^{17}	4.61×10^{17}	4.59×10^{17}	4.71×10^{17}
Angle From Horizontal (deg)	76	69	62	~79	55
Heading Angle (°N to W)	100.6	75.7	83.46	104.7	83
S-IVB Lunar Impact –Tumble Rate (deg/sec)	12	1	1	–	–
Selenocentric Latitude (°N)	–2.75	–8.09	–1.51	1.3	–4.21
Selenocentric Longitude (°E)	–27.86	–26.02	–11.81	–23.8	–12.31
Crater Diameter (calculated) (ft)	134.8	133.9	134.8	–	–
Crater Diameter (measured) (ft)	135.0	129.6	–	–	–
Distance To Target (nmi)	35.4	159	83	173	84
Distance To Seismic Stations (nmi)					
Apollo 12	73	93	192	71	183
Apollo 14	–	–	99	131	85
Apollo 15	–	–	–	593	557
Apollo 16	–	–	–	–	459
Azimuth To Seismic Stations (deg)					
Apollo 12	274	207	083	355	096
Apollo 14	–	–	069	308	096
Apollo 15	–	–	–	231	209
Apollo 16	–	–	–	–	278

[1] Compiled from Saturn V launch vehicle flight evaluation reports, Apollo mission preliminary science reports, and mission reports. Apollo 16 data based on seismic data due to loss of S-IVB tracking prior to impact; impact time is ± 4 seconds; impact site is $\pm 0.7°$ latitude and $\pm 0.3°$ longitude. Impact times for all vehicles are when impact signal was received on Earth.

S-IVB Solar Trajectory[1]

	Apollo 8	Apollo 9	Apollo 10	Apollo 11	Apollo 12
S-IVB Closest Approach To Moon					
GET	069:58:55.2	–	078:51:03.6	078:42	085:48
KSC Date	24 Dec 1968	–	21 May 1969	19 Jul 1969	18 Nov 1969
GMT Date	24 Dec 1968	–	21 May 1969	19 Jul 1969	18 Nov 1969
KSC Time	05:49:55 a.m.	–	07:40 p.m.	04:14 p.m.	01:10 a.m.
KSC Time Zone	EST	–	EDT	EDT	EST
GMT Time	10:49:55	–	23:40	20:14	06:10
Lunar Radius of Closest Approach (nmi)	1,620	–	2,619	2,763	4,020
Altitude Above Lunar Surface (nmi)	682	–	1,680	1,825	3,082
Velocity Increase Due To Lunar Gravity (nmi/sec)	0.79	–	0.459	0.367	0.296
S-IVB Solar Orbit Conditions					
Semi-Major Axis (nmi)	77,130,000	74,848,893	77,740,000	77,260,000	–
Eccentricity	–	0.07256	–	–	–
Aphelion (nmi)	79,770,000	80,280,052	82,160,000	82,000,000	–
Perihelion (nmi)	74,490,000	69,417,732	73,330,000	72,520,000	–
Inclination (deg)	23.47	24,390	23.46	0.3836[2]	–
Period (days)	340.8	325.8	344.88	342	–

[1] Compiled from Saturn V launch vehicle flight evaluation reports.

[2] Measured with respect to the ecliptic which is 23.5° to the Earth's equator.

Support Crews[1]

Apollo 7
Lt Cdr Ronald Ellwin Evans, USN
Maj William Reid Pogue, USAF
John Leonard Swigert, Jr

Apollo 8
Vance DeVoe Brand
Lt Cdr Thomas Kenneth Mattingly II, USN
Maj Gerald Paul Carr, USMC

Apollo 9
Maj Jack Robert Lousma, USMC
Lt Cdr Edgar Dean Mitchell, USN/ScD
Maj Alfred Merrill Worden, USAF

Apollo 10
Maj Joe Henry Engle, USAF
Lt Col James Benson Irwin, USAF
Maj Charles Moss Duke, Jr, USAF

Apollo 11
Lt Cdr Thomas Kenneth Mattingly II, USN
Lt Cdr Ronald Ellwin Evans, USN
Maj William Reid Pogue, USAF
John Leonard Swigert, Jr

Apollo 12
Maj Gerald Paul Carr, USMC
Cdr. Paul Joseph Weitz, USN
Edward George Gibson, PhD

Apollo 13
Maj Jack Robert Lousma, USMC
Vance DeVoe Brand
Maj William Reid Pogue, USAF

Apollo 14
Lt Cdr Bruce McCandless II, USN
Lt Col William Reid Pogue, USAF
Maj Charles Gordon Fullerton, USAF
Phillip Kenyon Chapman, ScD

Apollo 15
Karl Gordon Henize, PhD
Joseph Percival Allen IV, Ph. D
Robert Alan Ridley Parker, Ph. D

Apollo 16
Maj Donald Herod Peterson, USAF
Anthony Wayne England, PhD
Maj Henry Warren Hartsfield, Jr, USAF
Philip Kenyon Chapman, ScD

Apollo 17
Lt. Col. Robert Franklyn Overmyer, USMC
Robert Alan Ridley Parker, PhD
Maj Charles Gordon Fullerton, USAF

[1] Compiled from various documents and memoranda in the Rice University archives. For Apollo 7, Bill Pogue replaced Maj Edward Galen Givens, Jr, USAF, who was killed in an automobile accident in Pearland, TX on June 6, 1967. Military ranks are implied from available information and B. Hello (Rockwell) memo, 10 December 1969.

Glossary

ALSEP	Apollo Lunar Surface Experiments Package
Altitude	The perpendicular distance from the reference ellipsoid to the point of orbit intersect (measured either in feet or nmi)
Apogee	The predicted maximum altitude above the oblate Earth model (nmi)
APS	Ascent Propulsion System (of LM)
APS	Auxiliary Propulsion System (of S-IVB)
ARIA	Apollo Range Instrumentation Aircraft
AS	Apollo–Saturn mission designation
ASI	Augmented Spark Igniter (in a rocket engine)
ASPO	Apollo Spacecraft Program Office
ASPO	Apollo Spacecraft Project Office
BMAG	Body-Mounted Attitude Gyroscope
boilerplate	A research and development vehicle that simulates a production variant in size, shape, structure, mass and centre of gravity
BPC	Boost Protective Cover
BS	Bachelor of Science degree
CAPCOM	Capsule Communicator
CDR	Commander (of an Apollo spacecraft)
CM	Command Module (of an Apollo spacecraft)
CM-RCS	CM RCS
CMC	CM Computer
CMP	CM Pilot
CSM	Command and Service Module(s)
DAP	Digital Autopilot
DPS	Descent Propulsion System (of LM)
EASEP	Early Apollo Surface Experiments Package
ECS	Environmental Control System
EDS	Emergency Detection Subsystem
EDT	Eastern Daylight Time
ELS	Earth Landing System
EPS	Electrical Power System
EST	Eastern Standard Time
ETR	Eastern Test Range (at Cape Canaveral)

F-1	a rocket engine
G&N	Guidance and Navigation
Geodetic latitude	The spacecraft's position measured positive north, negative south, from the equator to the local vertical vector (deg)
Geodetic longitude	The spacecraft's position measured positive east from the Greenwich meridian to the local vertical vector (deg)
GET	Ground Elapsed Time
GH_2	Gaseous hydrogen
GMT	Greenwich Mean Time (Universal Time; UT)
GSE	Ground Support Equipment
H-1	a rocket engine
IMU	Inertial Measurement Unit
Inclination	The acute angle at the intersection of the orbit plane and the reference body's equatorial plane (deg)
IU	Instrument Unit
J-2	a rocket engine
KSC	Kennedy Space Center (in Florida)
Launch vehicle	A first stage booster and such upper stages as are required to put a payload on specific trajectory
LC	Launch Complex
LES	Launch Escape System
LET	Launch Escape Tower
LFA	LM Flight Article
LH_2	Liquid hydrogen
LM	Lunar Module (of an Apollo spacecraft)
LMP	LM Pilot
LOX	Liquid oxygen
LRV	Lunar Roving Vehicle
LTA	LM Test Article
LUT	Launch Umbilical Tower
LVDC	Launch Vehicle Digital Computer
Mach 1	The speed of sound
mascon	A concentration of mass at shallow depth in the lunar crust, mostly coinciding with mare-filled basins.
Max Q	The point in an ascent at which the launch vehicle and spacecraft are subjected to the most severe aerodynamic load
MBA	Master of Business Administration degree
MCC	Mission Control Center (in Houston)
MET	Modular Equipment Transport
MS	Master of Science degree
MSC	Manned Spacecraft Center (in Houston)
MSFC	Marshall Space Flight Center (in Huntsville)
MSFN	Manned Space Flight Network
MSS	Mobile Service Structure
NASA	National Aeronautics and Space Administration

nmi	nautical mile
Perigee	The predicted minimum altitude above the oblate Earth model (nmi)
Period	The time required for the spacecraft to complete 360 degrees of orbit rotation (min)
pogo	A longitudinal vibration, named after the child's pogostick toy
PGNCS	Primary Guidance, Navigation and Control System ('navigation' was the determination of position and velocity; 'guidance' was velocity vector control; and 'control' was control of rotational orientation about the center of gravity – i.e. attitude control)
PhD	Doctor of Philosophy degree
psi	pounds per square inch
psia	psi absolute
Q ball	an angle-of-attack sensor (on a rocket)
RCS	Reaction Control System
RP-1	Rocket Propellant (refined kerosene)
SA	Saturn–Apollo mission designation
ScD	Doctor of Science degree
SCS	Spacecraft Control System
S-I	The first stage of the Saturn I
S-IB	The first stage of the Saturn IB
S-IC	The first stage of the Saturn V
S-II	The second stage of the Saturn V
S-IV	The second stage of the Saturn I
S-IVB	The second stage of the Saturn IB and third stage of the Saturn V
SLA	Spacecraft LM Adapter (of S-IVB)
SM	Service Module (of an Apollo spacecraft)
SM-RCS	SM RCS
Space-fixed azimuth	The azimuth of the projection of the inertial velocity vector onto the local geocentric horizontal plane, measured positive eastward from north (deg)
Space-fixed flight path angle	The angle measured positive upward from the body-centerd local horizontal plane to the inertial velocity vector (deg)
Space-fixed heading angle	The angle of the projection of the inertial velocity vector onto the local body-centerd horizontal plane, measured positive eastward from north (deg)
Space-fixed velocity	The magnitude of the inertial velocity vector referenced to the body-centered inertial reference coordinate system (ft/sec)
Space vehicle	A term used in launch preparations, referring to the integrated launch vehicle and spacecraft
Spacecraft	A vehicle comprising a payload and all of the systems necessary for its mission in space
SPS	Service Propulsion System (of SM)
RTCC	Real-Time Computer Complex (in MCC)
US	United States
USAF	US Air Force

USMC	US Marine Corps
USN	US Navy
VAB	Vehicle Assembly Building

Bibliography

Akens, Davis S, editor, *Saturn Illustrated Chronology: Saturn's First Ten Years, April 1957 Through April 1967*, NASA George C. Marshall Space Flight Center, Alabama, August 1, 1968 (MHR-5)

Apollo 4 Mission Report, Prepared by Mission Evaluation Team, NASA Manned Spacecraft Center, Houston, Texas, January 7, 1968 (MSC-PA-R-68-1)

Apollo 5 Mission Report, Prepared by Mission Evaluation Team, NASA Manned Spacecraft Center, Houston, Texas, March 27, 1968 (MSC-PA-R-68-7)

Apollo 5 Mission Report, Supplement 2: Apollo Mission 5/AS-204/LM-1 Trajectory Reconstruction and Postflight Analysis, by TRW for Mission Planning and Analysis Division, NASA Manned Spacecraft Center, Houston, Texas, April 22, 1968 (TRW Note No. 68-FMT-642), and May 13,1968, MSC-PA-R-68-7, NASA-TM-X-72405)

Apollo 6 Entry Postflight Analysis, Landing Analysis Branch, Mission Planning and Analysis Division, NASA Manned Spacecraft Center, Houston, Texas, December 18, 1968 (MSC-68-FM-299/NASA-TM-X-69719) (N74-70888)

Apollo 6 Mission Report, Prepared by Mission Evaluation Team, NASA Manned Spacecraft Center, Houston, Texas, May/June 1968 (MSC-PA-R-68-9)

Apollo 6 Press Kit, Release #68-54K, NASA Public Affairs Office, Washington, D.C., March 21, 1968

Apollo 7 Mission Commentary, Prepared by Public Affairs Office, NASA Manned Spacecraft Center, Houston, Houston, Texas, October 1968

Apollo 7 Mission Report, Prepared by Apollo 7 Mission Evaluation Team, NASA Manned Spacecraft Center, Houston, Houston, Texas, December 1968 (MSC-PA-R-68-15)

Apollo 7 Press Kit, Release #68-168K, NASA Public Affairs Office, Washington, D.C., October 6, 1968

Apollo 8 Mission Report, Prepared by Mission Evaluation Team, NASA Manned Spacecraft Center, Houston, Houston, Texas, February 1969 (MSC-PA-R-69-1)

Apollo 8 Press Kit, Release #68-208, NASA Public Affairs Office, Washington, D.C., December 15, 1968

Apollo 9 Mission Report (MSC-PA-R-69-2 May 1969) (NTIS-34370)

Apollo 9 Press Kit, Release #69-29, NASA Public Affairs Office, Washington, D.C., February 23, 1969

Apollo 10 Mission Report (MSC-00126 November 1969)

Apollo 10 Press Kit, Release #69-68, NASA Public Affairs Office, Washington, D.C., May 7, 1969

Apollo 11 Mission Report (MSC-00171 November 1969/NASA-TM-X-62633) (NTIS N70-17401)

Apollo 11 Preliminary Science Report, Scientific and Technical Information Division, National Aeronautics and Space Administration, Washington, D.C., 1969 (NASA SP-214)

Apollo 11 Press Kit, Release #69-83K, NASA Public Affairs Office, Washington, D.C., July 6, 1969

Apollo 12 Mission Report (MSC-01855 March 1970)

Apollo 12 Preliminary Science Report, Scientific and Technical Information Division, office of Technology Utilization, National Aeronautics and Space Administration, Washington, D.C., 1970 (NASA SP-235)

Apollo 12 Press Kit, Release #69-148, NASA Public Affairs Office, Washington, D.C., November 5, 1969

Apollo 13 Mission Report (MSC-02680 September 1970/NASA-TM-X-66449) (NTIS N71-13037)

Apollo 13 Press Kit, Release #70-50K, NASA Public Affairs Office, Washington, D.C., April 2, 1970

Apollo 14 Mission Report (MSC-04112 May 1971)

Apollo 14 Preliminary Science Report, Scientific and Technical Information office, National Aeronautics and Space Administration, Washington, D.C., 1971 (NASA SP-272)

Apollo 14 Press Kit, Release #71-3K, NASA Public Affairs Office, Washington, D.C., January 21, 1971

Apollo 15 Mission Report (MSC-05161 December 1971/NASA-TM-X-68394) (NTIS N72-28832)

Apollo 15 Preliminary Science Report, Scientific and Technical Information office, National Aeronautics and Space Administration, Washington, D.C., 1972 (NASA SP-289)

Apollo 15 Press Kit, Release #71-119K, NASA Public Affairs Office, Washington, D.C., July 15, 1971

Apollo 16 Mission Report (MSC-07230 August 1972/NASA-TM-X-68635) (NTIS N72-33777)

Apollo 16 Preliminary Science Report, Scientific and Technical Information office, National Aeronautics and Space Administration, Washington, D.C., 1972 (NASA SP-315)

Apollo 16 Press Kit, Release #72-64K, NASA Public Affairs Office, Washington, D.C., April 6, 1972

Apollo 17 Mission Report (MSC-07904 March 1973) (NTIS N73-23844)

Apollo 17 Preliminary Science Report, Scientific and Technical Information office, National Aeronautics and Space Administration, Washington, D.C., 1973 (NASA SP-330)

Apollo 17 Press Kit, Release #72-220K, NASA Public Affairs Office, Washington, D.C., November 26, 1972

Apollo Lunar Surface Experiments Package (ALSEP): Five Years of Lunar Science and Still Going Strong, Bendix Aerospace

Apollo Program Summary Report, National Aeronautics and Space Administration, Lyndon B. Johnson Space Center, Houston, Texas, April 1975 (JSC-09423) (NTIS N75-21314)

Apollo/Saturn Postflight Trajectory (AS-502), Boeing Corporation Space Division, July 31, 1968 (D5-15773/NASA-CR-91983)

Apollo/Saturn Postflight Trajectory (AS-503), Boeing Corporation Space Division, February 19, 1969 (D-5-15794), (NASA-CR-127240) (NTIS N92-70422)

Apollo/Saturn Postflight Trajectory (AS-504), Boeing Corporation Space Division, May 2, 1969 (D-5-15560-4), (NASA-CR-105771) (NTIS N69-77056)

Apollo/Saturn Postflight Trajectory (AS-505), Boeing Corporation Space Division, July 17, 1969 (D-5-15560-5), (NASA-CR-105770) (NTIS N69-77049)

Apollo/Saturn Postflight Trajectory (AS-506), Boeing Corporation Space Division, October 6, 1969 (D-5-15560-6), (NASA-CR-102306) (NTIS N92-70425)

Apollo/Saturn Postflight Trajectory (AS-507), Boeing Corporation Space Division, January 13, 1970 (D-5-15560-7), (NASA-CR-102476) (NTIS N92-70420)

Apollo/Saturn Postflight Trajectory (AS-508), Boeing Corporation Space Division, June 10, 1970 (D-5-15560-8), (NASA-CR-102792) (NTIS N92-70437)

Apollo/Saturn Postflight Trajectory (AS-509), Boeing Corporation Space Division, June 30, 1971 (D-5-15560-9), (NASA-CR-119870) (NTIS N92-70433)

Apollo/Saturn Postflight Trajectory (AS-510), Boeing Corporation Space Division, November 23, 1971 (D-5-15560-10), (NASA-CR-120464) (NTIS N74-77459)

Apollo/Saturn Postflight Trajectory (AS-511), Boeing Corporation Space Division, August 9, 1972 (D-5-15560-11), (NASA-CR-124129) (NTIS N73-72531)

Apollo/Saturn Postflight Trajectory (AS-512), Boeing Corporation Space Division, April 11, 1973 (D-5-15560-12), (NASA-CR-144080) (NTIS N76-19199)

Apollo/Skylab, ASTP and Shuttle Orbiter Major End Items, Final Report (JSC-03600), March 1978

AS-202 Press Kit, NASA Public Affairs Office, Washington, D.C., August 21, 1966 (66-213)

Bilstein, Roger E., *Stages To Saturn: A Technological History of the Apollo/Saturn Launch Vehicles*, Scientific and Technical Information Branch, National Aeronautics and Space Administration, November, 1980 (NASA SP-4206)

Boeing Company, *Apollo/Saturn V Final Flight Evaluation, AS-502*, for the NASA George C. Marshall Space Flight Center, Alabama, July 31, 1968 (D5-15773/NASA-CR-91983) (N92-70435)

Boeing Company, *Final Flight Evaluation Report: Apollo 5 Mission*, for the office of Manned Space Flight, National Aeronautics and Space Administration, Washington, D.C., October 1968 (D2-117017-2/NASA-TM-X-64327)

Boeing Company, *Final Flight Evaluation Report, Apollo 6 Mission*, for the office of Manned Space Flight, National Aeronautics and Space Administration, Washington, D.C., February 1969 (D2-117017)

Boeing Company, *Final Flight Evaluation Report: Apollo 7 Mission*, for the office of Manned Space Flight, National Aeronautics and Space Administration, February 1969 (D-2-117017-4 Revision A)

Boeing Company, *Final Flight Evaluation Report: Apollo 8 Mission*, for the office of

Manned Space Flight, National Aeronautics and Space Administration, April 1969 (D-2-117017-5)

Boeing Company, *Final Flight Evaluation Report: Apollo 9 Mission,* for the office of Manned Space Flight, National Aeronautics and Space Administration, (D-2-117017-6/NASA-TM-X-62316)

Boeing Company, *Final Flight Evaluation Report: Apollo 10 Mission*, for the office of Manned Space Flight, National Aeronautics and Space Administration, (D-2-117017-7/NASA-TM-X-62548) (NTIS N70-34252)

Brooks, Courtney G., James M. Grimwood, and Lloyd S. Swenson, Jr., *Chariots For Apollo: A History of Manned Lunar Spacecraft*, The NASA History Series, Scientific and Technical Information Branch, National Aeronautics and Space Administration, Washington, D.C., 1979 (NASA SP-4205) (NTIS N79-28203)

Cassutt, Michael, *Who's Who in Space: The International Edition*, MacMillan Publishing Company, New York, 1993.

Compton, William David, *Where No Man Has Gone Before: A History of Apollo Lunar Exploration Missions*, The NASA History Series, Office of Management, Scientific and Technical Information Division, National Aeronautics and Space Administration, Washington, D.C., 1989 (NASA SP-4214)

Cortright, Edgar, Chairman, *Report of the Apollo 13 Review Board*, National Aeronautics and Space Administration, June 15, 1970

Davies and Colvin, *Journal of Geophysical Research*, Vol. 105, American Geophysical Union, Washington, D.C., September 2000

Ertel, Ivan D., and Roland W. Newkirk, *The Apollo Spacecraft: A Chronology*, Volume IV, January 21, 1966 - July 14, 1974, Scientific and Technical Information office, National Aeronautics and Space Administration, Washington DC, 1978 (NASA SP-4009)

Ezell, Linda Neuman, *NASA Historical Data Book, Volume II, Programs and Projects 1958-1968*, The NASA Historical Series, Scientific and Technical Information Division, National Aeronautics and Space Administration, Washington, D.C., 1988 (NASA SP-4012)

Ezell, Linda Neuman, *NASA Historical Data Book, Volume III, Programs and Projects 1958-1968*, The NASA Historical Series, Scientific and Technical Information Division, National Aeronautics and Space Administration, Washington, D.C., 1988 (NASA SP-4012)

First Americans In Space: Mercury to Apollo–Soyuz, National Aeronautics and Space Administration (undated)

Harland, David M., *Exploring the Moon: The Apollo Expeditions*, Springer–Praxis, Heidelberg and New York, 1999

Heiken, Grant H., David T. Vanimann, and Bevan M. French, editors, *Lunar Sourcebook: a User's Guide to the Moon*, Cambridge University Press, England, 1991

Johnson, Dale L., *Summary of Atmospheric Data Observations For 155 Flights of MSFC/ABMA Related Aerospace Vehicles*, NASA George C. Marshall Space Flight Center, Alabama, December 5, 1973 (NASA-TM-X-64796) (NTIS N74-13312)

Johnston, Richard S., Lawrence F. Dietlein, M.D., and Charles A. Berry, M.D., *Biomedical Results of Apollo*, Scientific and Technical Information office, National Aeronautics and Space Administration, Washington, D.C., 1975 (NASA SP-368)

Jones, Dr. Eric, *Apollo Lunar Surface Journal*, Internet: http://www.hq.nasa.gov/office/pao/History/alsj/, National Aeronautics and Space Administration, 1996
Kaplan, Judith and Robert Muniz, *Space Patches From Mercury to the Space Shuttle*, Sterling Publishing Co., New York, 1986
King-Hele, D. G., D. M. C. Walker, J. A. Pilkington, A. N. Winterbottom, H. Hiller, and G. E. Perry, *R. A. E. Table of Earth Satellites 1957–1986*, Stockton Press, New York, NY, 1987 (a compilation of installments originally issued by the Royal Aircraft Establishment in England)
Lattimer, Richard L., *"All we did was fly to the Moon": Astronaut Insignias and Callsigns*, The Whispering Eagle Press, Florida, 1985
McFarlan, Donald and Norris D. McWhirter et al, editors, *1990 Guinness Book of World Records*, Sterling Publishing Co., New York
NASA Facts: Apollo 7 Mission (E-4814)
NASA Facts: Apollo 8 Mission
NASA Facts: Apollo 9 Mission
NASA Facts: Apollo 10 Mission
NASA Facts: Apollo 11 Mission
NASA Facts: Apollo 12 Mission
NASA Facts: Apollo 13 Mission
NASA Facts: Apollo 14 Mission
NASA Facts: Apollo 15 Mission
NASA Facts: Apollo 16 Mission
NASA Facts: Apollo 17 Mission
NASA Information Summaries, Major NASA Launches, PMS 031 (KSC), National Aeronautics and Space Administration, November 1985
NASA Information Summaries, PM 001 (KSC), National Aeronautics and Space Administration, November 1985
National Aeronautics and Space Administration Mission Report: Apollo 9 (MR-3)
National Aeronautics and Space Administration Mission Report: Apollo 10 (MR-4)
National Aeronautics and Space Administration Mission Report: Apollo 11 (MR-5)
National Aeronautics and Space Administration Mission Report: Apollo 12 (MR-8)
National Aeronautics and Space Administration Mission Report: Apollo 13 (MR-7)
National Aeronautics and Space Administration Mission Report: Apollo 14 (MR-9)
National Aeronautics and Space Administration Mission Report: Apollo 15 (MR-10)
National Aeronautics and Space Administration Mission Report: Apollo 16 (MR-11)
National Aeronautics and Space Administration Mission Report: Apollo 17 (MR-12)
Newkirk, Roland W., and Ivan D. Ertel with Courtney G. Brooks, *Skylab: A Chronology*, Scientific and Technical Information office, National Aeronautics and Space Administration, Washington, D.C. (NASA SP-4011)
Nicogossian, Arnauld E., M.D., and James F. Parker, Jr., Ph.D., *Space Physiology and Medicine*, (SP-447), National Aeronautics and Space Administration, 1982
Project Apollo: Manned Exploration of the Moon, Educational Data Sheet #306, NASA Ames Research Center, Moffett Field, California, Revised May, 1974
Project Apollo: NASA Facts, National Aeronautics and Space Administration

Postlaunch Report for Apollo Mission A-101 (BP-15), Mission Support Division, NASA Manned Spacecraft Center, Houston, Texas, June 18, 1964, (MSC-R-A-64-2)

Postlaunch Report For Mission AS-202 (Apollo Spacecraft 011), Mission Support Division, NASA Manned Spacecraft Center, Houston, Texas, October 12, 1966 (MSC-A-R-66-5)

Report of Apollo 13 Review Board, National Aeronautics and Space Administration, June 15, 1970

Saturn AS-201/Apollo Postflight Trajectory, NASA George C. Marshall Space Flight Center, Alabama, June 6, 1966 (NASA-TM-X-53472) (N92-70421)

Saturn AS-202 Postflight Trajectory, Chrysler Corporation Space Division, October 1966 (Chrysler TN-AP-66-105) (NASA-CR-98403) (N92-70438)

Saturn AS-205/CSM-101 Postflight Trajectory, Chrysler Corporation Space Division (TN-AP-68-369) (NASA CR-98345) (NTIS N92-70426)

Saturn IB Flight Evaluation Working Group, Results of the First Saturn IB Launch Vehicle Test Flight, AS-201, NASA George C. Marshall Space Flight Center, Alabama, May 6, 1966 (MPR-SAT-FE-66-8/NASA-TM-X-61957)

Saturn IB Flight Evaluation Working Group, Results of the Second Saturn IB Launch Vehicle Test Flight, AS-203, NASA George C. Marshall Space Flight Center, Alabama, September 22, 1966 (MPR-SAT-66-12/NASA-TM-X-61958)

Saturn IB Flight Evaluation Working Group, Results of the Third Saturn IB Launch Vehicle Test Flight, AS-202, NASA George C. Marshall Space Flight Center, Alabama, October 25, 1966 (MPR-SAT-FE-66-13) (Volume 1, NASA-TM-X-59131) (Volume 2, NASA-TM-X-61695)

Saturn IB Flight Evaluation Working Group, Results of the Fifth Saturn IB Launch Vehicle Test Flight AS-205 (Apollo 7 Mission), NASA George C. Marshall Space Flight Center, Alabama, January 25, 1969 (MPR-SAT-FE-68-4)

Saturn Flight Evaluation Working Group, Saturn SA-1 Flight Evaluation Report, NASA George C. Marshall Space Flight Center, Alabama, December 14, 1961 (MPR-SAT-WF-61-8)

Saturn Flight Evaluation Working Group, Results of the First Saturn I Launch Vehicle Test Flight, SA-1, NASA George C. Marshall Space Flight Center, Alabama, April 27, 1964 (MPR-SAT-64-14)

Saturn Flight Evaluation Working Group, Results of the Fifth Saturn I Launch Vehicle Test Flight, SA-5, NASA George C. Marshall Space Flight Center, Alabama, September 22, 1964 (MPR-SAT-FE-64-17)

Saturn Flight Evaluation Working Group, Results of the Sixth Saturn I Launch Vehicle Test Flight, SA-6, NASA George C. Marshall Space Flight Center, Alabama, October 1, 1964 (MPR-SAT-FE-64-18)

Saturn V Launch Vehicle Flight Evaluation Report AS-501: Apollo 4 Mission, NASA George C. Marshall Space Flight Center, Alabama, January 15, 1968 (MPR-SAT-FE-68-1/NASA-TM-X-60911)

Saturn V Launch Vehicle Flight Evaluation Report AS-502: Apollo 6 Mission, NASA George C. Marshall Space Flight Center, Alabama, June 25, 1968 (MPR-SAT-FE-68-3/NASA-TM-X-61038)

Saturn V Launch Vehicle Flight Evaluation Report AS-503: Apollo 8 Mission, NASA

George C. Marshall Space Flight Center, Alabama, February 20, 1969 (MPR-SAT-FE-69-1)

Saturn V Launch Vehicle Flight Evaluation Report AS-504: Apollo 9 Mission, NASA George C. Marshall Space Flight Center, Alabama, (MPR-SAT-FE-69-4/NASA-TM-X-62545) (NTIS 69X-77591)

Saturn V Launch Vehicle Flight Evaluation Report AS-505: Apollo 10 Mission, NASA George C. Marshall Space Flight Center, Alabama, (MPR-SAT-FE-69-7/NASA-TM-X-62548) (NTIS 69X-77668)

Saturn V Launch Vehicle Flight Evaluation Report AS-506: Apollo 11 Mission, NASA George C. Marshall Space Flight Center, Alabama, (MPR-SAT-FE-69-9/NASA-TM-X-62558) (NTIS 90N-70431/70X-10801)

Saturn V Launch Vehicle Flight Evaluation Report AS-507: Apollo 12 Mission, NASA George C. Marshall Space Flight Center, Alabama, (MPR-SAT-FE-70-1/NASA-TM-X-62644) (NTIS 70X-12182)

Saturn V Launch Vehicle Flight Evaluation Report AS-508: Apollo 13 Mission, NASA George C. Marshall Space Flight Center, Alabama, (MPR-SAT-FE-70-2/NASA-TM-X-64422) (NTIS 90N-70432/70X-16774)

Saturn V Launch Vehicle Flight Evaluation Report AS-509: Apollo 14 Mission, NASA George C. Marshall Space Flight Center, Alabama, (MPR-SAT-FE-71-1/NASA-TM-X-69536) (NTIS N73-33824)

Saturn V Launch Vehicle Flight Evaluation Report AS-510: Apollo 15 Mission, NASA George C. Marshall Space Flight Center, Alabama, (MPR-SAT-FE-71-2/NASA-TM-X-69539) (NTIS N73-33819)

Saturn V Launch Vehicle Flight Evaluation Report AS-511: Apollo 16 Mission, NASA George C. Marshall Space Flight Center, Alabama, (MPR-SAT-FE-72-1/NASA-TM-X-69535) (NTIS N73-33823)

Saturn V Launch Vehicle Flight Evaluation Report AS-512: Apollo 17 Mission, NASA George C. Marshall Space Flight Center, Alabama, (MPR-SAT-FE-73-1/NASA-TM-X-69534) (NTIS N73-33822)

Spudis, Paul D., *The Once and Future Moon*, Smithsonian Institute Press, Washington, D.C., 1996

Technical Information Summary AS-501: Apollo Saturn V Flight Vehicle, NASA George C. Marshall Space Flight Center, Alabama, September 15, 1967 (R-ASTR-S-67-65)

The Early Years: Mercury to Apollo-Soyuz, PM 001 (KSC), NASA Information Summaries, National Aeronautics and Space Administration, November 1985

Thompson, Floyd, Chairman, *Report of the Apollo 204 Review Board*, National Aeronautics and Space Administration, April 5, 1967

Toksoz, M.N., A.M. Dainty, S.C. Solomon and K.R. Anderson; *Structure of the Moon*, published in *Reviews of Geophysics and Space Physics*, Vol. 12, No. 4, American Geophysical Union, Washington, D.C., November 1974

Trajectory Reconstruction Unit, *Saturn AS/205/CSM-101 Postflight Trajectory*, Aerospace Physics Branch, Chrysler Corporation Space Division, December 1968

Wilhelms, Don E., *To a Rocky Moon: a Geologist's History of Lunar Exploration*, The University of Arizona Press, Arizona, 1993

Index

Printing: Mercedes-Druck, Berlin
Binding: Stein+Lehmann, Berlin

Lightning Source UK Ltd.
Milton Keynes UK
UKHW030410060320
359809UK00012B/371

9 780387 300436